Heterostasis:

The Wisdom of the Nonverbal Mind

Book 1 – Theoretical

Book 2 - Experimental

By

Dr. Robert K. White

Heterostasis: The Wisdom of the Nonverbal Mind
By: Dr. Robert K. White

 Library of Congress Catalog Number TX 7-591-692

Printed in the United States of America

First Edition August 2012

Cover Illustration by Doris J. White

Publisher:
Dr. Robert K. White
P.O. Box 214
Ogdensburg, NJ 07439

Table of Contents

Book One – Theoretical

Chapter 4

Chapter 5

Book Two - Experimental

Chapter 1

Chapter 2

Acknowledgements

If experience plays the role I claim it does in shaping our nature, then the two men who could take whatever praise or blame may follow from the publication of this book are: First, my father, J. B. White, and second, from the University of Texas, Dr. George V. Gentry. They were more than mentors.

My father was a very religious man, but he believed completely in freedom of speech, as did Dr. Gentry. My father and I argued from the time we both got home on weekdays in the evening until bedtime, and all day Saturday and Sunday on any topic. Eventually, my father would say in desperation, "Bob, why don't you go outside and argue with a telephone pole?" My father simply taught me to argue and debate any topic, in general, but Dr. Gentry taught me what was worthy of debate.

Dr. Gentry, a full professor in the Philosophy department, usually spent at least one and sometimes two days a week at Balcones Research Center, in my office arguing on any topic we chose. Dr. Gentry, a second Generation Pragmatist, groomed me well to be a third generation pragmatist.

I am deeply indebted to the two students who ran experiments by themselves. David Mosley ran experiments M84, M87 (Book 2, Chapters 1 & 2), and Michael Demarco ran D87 (Book 2, Chapter 3), while he was still a high school student. One year as a high school student, he called William Paterson University and told them he wanted to volunteer to work on a science project for the summer. The school called me and I offered him a chance to run rats. He was most professional-far more than would be expected for his years. This design of this experiment was complex to execute, and its results were revolutionary.

In the text, I make some comments about the special challenges David Mosley faced with distinction.

The animals 90 Punish (Book 2, Chapter 4) were run by the students of my experimental class that semester.

I want to express my appreciation and affection for Maria A Dominguez for her dedication to the cause. Maria has been my brilliant assistant for over 30 years. Maria's job description, if we had ever written one, was simple..."doing whatever needed to be done...," from running rats, to keeping records (unfortunately the data for her brilliant hybrid Skinner box/maze experiment was accidentally destroyed) to cleaning, or feeding the animals in the colony room. Maria has long been a firm believer in Heterostasis, and she has always been ready with her tape recorder to discuss with me the latest development to the theory. She must have recorded hundreds of hours of these lecture/discussion sessions.

Then I must thank Mathew Harelick for writing over one hundred powerful original Pascal computer programs, which made it possible for me to carry out all of the massive data processing. Otherwise, they would have been impossible to compute.

Next, I must find a way to express in words my deep appreciation to Doris Jean (D. J.) White, my daughter in law and executive assistant, who in six months helped me to transform over 1500 pages of text into a coherent 800 page book. She is brilliant, a true "jack-of-all-trades"--a wonder to behold---a modern Atlas. Given access to Google and a computer, she can move the world. Mark, marrying D.J. may have been one of the smartest moves you ever made.

I want to thank my former student, long time college professor, and the last of a dying breed – a true Renaissance scholar – Dr. Jim Walker for being the first and only professional scientist to minutely read my book. Jim read all three books, over 800 pages of text and figures, and shared his observations with me each day. The book is a far more coherent text thanks to his diligence.

I want to thank Dr. Don Vardiman who was one of my first and finest students and lab assistant nearly fifty years ago at Texas Tech. Don has once again volunteered to assist me by discussing the book and finding ways of getting my book before the public. I also want to thank Don for suggesting his long time friend, Dr. Jim Walker read my book.

Dr. Robert K. White
August 2012

Acronym	Meaning
AA	Association Area
AIR	Anomaly Induced Retracing
AL	Adaptation Level
ANOVA	Analysis of Variance
ANS	Autonomic Nervous System
AOA	Alpha-Omega-Alpha
B and B or B&B	Brown and Blodgett
BFE	Between First Entry
BFEL	Between First Entry Latency
BFEM	Between First Entry Mean
BFEU	Between First Entry Unit
BFEULM	Between First Entry Unit, Latency, and Mean data set
BP	Blind Path
BPG	Blind Path Growth
BSRF	Brain Stem Reticular Formation
BT	Behavioral Telescope
C-A	concrete-abstract
CB	Cognitive Behavior
CC	Classical Conditioning
CCC	Causal Conceptual Conditioning
CP	Classical Pathways
CR	Conditioned Response
CS	Conditioned Stimulus
CUC	Certainty-Uncertainty Continuum
CV	Control Variable
DV	Dependent Variable
E	Experimenter
EI	External Inhibition
EMb	External Milieu of the body
EMm	External Milieu of the mind
EP	Essential Path
EV	Essential Variable
F11	Female 11
F3	Female 3
FE	First Entry
FEL	First Entry Latency
FELL	First Entry Speed
FEU	First Entry Unit

FEULL	First Entry Unit, Latency, Speed data set
FM	Full Maze
FR	Freedom Reflex
FREQ	Frequency
FREQL	Frequency Latency
FREQM	Frequency Mean
FREQU	Frequency Unit
FREQULM	Frequency Unit, Latency, and Mean data set
FRS	Functional Representation System
FUN	Functional Unit Names
G-H	nonverbal God - Human
GNMM	Generic Nonverbal Mammalian Mind
H	no mind one wisdom form of life
HH	no mind two wisdom form of life
H-H	human-human
HHH	one mind three wisdom form of life
HHHH	one mind four wisdom form of life
IMb	Internal Milieu of the body
IMm	internal Milieu of the mind
IV	Independent Variable
LTP	Loop Testing Procedure
MMM	Match-Mismatch Mechanism
M-N	mammalian-nature
MPS	Maze Probability Space
MSR	Maximum Steps per Run
MTR	Maximum Time per Run
NF	Node Frequency
NFM	Node Frequency Mean
NV	Nonverbal
NV/V	Nonverbal/verbal
OC	Operant Conditioning
OQP	Object Quality Problem
OR	orienting reflex
P	Parameter
PL	Path Length
PLTM	Path Length in Time
PLU	Path Length in Steps
PLUM	Mean Path Length in Units
PPA	Primary Projection Area
PSNS	Parasympathetic Nervous System
RD	Redundancy

RIR	Redundancy Induced Retracing
RT	Row-Type
S	Subject
S/N	Signal/Noise
S-D	Synchronization-Desynchronization
SNS	Sympathetic Nervous System
SOS1	Self-Organizing System 1
SOS2	Self-Organizing System 2
S-R	Stimulus-Response
SRS	Structural Representation System
TRF	Thalamic Reticular Formation
TT	transfer of training
TTP	Trial Testing Procedure
UR	Unconditioned Response
US	Unconditioned Stimulus
US	Unit Speed
USM	Unit Speed Mean
WGTA	Wisconsin General Testing Apparatus
WPW	Whole-Part-Whole
Y-D OR YD	Yerkes Dodson Law

"Before Einstein devised the theory of relativity, Newton's laws reigned supreme. In each case, it was never nature that changed, only our perception of it; and in each case, the new and more realistic perception depended on a larger, more encompassing perspective." (Erdmann and Stover, 1991, P. 7)

THE ANNOUNCEMENT

The purpose of this book is to announce the scientific discovery of yet another mind, the third mind discovered by a psychologist within the last forty years. Sperry, in the 1980's, received a Nobel Prize for having discovered that humans had two minds, a verbal and nonverbal mind, one in each hemisphere. However, this time the newly discovered mind is not a human mind but the nonverbal mind of a rat. A rat's mind is a radically new kind of mind profoundly different from both of Sperry's two physiological minds, but even more different from Plato's original transcendental concept of mind.

A timeline for the nonverbal mind

The primary thesis of this book is that the nonverbal mammals have a nonverbal mind that began emerging over fifty million years ago. This discovery could prove to be a revolutionary discovery, one that knocks humans off essentially all the pedestals they erected to celebrate their supposedly rational superiority.

The central tenets of classic human conceit is that only humans have a mind; that the verbal mind is the only mind; that the verbal mind did not evolve but suddenly emerged with the emergence of the human mammal. All these classic assumptions are false.

FIGURE0.01

FIGRURE0.01 is a timeline showing a theory of how the machine theory of mind emerged.

30. 25 Billion yrs. ago The Big Bang The cosmos began as a big pot of Hydrogen and Helium. Neither Humans nor speech was anywhere to be found.

29. 4.5 billion yrs. ago after over 15 billion years passed, our solar system emerged but with earth's atmosphere consisting of toxic gases.

28. 2.3 billion yrs. ago O2 emerged which produced the first great extinction. All the life forms that were adapted to the "toxic" earth environment became extinct, which made way for the O_2 forms of life.

27. 570 million yrs. ago Cambrian extinction occurred. The first great extinction of the O_2 forms of life.

26. 208 million yrs. ago Permian extinction occurred

25. 144 million yrs. ago First mammal emerged only 400 million years after the first great extinction.

24. 70 million yrs. ago primates emerged

23. 58 million yrs. ago Rats and Mice emerged, which are the subjects of the experiments in this book.

22. 5 million yrs. ago chimps emerged

21. 2.6 million yrs. ago first humans emerged
20. 1.5 million yrs. ago Homo erectus emerged
19. 1 million yrs. ago the mammalian peak
18. 40,000 yrs. ago Modern humans emerged
17. 450 B. C. Socrates, Plato, and Aristotle

- 450 B.C. is an important date because the first formal theories on the nature of biological life began. Plato and Aristotle began the two major schools of philosophy: Plato was dualist, a nativist, and a rationalist, while Aristotle was a monist and an empiricist.

16. 1600 A.D. Copernicus and Galileo revolutionized Aristotle's astronomy.

- Copernicus supplied the theory, while Galileo supplied the facts. This was the first revolutionary event in the history of science. The significance of these events is technically confusing, while Copernicus and Galileo disproved Aristotle's theory, the significance of their work actually supported Aristotle's monistic and empirical side in the end.

15. 1637 A.D Descartes The kings garden metaphorical insight.

- The Expression of the industrial revolution took different turns in Britain and in France. Machines in Britain were use for manufacturing, such as clocks for navigation, pumps for evacuating water out of mines, and for manufacturing plants or factories in general. Their innovation with clocks made the British navy the mightiest in the world.
- In France however, machines were used to make elaborate dolls that amused the king and royalty. The king of France, Henry IV invited Descartes to walk through his garden of dolls, each of which had been animated with a hydraulic system. However, Descartes' walk through Henry's garden revolutionized philosophy. Descartes reasoned if mechanical motion can make dolls seem alive and human, then perhaps the body is a machine. So, Cartesian dualism was born; Descartes was willing to assume that the body was a machine, but he stopped short on the mind. The mind was not a machine but the body was a machine.

14. 1755 A. D. Kant

Kant also produced a philosophical revolution concerning the nature of the mind. He contributed to both sides, both the rational and the empirical with his basic brick makers metaphor, which will be discussed several places in the text.

13. 1858 A. D. Darwin's Evolution

Darwin's revolution gave humans a new paradigm for thinking about the origin of their bodies. Evolution theory replaced the scriptural explanation of creation, Moses vs. Darwin. However, Darwin's theory, which explained the origin of the body failed to explain the origin of the mind.

12. 1861 A. D. Brocca's Area is discovered
The human compound nonverbal/verbal mind emerged about fifty million years after the emergence of the nonverbal mind, and was but a new model of the nonverbal mind.

11. 1864 A.D. Wernicke's area is discovered

10. 1902 A. D. Titchener's Psychology

9. 1904 A. D. Pavlov received a Nobel Prize for his physiological theory of digestion, not his theory of learning.

8. 1932 A.D. Cannon presents his revolutionary theory of Homeostasis Physiology in his book "The Wisdom of the Body."

7. 1949 A.D. Morazzi and Magoun discover the Brainstem Reticular Formation BSRF).

6. 1954 A.D. Jasper discovers the Thalamic Reticular Formation (TRF).

5. 1957 A.D. Gastaut replicates the work of Pavlov using Human EEG data.

4. 1961 A. D. Roger Sperry does split brain research

3. 1962 A.D. Kuhn publishes his revolutionary book on the nature of scientific revolutions.

2. 1981 A.D. Sperry receives the noble Prize for his discovery of the two human minds

1. 1984 A.D. The discovery that mammals have a nonverbal mind

Chapter One: The origin of the mind

The classic philosophical theory of mind, which was advanced by Plato over two thousand years ago, is still the most widely accepted theory of mind, and has a strange set of features whose examination is long overdue.

- The mind is an organ only humans possess. Proof: The mind confers speech, which only the human species possesses.

- The next feature is one little talked about and has no conventional name so I will call it the Standalone or Nativistic theory of the mind. The question is what was the origin of the mind? Did it evolve by Darwinian evolution, as did all other organs of the body? Are all the organs of the body derived from one primordial cell that first emerged billions of years ago? By calling it a standalone or nativistic organ, I refer to the fact that theorists today still deny that it evolved by natural selection. It did not evolve but was a special creation that suddenly emerged in humans without any precursors in the nonverbal mammals.

- The animation function – The strangest feature of the classic mind is that it is conceived of as a spirit by many. Spirits are in the image of god and are perfect eternal beings. Living and dying are properties of matter but spirits are never born or die, and are said to slip into a body at birth along with the first breath and exit the body with the last breath. This is one of the main functions of the spirit – to animate the body. A person is alive when the soul is on the premises and dead when it leaves.

- The rational function also has long been identified as one of the central functions of mind. Unfortunately, there are several meanings of rationality, which will require much of our time in this book to distinguish. The classic rationality is a verbal form, while the important new kind is nonverbal rationality, which mammals exhibit. A major part of my task will be to distinguish between verbal and nonverbal rationality.

However, starting from a scientific position, we must do a lot more work that was required of Plato or Chomsky.

For starters, two ideas are required to explain the concept of mental evolution: first, the emergence of mind and second, wisdom.

Theoretically or conceptually, this book begins as a variation on the name of Walter Cannon's great physiological textbook, *"The Wisdom of the Body"* versus *'The Wisdom of the Mind."*

This first chapter presents a theory of the origin of both the nonverbal mind, NV-mind, and the human compound NV/V mind.

This book is the product of a series of four experiments in which I discovered that mammals have a nonverbal mind. When I started conducting the first of these experiments in 1984, I knew only that neither I nor anyone else knew why or how rats run a maze, a fact that I discovered in the early 1960's. In the intervening twenty years, I had been assembling the technology required to design and execute what I hoped would be a proper experiment. Yet, I soon discovered my new technology was closer to being ready to do its job than was my ability to design a proper experiment. In this series of experiments, my rats kept emitting anomalous behaviors, which forced me to rethink repeatedly the way I understood or had misunderstood the rats' nature and their nonverbal behavior. Each time, the rats forced me to redesign the next experiment in the series. Nevertheless, by the time of the fourth experiment in 1990, I had finally achieved a sufficiently valid model of the rat's nature to be able to design a proper experiment involving Heterostasis.

In this series of experiments as each new anomaly, one after another, forced itself into my consciousness, my understanding of animal behavior and unavoidably also of human nature evolved so fast and so enjoyably that, I was almost unaware of the passing years, but after fifty years, I finally realized I was running out of time!

I find it hard to believe how lucky I have been, my original intention was to work within the context of Hullian (1943) learning theory, yet to my great surprise along the way I made many unexpected discoveries:

- First, I discovered that rats, mice, and probably all mammals have a nonverbal mind, NV-mind; or more technically, the GNMM.
- Second, in addition to Homeostasis, the wisdom of the body, nature designed a new kind of physiology, Heterostasis, the wisdom of the mind
- Third, the function of the generic nonverbal mammalian mind was to enable mammals to be self-organizing systems (SOS) that can spontaneously adapt to change.
- Fourth, the function of nonverbal communication was to enable the NV mammals to communicate with nature by converting anomalies into knowledge (ideational structures.)
- Fifth, the NV mind converts anomalies into knowledge, knowledge into models, models into scripts, plans, or means by which it can achieve its domain -specific end.

Ideas emerge

- Sixth, yet the core technology is the power of the NV mind to conceive sets of domain specific, purpose-specific, species-specific ideas (or Platonic ideals, or forms)

- Seventh, with the evolution of Heterostasis, the first rational technology emerged, which enables the nonverbal mind to create context dependent concrete ideas, either singularly or generally as a member of a set. This set or model can be rationally used and extended in many ways, which will be identified in the text.
- Eighth, this book is about the origin and the nature of ideas within the nonverbal mind, which is a mental speciation machine analogous to Darwin's Biological Speciation machine, natural selection. However, this mental machine creates species of mental beings or Ideas by rational selection not natural selection.
- Ninth, the nonverbal mind creates nonverbal ideas in order to create rationally a valid stream of nonverbal behavior. Nonverbal behaviors are a manifestation of a narrative constructed of nonverbal ideas conceived to provide mammalian life with a real time consciousness of its external world.
- Tenth, the mind is home to a varied population of ideas. Philosophers, theologians, and theorists of all kinds have had their say about the nature and origin of ideas. Unfortunately, most have been dualists; a set of humans who lost faith with matter, nature, and nature's god, and took the easy way out and assumed that Ideas are not natural but are the product of supernatural beings generally called spirits, or more simply--spirit god or gods.

The discovery of the nonverbal mind qualifies as the discovery of probably the largest bank of scientific anomalies of all times. Fortunately, each of these anomalies does not require the invention of a new paradigm because all these original words were presumably members of the old paradigm. Presumably, all the new conceptual elements, which will be required to understand all the anomalies waiting to be revealed, will furnish all the new ideas/words required to establish a coherent new set of ideas/ words, required to establish a solid foundation for a new paradigm or mind.

Even though Sperry (1961) received a Nobel Prize for his discovery and his claim that humans have two minds, one in the left and one in the right hemisphere, yet there are still those who ignore his theory and some who accept his theory.

For my first challenge to Sperry, I propose that humans have only one mind. Second, I challenge Sperry's concept of the nonverbal mind by proposing it is simply a component of the one human mind.

The purpose of this book is to present the concept of the generic nonverbal mammalian mind, the GNMM. I propose that there is a profound difference between Sperry's (1961) nonverbal mind and my rodent's nonverbal mind. In fact, I propose that humans have only one compound verbal/nonverbal mind, NV/V, only the nonverbal mammals have a pure nonverbal mind. The nonverbal mammals have two hemispheres in both of which there is only one nonverbal mind.

In this first chapter, I propose that the mind evolved out of at least two major components of the brain. The question is out of what tissues evolution constructed the nonverbal mind. The two components I propose are a horizontal cortical component and a vertical arousal function.

Sperry's (1961) view has been that the mind is an organ within the human cortex, which is divided into two components, two hemispheres, each host to a different mind. By contrast, I refer to the cortical component of the mind as the horizontal component of the mind. The second component, the vertical arousal component is actually the oldest component, older than the neocortex.

The Generic Nonverbal Mammalian Mind, GNMM was constructed of at least two sets of tissues, the horizontal, and the vertical. This book is divided into two books. Book 1 is the theoretical section, and Book 2 is the experimental section. The theoretical section has a preface and five chapters.

- Chapter 1 discusses; first, the origin of the NV mind and then the NV/V mind
- Chapter 2 describes the structure and function of the horizontal mind and five theories of mind.
- Chapter 3 discusses the structure and function of the vertical arousal component of the mind
- Chapter 4 describes the emergence of the nonverbal mind and the environmental conditions that caused nature to find a better way.
- Chapter 5 asks what technology armed the nonverbal mind. The focus is upon a new theory of conditioning, which extends Pavlov's set of seven stages or types of training periods.

The evolution of the nature of the mind

While the purpose of this book is to describe the structure and function of the nonverbal mind, the first step will be to present a historical review of the evolution of the nonverbal mind, six theories of mind will be compared.

Unfortunately, nature's pursuit of more powerful life forms has been advancing on parallel tracks; one called mind and another called wisdom. This complicates the story because mind and wisdom are often confounded.

The simplest set of basic actors that I have been able to identify consists of four wisdoms and six theories of minds with the four wisdoms consisting of two classes: the rational and pre-rational. The first two stages of wisdoms are pre-rational, while the last two are rational.

The four wisdoms/four stages of mental evolution are:

Stage one. The no-mind; one wisdom form of life H

Stage two. The no-mind; two wisdom form of life HH

Stage Three. The Non-Verbal, NV, three-wisdom form of life HHH

Stage Four. The Non-Verbal/Verbal, NV/V four-wisdom form of life HHHH

The implication of stage 3 on two models of human nature

Model 1: the classic closed passive model
Model 2: the open active model

The stages of evolution listed above reflect the fact that evolution advanced the adaptive behavioral capacity of animals simultaneously on two tracks: wisdoms and minds. Thus, two kinds of technologies must be distinguished. Sperry (1961) popularized the idea that there are two kinds of minds - one nonverbal and one verbal. However, I propose his strategy is not a useful way to explain the origin of the mind.

In this book, I will describe four kinds of wisdoms: Homeostasis, Hedonism, Heterostasis I, and Heterostasis II. However, there are no limits on how many wisdoms the human rational mind can construct. Each kind of rational wisdom simply requires a new set of symbols: mathematical and music symbols are two examples of two new mind-created forms of wisdoms.

My primary responsibility is to describe the structure and function of the nonverbal mind. Yet I must begin by distinguishing: two concepts of mind:

- First, the simple NV-mind,
- Second, the compound NV/V mind.

While presumably, humans have been informally talking about the Verbal, V-mind for thousands of years, a philosopher, not a scientist, conceived the first theories of the V-mind.

I propose that the V-mind conception of mind is a myth. The nonverbal, NV mind is the primary mind, and the basic element of any working mind, and served as the only mind the nonverbal mammals had and have, and indeed served them well for millions of years; allowing them to rationally and nonverbally to create the knowledge required for planning and executing their nonverbal behavior.

The scientifically conceived concept of the nonverbal mammalian mind described in this book has nothing in common with the theory of the verbal mind as conceived by Plato (1952) as a geometrical metaphor.

For thousands of years theorists have conceived of the human mind as a verbal, V-mind. The thesis of this book is that the human mind evolved in two stages. In the first stage, the nonverbal mammals evolved a nonverbal mind, a NV-mind. In the second stage, the human mind evolved as a compound NV/V-mind.

One way I could begin would be to list all the features of the verbal mind conceived by Plato and then show that the nonverbal mind is not like that at all. The paradox is that the major features of the nonverbal mind have always been obvious, in plain sight, yet these nonverbal features of the behavior of the nonverbal mammals have always been invisible. Why, because these features were nothing at all like what experimental psychologists had been led by theorists to expect. Historically the classic concept of mind has been myth/metaphor-driven.

The features of the fact-driven material nonverbal mind, I propose is the primary mind, contradicts every feature Plato (1952) proposed for his metaphor-driven verbal rational mind. Additionally, the classical Platonic spirit conception of the verbal mind is but a secondary verbal mind.

The revolutionary contribution of or--the magic of the verbal mind is that it enables the nonverbal mind to create new sets of symbols. The human verbal mind should be called the symbol generating and using mind. Each new kind of wisdom requires a new closed set of symbols, which are mapped one to one with all the variables required to create a coherent hierarchically organized model of some new space. Nevertheless, in this book, the focus is upon the fact that the nonverbal mind was both the first rational mind and the author of the first rational wisdom and additionally, created the first and the primary set of nonverbal symbols

Verbal behavior such as English is but one manifestation of the many sets of rational symbol systems, which the NV/V mind created. Numeric symbols or musical symbols are just an emergent creation of the NV/V mind as are the set of closed letters of the English alphabet, which make human speech possible.

However, in this book, I will present several new and strange theories and concepts, but the strangest new concept which the reader may have trouble understanding is that the nonverbal mind is the source of the third wisdom, which has this enormous closed set of primary symbols. The nonverbal mind inherited one copy of one set of the primary set of symbols (the sets of sensations) out of which perceptions are created. This set of first- order symbols are the ground upon which all the higher order sets of symbols are constructed. The letter *A* for example is a higher order example of a symbol, which consists of a set of italic pixels in our case arranged in the standard form of the letter, which is the first symbol of the English alphabet. Each symbol element of the set differs generally in the form, the cookie cutter forms, which the nativists insisted were given by some transcendental being.

Wisdoms add or extend the regions within which an organism can consciously function, just as headlights, air conditioners, or automatic starters add to the range of places where a car can function.

Stage one: The no-mind; one wisdom form of life H

In stage one; evolution employed the simplest kind of physiological technology to create the first forms of animal life. Cannon (1932) first described this physiology, which he called, *The Wisdom of the Body*. Cannon was obviously impressed with the technology of Homeostasis for he considered it the one and only technology the body required. He was unaware that there were different kinds of technology; three kinds are described in this book, one for the first time.

No one of these technologies is any wiser than any one of the others; each technology provides life with a new and different survival technological function. Each kind of wisdom plays a unique role, and in mammals, which are a three-wisdom species, each kind of wisdom supplements the other two.

All the no-mind species adaptive skills were instinctual. All the knowledge they used for their whole life was a gift of their genome; the knowledge obtained from one random dip into their species' gene pool. Homeostasis endowed life with only one power, the power to regulate the essential variables (EVS) of its internal body milieu, IMb.

Homeostasis

The first pure homeostatic kinds of life designed by biological evolution were closed forms, which were simple in several respects. First, they were designed as a one-niche species, which could exist only within the one niche in which they evolved.

Second, they were designed employing only one kind of physiology, Cannon's (1932) Homeostasis, the wisdom of the body, which employs only regulators, which are the simplest kind of physiological technology.

Third, Homeostatic regulators had only one adaptive trick; Each one had only the power to keep one of the set points of the body's Essential Variables constant, such as the need for food, water, air, blood pressure, etc.

Fourth, regulators are a "dumb" technology. They are error regulated and error regulating devices; they use error to regulate error so that the regulator is "on" only when there is an error. If the set point of a room is 70 degrees and the room temperature falls below 70 degrees, then the room thermostat turns on the heater, which warms up the room, and the heater stays on only as long as the room temperature is below the set point. What makes regulators dumb is that they are turned on only by error; they cannot act to prevent an error, they can only reduce or eliminate an error, which is called negative feedback.

Fifth, these no-mind species were simple in that they were designed to survive by instincts. Genomic knowledge completely supplied all the physiological knowledge required for these no-mind forms of life to survive. The genetic knowledge provided entirely, not only for the cellular construction of all the organs of all the bodies, but also provided the knowledge, which defined the cause-effect structure of all the instincts required for the execution of every requisite physiological function. These no-mind species were pure "nature" homeostatic beings. Nurture had no role in their survival.

The great extinctions

At various times and places the surface of the Earth can be and often has been relatively calm and constant for over millions or hundreds of millions of years, which is ideal for the no-mind animals who require constancy. However, on the other hand the fact that the surface of the Earth inevitably on occasion can be completely changed almost instantly by one of several possible catastrophic events, such as a meteor impact or a volcanic eruption, etc. Unfortunately, for the simple no-mind animals, while these catastrophic events are rare events they nevertheless inevitably occur.

In fact, the geologic record shows that these catastrophic events have occurred several times and each time they have destroyed most of the life forms.

These events are called "the great extinctions." Nevertheless, each time select species did not die; ones that had an ability to adapt survived.

Specifically after the great extinctions, evolution favored in the next step those species, which survived, so that in effect, evolution slowly created yet another technology to organize behavior, and this time evolution chose some chemical neurotransmitters to employ to induce rational behaviors.

Stage two: The no-mind; two wisdom form of life HH

A second kind of physiological technology invented by evolution is one that has gone by several names over the years. In this book, I will resort to perhaps the oldest name for this technology: Hedonism, which has been informally discussed by several authors but never put within the proper physiological/psychological evolutionary context it deserved and for good reason. In the context of Christianity, hedonism seemed as if evolution had taken a wrong turn. However, in the light of new data with Hedonism the wisdom of the emotions, evolution created some of its strangest but most powerful tricks.

William James's concept of the non-rational

William James (1950) called Hedonism non-rational by which he meant that pleasure and pain was an expression of a new technology, which today we know are specifically chemical tricks, which were neither rational nor irrational. With Hedonism, evolution invented and is still inventing a whole host of chemical neurotransmitters that trick the organism to do the right thing. Hedonistic or non-rational mechanisms trick animals into doing the right thing for the wrong reason. It is important to understand both the nature of the tricks and at least the kind of technology that made the non-rational behavior possible, for it sets the stage or provides a context for understanding rationality, which evolution does not invent until it creates a mind. While non-rational tricks induce an individual to do the right things for the wrong reason, not until evolution invented the Heterostatic mind did life make decisions rationally, doing the right thing for the right reason, which means understanding the primary local cause-effect conditions, which justify and induce behaviors.

Pleasure, for example, induces animals to eat the foods they need simply based on taste. Hedonism induces organisms that know nothing about nutrition to ingest the proper foods; they simply eat something because it tastes good. However, the wisdom of hedonism is that it induces animals to do the right thing even though they have no mind and are not yet rational. Pleasure and pain are the two best-known traditional examples of Hedonism.

Sex, of course, is the ultimate trick. Why is sex so pleasurable? Presumably, individuals may have avoided sex, yet nature just made sex even more pleasurable; nature ultimately made sex so pleasurable that most individuals find it to be irresistible.

Affective Neuroscience

Hedonism was one of the first names applied to behavior, which was a function of pleasure and pain. Today a whole new kind of neuroscience has emerged called Affective Neuroscience which has shown that the endorphins, for example, which are the Neurochemical foundation for pleasure is not unique but is but one of many neurotransmitters that have so far been discovered. (Panksepp, 1998)

The Neuron's role

The neurons talk to one another though synapses, which communicate by way of two-way chemical messages. The primary signal of neurons consists of neural impulses. However, these impulses must be transformed into a neurochemical code to cross the synaptic gap between neurons. This new science has shown that there are many kinds of neurotransmitters called neuropeptides, and each kind is responsible for a different emotion or a new hedonistic tone or non-rational trick. (Panksepp, 1998)

One of the most interesting and revolutionary new technologies that evolution has produced is this chemical technology, where the chemical creation is a function of neurons.

Neurons and Hedonism

If I were forced to choose which physiological organ was the most revolutionary, which could most likely work causes the greatest numbers of surprising kinds of behaviors. I would have to choose the neuron. This cell is but one of the three cell types in the human body, yet there are over one hundred billion neurons in the human cortex alone. Further to add to the complexity of this cell, it should be noted that each neuron could have over one thousand branches or dendrites, which make contact with other neurons. Each neuron is like a tree with over one thousand branches effectively making the human cortex like a great jungle with over one hundred billion trees, each with one thousand limbs, which interconnect with the limbs on innumerable other trees. These intertwined limbs form a great canopy over all the trees.

Now, there are two additional facts that make this picture interesting and important. First, where these branches connect, there is a special connection, a synapse, through which all trees technically communicate with one another mediated by chemicals, the neurotransmitters. The Affective Neurophysiologists have, in the last thirty or more years, identified over twenty-four neuropeptides and have shown that each of them has a unique capacity to induce a different kind of emotion in a subject. (Panksepp 1998)

These neuron-generated chemicals are all evolutionary advances in the class of behavior, which James (1950) called non-rational, which trick an organism into doing the right thing for the wrong reason.

No one yet has any idea how many of these special neurotransmitters there are functioning, growing, and evolving within the great canopy covering the billions of neurons, the tree jungle in each mammalian cortex.

Olds and Milner

Fifty years ago, the emotions and hedonism were two independent topics. Olds and Milner (1969) first identified the "pleasure centers" as they called them in the septum. However, today they have been identified as the special neurotransmitters, endorphins, which grow in our brain's "forest." Today, endorphins are just one of twenty-four or more of these special neurochemicals growing in the canopy covering our cortex. (Panksepp, 1998)

This is an example of two points: first neurons are responsible for two of the most revolutionary features of mammalian behavior: rationality and hedonism. Second, this is an example of one of the most interesting properties of the cosmos: its capacity to create "emergents." The next step is one of the most surprising creations evolution ever made. Specifically, after the great extinctions, evolution's next step created yet another technology to organize behavior, and this time evolution employed many major chemical neurotransmitters to employ to induce rational behaviors.

In one sense, the nonverbal rational mind could not have emerged without a new set of neurotransmitters, which supported the new configuration of neurons. Specifically, with the evolution of the mind, evolution employs these chemicals to create new forms of behavior: rationality with two kinds both the nonverbal and verbal kinds.

Stage Three: The NV, three wisdom form of life HHH

The third technology evolution created and perfected was Heterostasis, the wisdom of the nonverbal mind. The emergence of Heterostasis made the nonverbal mind the most revolutionary organ possible. Heterostasis, the wisdom of the mind, was the secret weapon responsible for the evolution of the whole line of mammals; first, all the nonverbal mammals; second, the eventual evolution of the verbal mammals; third, the rational evolution of all the technologies by which humans have redefined the environment of the planet Earth which is our home. This book was designed to describe this third kind of neuron technology, the generic mammals with one mind and three wisdoms: Homeostasis, Hedonism, and Heterostasis I. The particular mammal, which is the focus of this book is the rat.

The nonverbal mind

The central issue of this book is what kind of a mind is a nonverbal mind? What new adaptive features does the third wisdom convey on the long evolutionary story of the reign of the nonverbal mammals?

The nonverbal mind was the organ that enabled mammals to populate the planet and provided for the eventual emergence of the human four-wisdom mammals. The nonverbal mind is the primary mind and prepared the way for the human mind.

The primary function of the nonverbal mind was to supply all the knowledge, all the somatic skills employed by all the nonverbal mammals, for over fifty million years and speech was no part of that function! The primary function of the nonverbal mind was a creative function, which enabled the nonverbal mammals to self-organize in the presence of change and to enjoy exploring and making models of anomalous spaces.

What is mind? What is the root function of the primary mind?

There are dozens of important questions about mind that we must answer, but let us start with the first big one. What is a mind? Everyone presumes to know the answer to this question; for thousands of years, humans have deceived and flattered themselves into believing that since they were the only species of mammals with verbal behavior; they were therefore the only mammal with a mind.

Now, I propose that all mammals have a mind and that human speech is but an expansion of the functionality of the nonverbal mind. The question is how does the emergence of the nonverbal mammals fit into the big picture? How did the emergence of the generic mammalian mind change the future of the cosmos? The primary mind has no verbal speech, so what is the root function of a nonverbal mind? Now our question has evolved; why did evolution create the first mind as a nonverbal one?

The one-mind emergence

With the emergence of the cortex, the design of the nature of life took a one hundred eighty degree turn. The classic no-mind species were designed essentially as nature-type beings, i.e. as essentially robots dominated by instinctive behavior.

In the next stage, mammals and the cortex emerged together bringing with them the nonverbal mind and a new of kind of organ. These new forms of life were designed with a minimum of instinctive behaviors; they were designed to be self-organizing beings born free and mentally responsible for their own survival. Evolution began to design a radically different kind of adaptive behavior; mammals have been described as a family of species with a tabula rasa nature.

If the nonverbal mind was not designed to talk what was its purpose? What skills are required by a self-organizing system? Whereas evolution gave the no- mind animals instinctive solutions for fixed problems, the new life forms, the nonverbal mammals, were given a mind designed for exploring and creating new domain-specific nonverbal skills, scripts, and experiences.

For millions of years, the nonverbal mind served the nonverbal mammals well. The critical function was not to enable the one-mind mammals to verbally communicate or chat with one another but to enable life to deal rationally with change. Mammals were able to enter an anomalous domain of terrestrial space and transform every anomaly into a known class by evolving a concept, a mental form that can recognize or assimilate each anomalous thing of the same class. The purpose of the mind was to enable each mammal to act in the presence of change not to be destroyed by change. The facts seem clear that evolution leaned over backwards and created a mind that could do more than just adapt to change, but a mind that required change; a mind that feeds upon change; that required change in order to attain an optimal degree of arousal each moment of every waking day.

What is an anomaly

The basic insight that Heterostasis theory requires is that anomalies are objects or events the primary mind cannot assimilate. An anomaly is an event for which the primary mind has no concept, form, or cookie cutter. An anomaly is an event that cannot be assimilated or assigned membership in some set class. Anomalies, in principle, are arousing, exciting, or too exciting for they invalidate the conceptual system, which elicited them but cannot assimilate them. Any event or any precept that the mind cannot experience, and by experience, I mean assimilate, is a precept that the primary mind cannot match with some form already in its repertoire.

The nonverbal mind transforms "change- objects" into coherent sets of nonverbal ideas and concepts give birth to new ideas and new conceptual forms, which enable life to grow. They expand its mental gene pool; to give life a new set of conceptual tools, ones specifically designed for each domain, a string of coherent hierarchically organized ideas consisting of rational sets of categories unique for each species, each purpose, each domain, etc.

How many minds are there in the human brain? My considered answer is, "ONE." The nonverbal mind is the only organ capable of the execution of all the basic functions, which cumulatively form a nonverbal mind. Now, by looking back, it is possible to recognize that neither model of the mind was or is a full-function mind. Neither the nonverbal nor the verbal mind is a full- function mind, but was simply in its day a new model. In the case of the verbal mind, the 12-station model with the new functions added by the thirteenth station provided the new whole with new features that amplified the power of the nonverbal mind. Why do I say there is only one mind?

How many pairs of eyes does each mammal have? In fact, while the brain's plasticity can reassign many of the functions of the mind, yet normally there is a standard localization function and only one copy of each of the standard structures, and all those basic standard structures are a member of the nonverbal mind. The only original organs completely owned by the verbal mind are the Broca and Wernicke's areas. (Donald, 1991)

Kant on the origin of Order or Concepts: The Humpty Dumpty, HD problem

The simplest way to understand Kant's significance is to explain his theory of mind by employing a metaphor, which I call the Humpty Dumpty problem. According to the nursery rhyme, after Humpty Dumpty had a great fall, "...All the king's horses and all the kings' men couldn't put Humpty Dumpty back together again." The nonverbal mind, however, must be understood as better than all the King's horses and all the King's men.

The model of perceptual knowledge or empirical knowing requires that we conceive of all knowing as a three-stage function, a Whole, Part, Whole (WPW) function.

The object x, in the physical world are physical wholes. W1

The senses carry millions of bits of information about the object X, P

The mind combines all the parts into a perceptual whole, W2, W1, P, and W2.

The nonverbal mind creates experience by creating a narrative employing perceptual wholes or images of the physical things in the world. This book is about how the mind creates images or ideas.

The mind collectively creates experience. Later in this chapter, I propose that Aristotle's model processes the objects of the environment in a 12 step rational process, but for the moment let's look at a quick and dirty three-step process: an overview of the nature of experience. Humpty Dumpty originally sitting upon the wall was a Whole, W1, but after the great fall he broke into so many parts, P, that no one could put all the pieces back together again. In brief, each sensory act employs a three-step process, WPW.

The physical world can be understood as sets of hierarchically organized material structures with levels of atomic and chemical layers. But for the moment, let's focus on the outer most order of things, the world of things. At this level, each class or set of objects can be naturally conceived as wholes, which generally consist of parts. A tree, for example, has branches, leaves, trunks, and roots. But let's focus on the macro, the whole as the first whole, W. When an organism looks at or attends to some object then the five senses parse the massive sets of stimuli emanating from each object into massive sets of neural impulses, parts P, which enter the brain. Somehow, these neuronal parts are reassembled into a perceptual whole, W. One of the great mysteries of consciousness is how these parts are reassembled into a whole. This three-step sequence I call the WPW.

The WPW problem

W-The physical world is a world of wholes, W.

P- The sensory world parses the wholes above into parts, or Pieces P. The empirical world enters the brain as a vast collection of neural impulses or pieces. The five senses function as shredders so that objects never enter the brain. What enters the brain must pass through the senses in order to enter the afferent nerves as neural impulses.

In terms of a computer metaphor, the TV screen consists of large set of pixels. TV cameras, like the eye, parse the whole object into pixels.

W- The question is how Humpty Dumpty gets back together again. The fact is, however, THE WORLD OF CONSCIOUSNESS CONSISTS OF wholes, so Humpty Dumpty is reassembled but the puzzle does not go away. Where do the orders, the classes, the forms, the laws, the scripts, or the narratives that bring and assemble the parts back into some whole, come from?

There is much dark humor and irony in the history of the philosopher's discussions of where the order in the mind comes from. Locke (1914) said, "Nothing is in the mind that was not first in the senses." Kant (1952) said, "Nothing is in the mind that was not first in the senses, except what was first in the mind itself."

The empiricists argued that the mind was like a tablet of white paper. Nature alone wrote upon the tablet. The empiricists started with the senses, which are like an advanced paper shredder, which sense the millions of stimuli, which enter the five senses.

Kant (1952) argued that the mind imposed order upon all the small sensory pieces, the sensations. The ordering capacity built into the nonverbal mind was a surprisingly complex bit of technology.

The chief rational function of the nonverbal mind is quickly to restore mental stability or order to the mind in the presence of change-induced uncertainty. This calls for the nonverbal mind to be designed as a machine, a revolutionary kind of empirical machine that can conceive order as a coherent set of concepts by identifying a set of unique features for each class of objects, which makes each thing a recognizable predictable object.

The question is how and where the universal self-organizing System, SOS, called the nonverbal mind was conceived, so it will function properly for each possible niche and conceives a domain specific closed set of categorized objects. The nonverbal mind conceives closed sets of domain-specific ideas or models for each terrestrial domain.

When change occurs and nonverbal mammals are in an anomalous domain then the creative portion of the mind springs into action and begins to execute the creative function, what Plato (1952) called, the "god function". The nonverbal mind emerged because it had the intelligence to act as a self-organizing system, SOS. How does an SOS's mind work? How does the nonverbal mind deal with change?

An anomaly invalidates the current life space, and the nonverbal mind must begin to conceive a new life space, a new narrative, or a new script. (Lewin, 1943) The nonverbal mind must construe a new life space to prepare life for change and must begin to conceive, construct, or construe a set of categorizing forms, which creates a coherent domain-specific hierarchically organized model with a set of concepts incorporating all the objects of that space which only moments before was an anomalous domain.

New Territory

Our purpose to understand the nonverbal mind as a self-organizing system takes us into a wide range of topics or concepts, which never before have been considered within the domain of science.

What is truth and what is real

There are two kinds of truths: rational and empirical or verbal and nonverbal. The former rational truth was Plato's (1952) ideal. Empirical truth is the nonverbal scientific kind. Plato insisted that empirical truth or knowledge of nature or matter was impossible to obtain, and he was just as certain that, rational truth could be had just by thinking while sitting in his "ivory tower."

However, all the nonverbal mammals have an absolute need for empirical nonverbal truth. As self-organizing systems, they, in fact, have the greatest possible need for the nonverbal truth far more than a philosopher does. What need has a rat for empirical truth? Where does the nonverbal mind find truth? We must begin by redefining the nature of communication.

Communication

Where can nonverbal self-organizing systems get the knowledge they need to survive? The only place a self-organizing mammal can get the knowledge it requires is by interacting with the physical features of anomalous domains by probing the cosmos.

The nonverbal mammals get their kind of truth not by reading books, from parents, from schools, or from Google but by creating their own through the actions of their self-organizing system. When evolution endowed the nonverbal mammals with the rational power of the nonverbal mind, they were now SOS and were on their own. The nonverbal mind was designed to function in a world of things. An SOS has only one hope of surviving--by creating its own nonverbal model of anomalous domains while exploring terrestrial space such as the matrix maze.

So what and where is the nonverbal communication, which generally is assumed to function as the equivalent of the verbal speech mechanism of the human mind? To whom or to what could the one-mind mammals communicate while self-organizing and exploring a maze?

The nonverbal mammals cannot obtain knowledge by communicating with others of their verbal community; nonverbal mammals know nothing about meme communication and publically communicate with no one.

The nonverbal mind communicates only with the material or physical beings of the material cosmos or Nature through nonverbal behavior! What lesson must be learned here? We must expand our understanding of the nature of communication. First, we must explain how nonverbal exploratory behavior qualifies as communication/NV speech.

Nonverbal speech is a concept so foreign to our human ears, perhaps because the ears are not as central as in NV/verbal communication.

There are two areas where NV/speech should profoundly alter our understanding of our own behavior: First, The role of nonverbal speech in ordinary verbal speech; Secondly, the role of NV communication in that new and revolutionary form which humans have practiced for about four hundred years - experimental science.

The 12 station mind mammals feed on anomalies, which induce uncertainty, which activate the nonverbal mind. "Uncertainty" is the food on which the nonverbal mind feeds. The nonverbal mind transforms uncertainty into knowledge.

The nonverbal mammal's nonverbal uncertainty-reducing powers show that evolution has designed a new organ with the power to execute a new kind of evolution or a new kind of communication.

Dawkins

Biological evolution is limited to biological speciation, which allowed Dawkins (1989) to argue and defend his concept of “The Selfish Gene.”

However, the genes, which designed the nonverbal mind, can hardly be conceived as selfish genes. In Dawkins’ scheme of things, the future of life was controlled by the genome, and thus the genes controlled their future. Dawkins, however, missed one critical fact. With the emergence of the nonverbal mind, nature took the responsibility for the design of the future away from the genome and gave this responsibility to the mind. The selfish genes, if they had been smart, should have never let the nonverbal mind get off the ground. Now once the mind is out of the bottle, there is no way to put it back in. The nonverbal speciation mechanism has such a deep design potential that no one can predict its limits. In Dawkin's (1989) context, we must conclude that the genes that designed the nonverbal mind, it would seem, did not know their business! Or now we could make a case for their being a set of unselfish genes.

The nonverbal mind ultimately put each mammal in control of its own destiny. A self-organizing system's chance of surviving is a direct function of its ability to extract the potential knowledge stored within the nonverbal features of each domain that is relevant to its goal.

The ultimate purpose of mammals is to extract as much of the cosmos's potential as fast as they can.

"Man's ability to control the environment stems from his ability to perceive and to understand it, to observe the operation of cause and effect, to predict what might be the consequence of this or that activity. All control processes, manual or automatic, derives ultimately from such conscious processes." (Kelley, 1968, P. 35)

Kelly was right on, except for one point; note his quote begins with the word, "man." What should he have said: All mammals' ability!

The survival of the nonverbal mammals who are born free of the instinctive genomic knowledge that drove the no-mind animals is intimately dependent upon their nonverbal mind's ability to exhaustively read, understand, model, and predict each material cosmic domain as is physically/mentally possible. The survival potential of all self-organizing mammals is simply a function of their ability to extract nonverbally or read empirically the nonverbal knowledge stored within each domain that is relevant to their purpose. As scientists, we rat runners must learn that heterostasis not homeostasis drives the mind. The nonverbal mind with its creative rational power can always be counted upon serve any mammal plunged into an anomalous terrestrial space and count upon the self-organizing power of their nonverbal mind to interrogate nature. Heterostatic behavior allows the mammal to know the domain-specific niche in which it is immersed. Heterostatic behavior emits the rational powers that enable their nonverbal mind to extract and organize the potential knowledge contained within each anomaly into knowledge structures, which will enable them to achieve their domain-specific goal.

It should be obvious that since the primary mind is nonverbal, it can only communicate with a nonverbal hierarchically organized terrestrial world. The nonverbal mind was created in the image of the nonverbal cosmos.

Mental Evolution: biological Faults

One of the basic facts that is so tacitly obvious that no one bothers to state it overtly is that biological evolution is severely limited in several ways:

1-Biological evolution does only biological speciation,

2-Biological evolution brings about change in the gene pool slowly,

3- Mental evolution, at least by the advent of the verbal mind, potentially has no constraints other than factual: the mind potentially can invent or create any kind of being, biological, mental, or otherwise. The emergence of this general potential qualifies as one of the great instances of emergence.

4-The final limitation, which is more or less the focus of this book, is that biological evolution cannot create ideas, which is one of the main reasons why the nonverbal mind was created.

Science is nonverbal

Heterostasis can be understood many ways: first, either as a nonverbal mental function; or second, as the rational function of the nonverbal mind; or third, as a unique pattern of rational behavior emitted by the nonverbal mind, which creates coherent sets of hierarchically organized ideas, which are also an expression of mental evolution. However, please note that mental evolution is a rational function, which employs rational selection not natural selection. Also, please note that I refer to this mental function by evoking several metaphors:

- First, from Plato (1952), I borrowed the metaphor of the "god function."
- Second, from Darwin (1859), I borrowed the metaphor of evolution, mental vs. biological.
- Third, from information theory, I borrowed the metaphor of the concepts of uncertainty but not the concept of information. However, each of these words which were first defined in the context of verbal communication, were polluted, as noted above and must be completely reconceived and redefined in order to be rehabilitated. Information will be replaced with knowledge.
- Fourth, from physiology I borrowed the concepts of metabolism: biological and mental and two kinds of digestion: biological versus mental digestion.
- Fifth, from Claude Bernard (1957), I borrowed the distinction between the internal and external milieu, which I extend to the mind.
- Sixth, from Ashby (1960), I borrow the concept of a large neuronal system that uses inhibition to cut the large system into pieces or into local stabilities.

The primary reason biological evolution created the nonverbal mind was because biological evolution could no longer create or endow nativistic creatures with all the knowledge they needed to be stored in their genome that was required for a life of change. While the nonverbal mind with only three wisdoms was limited in its ability to create knowledge, by the time of the evolution of the verbal four-wisdom mind, it had achieved a place where it could now evolve without limits. The fourth wisdom may have the potential to extract all the potential nonverbal knowledge stored within the cosmos. The nonverbal mind was designed to enable each three-wisdom mammal to self-organize, to create on its own, the domain specific knowledge symbols required to take the uncertainty out of the anomalies of any domain.

The nature of wisdom: two verbal skills:

- Seventh, perhaps one of the special skills that made humans unique was the evolution of two not just one new speciation technology. The fourth wisdom obviously endowed humanity with the ability to create new concepts within closed sets of forms, new memes, but the secret fact is that nature endowed the human mammals with perhaps the ultimate advance, the ability to create new wisdoms and new alphabets.

The nonverbal mammals were the first species forced out of their closed comfortable niches and yet successfully made their way in a strange world. The nonverbal mind was designed to survive in a world of change, a world full of anomalies, full of strange new things never before seen. Yet the nonverbal mind takes all this novelty in stride for digesting this class of unknown nonverbal things is the rational nonverbal mind's reason for being.

The emergence of the nonverbal mind upon the face of the earth and probably on the face of the cosmos altered the possible futures for the cosmos for the better or worse for biological life without a mind had reached a dead end. It was as if life had run out of fuel and a new source of knowledge, the fuel of the mind, was required to get life moving again. The nonverbal mind bypassed genomic-knowledge which historically served all the no-mind animals so well, ignoring the fact that most of them, perhaps seventy-five percent of them, were now extinct.

Nature had to find a better way to supply life with knowledge. The mammals and the evolution of the cortex with its newly emerging neuronal cells proved to be the solution. Thus, the nonverbal mind emerged.

Heterostasis I

My goal of persuading the scientific community to distinguish Homeostasis and Heterostasis is not going to be easy and in the short term may be impossible. In this section, I will identify several ways that the two can be distinguished. Kluver come close in the quote below; I would have preferred him to say-- on a Heterostasis in the milieu externe.

"I submit that the Harvard emphasis on homeostasis in the milieu interne should be counterbalanced by a Chicago emphasis on a homeostasis in the milieu externe." (Kluver, 1951, P. 150)

The general differences between H and H

Human nature is a product of at least four kinds of wisdoms, Homeostasis, Hedonism, Heterostasis I, and Heterostasis II. Human nature is not the product of just one mind, or one-wisdom, or one-soul but is a function of at least four material wisdoms or technologies. Instead of calling all the various forms of behavioral technology by the one name, Homeostasis, I call each of the special kinds of behaviors, which these four technologies emit by a different name: Homeostasis, Hedonism, and Heterostasis I and II. In my opinion, there is nothing to gain but much to lose from failing to distinguish each of these special kinds of behaviors; it doesn't seem reasonable to assume that homeostatic instinctive behavior and the mind-based rational behaviors should be called by the same name. With the design of each of these new technologies, evolution created an ever more complex human nature.

Time-Sharing vs. Batch Processing

What is a mind? The concept of mind actually is simple to understand if we ask the right question and in the right context. What is mind? The context, which makes the mind the simplest and easiest to understand, is a classic computer distinction. If we broaden our question and ask what is the difference between Homeostasis and Heterostasis, then the operational question in computer jargon is; what is the difference between time-sharing and batch processing: Homeostasis consists of a set of batch processing systems, while Heterostasis consists of a time-sharing system.

Homeostasis: a batch processing solution

Nature designed Homeostasis as first described by Cannon (1932) in his book *The Wisdom of the Body* as a set of relatively independent physiological functions, each with its own set point and its own regulator. All the regulators are relatively independent mechanisms, which can function at the same time during both sleeping and waking states. Homeostasis is the wisdom of the body.

Heterostasis: a time-sharing system

The first computer system the state of New Jersey furnished all the state college campuses was computer time on a centrally located very fast time sharing computer for its day. Additionally, each user at each school was given a Selectric typewriter, a very slow system, with a modem, which connected the typewriter to the fast central computer. The central time-sharing system would read the text for each reader several times a second, so that each user felt he was the only user online, that the computer was only working for him.

This one mind, moreover, must be most versatile for it must solve every waking problem as a general-purpose mechanism, which can and must do everything that must be done, but on a time sharing basis. The mechanism of the mind, which emits Heterostasis behavior, consists of only one control center, which must be capable of executing or processing every heterostatic function of which the mind of every mammalian species is capable. There is only one mind, and every occasion, which requires the attention of the mind, must compete with all the other crises, which are calling for time on the time- sharing system.

What is change?

There are too many concepts of change, yet there is one concept, which meets our specification. By change, we refer to contextual anomalous events, events so new and strange that they cannot be assimilated into the current narrative. There is nothing about the anomalous event to remind the organism of any other event ever observed.

Consciousness and attention

For example, consciousness and attention are most easily defined with respect to the differences between Homeostasis and Heterostasis.

Since each Homeostatic error is reduced by a set of dedicated regulators, one for each EV, since these regulators all function constantly, and automatically; they have no need for consciousness or attention. Who is aware of what their liver or kidney is doing right now?

However, since the Heterostatic mechanism is time- shared, the mind must have a way to decide who and when a problem gets access to the controller. No problem of the EMm can usurp the full time of the controller for there is only one waking consciousness to go around.

I propose that anomalies are nature's way of assigning priorities. As long as a model is valid, it will function automatically and errorless outside of consciousness; there is no reason to shift consciousness or any justification for consciousness. As Angell noted, "consciousness recedes from a habitual task." Yet an anomaly turns on consciousness, and indeed the unusual grabs consciousness and usurps consciousness but recedes as a task is mastered. In the Heterostatic context, since there is only central processing register where every function is processed then there must be some way to assign propriety to each function. What is an anomaly? What is External Inhibition? The priority of a task is a function of the quantity of the uncertainty it induces.

Heterostasis, the nonverbal communication with nature, or god

Heterostasis, on the somatic side, is a two-sided coin. Heterostasis can best be described as a "seek and destroy operation." Heterostasis is triggered by an anomaly: its reason for being is to destroy anomalies; to find causes for surprising effects; to destroy uncertainty; to communicate with the nonverbal features of a material domain; to probe for causal answers for each anomaly.

All scientists, nonverbal and verbal, in executing an experiment are asking the cosmos a nonverbal question. Every experimenter in designing an experiment has a nonverbal question in mind that he/she puts to nature and the point of the experiment is to discover nonverbal answers for the nonverbal questions, which the experiment asks.

Perhaps one way to call attention to the central NV/communication role in Science is by calling attention to both the central role of numbers in the independent variables, IV, and the dependent variables, DV. The questions scientists ask of nature involve variables, which operationally CONSIST of numbers, and likewise the dependent variables output are numbers. Yet at the next level down these variables, the number questions are embedded in a design that operationalizes causality. I could wish to be more eloquent. The capacity of an experiment question to probe nature is probably more eloquent than Shakespeare is.

Why are experiments always nonverbal, because nature is nonverbal! Only the second half of a scientific experiment is verbal; each human experimenter is required to export his nonverbal knowledge to share his/her scientific results with the community. The verbal mind makes public what the nonverbal mind has done, observed, and knows.

However, with the emergence of Heterostasis I as the technology of the nonverbal mind, each individual mammalian species was not only enabled but also required to communicate nonverbally and directly with nature to create the domain-specific and species-specific knowledge required for them to complete successfully each of their self-defined domain-specific goals.

How does nonverbal behavior, which the nonverbal mind emits, enable mammals to create domain-specific models of each terrestrial space, and thus understand, and be prepared for whatever events come next in a series of events or scripts? The rational nonverbal mind enables each mammal to know where it is within the one hundred unit maze with respect to all the other units of the maze; or to understand how and where every unit of the maze fits into the whole (maze); the place of each part in the whole.

Therefore, rats and other mammals were given by evolution a nonverbal mind to interrogate nature and discover her hierarchical causal conceptual structures. In the first experiment of Book 2, the rational behavior invented by the nonverbal mind is a complex product of all the three wisdoms: HHH1.

The rational role of retracing in rational behavior

In a nonverbal communication encounter, which I will call an active mental event or a mind-matter exchange, knowledge not energy is exchanged. The difference a mind makes is something objective, something that can be seen, but it is harder to describe objectively. When a rat explores a maze, for example, the rat never moves randomly or ever blindly bumps into the walls of the maze, or makes random turns in the maze. Heterostasis in the experiments described in this book is the rational nonverbal pattern of interactions, the operational side of the nonverbal communication constantly occurring between the rat's nonverbal mind and the nonverbal structure of the maze component of the cosmos.

Heterostasis is a special rational form of behavior, which is limited in rats to two selective classes or forms of responses, but each of these responses serves a special role in the mind's interrogation of the physical world objects such as the maze. These two basic forms of behavior are designed specifically to have a particular knowledge-gleaning role.

Phase one: Exploratory behavior

The first segment of Heterostatic behavior is called exploratory behavior and is designed to discover anomalies, to locate objects or events, which induce uncertainty. When a nonverbal mind has created a valid model and the mind is getting bored, then it is time to turn on the exploratory behavior, which proceeds forward into the unknown, until an anomaly is encountered or discovered.

Phase two: Retrace behavior

Retrace behavior is designed to do just the opposite, to locate causes, which will reduce the uncertainty induced by each anomaly. Retrace behavior, by searching in the region before the anomaly, searches that part of the maze where logically the causes for each anomaly should be found. The rat's nonverbal mind knows the nature of the causal relation between these two kinds of Heterostatic responses - explore and retrace. First, you explore until an anomaly, which is an effect without a known cause is found then you retrace to find the cause for the latest anomaly. Once the nonverbal mind knows the causes of the anomaly, all those causal events, which predict the surprising events then it can convert the anomaly into a predictable class of events.

Homeostasis is the physiological function which nature designed to regulate the essential variables, EVs, of the internal milieu, to keep the EV's of the internal milieu at their set point or within the limits with which life is compatible.

Heterostasis, on the other hand, is a physiological function nature designed to control/regulate the EV of the external milieu. A valid model is one that is incapable of inducing surprise. A valid model is one in which the narrative in consciousness and the events of the External milieu of the mind, EMm, are in a one to one relation, so that the mammal is safe to act upon the content of its consciousness, i.e. can act without being surprised.

Lashley (1963) spent his scientific life searching for what he called an engram, which he defined as a physiological structure somewhere within the mind, probably within the cortex, which encoded learned changes.

Plato (1952) searched for the A, the eternal component of an idea, which was to his mind a sacred function, which could not be executed in any "Thing," any being made of matter.

One of the most exciting discoveries of my research is that this "so-called sacred function" can also be executed in the nonverbal mind of a rat. Moreover, I will show that one of the two halves of the heterostatic cycle, Exploring or Retracing, is the occasion during which the sacred function is executed by the nonverbal mind. Which behavioral element accounts for the evolution of a form?

Experimentalists find the retracing behavior of rats the most hated element of rodent behavior. Yet paradoxically, ideas are born within the nonverbal mind only during retracing, which provides the context within which ideas emerge.

While the exploratory phase of heterostasis behavior is the happy phase where each mammal explores without a care in the world, the unexpected encounter with an anomaly forebodes events or encounters with consequences that cannot be predicted. During the retrace phase, the rat is vulnerable for while searching for a cause they are behaving in a space where theoretically anything could happen. The retrace zone is a potentially dangerous place where it is impossible to predict what could happen.

Yet again, I must remind the reader that the whole heterostatic cycle begins with an anomaly.

Kinds of anomalies: absolutes and relative

One of the paradoxical aspects of heterostasis behavior is to understand the role of Retrace behavior, which is the most important element of rational behavior. When an experimental psychologist is running a rat in a maze, and believes that a rat cannot learn anything until it gets reinforced in the start box, and all the rat seem to do is retrace, some become unbelievably frustrated. The retrace behavior, they see as both a waste of their time and the rat's time. Blodgett (1929) was so proud of inventing the nonretrace doors, which in his opinion prevented rats from wasting their time and the Experimenter's time. The behaviorists assumed that the rat's behavior was naturally random behavior, and that retracing was the ultimate sign of random behavior. The retrace doors make retracing between rows impossible.

One of my first insights was that I should let the rat do whatever it wanted to do and just track it through the maze both out and back to the start/end box. What I finally was able to prove is that it is during retracing that rats learn. The retrace doors were in effect making it impossible for the rat to learn as it normally does. Since rat's have a nonverbal mind, when the experimenter blocks the animal's ability to retrace, it is a very counterproductive procedure.

First, retrace behavior and random behavior have several differences, but the most obvious and consistent difference is that random behavior is a form of Brownian motion with an average number of steps between nodes, a random number. The point is that random behavior has a special look about it. It looks like Brownian motion. It is a pure embodiment of random behavior.

On the other hand, retracing behavior is a critical element of rational behavior, which is never random behavior. In book 2, I will describe several ways that the retrace behavior of the rat is rational behavior.

First, for example, I will show that rational rats explore and retrace rationally from node to node, which random rats almost never do.

Second rational behavior creates retrace loops, which always retrace back to the alpha-one unit. The degree of retracing objectively measures how much uncertainty was induced by an anomaly. The length of a retrace loop provides an objective measure of just how much surprise or uncertainty was induced by any given anomaly.

Two kinds

However, there are, in principal, two kinds of anomalies: absolute and relative. A hierarchically organized structure such as the matrix maze consists of a four dimensional matrix. An absolute anomaly destroys the validity of the whole model X, which allowed the absolute anomaly to occur in the first place. By contrast, a relative anomaly, in the best case, plays a positive role by blocking the forward exploration until the uncertainty is reduced. Relative anomalies in an important sense direct the creative process; they direct the behavior that discovers the cause for each anomaly. Each cause is a new idea, a new parameter of the maze. Each new relative anomaly merely shows that what the rat may have thought to be an X-dimensional structure actually has another yet unknown dimension.

Retracing behavior plays a critical rational role by retracing all the way back to the Alpha unit so that on the way back the nonverbal mind can verify or update any of the parameters that are invalid.

What is the advantage of having a mind, one nonverbal rational mind?

The nonverbal three-wisdom beings have three kinds of adaptive mechanisms: homeostasis, the wisdom of the body: Hedonism, the wisdom of the emotions: and Heterostasis I, the wisdom of the nonverbal mind. Thus at Stage 3, the nonverbal mammals can: First, regulate their internal milieu; Second, regulate their feelings and emotions and; Third, control the essential aspects of their terrestrial external milieu. (Bernard, 1957)

The Heterostatic skills endow mammals with the general-purpose sensory-motor physiological technology required to control rationally their behavior within the External milieus of their mind, EMm. In addition, of course, they still have the capacity to regulate all the EV's of the internal milieu of their body, IMb. In addition, while no one talks about it, mammals have an elaborate chemical technology for regulating their emotions. The one-mind species have become a self-organizing form of life that can adapt to environmental change at many physiological levels. Biological evolution has endowed all these one-mind nonverbal mammals with the wisdoms to create all the domain-specific ideas or knowledge required for them to model anomalous spatial domains and become self-organizing beings.

Given the great vulnerability of the no-mind species, evolution fortunately began to design a new kind of species with more robust survival skills: a set of beings that are more complex, a new kind of life and ones that could adapt to new external circumstances, whose survival is a function of rational selection rather than natural selection. The adaptation time of a mammal can take only minutes, while change in natural selection can take millions of years.

The emergence of this new rational form of life was a manifestation of the emergence of a new rational organ, the nonverbal mind. This new organ went a long way in overcoming the no-mind animal's fatal adaptive flaw, its instinctive behavior no longer worked if its niche changed. A change in a native niche was a death sentence. If a native niche changed, the species that were adapted to it went extinct.

One of the central ideas of this book is that in order for a species of life to regulate and/or control their evolving or changing external milieus requires a mind. Heterostasis endows life with the power to change as their niche changes. As a niche changes, the nonverbal mind enables mammals to evolve new sets of domain-specific knowledge or models that allows them rationally to continue to conceive domain -specific goals and the means to achieve those ends even as their niches change.

While the no-mind animals, all the pre-mammals such as turtles, with homeostasis as their only form of physiological wisdom, have only one kind of adaptive wisdom; the kind they are born with and with which they can only regulate the Essential Variables of their internal milieu and external milieu with genomic knowledge. However, the one-mind mammals with Heterostasis 1, in addition to Homeostasis, can not only now regulate their internal milieu of their bodies, IMb; additionally, they have already begun the race to control more than the terrestrial components of their external milieu. The evolution of the nonverbal mind expanded the limited adaptive capacity of the no-mind species. (Bernard, 1957)

The essence of the difference is that the complexity of all the possible external milieus that these one-mind mammals could encounter in their lifetime is so varied and complex that the survival of this new kind of life required a most powerful supplementary knowledge source, the nonverbal mind armed with Heterostatic behavior.

As mammals acquired more powerful adaptive skills and technology; I must keep repeating that their success is the joint product of their cumulative wisdoms. In particular, the technology of Heterostasis works only because genomic knowledge is supplying, completely and constantly, all the physiological knowledge required for the cellular construction of all the organs of all the bodies, including the brain for the one-mind mammals. However, this genomic knowledge provides almost no assistance when it comes to adapting the behavior of any mammalian species to any one of the infinite number of available external milieus. In short, the genomic knowledge is adequate normally for solving all the physiological problems required by any species to regulate the internal milieu of the body but is useless for assisting in designing any of the novel somatic behavior patterns required for adapting to any novel external milieu of their nonverbal mind.

The nonverbal mind species are active beings. Technically, as we shall see with Female 3, each individual adult mammal is responsible for its own survival; each is responsible for conceiving and executing its own goals and plans of behavior. In addition, each plan of behavior must be rationally conceived to match the contingencies of each specific niche. Each adult one-mind mammal ultimately has the responsibility for its own life; each can only survive by rationally self-organizing a plan of behavior that is appropriate with both its own needs and the complex objective parameters of the External Milieu of its current domain. Thus was born the first self-organizing system. The domain-specific knowledge required to adapt to each species-specific unique external milieu has to be entirely and independently conceived or rationally evolved by each adult individual mammal's nonverbal mind.

"We can also see another possible goal in the behavior: the child is achieving knowledge, attaining a more differentiated cognitive map of his environment and thus satisfying an exploratory tendency or motive of curiosity."(White, 1976, P. 327)

Stephen

One of my first personal experiences with Heterostasis was a consequence of observing my son Stephen, when he was at the age where he could pull himself up off the floor and hold onto a piece of furniture and use it to walk. One day he was holding on to his high chair and walking around the chair. I wasn't paying any attention to what he was doing, but all of a sudden, I heard a loud noise, and Stephen started to scream as if he was injured.

I dropped what I was doing and rushed to his rescue. He was all tangled up with the chair, so first I pulled him out of the chair. Then I set the chair back up and allowed him to continue his exploratory play behavior. I had no more gotten back to my chair and sat down when again, I heard a loud noise, and Stephen screamed again. Again, I rushed to his rescue. However, he had turned the high chair over again. Suddenly I realized he was not hurt, but only excited, in fact he was pleased, he had discovered for the first time that he could "turn chairs over." The first time was an accident, but the second time was deliberate, in fact, he couldn't have more pleased with himself. Soon he was pushing on all of the several large pieces of furniture in our house seeing if he could turn them over.

This behavior illustrates the way the Heterostat learns about himself and his world.

However, mammalian young are born small and helpless requiring a long period of childhood and so mammals form families. The role of the mind in enabling the advent of families is a topic beyond my expertise. Alternatively, perhaps the questions should be; which came first, the mind or the family; the breast or the mind? Should mammals be called mammals, or mindmals?

The fundamental difference between the no-mind and the one-mind species is that the no-mind species were designed to survive in a world of constancy. While the one-mind species were designed to deal with, to survive, and thrive, within a world of change. The former closed worlds are worlds without uncertainty. However, the new second kinds of worlds are open worlds, worlds full of change and uncertainty that both threaten and challenge survival, and must immediately be reduced.

As noted above, Claude Bernard (1957) distinguished only two of the four kinds of essential milieus the internal and external milieu of the body. Walter Cannon (1932) first explained that the physiology for regulating the internal milieu of the body was Homeostasis, which he called the wisdom of the body. In this book, I introduced two additional kinds of mental milieus, the Internal Milieu of the mind, IMm and the External Milieu of the mind, EMm.

Cannon (1932), however, at places in his book, talked as if homeostasis was the one-and-only, all-purpose physiology, and I am surprised at how many physiologists and psychologists still accept this assumption. They are obviously unaware of how much more complex is the physiological technology required to adapt to the external and the internal milieus of the mind than the internal milieus of the body. They are still unaware of how great are the differences the nonverbal mind makes.

Nevertheless, the one phenomenon that surprises and admittedly annoys me are the considerable number of book authors, who are so cavalier about writing a book with the word mind in the title, which I am sure sells books, and never even use the word on the inside of their book.

The greatest opposition to the concept of Heterostasis, I expect to come from the experts who have adopted Cannon's idea that Homeostasis is the end all and be all adaptive technology. Cannon (1932) even presented some phenomenon in social psychology as examples of Homeostasis.

However, there is a ray of hope; there have long been those who intuitively understood that Homeostasis was not the end all, be all of physiological motivating forces of human nature.

One of the authors who defined heterostasis was Carl Menninger. I have included two quotes below to illustrate the way he conceived Heterostasis.

"There is a strong urge within the organism to effect change, to initiate some of the very disturbances which the regulatory processes of the organism are patterned to resist. It is the conflict between the wish for change- newness, variety, opportunity- and the fear of other consequences of change which makes for the complexity of human behavior." (Menninger, 1963, P. 85)

"One danger in this way of thinking is that it tries to explain the applicability or inapplicability of homeostasis by the principle itself. This is partly because there are certain phenomena, such as growth, which more aptly illustrate the entirely different principle of heterostasis, i.e., the progressively moving away from the status quo, the search for new and unsettled states, in contrast to the automatic return to the comfortable and relatively tension-free previous state of balance."(Menninger, 1963, P. 85)

The experts who continue to call Heterostasis, Homeostasis all make the same mistake, the two physiological functions, technically are confounded. In the next section, I will illustrate this most classic and prevalent kind of confounding.

An example of Confounding: Thermostats

The most common reason most scientists don't see Heterostasis is because they see it but classify it as homeostasis. This is a common example of confounding, which professional scientists make when explaining Homeostasis. They advance the idea that the regulation of the body temperature and the room temperature are both examples of Homeostasis. From my perspective, Homeostasis technically should be defined as the regulation of the internal milieu of the body, while Heterostasis is the control of the Internal and External milieu of the mind, IMm and EMm.

Temperature regulation in the body is genetically programmed. The set point and all the organs, which are a part of the body temperature regulator, have a biological origin, and are a part of the internal milieu. However, the thermostats on the walls of our homes, which regulate room temperature are not biological organs and were not designed by biological evolution. All the parts of the thermostats on our walls and the electricity to run the unit, and the heaters, etc. are dry technologies and were designed by the human mind.

While both the body and the room temperature regulators employ the same engineering logic, all regulators employ the same functional logic, but the technology which regulates IMb is designed by our genomes, while the technology which regulates our room's temperatures are mind –designed machines. No mammals, other than humans, have homes with heating and air conditioning units and thermostats to regulate them. To call the two profoundly different technologies which embody the two different kinds of temperature regulating mechanisms the same, Homeostasis, perhaps does no great harm, but to do so means that the difference the emergence of the nonverbal mind has made and is still making, will go unseen. Examining the modern technology which dominate our lives: cars, smart phones, planes, computers, etc. all are technologies which are the product of the emergence of the generic nonverbal mind and have made the most revolutionary changes in the nature of our life than the emergence of any other organ.

The generic mind

I call the nonverbal mind the generic mind; first for its general capacity to master all kinds of hierarchically organized terrestrial spaces; and secondly, for its general capacity to work for all the many and varied types of mammals. Mammals for which this generic mind has worked so well for millions of years in spite of all the many kinds of domains they have occupied.

Rationality: nonverbal and verbal

Both the mammalian nonverbal mind and the Platonic model of the human verbal mind are rational. However, the word rationality has a completely different meaning for each of these two minds. For Plato (1952) rationality was a verbal function. Plato was not focused upon thinking about how the individual nonverbal beings, humans, or nonhuman's minds worked. Plato was an ivory tower man, a talker not a doer. For the nonverbal mammals doing is everything; talking is not an element of their life; but then for Plato doing was not a major part of his life.

For nonverbal mammals, rationality is a nonverbal function. Defining nonverbal rationality is a serious challenge. However, rationality, in general, is defined as the ability to act as a self-organizing system, to engage in both types of mental functions, both the god and the man function; which both involve complex acts, which require knowledge stored within hierarchically organized sets of categories which enables the Internal Milieu of the rat's mind to return to its set points.

The secret of the nonverbal mind is its chief mental nonverbal function, which enables mammals to self-organize their behavior in new niches. The secret is their rational ability to transform each particular anomalous domain into one in which all the objects of the domain are categorized, but not into separate or independent categories; but into a coherent sets of hierarchically organized categories, from which plans or scripts can be conceived and executed. The magic of the nonverbal mind, what makes mammals unique, is certainly not the power of verbal speech. The root rational power of nonverbal speech is its rational nonverbal evolutionary speciation function. Each new nonverbal idea or new nonverbal concept or form, Plato (1952) insisted were spirits, and were the product of the mind of god. Rationality is a set of categorizing skills or rational wisdoms. Each kind of new wisdom expands the adaptive capacity of life. The first two wisdoms were not rational wisdoms for they evolved before the nonverbal mind emerged. There are many possible rational wisdoms, each expands the adaptive capacity of nonverbal behavior and humans by employing the primary intelligence of the nonverbal mind to conceive a new rational skill.

Two-mind functions

The nonverbal wisdom, the primary wisdom, is the first-rational wisdom, but is the only wisdom the nonverbal mammals have. The general topic of wisdoms and symbols; the evolution of the higher order wisdoms and symbols belongs under the topic of the two-mind behavior, and will be discussed in more detail in later sections. For the moment, however, I want to start our conversion about the topic of the many kinds of wisdoms and symbols by commenting on the secondary role of the verbal mind and its creation of the nonverbal symbol, the nonverbal wisdom, the nonverbal consciousness, etc.

Only the human's with a compound mind, a NV/V mind, can expand this creative evolutionary function of wisdoms and symbols into the higher-order domains. All kinds of new domains: music, art, dance, math, etc., but each new kind of wisdom, which expands the rationality of a species into new regions of the cosmos, requires the invention of a new unique set of symbols. Music required the invention of a musical scale: a new set of domain-specific symbols, which can create a set of musical "words," ideas, or "things" which can represent every possible class of musical sound which nature can produce. Only a two-mind species has the rational capacity to raise the categorizing skill of the nonverbal mind to higher powers: to conceive the new sets of symbols, secondary symbols, which enable the mind to expand rationality into new non-spatial domains.

The need for animal experiment

Why should science study the rat's behavior in the Lashley III matrix maze? Because the simple fact is that there is probably no better place for mankind to go to begin to master the way the primary level of the vertical mind works than by studying the behavior of a nonverbal organism adapting to anomalies, novelty, and redundancy. The experiments of this series offer the students of the NV-mind a place to start to study these kinds of behavior described in these experiments.

Hierarchy in rats

Rats are exemplars of the prototypical mammalian organism. Rats were the first organism to evolve a pragmatic nonverbal mind with the capacity to create a hierarchically organized set of primary symbols isomorphic to a hierarchically organized domain such as the Lashley III matrix maze. Such a system creates sets of hierarchically organized primary symbols to mediate its behavior in response to anomalies, novelty, and redundancy.

Humankind, if it ever became aware and interested in the symbol generating NV- mind, will have to first overcome its hubris and admit to sharing with mammals the NV-mind, which creates the primary set of nonverbal symbols. Mindlessness is the denial of the primary role the nonverbal symbols play in higher order symbol systems, the verbal, mathematical, musical, etc.

Two examples of the blinding conceit, which still animates/shackles so many human's minds, are quoted below.

"Animals, in fact, can be very smart, emotionally attached, and caring – no less than humans - but in all their actions they are driven by reflexes and conditioned reflexes. They always execute a plan planted in their mind by training, previous experience, or hard-wired into their DNA." (Vyshedskiy, 2008, P. 47)

There are no self-organizing mammals possible in his estimation. However, Maslow reveals almost the same kind and degree of human conceit!

"There is no reason whatsoever why we should start with animals in order to study human motivation." (Maslow, 1974, P. 241)

One of the most fundament beliefs, which made this book possible is and was my belief that statements like the two above had absolutely no scientific or empirical justification.

For the first generation pragmatists the ultimate meaning of a word was an action. The first generation was just as deeply committed to the verbal model of human nature as was possible: where they chose to make their stand was to raise a question about how words got meaning. It is impossible for me to speculate about they would have said about the idea of the nonverbal mind, and the idea of a nonverbal idea, and nonverbal meaning, etc.

To this third generation pragmatist, pragmatism has moved miles beyond the verbal context into which the first generation was locked. I am now locked into the topic of the nonverbal mind, and all the Pragmatic questions suddenly are nonverbal ones. Heterostasis is a nonverbal form of pragmatism, which requires a more rigorous behaviorism than practiced by the classic behaviorists. Every idea must be defined, not by running to Webster, but rather to some nonverbal data in some experiment, that measures the behavior of the organism. For example, in the series of experiments to be reported, I will show that the rat has a nonverbal ideas or concepts of a unit, row, and two kinds of rows, and several exemplars of each kind of unit. However, now the rule is that all nonverbal ideas, etc. are defined by the behavior of the rat.

Ideas evolve and grow and in most complex ways. A part of my task is to present data that will make it difficult for the reader to agree with Plato. Plato believed that ideas are spirits or nonmaterial beings, but I propose that ideas begin life nonverbally as material beings created within a material organ; a nonverbal mind busy coming to terms with hierarchically organized domain full of nonverbal ideas created by the pragmatic nonverbal interaction between physical objects and a nonverbal mind. The nonverbal mind works to enable nonverbal mammals, as SOS, to conceive a model that is isomorphic with a down to earth physical reality. How can I show this, by allowing my rats to make my case, by showing that the primary ideas of the matrix maze can be seen in their behavior. Our purpose is to establish the nonverbal mind to its rightful place in all those scientific disciplines today that concern themselves with a physical model of human nature.

My task is to show how the primary mind evolves ideas or models of things external to the brain. Each class of thing pragmatism claims is defined by the behavior of the organism. Heterostasis cannot evolve as a one sided science as did the classical behaviorist who focused primarily upon the output side of the organism. Heterostasis, as a pragmatic science, is obliged to show that each mammalian species' idea of each class of thing is defined by the organism's behavior while interacting with the things of the cosmos. As a pragmatist, I will modify one of Skinner's concepts and claim that every idea must be defined by its behavioral signature.

Primary symbols

What is a symbol? In the standard theory of language, the conventional symbols are alphabetic, and are parts of words, the elements of words. Words are spelled with alphabetic symbols.

However, in the nonverbal world, the myth would propose that there are no symbols. Yet I propose that pre alphabetic symbols are a primary element of the verbal mind, the elements out of which images are constructed--sensations are the symbols of the nonverbal mind, which create images or perceptions.

Images in humans are absolutely tied to the verbal behavior and the verbal mind. This classical concept of language does not recognize the rational role of the nonverbal mind.

This conceit is but one small part of the big bag of conceits, which sustain the classic myth of human nature: humans are the only species with a mind; the only symbol using species, the only rational species. Etc. All mammalian species with the generic nonverbal mammalian mind employ the primary set of systems, the sensations out of which images are constructed, and which create the narrative of the nonverbal consciousness.

The classic western myth of human nature advances the notion that humans are special; are the only species endowed with a mind; there is only one mind, and it is a verbal mind. Symbols are the inner most element of the hierarchical structure of the mind. The full-blown myth, of course, holds that a nonverbal mind does not exist.

My purpose is to make the invisible nonverbal mind visible and to make the nonverbal mammal's invisible mind recognizable.

The failure of the human mind to recognize the elements of the primary mind shows that the verbal introspective skills employed by the philosophers for two thousand years have proven inadequate to the task of understanding the nature of rationality, the nature of mind, and the nature of consciousness. The introspective methods have not worked, nor have the methods of the behaviorists worked. It is clearly time to try a new method. The paradox is that the invisible primary mind has become visible for the first time in the spontaneous exploratory play behavior of rats in the matrix maze. The invisible nonverbal mind now can only be seen in the public objective data generated by rats exploring the matrix maze.

Once the rat data works its way into our human consciousness, once science has evolved a conceptual system capable of understanding this prototypical mental behavior, then it should be an easy matter to begin to see the same behaviors in humans, and to understand what is going on in our human consciousness. Perhaps our own mind cannot see itself until it learns to see the mind of the rat, but having accomplished that step, then perhaps our mind can begin to become conscious of itself. Mind consciousness may well be the last kind of consciousness to evolve.

History has proven one thing; the human verbal mind cannot see the human nonverbal mind by the method of introspection. The only way the nonverbal mind can become visible to the verbal human mind is by practicing a more powerful form of methodological behaviorism than was practiced by Watson (1972). Watson was half right- half wrong. While his conceptual behavior in effect was able to destroy the potential effectiveness of methodological behaviorism for the first hundred years, yet methodological behavior is the right prescription; the only problem is that Watson's prescription was not a strong enough dose of medicine. The purpose of this book is to perfect a super dose of methodological behaviorism, one strong enough to make the heterostatic behavior that is the sign of mind and consciousness, public and objective data for the first time.

The purpose of this book is to make it possible for the human verbal mind to see the nonverbal mind by observing the nonverbal behavior of the primary mind of the rat. The primary mind was never observed by introspection, but the public and objective data collected by the behaviorist was less effective than was introspection. Only a more rigorous form of behavior created by the new method employed in the four experiments in this book can reveal the heterostatic data of rats. Paradoxically the behavior of the rat in the maze that the behaviorist called random behavior is not random behavior, but is the sign of the rational nonverbal mind. Anyone interested in his or her own mind, I am proposing, must be willing to take a lesson from a rat. Our long lost ancestor, the rat has much to teach us about ourselves. The primary mind must first be observed in the behavior of another species. The verbal mind is too smart for its own good. The verbal mind has been taking the credit for all the functions of the nonverbal mind.

However, once the verbal mind has learned to observe the nonverbal mind in the behavior of the rat, the magic spell will surely have been broken and the nonverbal mind of humans will finally become visible for all to see. Once the verbal mind not just sees but acknowledges the primary role of its silent partner, the nonverbal mind, the concept of human nature will have taken a giant step forward.

The chief function of the nonverbal mind is to create causal conceptual systems, (this will be discussed in detail in chapter five), and this task is what mental growing is all about. The primary mind creates the nonverbal primary symbols which give life and meaning to the neural impulse patterns entering the brain. The primary mind employs the primary nonverbal symbols to create in consciousness a representation of the external milieu that seems so real and that has functioned so successfully for so many millions of years that the existence of the symbols has escaped notice. Yet understanding the role or functions of the primary symbols are the critical step, not just a token acknowledgment of their existence. The only small step, humans have taken in the direction of modesty is that they attributed their rational superiority over other forms of life to the yet unidentified, set of verbal skills of their verbal mind. Yet, the next big step, which this book is promoting, is one that humans have not been willing or able to even understand, much less to take, that the emergence of the nonverbal perceptual world, collectively called consciousness is wholly the creation of the verbal mind's nonverbal partner. The classic human myth attributing our entire rational superiority to our obvious verbal skills is like a jockey who takes all the credit for winning the Kentucky Derby, as if his horse didn't even exist. Our consciousness, our rationality, our verbal narrative are all the work of a powerful invisible staff of experts who do all the work, but never get the credit. Everyone praises President X for a great speech, when on some occasions our president should only be praised for reading a great speech well. How many great humans are great only because they have at their disposal a vast army of verbal eunuchs?

I propose that each human who aspires to greatness, had better hope that he or she has a powerful nonverbal mind working in him or her, and on the job twenty-four hours a day, 365 days a year, etc.

Mental evolution is the product of our ability as slaves sentenced to a life within a cave to create nonverbal streams or sets of nonverbal metaphors, which gives meaning to the neural impulse, or shadows, which enter into the cave through the afferent nerves. (Plato, 1952)

Sternberg's problem

Sternberg (1998) defines intelligence as "the ability to use metaphors," but which kind? Sternberg is talking about verbal metaphors, or in other words, he is talking about secondary verbal intelligence. However, in the text just above and everywhere in this book, I am talking about the nonverbal primary intelligence that creates the nonverbal perceptual metaphors of the primary mind. The primary intelligence of the rat and the secondary intelligence of humans are two different kinds of intelligence. Sternberg is another expert for whom the nonverbal primary mind is invisible.

To explain the general concept of wisdom requires a general concept of symbols. I anticipate one more problem, and that is selling the concept of the primary symbol or the nonverbal symbol. The set of primary symbols are those sensory qualities out of which all other symbols are constructed, namely the sensations. The sets of sensation symbols, which are at the disposal of the nonverbal rational mind, are the foundation for both the verbal and nonverbal consciousness, indeed for all possible wisdoms.

In the section on Aristotle's model (1952), I will show that all perceptions, all the exemplars that enter consciousness of every class of things before they enter consciousness are constructed from the elaborate sets of sensations, which are created by the five senses within the Primary Projection Areas by the first seven stations of Aristotle's twelve-station action cycle. The image of the rose in the human verbal consciousness, which is the meaning of the word rose, is assembled by the sixth sense from all the sets of sensations created by all the Primary Projection areas, PPA. These sensations are then delivered to Station 6 for the mind's animator to assemble all the pigments on its palette and are ready to become an image of a rose.

The nonverbal mind speaks the primary language, which I call "nonverbal," and its sensations are the building blocks of the images of consciousness, and are the first and the primary symbols and are the raw material out of which the nonverbal consciousness is constructed by the verbal eunuch partners of the verbal mind.

Image based consciousness

Rationality for the one-mind mammals means that the one-mind mammals are limited to transforming anomalous objects into a rational nonverbal world of recognizable things, which are experienced as exemplars or images of some nonverbal class otherwise known as assimilation.

Where does assimilation occur in a stage of verbal behavior or in a nonverbal stage of verbal behavior?

If, we get in the habit of thinking about rational behavior as occurring in two phases that practice alone will simplify our scientific life.

For example, for the nonverbal mammals, which do not have either components of the verbal mind in their left hemisphere, our question is pointless. If their mind is capable of the act of assimilation, which it is then that assimilation has to occur nonverbally.

However, the question only has meaning with respect the verbal mammal, and in my opinion is an important and interesting question; Sperry, I think, answered this question, when he reported that a quarter was recognized (assimilated) in the right hemisphere. In other words, the quarter had to have received a nonverbal assimilation. (Gazzaniga and Sperry, 1972)

However, a rational mammal with a rational nonverbal mind must experience assimilation nonverbally. The nonverbal mind does not and cannot talk verbally, but it thinks and talks in nonverbal classes, in categories embodied in primary images constructed of sensory symbols. The nonverbal mind uses images to plan its nonverbal behavior by relating to objects in terms of categories, classes or images. Each class is the product of the nonverbal mind's having evolved a new nonverbal concept, a new mold, or form, or category; but made flesh by its incarnation as images, beginning with sensations, and then into 3D-space, and then into a virtual reality.

The nonverbal consciousness

The rational powers of the nonverbal mind to evolve on-line and in real time new sets of domain specific classes are but one element of the nonverbal rational armament. However, this power to conceive would probably be useless if not for the nonverbal consciousness. The nonverbal mammal, with the nonverbal mind, is not an ivory tower thinker; but lives, or so it believes in a real world, which it experiences in every second of its waking consciousness. All it has to do is to open its eyes and all those rational categories become animated or alive before its eyes, and are as real as it is physical reference in nature are real. The nonverbal mammal, like Descartes (1946), can no more doubt the existence of the images of objects with which it shares consciousness than it can doubt its own existence.

In its waking consciousness, each nonverbal mind lives with all the animated embodiments of those rational categories that only moments before were but ideas just conceived, and waiting to be born in some neuronal network within some station of Aristotle's model which will be described later.

Compound or sets of ideas

However, even in rats, the conceptualizing ability long ago achieved such an extraordinary power that the rat's mind does not, or probably could not, conceive one lone idea. The rational power of the rats' nonverbal mind already is a master at conceiving coherent sets of hierarchically organized nonverbal ideas. The fun of exploring for F3, was exercising her nonverbal mind's "god-function."

Mental evolution; mental speciation--the god-function: What causes the order in experience: the categories or forms of the mind

Paradoxically, the functions, which were for Plato (1952) the "god-functions", and which provided, in his opinion the power to engage in mental evolution, the power to conceive and evolve hierarchically organized sets of ideas, is at the heart of the mammalian nonverbal rationality. Even Kant (1952) insisted that the creation of order or concepts was nativistic. Philosophers have consistently taken the view that "the god-function" was too great a power to have been entrusted to humans, even though the human mind was also a spirit.

There are important reasons why, for mammals, mental evolution, which was Plato's (1952) god-function is the key to rationality. The chief function of the nonverbal mind is the power to digest anomalies, the power to explore, discover, and transform each new anomaly into a new concept, a new idea, and a new image, which makes manifest the knowledge component inherent in its conceptually coherent model of the real world.

This stage of the development of my argument that the nonverbal mammals have a nonverbal mind brings us out of the fog, which characterized the last two thousand years, and allows us to see if not forever, at least to see things that couldn't be answered for the simple reason they couldn't even be thought or asked.

Intelligence

Historically, the nature of human verbal intelligence has been conceived as a question that could be answered with a pencil and paper test. Yet for someone working not with humans but with animals, the pencil and paper test doesn't work. Moreover, suddenly the intelligence of the nonverbal mind must be conceived as nonverbal, not as verbal. I must say again, the mind did not evolve in order to enable humans to talk with one another. Nonverbal intelligence evolved to enable mammals to deal with change, with anomalies, and with uncertainty, to act rationally in the midst of change. These kinds of problems require a mind with the biggest guns.

In the first stage, evolution created a series of pure genomic beings with only the capacity to adapt in one milieu, in one of Bernard's (1957) two milieus, the internal milieu.

After the great extinctions, evolution took on a new more ambitious task, to design a new form of life that could adapt in both of Bernard's (1957) milieus: both the internal and the external milieu. Walter Cannon (1932) called the skills required for adapting within the internal milieu, the homeostatic adaptive skills, and the wisdom of the body. The second wisdom, Hedonism was a nonverbal form of wisdom, which induced animals to do the right thing for the wrong reason.

Metabolism: biological and mental

Perhaps, one of the simplest ways to understand the differences between the no-mind and the one- mind species is to conceive of both forms as open systems. Both kinds of life have their own unique special kind of metabolic system and metabolic rate. The no mind species is open only for homeostatic needs, such as food, water, and air, and can be conceived as a great nutritional digestive system; it digests nutrition and energy.

The nonverbal mind species, on the other hand, are open systems in two ways and each system digests a different kind of input. The one-mind species is still a biological system that has all the standard homeostatic needs. In fact, in some respects its brain/mind system has even greater needs for some homeostatic substances such as oxygen than any no-mind species

What changes with the emergence of the third wisdom is that the one-mind species has two sets of needs, the traditional homeostatic or body needs, and it has a brand new set of mental needs, the Heterostatic needs. The mind feeds upon and digests novelty and uncertainty. Moreover, it actively seeks uncertainty, just as the body needs food to digest, just so the nonverbal mind needs uncertainty to digest. Perhaps we need to amend the Lord's Prayer:

"Give us this day our daily bread for our body and our daily uncertainty for our nonverbal mind."

The nonverbal mind digests anomalies by turning them into knowledge. The mind-generated knowledge means that the nonverbal mammal's survival is a function of two kinds of knowledge: (1) genomic- knowledge and (2) nonverbal mind created knowledge. With the emergence of the nonverbal mammals, evolution changed its focus and began to specialize in the design of open species, which no longer have to survive only on their genomic knowledge. Evolution began to endow mammals with a nonverbal mind, which made them self-organizing beings, which can create in real time the domain specific nonverbal knowledge required to achieve some domain-specific goal in some new domain.

The nutritional value of experience

The heterostatic mind feeds on novelty or anomalies. I will discuss this complex topic several times in this book, and each time for a different reason. However, one of the most important of these topics is called the Yerkes-Dodson law, which I will discuss in Chapter 2 at some length.

We need to talk about the nutritional value of experience. The nutritional value is always expressed in its ability to maintain consciousness. Why do we do new things, we have no choice. One of the central responsibilities of the nonverbal mind is to maintain consciousness at an optimal level of arousal.

When an anomaly is encountered, the arousal system of the mind, which will be discussed in chapter three knocks out all the learned behaviors active when the anomaly occurred. A strong anomaly destroys the model that allows the anomaly to happen. Under those conditions then, it is not that this provides an occasion for a mammal to learn,, on the contrary this condition provides an occasion, when a mammal must learn, in order to reestablish its capacity to function as a self-organizing system.

On many occasions, domains force mammals to learn. For example, in learning a concept, the mastery of the concept must pass through many grades, but learning always begins in the first grade. All the distance between a particular and a general and between a concrete and abstract are all *food for thought*. Every graded learning situation provides food for the mental life.

The neuron

Strangely, the hardest aspect of the story of the mental evolution of mammalian nature to sell is the idea that the technology of our biological machines changes even though the homeostatic needs stay the same. Most humans are aware to some degree of how the technologies of our manmade machines change and how they change our lives. How our cars, for example, change even though they still remain units of transportation. My point is that the emergence of the nonverbal mind constructed of billions of neurons changed the way the simplest of human tasks are executed. Homeostasis, Hedonism, and Heterostasis1 are three new physiologies, which only become possible because the kind of cells of which the bodies of all the mammalian species are composed changed. My concern is with the some of the differences the emergence of neurons make. Actually, the emergence of both Hedonism and Heterostasis1 were only possible because of the emergence of the neuron.

Heterostasis is a new kind of behavior, rational behavior, which emerged as the nonverbal mind emerged in parallel with the expansion of the mammalian neocortex or so the story goes. However, at least three sets of tissues had to evolve together. This cortical expansion is directly a function of the number of neurons that make up the nonverbal mind.

The emergence of the nonverbal mind was possible only because the neuron had set the stage for a completely new kind of technology. My thesis is that the technology of the nonverbal mind has provided mammalian life with two new causes: the emotions and reason. Historically, these two technologies were conceived as enemies; the classical view was that emotions made humans potentially irrational, but the data of this experiment showed that evolution sidestepped this belief and created an unexpected two- step evolutionary emotional trajectory: step one, the emotions made life non-rational, which generally may look irrational. Then in the second step, with mammals, the emotional energies were redirected in a way that made mammals rational. It is a long story but worth the telling and understanding.

The concept of the emergence

It is impossible to exaggerate the importance the role of the concept of emergence for the evolution of the nonverbal mind.

One of the chief emergent features of biological life is the emergence of the newest and third kind of cells: neurons to go along with germ and soma. Our concern is with the emergence of the neuron, which has all kinds of emergent features never before seen within the cosmos, each of which itself opens the door for still newer features of life to emerge.

Rationality: mental speciation vs. biological speciation

The different definitions of rationality are:

The central rational function of the mind is a new kind of evolution: mental evolution, or mental speciation

The central rational function is the new power of an organ, the mind, to create an idea or a complex set of domain-specific ideas, which was Plato's (1952) concept of the god-function

The power to conceive concepts; to self-organize

The power of mental rather than biological speciation

The evolution of hierarchically organized coherent sets of ideas

Just as the visceral organs of the body were designed to organize or stabilize the internal milieu of the body, just so the new adaptive organ, the nonverbal mind, was designed as a general purpose system capable of organizing or stabilizing mammals in any possible earthly or even cosmological external milieu, such as the space station. The generic nonverbal mind has served well in the body-mind systems of every mammal that ever lived.

My purpose is ultimately to explain the nature of an idea; to explain the evolutionary process by which large neuronal structures, which form the nonverbal mind, can rationally conceive coherent sets of ideas, not by natural selection, but by rational selection! The emergence of a new kind of organ, one with a rational function, which can conceive domain-specific sets of Ideas, which gave mammals a profound adaptive advantage over the no-mind, or Stage-one animals.

The nonverbal mind by interacting with the specific local forms of matter can conceive coherent sets of nonverbal ideas. Only the verbal mind can conceive metaphor-based ideas, but only the nonverbal mind can conceive empirical or fact-based sets of ideas.

Empirical knowledge is a function of the nonverbal mind: each scientific fact is a reproducible event, whose existence and validity can only be established by the nonverbal mind.

The generic nonverbal Mind

What especially makes the generic nonverbal mind interesting and important is the idea that it is generic. In spite of all the obvious major sensory and motor differences between the various kinds of mammals that evolution has ever produced: whales, bats, moles, rats, and humans; there is some basic common rational denominator, some generic nonverbal universal logic, or generic intelligence built into the nonverbal mind. This generic logic works equally well in all the various types of bodies and the diverse niches occupied by all the variety of mammals that has ever lived. My purpose is to identify this generic function, which has served all mammals equally well in spite of all the differences in the composition of their sensory-motor functions as required by all their different niches.

This universal feature of the adaptive logic of the nonverbal mind served all mammals equally well by conferring upon them the rational capacity to self-organize which requires an organ that can digest anomalies and create coherent sets of hierarchically organized ideas under the most diverse of circumstances.

Western humans have long conceived of themselves as a special or unique species, and generally, this conceit was justified upon the obvious belief that only humans had language, or verbal behavior.

This obvious human superiority complex nevertheless failed completely to address the central issue of the superiority of mammalian survival over all other species. The profoundly superior survival skill of mammals is obviously independent of this verbal adaptive skill, which is unique to humans.

What is a mammal?

The classic essential mammalian features recognized by taxonomists are the existence of the organs, which give mammals their name: the mammary glands.

However, the adaptive advantage evolution conferred upon mammals is something far more serious than mammary glands; all mammals, male and female, have a nonverbal mind, which has not gotten its due attention from specialists of any kind. Hopefully someday, not too far in the future, they will recognize the existence of the nonverbal mind and acknowledge its role in mammalian behavior.

E. O. Wilson (1975) says, "Mammals can only be understood in terms of milk!"

With the pre-mammalian forms of life, evolution endowed each species with the specific adaptive technology required for instinctive adaption to one closed niche. However, evolution gave mammals something even more powerful; a new technology which can adapt to change and create domain-specific, purpose-specific, and species-specific nonverbal conceptual systems. The nonverbal mind enabled all kinds of mammals to move into and adapt to almost every open niche upon the face of the earth and so mammals have become whales, bats, etc. and live in all kinds of terrestrial domains.

By giving any individual species a genetic designed fixed form, you have prepared that species to live in one closed niche, but give a species the rational capacity to create or evolve hierarchically organized conceptual systems and they will be prepared to live in any closed niche and many, if not all, open niches.

The master mind

The nonverbal mammal's nonverbal language behavior is a nonverbal twelve-station function. The nonverbal mind is the mastermind; for millions of years it served as the only mind, the all-purpose mind, and faithfully served all the needs of all the nonverbal mammals. As you read about F3 running the maze, you will see how the nonverbal mind spontaneously emits Heterostatic behavior and how these rational units of behavior interact with the parameters of space. F3 will probe nature, forcing raw nature to reveal its causal structure. In addition, each time nature yields, or gives up another bit of empirical knowledge about the mystery of the puzzle; the nonverbal mind digests this bit and updates its model of the space in which the rational nonverbal mind is evolving. Each new piece of the puzzle rationally enables the mind to direct the rat to modify its behavior, now to retrace, now to explore, etc. Sooner than I ever expected, the game is over, the puzzle solved, and the rat's behavior reveals that the nonverbal mind had created a model of the hierarchically organized maze, which F3 exploits to create a plan by extrapolation for running the maze without errors.

However, the private nonverbal narrative, the third wisdom, created by the nonverbal mind is a private wisdom, one without an export function to enable sharing with other rats. The nonverbal mind of each individual one-mind mammal must create the mind-generated knowledge required for the rat, or any other nonverbal mammal, to achieve its goals.

The role of the nonverbal mind is to probe the external milieu using the unique pattern of Heterostatic behavior and provoke nature into empirically revealing its spatial structure's causal nature, grade by grade. Each new bit of empirical knowledge enables the mind to revise and evolve an ever more powerful model of the maze.

My claim is that Heterostasis, adaption induced by anomalies, is a phenomenon that is common to all mammals, nonverbal and verbal. The growth of a mammal's perception of the world is one of the great achievements of the mind. There are different kinds of growths, but for the moment, my point is to distinguish between the growth of the body and the growth of the mind.

The emergent mind

The emergence of the nonverbal mind was one of the great transforming events in the emergence of the cosmos. There is a lot of sick humor about people changing their mind, which reflects our profound lack of our understanding of the nature of mind.

The mind emerged after the great extinctions almost destroyed the entire simple one-niche animals. The virtue of the mind was that it was designed to deal with change, or perhaps we should say, that the mind was designed to change. In a most profound sense, the difference between mammals and the non-mammals is that mammals are the only form of life that can change. Humans, however, like all other animals can do nothing consciously to change their bodies. There is only one thing we can change, and isn't it paradoxical that we make fun of those who change their mind.

I have often wondered how many humans, and how well, could solve an equivalent problem to the one solved by F3. I have long wished to cut a Corn Field maze so that I could run humans on an equivalent of the matrix maze. My son Mark suggested this to me many years ago.

The concept of the creator god

Plato's (1952) intuition about the most critical function of the mind was partially accurate. Mental evolution or mental speciation, which he recognized as the most critical of its functions, he reserved for god. Accordingly, I call this function the "God-function."

Plato's thinking at this place in his theory was actually conventional for his time. Both Moses and Plato began with this same assumption. When Moses wanted to explain the origin of the earth and all its species of life, he fell back upon a verbal god as the creator.

Darwin (1859) was the exception; he presented his theory of creation without appealing to any spirit god. The chief function of Darwin's theory was to explain "The Origin of the Species." The language theorists have not followed Darwin's lead, but in this book, I follow Darwin's lead and propose a natural mechanism to explain mental evolution. As I will show, the difference between Darwin's biological theory of evolution and my theory of mental evolution is that Darwin's evolution theory was based on a theory of natural selection, while mental evolution is a function of rational selection.

Stage Four: The NV/V: four wisdom form of life HHHH

Sperry (1961) was the first scientist to claim that human speech was the product of two-minds with the right hemisphere computing the first half of speech behavior then the left hemisphere executing the second half.

I propose that evolution created a verbal mind, a mind with the capacity of verbal speech by evolving a new somatic control region for controlling the somatic muscles controlling speech in the left hemisphere employing a division of labor. Evolution put some parts of the machine in the left hemisphere and the rest of the machine in the right hemisphere. Evolution did not give humans two equal minds but created one new and more powerful mind by evolving a new motor center in the left hemisphere; a new thirteenth station on the assembly line.

What is the verbal mind?

The Sperry researchers seem to have defined the two minds primarily based on the two hemispheres. If a function is in the left hemisphere then it is a verbal mind function, while if it is in the right, then it is a nonverbal function. What criterion is appropriate for distinguishing whether humans have one or two minds?

The role of Rolandic regions

The simplest way to distinguish what Sperry (1961) called the two minds is to focus attention upon the output structure of the mind, the somatic motor regions. The Rolandic region of the cortex is an ancient region, which has been controlling the somatic nonverbal output behavior of mammals from the beginning. However, the emergence of Broca's area is new and almost certainly evolved for controlling the special combination of somatic motor muscles, which control vocal behavior.

The ancient action-centric rational nonverbal mind was the original mind and when evolution divided this original nonverbal mind between the two hemispheres, no attempt was made to create two duplicate minds. In fact, evolution created only one copy of the new speech motor center and put this new neuronal structure in the left hemisphere. The hemisphere in which this new speech motor was located, Sperry claims is the primary mind, but in a large system, each part of the whole is equally important. However, the nonverbal functions located in the right hemisphere are the primary mind centers.

Sperry did not know about the generic mammalian mind, and so he distinguished his two human minds in total ignorance of where, how, and when the generic human mind evolved.

Sperry (1971) called the right hemisphere the nonverbal mind because Broca's area is in the left hemisphere and thus the right hemisphere, which in my opinion is the primary action-centric/thinking mind, cannot initiate speech because evolution put Broca's region in another and inaccessible hemisphere. Yet, that does not mean the right hemisphere has nothing to say. Sperry states:

"One of the main goals of our testing program has been to find out what goes on in the nonverbal, minor hemisphere, which can neither speak nor write." (Sperry, 1971, P. 188)

It would be interesting if the fossil records could make any estimate as to when and where Broca's area emerged and how fast it evolved.

In any case, the best justification for the study of rodent behavior is that it is most important to study the generic nonverbal mind to discover exactly what the nature of the root functions of the action-centric mind are. Are there two copies of the root functions of the nonverbal mind in the human brain? Of course not, because there is only one mind in the human brain. The human compound mind is but a more advanced copy of the generic mind. Each year Americans are well aware that car and computer companies come out with new models or new versions of either hardware or software. When the nonverbal mind added Broca and Wernicke's regions to the left hemisphere it did not create a second kind of mind, just a more advanced or upgraded edition of the old generic mind.

The implications of this definition are interesting because they reduce the scope of the verbal mind enormously. The verbal mind is now defined as those structures, whose functions control the complex set of somatic muscles, which control Broca's and Wernicke's areas.

While it may sound outrageous and perhaps demeaning, I suggest that the verbal mind probably does not even create the narratives it expresses through verbal speech. The verbal mind is but the export or shipping department for the nonverbal mind. I propose calling the nonverbal mind the "doing" mind because it controls all somatic muscle behavior, which enables all mammals including humans to do everything we do, except talk. The two different functions of mind respectfully, perhaps should be labeled "The Doing Mind" and "The Talking Mind." The role of The Talking Mind is to enable each human mammal to talk about or export a verbal report about what the "doing mind" was doing, or had done, or was going to do.

The HHHH mind

The one-mind, three-wisdom mammals HHH, have only the adaptive power conferred by the nonverbal knowledge created by their own nonverbal mind. On the other hand, humans are a one- mind with four wisdoms mammal. The fourth wisdom, human speech, endows humans with yet another source of vital knowledge to supplement the knowledge that comes in the genomic starter kit. Now they do not just have access to the knowledge created by just their own nonverbal mind, but now they may have access to all the accumulated wisdom of their whole language/knowledge community.

In the next section I will use Aristotle's (1952) famous ancient model of the mind to explain several important aspects of the nonverbal mind.

Joshua, a four wisdom mammal illustration

For over two millenniums, human nature was characterized as essentially a one-mind verbal function. This is another prime example of the superficial nature of the empirical observations of nature; another prime example of the behavior I flatteringly call naked-eye science, which only observes the more obvious set of facts, but then ignores all the facts of the opposing set.

For example, in young human mammals, nonverbal mind behavior is almost as obvious as is verbal mind behavior in adult humans. On the way to becoming adults, most humans eventually master verbal behavior, but if we stop and consider all the facts, we should note, but we never do, that verbal behavior is not the natural language of humans. All young humans naturally and spontaneously first master nonverbal behavior and have to then learn to master verbal behavior, some more easily than others, or to put it the other way around, some young humans master verbal behavior with some reluctance and difficulty.

If we consider all the empirical facts, humans begin life with the nonverbal mind as their primary mind, as their mastermind, and then later master the verbal behavior of their secondary mind. Eventually most acquire some degree of fluency in speaking the verbal language of their parents. However, while all humans are born with the structures of both phases of the mind, yet for a considerable time after birth, for all young humans, the nonverbal mind is the primary mind and the verbal mind is a mind in waiting.

The fact that the nonverbal mind is the primary mind is most obvious and incontrovertible in human children. Which language do human children master first nonverbal or verbal? Which is the most natural, which is the most difficult to acquire? The nonverbal mind's language functions are mastered naturally and spontaneously, while the verbal mind's language behavior is mastered second and perhaps with some degree of difficulty.

I watched my daughter in law, Julianne White assist her youngest son Joshua when he first began to learn new verbal words for all the standard fixtures and objects around their home. She attached flash cards to the sink and most of the other objects in their home. The order of the mastery of the two-language categories, I have no doubt, was nonverbal images followed by words. Joshua's nonverbal mind spontaneously mastered recognizing and using all the objects of his home. Joshua's language skills first succeeded in mastering the nonverbal image based categories, which evolved well before he was able to begin to associate and acquire the new verbal mind's words.

A child acquires words for each nonverbal category only well after they have mastered the nonverbal mind's nonverbal images/symbols. What is most obvious is that verbal words are not the natural primary language. The order of mental evolution is images of the nonverbal symbols first, then verbal second. The primary nonverbal language categories are mastered first. The generic nonverbal mind we share with all the other mammals is still our primary mind, and some of us are in no rush to give it up. In fact, even as adults, some famous scientists find that this primary nonverbal mind is still home to the creative god-function and is thus still the best place to go to think. Only after they have solved their problems with what one famous scientist called, "thought experiments" nonverbally do they bother with memes.

Nonverbal conceptual evolution in children is the foundation of **language behavior learning**. Classical psychology has proven incapable of conceiving of nonverbal concepts as evolving during everyday life; they had no more idea than did Plato where forms come from. No one could imagine that conceptual evolution occurs in the nonverbal mind of even human children, and not just animals.

The **primary function of the primary intelligence** of the preverbal child is inventing the conceptual systems that will eventually fit the forms for which the adults of their linguistic community have verbal names. The function of the primary intelligence of the nonverbal child is inventing conceptual systems that match the verbal classes of events contained in the paradigm of the adult verbal community into which they are born. When the names of the verbal community begin to fit the conceptual system invented by the child's nonverbal mind, only then can verbal intelligence begin to be measured. The business of the human child's nonverbal mind is evolving conceptual systems that allow them to experience the things of the social and physical world conventionally, as exemplars of the classes of things for which the adult community has names. The language forms of the adult community cannot begin to work for the child until the conceptual system of the nonverbal child begins to match the conceptual system of the verbal community. They cannot learn to talk until they learn to see, to recognize things conventionally. The main task of the nonverbal child's primary mind is to evolve a stable conventional conceptual system, to answer Plato's question where do forms come from.

Chomsky's problem

Chomsky's (1979) problem is that he does not know where the forms came from any more than Plato (1952) understood the origin of the forms of the mind.

The conceptual system of the nonverbal mind evolves by nonverbal conceptual conditioning and neither Plato nor Chomsky knew or even dreamed that there was a nonverbal mind. Conceptual conditioning is a nonverbal logic that evolves in the nonverbal mammals in each special domain in order to enable each mammal successfully to self-organize in each domain. The same demand in the case of the verbal mammals also exists, but additionally the human child's role in its family places an additional requirement. The human child must evolve a conceptual system that will match the words of its family's verbal community.

The extra responsibility of humans

The emergence of language behavior upon humans I will argue hardly seemed fair. The revolutionary new potential of language behavior thrust upon human's additional responsibilities for which they were not prepared. To this day even the brightest humans don't know what to make of this skill; they don't know where it came from, and they have no idea of how it functions. Giving human's speech was like giving a fourteen-year-old child a driver's license. Nothing seems more important than assisting humans to understand their own nature. It is difficult to put into words the many ways that humanity can destroy not just themselves but the planet also.

Knowledge vs. Information

Knowledge is metaphor free

Developing a theory of the nonverbal mind is especially difficult:

- First, because all the key words have been defined in the context of verbal speech, and thus have no useful meaning in the nonverbal context; all the words needed to describe and discuss the nonverbal behavior have long been taken and corrupted.
- In the human context rational truths are verbal truths, deductive truths, which simply mean that systems of words, which are internally consistent, don't contradict themselves, and are tautological even when they have no empirical justification are true.
- The nonverbal mammals are not capable of communicating with verbal truth, with tautologies, nor with verbal metaphors, which have no empirical meaning. The nonverbal mammals live immediately within an empirical world, a world of action that is the real world.
- Watch Plato, in section one of this chapter, play with metaphors whose virtue in his eyes is that the words are empirically meaningless, and thus they have only the meaning, which some metaphor of his choosing, confers on them.
- Watch each religion play with metaphors, whose virtue, in their eyes is that they are empirically meaningless, and thus they have only the meaning, which some metaphor of their choosing, confers on them.

- Watch Aristotle disavow metaphors. Aristotle rejected Plato's metaphorical geometrical world. Aristotle chooses to live in a world defined by his six senses, whose virtues, in his eyes, are that they are empirically meaningful, and thus they have only the meaning which nature confers upon them. The mind evolved to enable each mammal as a SOS to empirically interrogate nature and survive only so long as it did not stray too far empirically from nature. A self-organizing system is a misnomer; they are not self-organizing nor are they organized by metaphors but by nature.
- Too many humans role-play at life and fall instantly and without reservation into the word-driven, metaphor-driven world of some religious leader, political leader, or philosophical world.
- The human's gift of speech was a mixed blessing because acquiring the ability to enjoy the verbal metaphor driven form of speech was a mixed blessing/curse. Humans are probably on a collision course with destiny, and they have no empirical data-driven idea of what is up ahead. Metaphors are delusions, which the nonverbal mammals never suffer.

The primary thesis of this book is that rats and all other mammals have a nonverbal mind, which is a non sequitur for humans. Historically mind by definition is verbal, so for most humans the words "nonverbal mind" makes no sense.

My purpose in this section will be limited to making only some preliminary remarks about the extra burden that humans acquire by virtue of having acquired the power of verbal/ metaphorical speech. One symptom of the human condition can be revealed by simply distinguishing between information and knowledge.

There are words I will attempt to define and distinguish in this book: they are information, communication, knowledge, uncertainty, ideas, and metaphors

The believers in verbal-myths conceive of verbal behavior as 100% verbal. The linguistic experts have no clue that human communication behavior is perhaps 90% a nonverbal function and only 10% a verbal function.

The myth that the verbal mind is the only mind is one that Chomsky and Pinker still support, they have never imagined the existence of the generic nonverbal mammalian mind, GNMM; and that the verbal mind simply serves as the shipping department for the nonverbal mind. The linguists take full credit for the human verbal skills never suspecting that the nonverbal mind does probably 90% or more of the work. They are jockeys who don't believe in horses.

The verbal speech specialists have no idea what a nonverbal anything is; what is a nonverbal rationality, take each of the six words above; how many can you define nonverbally!

Communication: Information vs. Knowledge

Consider the word **communication.** There are two kinds of communication verbal and nonverbal. The verbal communication of humans employs **metaphor/information**, while the nonverbal communication of the nonverbal species employs **knowledge**.

The human speech function must be characterized as rational Human to Human, H-H, speech, while nonverbal speech will be characterized as mammalian empirical or mammalian to nature, M-N, speech.

Human-Human, H-H, communication

Next, consider the word information, which was formally defined as the Bell labs form of communication.

Shannon and Weiner (1951) working at Bell Labs devised a mathematical technology for objectively measuring how to get the most "information" on a channel of communication, i.e. on a telephone wire. They created the science of information, which focuses on measuring the content of a message being seen over a wire. They received a Nobel Prize for their work.

But in the process, they usurped three words on my list of important mind-words: **information, uncertainty, and communication.** The Bell Lab scientists took these three words gave them a technical and formal domain-specific, function-specific definition, which most non-mathematicians are never going to be able to understand.

Information reduces uncertainty

What is information? They defined it thus: "Information is that which reduces ***uncertainty****." Each bit of information reduces uncertainty by 1/2. Fine, but what's uncertainty?*

However, in their scheme of things creating a message is only possible in a closed set of symbols, and is restricted to selecting one letter at a time from a closed set. Thus, it is possible to formally compute the probability of every possible choice and thus create a string of a series of alphabetic characters that will reduce the sender's and the receiver's uncertainty. The company's business was sending messages (alphabetic characters over a wire) and this is the basic fact to understand. Their business had only so many wires, so the more messages they could send over every line the more money they made. (Shannon and Weiner, 1951)

Since words and sentences do not contain the same amount of information, many letters are redundant, the telephone company needed a way to identify those letters in a message, which carried the most information and those that were redundant. The object was to restrict the messages only to those letters that carried information and eliminate those that were redundant.

The original concept of information was defined in this context. Originally, the only thing they could send over the wire was sound- encoding- letters of the English alphabet. The message must consist, in other words, not of letters randomly selected from the alphabet pool, but letters rationally selected so that the particular letter sent over the wire was statistically the one, which carried the most information at each moment.

The act of communication was simply restricted to picking one alphabetic character at a time out of a set of symbols which is a closed system consisting of

26 capital and 26 lower case letters. The math was restricted to a logarithmic scale, which measured the information or rarity in each successive letter.

Again, Bell labs defined information as "that which destroys uncertainty." This concept works fine if the goal is to communicate using the fewest possible letters of the alphabet.

Mammalian-Nature, M-N communication

Now it is time to examine that mysterious kind of communication the M-N type.

The verbal community has long claimed they had the right to define and have a monopoly over all the communication words, since they alone had the only forms of communication.

However, verbal communication is but a relatively minor improvement of the somatic behavior of mammals. Broca's area is a somatic motor center just below the primary somatic cortex, the Rolandic cortex, and evolved to control the somatic muscles concerned with talking.

Knowledge/ideas reduce uncertainty

My concern is not with how to maximize the efficiency of a telephone company to send alphabet characters over a telephone line.

My concern is with how the nonverbal mind of a rat such as F3, which was designed to supply life with a better way to acquire the local knowledge they need to accomplish some domain-specific task, actually conceives knowledge while engaged in exploratory play behavior. The nonverbal mind for example, creates empirical knowledge by creating a hierarchically organized model of the Lashley III maze space. The model encodes nonverbal knowledge, which the NV mind can extract and compute a plan, which she can extrapolate and use to run the maze error-free. Knowledge reduces the uncertainty, which is a state of disorganization of the EEG rhythms within the cortex. The nonverbal mind must conceive a model of the maze by engaging in a mental speciation behavior.

The reduction of uncertainty in verbal behavior and in nonverbal behavior is completely different. One of the differences involves the selection of the ideal string of verbal characters.

Nonverbal communication is a knowledge exchange, while verbal communication is an information exchange. In M-N communication, knowledge is coded and stored as **ideas**. The only kind of ideas the nonverbal mind generates are nonverbal.

Knowledge/Ideas reduce uncertainty and makes models, maps, and plans, all of which are ideational structures and reduce ignorance or uncertainty.

Open and Closed sets

Information and knowledge are opposites.

Information is a property of a closed set, such as the 26 letters of the English alphabet. Information cannot be calculated in an open set.

Knowledge closes an open set. When a mammal encounters an anomaly it opens what had been a closed paradigm. It induces uncertainty about what next to do, about the nature of the domain in which the animal is exploring. The creation of an idea reduces the ignorance or uncertainty, and enables the organism once again to begin to explore and act with confidence. **Anomalies open closed sets, while knowledge closes open sets.**

Mammals evolved because nature's original strategy for designing species had failed on a grand scale on a number of occasions when cataclysmic events, the great extinctions, almost wiped out all the original type of species; those that evolved to live in only one niche that had been stable for millions or hundreds of millions of years. These original species, which were designed to live in only one stable environment were supplied at birth with all the genomic **knowledge** they would ever need to live their whole life span in that one stable niche. Evolution however made one false assumption; the great extinctions proved there are no permanently stable niches.

Mammals evolved because they were designed with a completely new survival strategy. Whereas the original species were designed for life in a constant or stable niche, mammals were designed to survive or thrive in the face of change. The difference between the old pure-nature species and the new almost pure-nurture species is that the latter were designed with a nonverbal mind; one that could rationally create a Life Space appropriate for almost any niche upon the face of the earth. (Lewin, 1943)

Mammals were born with a nonverbal mind that could create in any open domain the domain specific knowledge/ ideas required for them to close the domain and create all the knowledge needs required in that domain.

How do mammals with a nonverbal mind and not capable of getting knowledge by communication with others of their species manage to be self-organizing systems. The mystery has two parts.

- First, what is nonverbal communication, and

- Second, where can a nonverbal mammal go to get the nonverbal knowledge it needs to survive.

First, let it be noted that communicating over a Bell Telephone system is not a form of communication that any mammal other than a human is likely to use.

The primary form of nonverbal communication employed by the nonverbal mind is obviously not between humans, H-H, or equally obvious, not even between other mammals, M-M. The strange aspect of mammalian life is understanding: how, when, and where, the nonverbal mammals obtain the domain-specific, species specific knowledge that enables them to be self-organizing systems. What is the meaning of the word communication in nonverbal mammals?

There is only one kind of communication used by nonverbal mammals between mammals, M, and Nature, N-the **M-N** kind. The only way the individual nonverbal mammal gets the knowledge/ideas it requires to act appropriately within domains that initially were full of change or anomalies is with M-N communication.

The nonverbal M-N form of communication is the primary form of communication. The cosmos was not capable of verbal communication until the human 13-station mind emerged. Yet nonverbal communication had been the mainstay of the nonverbal mammals for over fifty million years. Only humans' conceit ever supported that humans alone had a mind.

Nothing seems to have a higher priority to essentially both sides of the debate. Science, religion, and philosophy all want to be on the side of truth. No matter how much religion bravely emphasized the sacredness of faith, yet they all clearly want to be on the side of truth and knowledge. Plato claims that his system in the divided line gives knowledge; the scientists claim their empirical behavior induces knowledge. Who will make the decision whether knowledge is defined as empirical or rational I have no idea. However, as a scientist I most fervently desire to be on the side of knowledge. So does rational behavior or empirical behavior produce knowledge? Both sides want to claim a monopoly on the term knowledge.

In this book, I will consistently advance the idea that empirical knowledge is the only kind of knowledge. Female 3's behavior in the maze generates knowledge, which is encoded in models that is then used to extrapolate.

Chapter two: the structure and function of the horizontal mind: Five Theories of Mind

I. Plato's dualism: the spirit- verbal mind

Plato's ancient concept of the human mind was a salad bowl of critical technical philosophical ideas or ideals that over the centuries have wrought all kinds of havoc.

Plato's Tri-Partite concept of mind

Plato (1952) was a philosopher, a professional thinker who described humans as a unique three-soul species. Plato's tri-partite concept of the human soul has an evolutionary aspect. The three souls are the vegetative soul, the animal soul, and the rational soul. Each form of life had a unique pattern of souls and each soul had a special set of features.

The tri-Partite souls:

- **The Vegetable soul**: Plants had this soul with the properties of metabolism and reproduction

- **The Vital soul**: Animals had two souls, but their unique soul was this soul, which had two attributes: motion and emotions

- **The Rational soul**: Only humans had a rational soul, but additionally only humans had all three souls.

Humans were the highest form of life with three souls. Humans had a unique soul, a rational soul, which endowed humans with verbal rationality.

Plato's (1952) concept of humans as verbal rational beings became the standard or classical model for the West's concept of human nature. The remarkable aspect of Plato's theory of the rational soul was its ontology. Plato claimed the human mind was a soul or spirit. However, Plato does not give animals the rational soul. There are many different rational souls, which is a concept Plato did not grasp.

Plato's theory, some could argue, was a precursor to my theory of four wisdoms, yet there are obvious differences. Plato gave plants a soul. I would not presume to make such a claim about the nature of Plants. The vegetable souls, it could be argued, were a kind of a Homeostatic soul. I suspect that Cannon would not have taken kindly to such an idea. However, I could not see how based on the scientific evidence provided, anyone could propose that the human rational soul is anything but a Heterostatic soul. I am supporting Plato's basic idea of humanity having three or more physiological mechanisms; however, I would call them wisdoms rather than souls as Plato does.

II Aristotle

However, a few humans from that time were already serious about carefully collecting all the available facts before they began their talking and thinking. Aristotle (1952) was one, who chose to build a set of empirical facts as a foundation to stand upon as a support for his thinking.

Aristotle's monistic mind

Plato (1952) and Aristotle (1952) presented opposite ontological concepts of mind. Plato described humans as dualistic beings, with the body made of matter but the mind was a spirit. Aristotle conceived of mind as monistic, as made of matter. Unfortunately, later philosophers in the West generally have followed Plato. However, in this book I propose a new monistic theory of mind, which is based upon Aristotle not Plato. In fact, in the section on Aristotle's Model, I recreate, or update, a model of Aristotle's monistic theory of mind, which gives a rough outline of human nature that is still empirically valid today, after all these years. Most of Aristotle's model long ago became part of the West's body of common knowledge, especially his phrases; "the five senses" and "the sixth sense”. However, while these words survived the ages almost intact, his concept of the meaning of these words has not fared so well.

Four versions of Aristotle's empirical model of mind

Today, the tree of mind has branched so many times that it is possible to distinguish many more kinds of mind than are practical or useful. However, four theories are useful:

> The six senses nonverbal model: My Primary purpose in this book is to propose a Aristotelian type theory of the nonverbal mind, which I will call an old fashion sixth sense model of the mind. This model serves as the foundation for a generic mammalian theory of mind. However, when Aristotle created his model he clearly did not conceive of it as a model for the nonverbal mammals. It is my responsibility to specify what additional functions must be added, for it to achieve such a purpose.

How can the origin of the human mind be explained?

- The classic theory of the origin of the human mind is called a nativistic theory by Chomsky, who insists that the mind did not evolve but suddenly emerged in humans.
- By contrast, I propose that the nonverbal, NV, mind emerged perhaps fifty million years ago in one of the first mammalian species, such as a rat.
- The nonverbal mind can be understood as a kind of Aristotelian machine.
- The NV mind manufactures behavior on an assembly line, with say 12 stations.

- The human Verbal, V, mind emerged as an extension of the NV mind; as a NV/V mind.
- The Verbal mind was simply an extension of the original 12-station model. With the emergence of the new station two new modules were added, the Wernicke and Brocca's areas. The human verbal mind is a 13-station machine.

Ontology: two kinds

Distinction: there are two kinds of beings: dualists and monists.

Ontology is the study of what is real; the central ontological question has been what human nature is? Historically there have been two camps of ontology: dualism and monism. Plato (1952) represents the former camp, while; Aristotle and Spinoza (1933) were philosophers who represented the monists. Plato conceived of humans as dualistic, the body was made of matter, but the mind was a spirit.

Aristotle and Spinoza were monists. Spinoza most importantly conceived of all of nature as god. His concept of a nature-god is called pantheism.

Plato's Dualism

Plato's (1952) chief concern was with ontology; what is real? Plato turned the common person's conception of what was real, upside down. Most people believe that real things are material things that can be touched or experienced by one or more of the senses. Nevertheless, Plato argued that material things are not real; for all material beings, even the most permanent things, including rocks and mountains, die or waste away and are mere appearances. All material things last only for a while. According to Plato, real things are eternal beings; they do not change, nor can they be experienced with the eyes, but only by reason. Essentially Plato took what seemed to be a most uncommon or strange position.

First, when he acknowledges that real things cannot be experienced or known with the eyes, he was tacitly recognizing that his assumption defied common sense. Most humans believe that the eye plays a special role in defining the real.

Second, Plato (1952) defined the ontological question as a rational geometrical question rather than as an empirical scientific one. For Plato thinking was driven by geometrical logic, so that truth was established by reason, rationally, not by the senses, empirically. The geometrical Metaphor of the Divided Line allowed Plato to turn reason upside down by arguing that only nonmaterial beings i.e. spirits, were real.

What is an exemplar?

There are two kinds of exemplars: physical and mental. An exemplar is an instance of a class. The important issue is what do we see? Fact: we cannot see any physical objects, all of which are exemplars of some species or some class! So what do we experience?

I have tried many times to make the point that physical objects do not enter the eye, the afferent nerve, or the occipital lobe. What happens is that each of the five senses transduce one of the several kinds of physical energies emitted or reflected from a physical object; the PPA then transduces the neural impulses into sensations, which the sixth sense uses to paint an image of the object in station one. What we experience are images! But, images are mental exemplars created by the nonverbal mind and are mental exemplars of some class of object. For example, normally all we can experience in time/space are concrete images conceived by the nonverbal mind to give meaning to each class of object. The look of the exemplar of a white Pine tree is an exemplar of our mind's concept of a white Pine. The mind creates an image that gives mental meaning to the rose or to a pine tree in space, but the rose we experience is, in effect, a likeness, a mental nonverbal metaphor for the rose in space. Every image, each mental exemplar, is a nonverbal metaphor, yet the red in our perceptions of the rose is a sensation. There are no sensations outside our head.

The metaphor of the divided line

In Plato's (1952) system, there are two kinds of beings: things and spirits. Things are physical beings; beings made of matter, which he insisted were not real, but merely imperfect copies of the eternal real forms, molds, or ideas of which material classes of things are only exemplars or copies.

No human has ever been so diligent and effective at both promoting human rationality and denigrating the rational powers of the human mind than was Plato.

Plato (1952) gave humans, the ultimate paradoxical metaphor, the famous/infamous Metaphor of the Divided Line, which has intrigued philosophers for thousands of years, and may be the most complex philosophical metaphor of all time; the most successful and the most harmful. With this metaphor Plato, in effect, put Stratton (1896) Glasses on the West's mathematical and philosophical community, which turned their world upside down. Over the last two thousand years, this Platonic upside-down world has irradiated far afield, for example, through the transcendental religions, and from there throughout the entire everyday verbal world, which the West shares.

Figure0.1A

Figure0.1A shows a simplified version of Plato's (1952) metaphor of the divided line. Plato’s Metaphor was a multilevel metaphor, and FIGURE0.1A only represents the half that is relevant to this book. Plato devised this metaphor to describe his metaphysical view of the world including both its ontology and epistemology aspects.

The three cuts

Plato conceives of the world as having been divided by lines, one vertical and three horizontal lines.

Vertical cut one

His metaphor first depicts the world as divided by a vertical line into two halves, right and left. On the left side, he describes the **ontological** nature of the world, while on the right side he describes the **epistemological** nature of the world.

Horizontal cut two

Plato is a dualist, and his metaphor depicts the two halves of such a world; a world divided horizontally into two halves, top and bottom. In the bottom half, Plato locates the physical world, while in the top half Plato locates the transcendental, or spirit world.

Horizontal cut three

With the third cut, Plato divides the two halves of the world into four quarters. In FIGURE0.1A, Plato shows the spirit half of the world divided into two parts, Quarters C and D, while the physical world is divided into Quarters A and B.

Being and Becoming

Two words on the outer upper edge of both the right and left halves succinctly summarize Plato's take on his story. On the left in bold letters, observe the two ontological words **BEING** and **BECOMING**. On the right side also in bold letters, observe the two epistemological words **KNOWLEDGE** and **OPINION**.

In other words, on the left side of the vertical line Plato, the dualist, imposes his dualistic view upon the world, using two strange, but interesting words which describe two ontological states: **being** and **becoming**. To which ontological state would you choose to belong?

All the objects in the lower half of the figure, A and B share the ontological status of Becoming, while all the objects in the upper half of the figure, C and D share the ontological status of Being. It is important here to learn to think like Plato. First, since all the objects of the physical world are of matter, Plato classifies them as in a state of **Becoming.** Secondly, all those objects that are not of or in the physical world, but belong in the spirit or transcendental world are in a state of **Being.** When Plato's says that matter is in a state of becoming, that matter is always changing and thus not real, while all the beings of the spirit world are in a stage of being; they are constant, or unchanging and therefore real.

Rows A and B; the Becoming kinds of objects

Row A and B or Quarter A and Quarter B are to Plato's eyes seen as Becoming kinds of objects. In the first row, Plato lists images and reflecting shadows, which are the most nebulous kinds of beings, while on the right side of the first row Plato proposed that the epistemological technology available for knowing these objects were most crude; namely conjecture and picture-thinking.

Row A is puzzling, yet it is clearly the foundation for Plato's big picture of the metaphor of the divided line. All the items of quadrants A and B are of the physical world, yet the two individuals of row A, images and shadows, seem the least real things it is possible to imagine. Nevertheless, shadows and reflections are essential features of all real things. All physical things cast shadows and have a reflection in water or a mirror.

ROW B: Becoming/beliefs

- All the exemplars of physical things reside here. Row B is home of all exemplars and ONLY exemplars. Therefore, Plato is able to reason since all physical things are exemplars; no physical things are real. Concepts, of which exemplars are only a poor copy, are the real things, so guess where the real things exist?

On the left side of the second row, Plato lists all the physical objects, all the exemplars that can be known by sense experience like trees, people, houses, etc. These objects can be empirically processed by Aristotle's 12-station model, but in Plato's view, we cannot **know** these exemplar type objects; the best we can do is to have **beliefs** about them. Why can exemplars not be known, because they are always changing! Knowledge's chief property is that it is always true, and exemplars that are always changing cannot qualify. All that can qualify to be known are eternal things that never change and nothing like that can exist in matter!

On the right side of the second row, Plato is painfully contemptuous of empiricism or sensory knowing, which Aristotle claimed to be the only epistemological skill that works. In this book, I side with Aristotle who claimed that the nonverbal mind of mammals only have one kind of knowledge, that provided by their six senses. Plato still insists, however that **becoming** of this type cannot be known, that only opinion can be had for these exemplar type objects.

On the other hand, when Plato attributes to the objects of the lower half of the world the status of Becoming, he attributed to those objects, in his opinion, all the contemptible attributes for which he has no regard. Exemplars, things that change, never got any respect from Plato.

Plato's big question asks which of his four Quarter's classes of objects are real. There are two upper quarter sets to choose from, which are 100% free of matter, and the lower quarter set of objects, which are 100% composed of matter. "Things" are objects made of matter, while "spirits" are objects not composed of matter. Spirits are "no things," thus they are eternal, etc. The culprit is matter.

Too bad Plato was born too soon. Today he could argue that matter is moving toward entropy, i.e. always changing, and cannot hold its form. According to Plato, objects in the state of becoming are changing and thus are not real, and matter is the problem.

Now we come to Plato's point. What is real? Plato makes it simple. Beings that change are not real, while beings, spirits, that don't change are real. Plato despised matter and thus despised all things. Things keep changing, which drives Plato mad, things, which he characterized as in a state of **becoming**, means that they are not real or not eternal. Only "no things," which consist of no matter, are eternal, and thus qualify as being in a state of perfection, or in the state of **being**.

Plato's goal is to denigrate "things" and elevate "spirits." Plato's goal is to turn the everyday ontological beliefs about the world upside down. In his scheme of what's real, the ontological objects that the ordinary human conceives as real, **and all things**; all beings made of matter: yet in his scheme, things are not real, because they always change. But on the other hand, all beings not made of matter are real! Presumably, with a straight face, Plato says that only beings that are nothing can't change; only beings that are nothing are real!

Plato employs several metaphors here to make his point, but I will not honor his metaphors by even discussing them, for what he is trying to say is senseless and no metaphor can enable "nothing" to make sense. Plato's ontology is profoundly wrong even before he begins to argue for his metaphors, but his metaphors cannot make what was wrong right.

Plato is trying to do two things.

- First, to convince the world that that the spirit world and its occupants are real, the physical world, and its occupants are not real. The basic logic is clean and simple, a spirit is a "no thing" a nothing, and a nothing is a nothing. Nothings have no nature; no features, no anything, and no metaphor can make a nothing into something.

- Second, that only a philosopher can deduce the truth about the spirit world.
  Plato can't or won't say how many spirit beings occupy the spirit world, what ranks they hold, how many of each rank there are, and who and how many does the thinking, how the decisions are made, does every nothing has the same rights as all the other nothings?

Row C and D: The Being kind of objects

Plato's ontological word **Being** is loaded. From his point of view when he attributes the property of **Being** to objects, he is attributing perfection to those objects. These **Being** objects, we might say, in Plato's eyes have arrived. They have achieved perfection, so they have no place to go or to do; they are right where they ought to be, and they are where they shall remain forever. Beings as objects, by definition can't change.

These objects to which Plato attributes the status of **beings** all share in Plato's view, the attributes of god. These classes of objects, in effect, share god's perfection including god's eternal nature, god's ideal nature, and god's perfect form. In other words, they are transcendental nonmaterial beings that can never change, they have no reason to change; they are perfect, they are incapable of change; nothings are just fine the way they are.

The Top-two rows or kinds of objects

The top two rows include both sides of the vertical line of rows C and D inFIGURE0.1A

ROW C- Being/Understanding

The left side of the third row Plato reserves for the class of mathematical objects circles, spheres, and triangles, etc. Plato locates all abstract mathematical figures together on this row. My discussion of how objects are transformed scientifically or empirically from concrete to abstract will be described in Book one, chapter five.

On the right side of the vertical line, Plato explains that the epistemological picture for the objects of mathematics is much better than for physical things. Mathematical objects have the advantage of being **understood**, which are the product of some degrees of philosophical discussion and debate. Plato has never been known for modesty, either personally or philosophically. **Understanding** presumably is an epistemological state that only philosophers can achieve.

ROW-D: Being/Dialectic

All the A concepts resided here!

The left side of the fourth row, Row D, Plato reserved for all his personal favorites, a class of the highest kind of classes. Row D is the home of all the special concept of objects, forms, or ideas that only god can create. His concept of forms as spirits changes everything. When I say the creation of spirit forms only god can conceive this is a statement, which is inevitable. This position is necessarily so, for they are not forms in matter but are spirit forms. In other words, as spirit beings they could not be the product of evolution, my expression is serious because there is no other alternative; the "god-function" produces all the objects of this row for there is no other way logically to explain their origin. If Plato had conceived of concepts or ideas as things, as being made of matter, then their origin could be explained by natural selection. Plato's God-function does not refer to a function that could be implemented in matter or a material mind.

All the beings of this class are spirits, i.e. by definition, they could not have evolved by natural selection, only material beings can evolve. Their origin is and always has been called Nativism. The A-forms are the eternal, perfect, ideal forms, which all the physical things attempt to copy; but of course, they never succeed. Exemplars, B, made of matter can never achieve the perfection of spirits, A, which are the product of the "god function." Plato is funny: here Plato is conceiving of even inanimate matter lusting after spirits, and furthermore, he is contemptuous of all physical things because they are not spirits. In our religious culture even animate objects, especially the human Christian's soul is conceived of as being a spirit soul, which is necessary, because by Plato's account matter cannot enter eternity-a spirit world is for spirits, and yet Christians fantasize about the resurrection of their body, and having their spirit soul occupy a material body form during all of eternity! They simply do not seem capable of taking Plato seriously. Eternity, the spirit world, is for perfect beings, spirits without any corrupting matter, which is incapable of being in a state of being. Matter, Plato stated in his metaphor of the divided line that matter-beings always belong to or are in a state of becoming. It is perverse for anyone to suppose that

Plato would support the idea that after death matter could suddenly change its stripe and cease to be a "becoming-being" and suddenly become a "being-being."

This eternal place is home to not only all spiritual forms for all the physical objects of row B, but also the abode of all the highest kinds of philosophical, psychological, and ethical concepts, such as ideals, forms justice, purity, and intelligence, etc.

However, note that on the right side of the vertical divided line, Plato is boldly optimistic about the epistemological prospects for the intellectual mastery of this class of objects.

Plato could not make his position more clear; the best thing he can say about the bottom two rows, A and B, is that experience at its best only gives us **opinion.** Plato saved all his accolades for Row D, for his most critical point was that the top two rows, C and D of the matrix, where our reason reigns, our reason gives us **knowledge, or more technically-the philosopher's reason can give them knowledge. Knowledge is not about the senses. It has no place in Aristotle's thinking. Plato's knowledge is rational not empirical.**

By contrast, note the second row where our senses reign. According to Plato, our senses cannot give us knowledge, only beliefs. So what is the conflict all about here?

THE DIFFERENCE is now clear, it is between Plato and Aristotle's epistemology: for Plato KNOWING is a rational achievement, while for the Aristotelian KNOWING is an empirical achievement.

Plato vs. the nonverbal mind

Plato goes out of his way to make his point that empiricism does not work, that material objects cannot be known; the best we can do is to cultivate a **Belief** or **Opinion** about things. Yet note the extreme contrast, which he expressed about our capacity to know transcendental beings. According to Plato, as long as a group of philosophers can discuss or have dialectics about spirit objects then knowledge is possible. Knowledge of spirits is created by reason; empirical experience **cannot** produce knowledge. Would that Plato were alive today, would he watch TV, drive a car, wear glasses--all the consumer products of today are the product of empirical knowledge, and Plato could claim that all these products prove him right--they are all becoming kinds of beings. In fact, they are designed with obsolescence deliberately built in. Plato's Philosophers, to the best of my knowledge, have never produced anything as a product of their rational knowledge.

That is as it should be. There can now be a distinction between empirical knowledge and Philosophical or rational knowledge. The former produces capitalistic things, while we are still waiting for philosophical rational knowledge to produce any being--a perfect thing that never breaks, or wears out, or changes in any way.

Plato's conceit seems to be the core of his arguments. As Plato, the philosopher, tells the story when philosophers get together and give some verbal question a dialectic workout, the philosophers can't fail; their behavior will produce knowledge! But, according to Plato, who never heard of science, empiricism has no comparable magic. Plato, two thousand years ago, in effect claimed that experimental scientific/engineering would forever be impossible.

Plato's metaphor is ideal to make two points: First, his point was that empirical knowledge is not reliable, valid, etc. but second, the verbal rational mind of philosophers has a special set of skills for **knowing** spiritual objects. This is the particular point where Plato put Stratton glasses upon those who would buy into his metaphor. This chapter was designed to start to undo Plato's two thousand year head start of selling the myth that humans are the only species with a mind, etc.

The tragic legacy of Plato: Science vs. Plato

In one sense, I feel guilty criticizing Plato whose philosophical work was produced over two thousand years ago. Perhaps, there is no point in criticizing Plato, my criticism is directed not at him, but to all those who still follow him today.

The irony is that precisely those points in which Plato takes such pride are precisely those places where modern science is careful to take a position opposite to that of Plato.

- Point one: Where Plato takes great pride in his belief that knowledge is the property of the domain of philosophers, science makes the point that scientific knowledge is not the province of any special group of people.
- Point two: Scientific knowledge is pluralistic; it has to be open to all, independent of race, religion, age, sex, etc. Science could not recognize as scientific knowledge any fact that is only true for some particular philosophy, race, religion, sex, etc.
- Point three: Plato wants truth to be eternal, and so it is with science. The laws of nature, the force of gravity, the speed of light, etc. are the same everywhere and at all times. They are the invariants of nature--a kind of phenomenon to which Plato was blind.

Spinoza's Monism, vs. Plato's dualism

We must now come back to our theme. What is an idea, ontologically; what is an idea and where do they come from? However, we must keep in mind that our concern is with nonverbal ideas not verbal ideas. As Plato said, Ideas come in two parts; A and B, the concept and the exemplar. But where do the A components of an idea come from? Is it mind or matter?

My task as a rat runner is to show that rats in the matrix maze, F3, for example, who have no verbal speech capacity conceive a set of As, with four coherent nonverbal ideas in less than thirty minutes. F3 showed that she could convert these ideas into a model, then into a plan, and then into a action pattern that allowed her to extrapolate in the third row and run the maze without error.

Spinoza

Spinoza (1933), on the other hand, proposed that god was monistic, that everything within the cosmos made of matter was god. God is what "Is." One of the advantages of a monistic god is that since everything that is real is made of matter and can be "known" by the senses, empirically, finally it will be possible scientifically to define the ontological differences between faith and knowledge. Only "things" can be known. All Platonic nonmaterial beings; all out of this world spirit beings, cannot be known. All Spirit or word-based religious concepts, which cannot be empirically known, can only be justified by faith!

Nonverbal based religions do not require faith. In a Spinoza (1933) based religion, god is known; no faith is required. Empirical religions are action based. Of course, in the context of the nonverbal mind, truth is defined simply and elegantly in term of feedback; each action sequence that does not create surprise is true. The truth of every nonverbal experience is characterized by a closed sensory experience. Everything that is monistically real can be known by the senses and by science. The point is that domain-specific sensory experiences that are true are closed sets of sensory experiences, or are consistent with the natural laws of Spinoza's god. Anomalies then play a critical nonverbal role; they in effect, are the natural sign of falsehood, or are nature's lie-detector test.

The nature of the nature-god

However, most importantly, Spinoza's monistic god has important empirical properties, which are the way the cosmos works and evolves. One of the most important consequences of a material god with empirical properties is that it can be known by science, just like all the other things or objects of the cosmos.

However, for our purposes these features of the cosmic god are permanent universal features of everything within the universe. Just as the eye evolved to know the properties of light, just so the nonverbal mind evolved the power to recognize or read the local nonverbal manifestations of these universal features or attributes of objects, the things of nature, whose features, are the universal features of the cosmos.

The cosmologists today are identifying these universal properties of the cosmos. For our purposes the two most important of these features are:

First: it has a hierarchical structure,

Second: it has emergent properties

Ontology: Monism vs. Dualism

Each community conceives of its god as the cause of all things, yet Spinoza's (1933) and Plato's (1952) god not only have a different nature they also exist in different worlds with which they share their nature.

However, Spinoza's physical god cannot be said to exist as a separate object in a physical world, but IS the physical world, while Plato's Spirit god occupies a spirit world. Both gods have their own kinds of laws, their own kinds of truth, and their own kinds of judgment days.

Actually, all the transcendental religions share the origin of their nature with Plato's philosophy. Both philosophy and religions concept of spirit-beings have their origin in Plato's metaphor of the divided line.

It is strange, almost unthinkable, but the nonverbal mind evolved when the world was without words, and without metaphors, and excuse me when I say, they evolved in a world Pre-Platonic. Only with the emergence of humans did the cosmos, for the first time, acquire verbal behavior with the capacity for metaphors, yet early on Plato's metaphor of the divided line used metaphors to turn the world upside down, but nonverbal Ideas evolved perhaps fifty million years before words and metaphors emerged.

Now, one critical change has occurred, my research with the nonverbal behavior of rats has provided me with detailed knowledge of the ontological nature of the nonverbal mind. Armed with this empirical knowledge of the monistic ontological machine, I can scientifically challenge the dualistic ontology of Plato with scientific facts. The nonverbal mind can rationally be called the material ontological machine; the mental organ that houses Aristotle's (1952) Sixth Sense. This organ serves several ontological functions but most importantly, it houses the technology that conceives consciousness that creates the life space, which induces nonverbal behavior. (Lewin, 1943) How can science take possession of this ontological domain that has always been a no-man's land up until now?

The answer lies in the difference first exhibited between Plato and Aristotle (1952). Aristotle began first to describe the mental machinery that biological evolution designed to enable all mammals to move into and take possession of those domains of space, which were full of uncertainty; that were ontological, conceptual voids. Not only did they survive, but also thrived by digesting the uncertainty which created the void. When the nonverbal mind moves into one of these voids of which nature has an infinitely large population and begins to digest the uncertainty to destroy the darkness and takes dominion over the space, as self-organizing systems were designed to do, then we see for the first time the core function of the nonverbal mind. The generic mind was not designed to grant humans the power of speech, which would emerge in its own time. For the moment, the nonverbal mind confers upon life the rational power to probe and digest the uncertainty in these domains one at a time. Slowly mammals expand their territory, expanding their competence to self-organize their way through the chaos of the nonverbal populations of space.

But the nonverbal mind provided, as it was designed to do, a better way for life to adapt to change, yet the core element of the adaptive rational function is the opposite of incarnation. The verbal or spirit believers have long used the concept of incarnation, where the word or a spirit becomes flesh. In their use of this word, thy tacitly acknowledge that in order for a spirit to be real it must change its ontological status. However, due to the glaring absence of a word in English for the opposite of incarnation, I coin the word inideation as a rival of incarnation. Incarnation means the word is made flesh, while in inideation the flesh is made word.

Book of the pantheistic god

Evolution designed the nonverbal mind for the infinitely long adventure of getting to know the physical world, Spinoza's god, hopefully and strangely, may it take forever. However, one thing is clear, evolution did not design the nonverbal mind to wait for some spirit god to give it the revealed truth, either in Bibles or as sets of ideas injected into the mind at birth nor written on tablets of stone. Rather, the creative mental adventure begins anew each time a mammal finds itself freed in some one of an infinite population of unknown terrestrial spaces and is able to unleash the full creative power of its nonverbal mind upon this space. The SOS's technology of the nonverbal mind was designed to probe and digest the anomalies of each new population by creating a set of concepts, which, as it were, turns on the conceptual light, which does not let us see with our eyes, but more importantly to see with our mind's eye.

The evolution of the nonverbal mind raised several conceptual challenges, which every theorist must confront.

- **Challenge one:**

My first task is to describe the stations of an ontological machine and to describe how such a structure functions. I have adopted the Aristotelian (1952) type model of mind, and will describe how a mind evolved first in the horizontal plane and then in the vertical plane. The mind is most complex. We will describe these two components of the mind, the Horizontal-Vertical mind in two chapters: how they work, separately, then how they interact.

For Plato (1952) only a spirit god could create eternal ideas with a pure spiritual nature. While for a materialist like Aristotle (1952), only a pure material nonverbal mind could create domain-specific closed sets of material-based ideas. The material mind creates all the real classes or sets of ideas, A of which things, B are exemplars.

- **Challenge two: the sensory-driven mind**

Plato's (1952) cause was to create a set of real eternal things. However, his concepts of real beings were spiritual beings, which never change. This is more plays on words. Why do spirits never change, because there is nothing to change? Indeed what kind of change could a nothing execute? Can a nothing change to something? If so, was it a real eternal nothing in the first place?

In Aristotle's case, on the contrary, real things are beings made of matter, and thus have the nonverbal mind's empirical sensory properties as attributes. So, the material mind's ontological function defines things as a set or syndromes of "sensory mental attributes."

- **Challenge three: Concepts: recognition function**

  Third, the ontological function of the material mind must also create a rational detection or recognition mental function, which can detect the presence of every class or set of sensory attributes, which match the respective set of sensory properties created by the ontological function of the nonverbal mind as concepts for defining each class of things.

- **Challenge four: the Certification Function**

The ontological mental function of the nonverbal mind creates a Certification Mental function, which determines which set of attributes qualify as exemplars and therefore qualify to enter the nonverbal consciousness. Exemplars are certified as real by being written into the image-based narrative, which is running in consciousness. On the practical side, the role of the ontological sense gets involved in real life in the predator and prey function. The goal of both the predator and prey is to avoid detection by the ontological detection function of their opponent. For example, evolution created the ontological functions of the nonverbal mind to detect, classify, and initiate action toward every set of neural impulses that enter the mind through the five senses that qualifies as empirically real. However, this recognition function of the ontological mind is both an advantage and disadvantage. Empirically established ideas or things would have to have their validity certified empirically.

Nevertheless, for the nonverbal mammals, the ontological function, what is real, is generally a matter of life or death. To conceive sets of ideas is one thing, but is useless without the assurance that every idea or component of a rational system is real, and is safe to act upon.

Invariant

On the monistic side, the ontological problem is no less acute for the nonverbal mammals than it was for Plato. Today's science has a well-established criterion for defining what is experimentally real. Scientific questions can only be settled with reproducible scientific facts. What makes an experimental result real? Science today has a sizable set of active probes to employ to force nature to reveal her invariants in each scientific class: physics, chemistry, biology, etc. and has designed a technology, which becomes more complex and powerful by the day. Science defines the real as a reproducible fact or a reproducible experiment. It is interesting that an invariant and a reproducible experiment are two names for the same thing. A reproducible experiment is one that can be reproduced by anyone, at any time, and any place that is, in other words, an invariant, or that which is independent of person, time, and place. Scientific experiments as invariants are the product of an advanced form of nonverbal mind probes.

To act in the real world is to commit one's very life upon the ontological validity of each element of consciousness, and if any one of the ideas of the whole is false, or is not real, then the action is not justified. The role of the ontological machine is to certify every idea that enters the nonverbal consciousness. The role of the nonverbal consciousness is to initiate and sustain nonverbal actions, but every element of the script that causes action must be nonverbally real!

Challenges to Plato

While Plato provided the foundation, upon which the West's concept of human nature formed and generally naively accepted, yet, Plato's assumptions from his philosophical school of thought have had their challenges and challengers.

- **Challenge six: What is Eternal?**

Plato wanted facts to be eternal but Plato had a very narrow view of the meaning of the concept of the eternal. Plato (1952) understood the eternal in a way that is meaningless in the real empirical world. However, Plato was not wrong in wanting the real to be eternal; he just did not have the experience to understand the difference between his out-of-this-world concept of eternal, and the real world's everyday meaning of eternal.

For example, water is H2O eternally. Science has never found water that was not H2O. Every scientific fact can be counted upon to work anywhere and anytime. Gravity works the same everywhere, on earth and out in space, as NASA has repeatedly verified for years. The simple fact is that today only science's concept of a scientific fact, as a reproducible event, is real or valid. However, Plato despised all empirical facts; yet only scientific or empirical facts are an eternal truth!

- **What is a god**

There are two kinds of gods.

The verbal god's verbal book religions

Each of the classic transcendental or spirit religions has its own divinely inspired book in which each verbal god reveals itself, and gives its followers the knowledge/truth they are to live by. Each transcendental religion is a book-defined religion. All we know and can ever know about each spirit god is limited to what each spirit god chooses to reveal about itself in its book, the nature of each book-god is bound by the words in its respective book. Every book religion or every book god has its own book and presumably describes its nature, and provides its own divine narrative.

These revealed transcendental religions assume that humans are not actually capable of self-organizing. On the contrary, the revealed religions are like the nativistic forms of life, they too live their whole life by a new kind of "genomic," or book knowledge. Most believers refuse to **recognize the insulting nature of the book religions.** While they may differ in how they received their program: upon a table of stone, injected into the mind at birth, or revealed in some book; yet all conceived human nature as passive: as computers which can only blindly execute the programs they are given by their spirit programmer in the sky.

The book-gods each write their own book, this is Plato's god-function. The book god's followers are only half-rational. The book god while executing the god-function writes the book, while the book god's half-rational followers simply passively live, and execute the words of the book.

The nonverbal book

Science creates a special kind of truth, a nonverbal reproducible truth, or empirical facts. Plato would have nothing to do with empirical knowledge. By definition, such knowledge did not or could not exist.

Knowledge for Plato had a rational not an empirical origin

Science has only one purpose, to describe what can be known with the senses, what is empirically reproducible. Within the sensory-driven concept of knowledge, the pantheistic god is all that can be known. Yet the pantheistic god is not of this world, but the other way around. Everything that can be known that exists in space and time is of god and belongs to everyone who is willing to live their life upon the foundation of fact.

According to Plato, half-rational sense-driven humans can never empirically know the spirit god or discover the real; nor can they create ideas individually or in sets or narratives of any size. So the verbal gods have no other option than to give humankind their own written versions of verbal truth in verbal forms, in their own books, or bibles.

Each GNMM can communicate with the material cosmos

The nonverbal pantheistic god's nonverbal book

Those scientific specialists that devote their life to the study of domain specific features of the nonverbal material world spend their life studying/reading the domain specific features of some domain of the cosmos. Only science generates empirical knowledge; only science is writing the story of the universe in one unified narrative. Science is doing what Plato wanted; scientists are writing the only "one truth" that is true for all humans in all times and places.

When nature finally created a family of mammalian species that had evolved a better way to organize life, it had adopted the scheme of giving each organism an organ capable of executing Plato's (1952) "god-function". Each mammal needed no book but rather, only needed that part of the cosmos within which it found itself. Each mammal had the mental mechanism to extract from nature the knowledge it required to self-organize its own behavior to satisfy its own self-defined goal.

The classic mind

The classical religions have effectively taken away from human mammals the ability to self-organize, and given them nothing in its place but a book full of words, which have no empirical validity yet give an adult community a set of adult Santa Claus myths, which explains how the adult Santa will give them their daily bread.

Mammals are born with the mental power to discover and write for themselves an empirically true narrative, which empirically works and contains a valid model of that domain; in the domain, they must achieve some goal; no verbal myths are required.

II. Sperry: The first scientific challenge

The evolution of the monistic conception of the mind

There have been two scientific challenges to the classic dualistic paradigm of mind; first, Sperry's, and then mine.

In 1985, Sperry received the Nobel Prize for scientifically proving that humans had two minds: a nonverbal and verbal mind, one in the left and one in the right hemisphere. Sperry's (1961) two human minds were a start; they clearly established that Plato was wrong, mind was not a spirit but a physiological organ located within the two hemispheres of the human brain available for science to study.

Since both of Sperry's (1961) minds were human minds he left Plato's concept that only humans have a mind still standing. Sperry also offered no suggestion as to how both minds emerged at the same time in the human brain. However, Sperry's two-mind theory made a powerful start at establishing a new paradigm.

Unfortunately, both Sperry and Plato conceived mind as a speech faculty, but Sperry (1961) changed one major component of Plato's theory. Sperry introduced the concept of the nonverbal mind. However, Sperry's concept of the nonverbal mind was misleading. Sperry showed that certain of the verbal functions, such as the recognition of objects, were computed in the nonverbal mind, but then the results transferred across the corpus callosum to the primary speech side of the brain. Sperry's phrase The Nonverbal Mind, unfortunately, gave the wrong impression about the role of the right hemisphere. His phrase did not mean that the nonverbal hemisphere did not play a central role in human speech behavior, but simply that the initiation of speech was always from the left hemisphere, presumably, because that is where Broca's area is located. However, Sperry conceived the nonverbal mind as playing a subordinate or secondary role in generating language behavior. Sperry's concept of humans as a two-mind species had one major fault, his downplaying of the role of the nonverbal mind.

In fact, from the point of view of the generic nonverbal mammalian mind, I must suggest that Sperry's distinction between the verbal and nonverbal mind is based on an artifact. What he showed was that the human speech mechanism was essentially an assembly line. Some components of the human speech assembly line were in the right and some in the left hemisphere, but I suggest that the components on the two sides of the brain still added up to only one verbal speech mechanism.

By my definition of terms, Sperry had no concept of a nonverbal mind. In the context of the generic nonverbal mammalian mind neither of the hemispheres had a Broca's speech center or in other words as I use the term, the nonverbal mind I refer to lasted for over a fifty million years, when neither hemisphere of the nonverbal mammals had a Broca's speech center. In my context, both hemispheres of the nonverbal mind were nonverbal.

Sperry's (1961) human brain with two hemispheres, he claimed, had two minds, one in the right and one in the left hemisphere. In my way of thinking, this is but an instance of the belief that in terms of evolution there have been two models of the mind. For the longest time, there was only the generic nonverbal mammalian mind, GNMM, then later a human verbal mind with a Brocca's area evolved, but the human mind is but a new model of the old generic nonverbal mammalian mind with a new station for Broca's area.

The Talk-Centric mind

Sperry's view of the mind was Talk-Centric as was Plato's and all the other philosophers and linguists down through the years.(Kuhn was the exception--see below) The nonverbal mind meant for Sperry (1971) only that the nonverbal mind was such a "subordinate" and "minor" component of the Talk Centric mind because it could not initiate speech.

The Action-Centric mind

I propose an alternative to this classic Talk-Centered conception of mind, an Action Centric mind. For millions of years the generic nonverbal mammalian mind was exclusively an Action-Centric mind and the only mind which all the prehuman mammals had. The action-centric mind, of course, had no verbal speech ability, no Broca's speech center, and I would expect that the first question that comes to most reader's minds is what does a nonverbal mind do that isn't verbal?

I propose that the nonverbal mind is the primary mind and its role even in the human mind is to execute perhaps ninety percent of all mental functions, which are indeed, nonverbal functions. My thesis is that the verbal mind and the verbal functions are secondary functions; the nonverbal functions have always been the primary function, and are still the primary functions even in the human verbal mind. The nonverbal mammals had already been around for millions of years before the verbal human mammal entered the scene. Therefore, perhaps we need to make a list of what rational functions nonverbal minds create that is useful to nonverbal mammals.

But a faster way to answer this question is to note that the sole function of the generic nonverbal mammalian mind is to organize the somatic behavior of the mammalian Rolandic cortex. The verbal mind is, relatively speaking, simply one of the newest features of the mind, which evolved to control Broca's area located in the left hemisphere below the Rolandic cortex.

Sperry (1961), in my opinion demeaned the nonverbal mind when he called the right hemisphere mind the nonverbal mind and the minor mind. Both these names: the nonverbal and the minor mind obviously reflect Sperry's view that the true or only function of the mind is verbal, and reflects his view of the mind as Talk Centric. By my calling the original generic mammalian mind the "Action Centric Mind," I intend to call attention to the fact that the primary role of the mind even in humans is nonverbal. Someone needs to measure or at least estimate what percent of human behavior each day is verbal and nonverbal. However, my point is that the mind does more, much more, than control human speech behavior, everything from putting on your shoes, to brushing your teeth, to playing golf, to playing the piano, or to driving your car. What other mammal does these things? The human mind enables all kinds of behavior that require a special vocabulary. Speech is but one of the special wisdom-based skills that the human mind with the fourth wisdom enables humans to enjoy.

Even with humans, the major percent of the behavior each mammal emits each day is still nonverbal, a function of the third wisdom. Many jobs are purely nonverbal, and do not employ the fourth-wisdom, talking behavior. For example, brushing one's teeth and talking are mutually exclusive.

While Sperry claims the execution of the verbal function is the sole function of the verbal mind, and therefore the primary mind, I would estimate that the verbal function represents no more than ten percent of the human mind's motor function.

However, Sperry's (1961) phrase the nonverbal mind has a meaning, which completely misses the point of this same phrase used with respect to the nonverbal mammals. Sperry's phrase refers to the fact that the right hemisphere side of the human mind does not initiate speech since the Broca's area is in the left hemisphere.

However, the rat's brain has no Broca's area in either hemisphere; no side initiates verbal speech. The nonverbal mammals cannot export the nonverbal narratives running in their nonverbal consciousness.

I call the generic mammalian mind the nonverbal mind, but for a new reason. Evolution created the nonverbal mind to enable mammals to evolve a new nonverbal self-organizing kind of behavior, not instinctive behavior, but a rational kind of somatic behavior, which turned mammals into self-organizing systems, which can spontaneously create on line and in real time the nonverbal knowledge that is required to function in an anomalous domain. The primary rational function of the mind is to make models, extrapolate, and execute scripts, which enable mammals to act without error.

For me the primary meaning of the name, the nonverbal mind, refers to the fact that the nonverbal mind has been on the job for millions of years. Evolution designed the nonverbal mind when there was only one mind in two hemispheres, and it was the nonverbal mind. This rational nonverbal mind earned this name for all of the nonverbal mammals in existence because all their behavior was nonverbal. For millions of years all mammals had only the Rolandic region for controlling their somatic motor behavior. In my case, the name nonverbal mind has a positive meaning. The exclusive role of the nonverbal mind was to control the Rolandic region, the somatic nonverbal region of the brain/mind.

Sperry's impact, as was noted earlier, was almost entirely upon our theory of human nature, not our theory of animal nature. The idea that mammals, such as rats, might have a nonverbal action-centric mind rather than the old Platonic (1952) vital soul was simply outside Sperry's frame of reference. Before Sperry, the facts about animal and human language behavior were still the same as they were in Plato's days. However, Sperry's (1961) scientific data offered, for the first time ever, a set of empirical facts about the mind and required some specific changes in our understanding of human nature.

For the philosophers and linguists, the mind since the time of Plato has been synonymous with verbal behavior. If science can objectively prove that there was such a thing as a nonverbal action mind, this would ruin their day.

However, if the nonverbal mind should additionally prove to be the original as well as the primary mind, this would shatter centuries of myths.

The linguists today generally ignore Sperry and explain human verbal behavior as if the human brain supported a one-mind pure verbal function, ignoring completely even Sperry's (1961) reference to this nonverbal mind. Generally, the linguists focus exclusively upon the laws of grammar as if they were a product of a verbal spirit, which magically generates the language function. They ignore the possibility that language behavior is the product of a material mind with a structure which generates the language function, so that to put it mildly, they have made some grievous errors, which, now it is both necessary and possible to begin to correct. Chomsky's theory of a universal grammar and Lashley's counter theory will be discussed later.

The one-mind theory of human nature

My basic assumption is that humans have one material mind distributed between the two hemispheres. Human language behavior is a joint product of the functions of both sides of the brain. My first task is to broadly describe and explain how nonverbal behavior is the product of the nonverbal mind, and then secondly, I will broadly describe how human verbal behavior is the joint product of a new thirteen-station model of the nonverbal mind.

In order to understand why the nonverbal mind emerged first-and enabled the nonverbal mammals for millions of years to self-organize their own domain-specific behaviors, we must take a step back and ask why evolution finally began to create a new form of life, the mammals, in the first place.

The origin of the human compound mind

What was the origin of the human mind? Where did the two models of the human mind come from? Did they both evolve like all other organs of the human body? The linguists' standard answer today is still "Nativism." The authorities on the verbal mind say that there can be no answer to that question other than to state that the existence of the verbal mind is a fact. There are two basic facts: first, humans have the gift of verbal speech; second, is the fact that no other species does. Obviously human speech emerged "out of nowhere." Nativism says that the verbal mind was not a part of evolution for there were no precursors of speech behavior observed in any other species. They ignore the existence of the primary nonverbal mind, and ignore the possibility that verbal behavior is only a more complex form of somatic motor behavior that emerged for the first time in humans, and they ignore the possibilities that syntax and grammar could have evolved as a nonverbal function. See my discussion of Lashley's theory of serial order in Chapter Four.

III. White: The second scientific challenge

The two-stage mind

Every question about the nature of mind inevitably is an ontological question, but not all ontological questions are equal. In this section, I am going to address the biggest ontological question of them all, where did the verbal mind come from; obviously, each side of the ontological question has their own answer. Historically, the question has been one sided. The classic philosophers were dualists who only believed there was one mind, the human verbal mind, and so obviously, they had only one-sided questions and answers. Their only question was where, when, and how did the verbal spirit mind enter the human mammal? The dualists, those that believe that mind is a spirit, had no choice but to embrace the classic nativistic argument, the mind did not evolve, but was magically created by a spirit god or a talking god, who must have at some late date said, "let there be speech" and saw that it was good!

While on the other hand, the monist theoretically can argue, given the proper set of empirical facts, that the verbal mind could owe its existence to evolution but how no one could scientifically say!

The scientific problem has been that historically no one knew of the existence of the generic nonverbal mind. However, this set of experiments settles that question hopefully for the first and last time, knowing about the existence of the GNMM sets the stage for the next step, and from here, it was only a relatively small step from the emergence of the nonverbal mind to the evolution of Broca's area in the left hemisphere.

Psychologists, probably do not all agree; but at least some of us, and hopefully all of us desire for Psychology to become a science. The biggest roadblocks are disagreements about the nature and origin of the mind. As long as mind's conception is seen as spiritual in nature, our ambition for psychology does not have a chance of coming to fruition. The natural or hard sciences rightly insist that only material beings attain scientific recognition.

Most scientists today take evolution for granted, yet hundreds or thousands of fundamentalist Christians and their school districts are preventing: first, evolution from being included in their textbooks; second, evolution from being included in their science classrooms. Yet there are famous scientists such as Chomsky (1997) who still insist that the verbal mind had a "nativistic" origin and thus are blocking the path for the acceptance of psychology as a science. The Nativists assert that mind is a special creation (of god) and could not have evolved naturally. When confronted with Sperry's (1961) concept of the nonverbal mind, some take the easy way out and simply ignore the data. (Pinker does not even reference Sperry in some of his books on language.)

The Taxonomy issue

Even those who are trying to confront the facts such as Sperry's have special problems particularly with taxonomy issues, the science of classifying things. For example, how do we know when we are dealing with a new mind, or when one mind upgrades to two minds; how do we distinguish between the emergence of a new species, or just a new model of an old species?

Sperry (1961) jumped too quickly to his conclusions that the human brain was home to two minds, a verbal mind in the left hemisphere, and a nonverbal mind in the right hemisphere. My discovery that there is a nonverbal mind, the first stage GNMM, proves another alternative must be considered: Sperry's two minds are two stages in the evolution of mind. How can we decide whether we are comparing two kinds of cars, or two different models of the same car? When does a new model of a Ford become a Cadillac?

The somatic motor system and its two control centers

The body contains two motor systems, the autonomic, and the somatic. The former controls the smooth muscles, while the somatic controls the striate muscles. The smooth muscles move things through the body, while the striate muscles move the body. The former are homeostatic functions, while the latter are Heterostatic functions.

The most telling point is that even in humans the nonverbal mind is the major control center for the somatic muscles. The nonverbal mind controls perhaps 98% of the somatic muscle's actions, while the newest model of the verbal mind controls only the somatic muscles from Broca's region, which only controls the somatic muscles of the vocal apparatus.

The two somatic motor centers in the mammalian cortex

Fact only mammals have a cortex.

Stage one: the one-way rational nonverbal communication with nature.

The only region, which concerns us now, is the large central area at the top of the brain, called The Rolandic cortex or otherwise known as the somatic motor cortex. This Rolandic motor region of the brain controls all the nonverbal somatic muscle behavior in the body of all mammals, all the nonverbal mammals, and the one verbal mammal.

Stage 2: Verbal language behavior evolved:

- **First, the nonverbal communication with nature in the nonverbal mind**

- **Second, the full blown form of nonverbal/verbal communication in humans**

In the human mammal, there emerged a division of the motor region. In the lower perhaps one tenth of the left hemisphere there emerged a specialized somatic motor area-Broca's area, which controls the somatic muscles, which control human speech. Broca's area is generally called a speech center, yet first, it is but a special form of somatic motor behavior. (Donald, 1991)

What do these two somatic motor regions tell us about mammalian language behaviors? Primarily there are two kinds of language behavior, Rolandic speech and Broca's speech, or nonverbal and verbal speech?

The emergence of the nonverbal mind, perhaps we should, say happened twice, first as a physiological fact, second now, when it is recognized by science.

Now we must consider this question - do humans have two minds or only one mind with a two-stage communication function; one with two stages in the production of speech? In other words, what is the difference between Sperry's (1961) two human minds, the verbal and nonverbal minds? Is the nonverbal mind a second mind or the first stage of the verbal mind?

The revolutionary question is what language function does the large region of the Rolandic cortex serve? Sperry doesn't say, but obviously, in his opinion it has nothing to do with speech.

However, the entire motor control in the original first model of mind is a rational function of the Rolandic region. What is called for now is a new paradigm of rational behavior; one, which recognizes that rational behavior, emerged as a two-stage function. What we must first consider is that the role of the nonverbal mind was to enable mammals to have rational control not for talking but for the rational control of Rolandic behavior. The first stage nonverbal rational communication, which controls the nonverbal behavior of all mammals emerged and served all the nonverbal mammals well for millions of years, then, with the emergence of the special speech centers in the left hemisphere humans now emerged with a new two-level or two-stage rational communication function. Now they can not only communicate with nature, but they can now share their knowledge of nature with one another.

In the first stage model, the nonverbal model of mind, the entire Rolandic area became as it were a speech center for the nonverbal form of speech, the only form of language spoken by all the nonverbal mammals. The problem with the classic communication paradigm is that Sperry's paradigm does not recognize any kind of speech except verbal. I must admit that this theory requires a revolutionary transformation of how we understand communication.

My radical proposal is that the there are two kinds of rational communication: the first stage is the nonverbal kind of rational communication with nature.

Second, this first type of communication was expanded from a stand-alone system, into a type of communication, where the second stage simply exported the knowledge generated within the first stage's interaction with nature.

The nonverbal mammals have only a Rolandic region; there is no Broca's region, yet they have the primary nonverbal communication function, which enables them to know nature, to know where they are, to conceive goals, and to conceive means to achieve their goals, etc. The nonverbal mind creates all the meaning, which words can obtain, but rationality first emerged as a nonverbal function, which enables mammals as SOS1 to organize their nonverbal behavior emitted from the Rolandic region of the mind..

The great fantasy ball

Think of the nonverbal mind as creating a great body of nonverbal knowledge, a great pool of knowledge much akin to the gene pool of biological evolution. Millions of years pass, and one day a new mammalian human species appears upon the planet earth, which can speak in a new higher-order way. This species has the capacity to engage in a new two-stage kind of communication. They have both the god-function, the power to know nature, yet they have this new power, the ability to export the knowledge of nature, which originated in the nonverbal communication function. However, while the first stage of nonverbal communication was mediated by images, the new second stage communication was mediated by words.

However, at the "first light" of this new stage of the emerging cosmos, the age of verbal communication, nature threw a great ball. Nature had a plan for all the images and all the words to meet one another at this great coming-out ball. All the boys, the nonverbal images created by the first phase of the mind, and all the girls, the verbal words from the second phase verbal mind, come to this great hall on the night of the ball. First, they were blindfolded and then two sets of communication forms, the nonverbal images, and the verbal words formed into two great concentric circles. The music started and the nonverbal boy-images go round to the right, and the girl-words go round to the left, and they go round and round waiting for the greatest bonding moment in the history of the cosmos. Suddenly the music stops, and the two circles are told to face each other and take their masks off, walk forward, and meet their new life-partners and so all the feminine words and manly images were randomly mated.

Today people still argue what happened on that night, in the marriages between the manly images and feminine words. Did the men, the masculine nonverbal images, give meaning to the women, the words formed of secondary symbols, or was it the other way around, did the words formed of symbol, the girls, give meaning to the manly men, the images formed of nonverbal symbols. Did the men give meaning to women, or did the words break the silence of the men?

The consequence of this marriage of the minds

Advantage one-progress

Being a nonverbal mammal was a great gift, for each was given a new adaptive power, no other animal had ever enjoyed the power to self-organize, to communicate with nature. Female three, F3, for example was able in less than thirty minutes to know the maze, but as a member of the set of nonverbal mammals, she was not able to share her knowledge around a campfire or to publish it in a scientific journal.

But with the marriage between the two stages of mind, progress was finally possible. Each two-stage human mammal can now share the personal knowledge they conceive with their nonverbal mind with others of their verbal community; knowledge now need not be lost with a two stage mind nonverbal knowledge can now be exported and shared with all the members of their verbal community, and possibly with all the members of the planet earth.

Advantage two-division of labor

Only the human cortex has a divided motor cortex; part nonverbal and part verbal, part male and part female, but the Rolandic region is by far the larger part of the motor cortex.

Broca's area, the speech center for verbal speech, emerged with the newest model of the verbal mind, while the Rolandic region emerged with the evolution of nonverbal speech.

The reader should be asking what nonverbal speech is.

The somatic nonverbal motor functions are the primary kind of speech; the only kind of communication the nonverbal mammals have ever had, but it served them well for millions of years. Controlling the Rolandic region is the function of a set of technologies, which I will discuss as components of Aristotle's (1952) model of the mind later in this chapter.

I must continue to note that all kinds of speech, both nonverbal and verbal, are functions of the two somatic control centers. First, the nonverbal, which is a function of the entire large Rolandic cortex and second, verbal, which is specifically a function of the smaller motor region, Broca's area. Which kind of speech is primary and which is secondary? Which kind of language evolved first? Which kind of speech occupies the largest percentage of the cortical motor centers?

Human verbal speech is a joint product of both speech centers or both models of the mind and both kinds of languages.

The emergence of the generic nonverbal mind turned mammals into self-organizing systems. The only language these self-organizing mammals speak is nonverbal. Their Rolandic cortex mediated all the communication done by these nonverbal mammals. How does the Rolandic cortex communicate? With whom does it communicate?

Hypothesis: The two-stage theory of rational communication

Hypothesis One: There are two one-stage minds in the brain

Hypothesis Two: There is one two-stage mind in the brain.

The emergence of the first-stage of the mind

Evolution in the first step evolved the generic nonverbal mammalian mind, the GNMM, which enabled the nonverbal mammals' rational control of their somatic muscles within the Rolandic motor cortex.

However, in step two with the emergence of the Broca's area, human's verbal behavior emerged. The verbal mind is but a new edition or a new model of the nonverbal mind; one with a new extension of the motor cortex, the Broca's region.

In my discussion of Aristotle's (1952) model, I will make some of these points more explicitly and effectively but for now let me make just one more point.

Plato and all his nativistic followers have managed to ignore for thousands of years that when humans talk it is talk about people, places, and things; all objects created by the first stage of the nonverbal consciousness, which are not the physical things in the world, but they are the product of the activity of the nonverbal mind. They are elements of the nonverbal consciousness created by the nonverbal mind. All we humans can know and talk about empirically are the creations of the nonverbal consciousness.

Both stages serve a different purpose: the first stage enables life to know: to communicate with nature; the second stage enables life to communicate with others of its own kind what they know about nature.

The purpose of the first-stage mind

The first state mind evolved to control the Rolandic cortex, or the somatic motor system. At this stage of evolution, all the somatic behavior the nonverbal mammals generate was planned and executed rationally, employing the knowledge stored in the model of some space, such as the maze. Rats employ this model-generated knowledge to create a plan for running the maze. The most impressive aspect of this behavior is that rats can conceive a model of the maze in the early part of the maze, and quickly conceive a plan to achieve their domain specific goal by extrapolation.

But for thousands of years, humans no doubt have seen and heard rats running through the walls and floors of our homes, but we never attribute this behavior to the nonverbal mind: to being an expression of nonverbal planning of the rat's self-organizing mind. Yet the data of this set of experiments support the idea that all mammalian somatic self-organized behavior is generated by the nonverbal mind.

When we humans begin to think about how our mind works, generally we start with Plato's (1952) one mind theory of human verbal behavior. No one gives a thought to the idea that down deep in our brain there is the first stage of mind rationally controlling all our nonverbal behavior. It never occurs to us that as mammals we share the generic nonverbal mammalian model of mind, which is the only mind the nonverbal mammals have. When we learn to crawl, then to walk, to run and jump, and to eat, etc., this nonverbal first stage-mind is active; all mammals are self-organizing systems, and our first stage primary model of mind enables us to master everything except talking. There are two kinds of organization: first, the organization of a script or what we want to say or do is a nonverbal accomplishment; then second, the nonverbal script must be translated into some new alphabet and organized into words. The nonverbal mind does the first step: the verbal mind does the second.

According to Plato (1952) and all his modern nativistic followers, humans have a verbal mind within our brain, which suddenly emerged out of nowhere and functions without interacting with a nonverbal mind. It may consist of too many parts to force into one hemisphere, and so part of its elements may function out of the right hemisphere and part function out of the left hemisphere.

My first task in a sense is to bring words down to earth, and give them a material nature. This book teaches the reader to think of the mind and behavior, both nonverbal and verbal, as two stages of somatic motor functions; evolution took millions of years to perfect the nonverbal motor behavior systems before it expanded somatic motor behavior into a new function, the evolution of Broca's area. The evolution of Broca's area makes the two hemispheres different, but the evolution of the new somatic feature of the mind, which modernized the mind, was the product of the speech motor centers in the left hemisphere, which ended up being at the tip of the iceberg. The structural arrangement of these two models of the mind allows, or actually encourages, the wrong implications to be drawn. The fact that Broca's region is on top simply reflects that the communication between the two motor centers is from the bottom up because the primary narrative generating functions are created by the nonverbal mind and this has been so for millions of years. For all this time, the behavior of each mammal is a function of the nonverbal mind functioning as a stand-alone nonverbal self-organizing system. All mammalian nonverbal behavior was and remains completely directed by the knowledge contained within the nonverbal script created by the nonverbal mind. What was new about the two-stage mind was that evolution discovered there was an adaptive advantage to sharing the nonverbal narrative and evolved a completely new somatic motor cortex, Broca's area. The verbal mind is not now and has never been the creative region of the mind. The structural arrangement of the two hemispheres simply reflects the fact that the nonverbal mind is a gold mine of knowledge. Evolution finally made it possible for this nonverbal mother lode of knowledge to be transformed and used not by just one individual and then lost forever, but on the contrary, with the evolution of Broca's region, evolution makes it possible that this private nonverbal wisdom not be lost but can be transformed and shared with the entire verbal community.

What Sperry (1971) called the major division of the brain is between the right and left hemisphere. Sperry reminds me of the early Pavlov (1960) who took the position that all the mental activity was happening within the cortex, in particular across the cortex.

But technically, as Sperry (1971) showed the nonverbal knowledge transfer happens from the right to the left hemisphere of the mind, yet if we adopt a historical perspective, then this right-left transfer of knowledge is actually a bottom up transfer of knowledge, which reflects the primary importance of the nonverbal knowledge created by the nonverbal mind. There is a great new social or cultural adaptive advantage in not requiring each individual to rediscover the wheel. In fact, this bottom up structural arrangement shows that the validity of the nonverbal narrative is too good to not share.

The classic verbal authorities who advocated the notion that the verbal mind was a one-mind mind, never took the time to consider the big question: where is the information the verbal mind is showering upon the public coming from? One of the main points of my theory, which I develop in my account of Aristotle's (1952) model of the mind, is to show that Sperry was probably incorrect. There are not two minds within the human brain. For starters, there is only one set of the five senses. One of the most important differences between the nonverbal and verbal mind is that it creates the nonverbal consciousness, which functions as the first narrative, and all this nonverbal conceptual creation of form happens only once, within the GNMM. The verbal mind just picks up the ball at that point, as the export department of the mind, and converts the nonverbal into memes and spreads the word.

In other words, the five senses belong to the nonverbal mind, and the sixth sense belongs to the nonverbal mind, so if a newspaper wants to provide its readers knowledge directly from the front, any front, then it must have a reporter on the ground at that front.

The nonverbal mind's designation is as a universal reporter. It has five senses, which can collect and process all possible physical energies, which can carry critical knowledge about what is happening on the frontlines in any domain of nature.

NASA discovered that there were things going on at various fronts in space, which created a need for additional sensors, which were different from the five senses. So NASA invented several new sensors, which allowed it to monitor essentially all possible significant physical energy emitted by every possible cosmic event, which humans must know about, but still these new senses had to ultimately flow through the sixth sense of the nonverbal mind which then translated them into the verbal code of some verbal community.

Conversely, the bottom-up model seen in the human mind still is the ultimate reporter, which will create a narrative upon any conceivable planet-earth event, but also on any event that may ever have happened at any time and in any place in the path of cosmic evolution.

The idea that the verbal mind is the primary mind is the product of an incomplete thought process. The nonverbal mind is the ultimate universal mediator between consciousness and the cosmos. The nonverbal mind is the ultimate reporter, the ultimate "eyes and ears on the ground." The nonverbal mind has what is an almost miraculous power to transform any and every cosmic energy-creating event, which happened anywhere in the past or present into consciousness. In the beginning was the nonverbal, but evolution invented the ultimate weapon against the nonverbal: it created the two stage human mind.

The verbal mind's functions do not duplicate the nonverbal mind's functions. In other words, the functions of the verbal mind take off where the functions of the nonverbal mind leave off. The two minds jointly form one complete assembly line. The nonverbal mind, as I have noted many times, does not do memes. Likewise, the verbal mind only does memes. The two make a perfect couple. It reminds me of a nursery rhyme. "Jack Sprat ate no fat, his wife ate no lean; between them both they licked the platter clean." The two minds, the nonverbal and the verbal, are a marriage made by evolution over billions of cosmic years.

The generic nonverbal mammalian mind, which carried mammals through millions of years of change, evolved for perhaps fifty million years before evolution took the next big step and designed the verbal mind. But the nonverbal mind did not go away with the emergence of the verbal mind. On the contrary, the human verbal skills are the product of the generic nonverbal mammalian mind plus the extension, which enabled humans to enjoy verbal speech. But one of the radical items in this theory is that the nonverbal mind is the primary mind and works invisibly below the surface, so that it might be called the unconscious mind.

The verbal function of the human mind is simply the newest feature of the mammalian mind that emerged as the export or shipping department for the nonverbal mind. The verbal mind's chief function is to transform the nonverbal narrative created by the nonverbal mind into memes and to prepare it for publishing, for shipping, or for export.

The great extinctions

After the five great extinctions, it was clear that evolution required a new and better way to provide life, mammalian life, with knowledge. How did this accomplishment occur? It occurred with a two-step evolutionary process.

In the first stage, evolution created the generic nonverbal mammalian mind and in over a fifty million year period produced a series of mammalian species that occupied most of the niches of the earth. The nonverbal mind has had a long evolutionary run and its emergence revolutionized everything. The emergence of the nonverbal mind made mammals special, into self-organizing systems. The emergence of self-organizing systems was probably a more revolutionary event than was the emergence of verbal behavior. My guess is that a creature that was capable of speaking, but not self-organizing would have little if anything to say. The central component of a rational self-organizing system's mind was the emergence of the ontological machine function, which I will describe in the context of my version of Aristotle's twelve-station model of the mind.

Wisdom of mind

What is the difference between the nonverbal and the verbal mind?

In the first stage, evolution produced the nonverbal mind, and in the second stage, evolution produced the verbal component of the two-phase mind. These two phases give humans the cosmos. Why do humans have a compound nonverbal/verbal mind; why are there two sexes? It takes two to tango; full-blown communication requires a two-stage communication machine, a nonverbal, and a verbal.

I propose that Sperry's (1961) conception of this story of the human two minds was conceived too hastily. My alternative proposal states that humans have one mind. The difference between these two narratives fails to consider the difference between a two-stage mind and two-models of mind. The human brain is not a two car-garage; however, it has room for the latest two-phase model of mind.

The difference between the two versions of mind was chiefly that the newest model of mind had a new adaptive somatic motor ability, the Broca's region that allows the nonverbal mind to go public for the first time. The nonverbal mind created a nonverbal narrative, a script that directed the behavior of the nonverbal mammals, but with the evolution of the Broca's region, this nonverbal script conceived in the right hemisphere can now be shipped across the Corpus Callosum to the left hemisphere where it is edited and prepared for processing by Broca's area. So finally, the nonverbal cosmos got a voice! By my version of the creation story, in the beginning, the pantheistic god had no voice so there were no words in the beginning. The human mammal's words were the first words spoken in, by, and for the cosmos. The human mammal is the first species with the capacity, finally, to write books that describe the cosmic god; that enable every inhabitant of the planet earth to know the nature of the cosmos, and their nature, and everything on the face of the earth that can be shared on the internet.

A new human nature paradigm

My purpose is to describe the structure and function of the nonverbal mind, which will create a new scientific paradigm of animal and human nature.

The nonverbal mind

The West's classic paradigm of animal and human nature has Plato's concept of the human verbal mind as its founding assumption. The classic Western paradigms of animal and human's natures have been around for so long and seem so right, so natural, that few humans today probably feel any great need for a paradigm shift any more than the church did when Galileo discovered the moons of Jupiter. Paradigm shifts can arouse intense emotions and some devout believers find them hard or impossible to accept. Indeed, it took the Catholic Church four hundred years finally to admit that Galileo's concept of the Heavens was correct while Aristotle's concept of the heavens was wrong.

A monistic paradigm of human and animal nature

In this section, I will begin to propose and develop a pure monistic, spirit-free, paradigm of human and animal nature: specifically, I propose to evolve a monistic alternative for each of the dualistic elements of the classic paradigm of animal and human nature, to propose a monistic opposite for every dualistic element of Plato's theory. My task is to turn the real world, the physical world, right side up for the first time in thousands of years: to recognize and reestablish material things as real. To show that only material things are real; to show how all those features of mind which Plato (1952) insisted were spiritual or dualistic functions can be scientifically shown to be either false or the product of monistic functions of the generic nonverbal mind.

This series of four experiments, fortunately, were designed so that the empirical data they produced enabled me to test and verify many, if not all of the aspects of the nonverbal features of the new monistic paradigm of mind. The test procedures designed for these experiments happened by reversing every feature of the behaviorist rat runner's classic trial-test procedures.

My new test procedures worked flawlessly. The behavior of the nonverbal mind finally became publically visible. This book is the product of these new loop-testing procedures. The revolutionary ideas, which drive the thinking in this book, are all the product of the new nonverbal generic mind, whose discovery came about thanks to this new testing procedure, which will be described in more detail on Chapter One of Book 2.

The implications of the nonverbal mind for the classic paradigms

The generic nonverbal mind described for the first time in this book offers answers many scientific questions that Sperry did not ask or answer about the mind.

- If rats and perhaps all mammals have a nonverbal generic mind, then finally Plato's theory that only humans had a mind immediately is a myth.

- The fact that could be the most troubling to the classic paradigm is that if all nonverbal mammals have a nonverbal mind, then finally the idea that mind is a verbal language faculty is a myth.
- In addition, if the rat's nonverbal mind evolved over fifty million years ago, then the idea that rationality began with humans is a myth.
- The idea that the generic mind evolved for millions of years before humans appeared on the planet, offers an easy solution to the problem of how the two human minds emerged. This explains how the generic nonverbal mind evolved first and then the plasticity function of the human brain at some point redistributed the verbal and nonverbal functions into the two human hemispheres.

IV-Kuhn's mind

One of my former students Bruce Lombardi gave me a copy of Kuhn's book for Christmas in the nineties and suggested I read it. Bruce had already seen the similarity between Kuhn's (1962) theory of science and my theory of Heterostasis in spite of the fact that Kuhn was talking about human scientific behavior and I was talking about rat behavior.

Kuhn was a philosopher of science, he was not the typical philosopher; he was more like Aristotle rather than Plato; he paid careful attention to the empirical facts. Kuhn's (1962) theory of science contains three concepts: First, there are two stages of science, second, there are paradigms, and third, there are anomalies.

There are two stages of science and each science can shift between two stages, a normal, and a revolutionary stage. There are two kinds of scientists, the normal, and the revolutionary. A paradigm is like a jigsaw puzzle. The normal scientist works within a paradigm and fits pieces into place into the puzzle. (Kuhn, 1962)

A paradigm or model

According to Kuhn (1962) each discipline, only becomes a science when it has a paradigm, which is a universally accepted model of how nature works in each of her many domains. Each science, additionally, has an accepted model of how their domain of nature is studied.

An anomaly

On rare occasions, a revolutionary scientist discovers an anomaly, a fact that causes change. While generally they are rare events, yet historically they do occur. An anomaly is an illegal event, a fact that logically does not fit into the current paradigm. Anomalies invalidate the paradigm from which it emerges and may force a total conceptual revision of the old paradigm. An anomaly invalidates a paradigm and shows that a new paradigm is necessary. Revolutionary scientists play two roles, first they discover an anomaly that destroys the old paradigm, and second they create a new paradigm to replace the old. (Kuhn, 1962)

One of my basic beliefs was that the emergence of the human verbal mind was not the end of the GNMM, but that on the contrary the GNMM had become the foundation for the verbal mind.

My beliefs are the human verbal mind emerged out of the GNMM: the GNMM is still functioning as the primary mind in humans, and that the verbal mind was a secondary level concept of mind. But how could I prove my point? Luckily, Kuhn (1962) had already partially done my work for me. As Bruce first pointed out to me, Kuhn's concept of an anomaly was essentially just another way to recognize a nonverbal mind function, which was the same as my concept of Heterostasis. However, Kuhn did not recognize the difference between Homeostasis and Heterostasis, and was not aware of the GNMM.

Rats and mice exploring an anomalous domain create a model of the domain, which I call a model instead of a paradigm. While the words are different, a paradigm and a model serve exactly the same functions.

The key event that triggers rats to stop exploring and to begin retracing is the discovery of an anomaly, which causes the rats to shift the focus of their behavior from exploring to retracing. An anomaly, in both the rat and human's mind or science, means the same: in both the maze and in human science. Each anomaly, depending upon its force, compels the nonverbal mind to change in one of several possible ways: to destroy or abandon the old model completely, to complicate the model by creating a new idea, or to begin a new grade. This will all become clear in my discussion of F3's behavior.

In Kuhn's (1962) two- stage theory of science, the normal and the revolutionary stage are almost identical with the nonverbal behavior emitted by a rat while exploring an anomalous maze! What was remarkable about Kuhn was that human verbal behavior did not mesmerize him; he actually was the first to observe the nonverbal component of human scientific behavior.

Humans have become so enamored with verbal behavior that the nonverbal-verbal continuum does not even exist in most scientists' consciousness.

It is most unlikely that Kuhn, as a philosopher, ever had any reason to recognize that all mammals have a mind or to recognize that the nonverbal mind's only function is to regulate nonverbal behavior. However, Kuhn (1962) was the first to see beyond verbal behavior and to recognize that the nonverbal mind was involved in scientific behavior, and was the first to recognize the primary role of nonverbal anomalies in scientific behavior.

If we pause to think seriously about the purpose of having a mind, we must eventually recognize that there are only two options, the human, and the nonhuman. I presume the reader knows by now: humans are the only mammalian species with the gift of speech and with the verbal mind. So if we ask what the primary role of the mind is: there is only one option, speech, and this has been almost the universally accepted option. If however, the data I present in this book is valid, then a very different conclusion is required. If there is a GNMM, if all the pre-human mammals had a nonverbal mind, then the nonverbal mind emerged and lasted for over 50 million years as the only source of **knowledge** for all the nonverbal mammals that ever lived. If the GNMM enabled all the tens of thousands of nonverbal species of mammals which evolution designed to survive and thrive without the equivalent of human speech then we must conclude that the primary function of the GNMM is some function other than verbal behavior.

My thesis is that the GNMM emerged to give life a better way to get the knowledge it required to survive and thrive. The GNMM's purpose is to seek and destroy anomalies, to convert uncertainty into knowledge.

Plato's game

If we agree to play Plato's game, then the world is upside-down.

First, there are two kinds of worlds, Plato's being and becoming worlds. Moreover, there are two kinds of ways of establishing knowledge/beliefs.

The two kinds of ontological beings and worlds:

- The sense-based world
- The reason-based world

In science, knowledge is only obtainable about the sense-based world; or action in a sense-based world is only justified by knowledge.

Action in a reason-based world can only be a function of faith.

If the human verbal mind's sole purpose was to provide for humans to become effective social beings, then I have no way to understand what the nonverbal mind was doing for fifty million years of existence in the nonverbal mammals. My position is that the nonverbal mind's role was to enable self-organizing nonverbal mammals to create on-line and in real-time the specific **nonverbal knowledge** their nature and their situations required.

Kuhn (1962) was the first to call attention to the fact that most important events even in human science were nonverbal events such as an anomaly.

Let us consider two of his most famous examples.

Two examples of paradigm shifts

The two most famous scientific paradigm shifts were first, in Astronomy, the shift from Aristotle's earth-centered theory of the heavens to Copernicus's sun-centered theory of the heavens. The astronomy paradigm shift was possible only because Galileo's (1989) new scientific telescope allowed him to observe the moons of Jupiter.

The second paradigm shift was Darwin's (1859) theory of evolution, which caused a revolutionary shift from the biblical story of creation to the speciation mechanism of natural selection. This biological paradigm shift happened primarily due to Darwin's observation of the finches in the Galapagos Islands.

My prediction is that F3 could quickly join these two anomalies. The first two anomalies, of course, are famous for causing a great intellectual turmoil in the events of their day. F3 actually has the capacity to induce an even greater degree of turmoil in our day.

Paradigms are theories that consist of a set of closed facts. Anomalies are facts, which do not, and are not, valid elements of the accepted paradigm, or they do not fit into the current jigsaw puzzle. Anomalies really break wide open a closed conceptual system.

Anomalies are rare events. The classic paradigms, which Galileo (1989) and Darwin's (1859) anomalies invalidated were created out of data I describe as naked-eye facts. However, the anomalies, which invalidated these two classic sets of ideas, were not naked-eye facts. Galileo's anomaly, the observation of the moons of Jupiter, was observed by Galileo because he had constructed a new and better telescope, which enabled him to observe facts about the heavens, which could not be seen with the naked eye.

Darwin's finches were as difficult to observe, as were the moons of Jupiter, because they were well off the beaten paths. Darwin had to travel halfway around the world in order to be able to observe these rare events.

My observation of F3's behavior, likewise, was only possible because like Galileo, I effectively designed a kind of telescope, which I call the "behavioral telescope" which enabled me to observe aspects of F3's behavior which had never been observed before because it was invisible to the naked eye.

Rats are prototypical scientists.

One of the most interesting features of rodent behavior is that a rat's nonverbal behavior, which I describe, and the human scientific mind described by Kuhn (1962), works the same in the revolutionary mode! The generic nonverbal mind is the primary nonverbal adaptive tool of all rats and presumably of all self-organizing mammals from rodents to human scientists. Scientific behavior is nonverbal behavior and the same brain/mind mechanism is engaged in all mammals by anomalies, both the one and the two stage mind mammals probe and interrogate their local environments by using the same brain mechanism. The rat is a prototypical scientist. This point becomes ever more interesting the more we understand how the nonverbal mind works. Both the human and the rat's nonverbal minds, when they are acting as a revolutionary scientist, exist to dominate anomalies.

Conceptual mutations

Absolute anomalies require that a completely new conceptual scheme be devised that will create a new coherent model of all the known facts in a domain. In the process, the old conceptual pieces of the puzzle generally get revolutionary new meanings.

Kuhn, however, missed one of the most interesting features of anomalies. There are two kinds of anomalies: absolute and relative. Absolute anomalies destroy paradigms, but relative anomalies generally require that a hierarchically organized structure grow by adding other levels or grades of complexity; or by adding a new hierarchically organized level of complexity to the model.

F3's behavior furnishes several examples of how relative anomalies work, and will be discussed in the first chapter of Book 2.

The implication of Kuhn's theory of mind

In the human context, it is not too surprising that humans supposed that mind was all about words; that the sharing of words was all there is to mental behavior.

Indeed no one need be surprised or embarrassed by the fact that humans for thousands of years and even today, even the top verbal scientists, fail entirely to see Kuhn's (1962) scientific behavior hidden within the world of words created by humans.

So what should a scientist say about the components of language when they write and lecture about the nature of verbal behavior, especially when the experts tell us that verbal behavior simply consists of two elements, words and rules, plus meaning.

"The premise of this book is that there are two tricks, words and rules. They work by different principles, are learned and used in different ways, and may even reside in different parts of the brain." (Pinker, 1999, p. 2)

What are the most important elements of human speech: the verbal or the nonverbal components?

Let us review Plato and all those other dozens or hundreds of supposedly brilliant philosophers who missed it, but down deep the nonverbal elements of scientific behavior have always been confounded with words.

What is human speech all about? Is it possible that the most important component of human speech is that it has the capacity to amuse and inform, as well as surprise? The classic language theorist makes the natural but erroneous error of thinking of speech more in the context of its Bell Telephone role than in context of Shakespeare or even more simply in the context of newspapers.

Kuhn's (1962) research proves the rat was a prototypical scientist, that the mind of the rat and all mammals up to and including humans, share the same nonverbal technology, the same Heterostatic arousal mechanism with respect to both the structure and function of the nonverbal mind.

Kuhn's (1962) theory of mind supports my theory of the generic nonverbal mammalian mind in two of the most substantial ways and provides more support than I could have wished. The fact that both anomalies and paradigms play interacting roles in both species handed me just the proof I required to show that all mammals share the GNMM, and show that the emergence of the human verbal mind did not destroy the GNMM.

Sperry's (1961) research only focused upon the classic speech functions of mind so the fact that Sperry never observed the role of anomalies in human speech behavior should not be too surprising.

Synopsis

I propose that the generic nonverbal mammalian mind is the primary nonverbal mind for all mammals.

The primary role of the nonverbal mind is to make mammals nonverbal self-organizing systems by enabling each mammal to be self-sufficient in the sense that it can glean all the knowledge it requires to achieve any task it can conceive in any domain of earth.

How can an anomaly be detected

The normal search function of the Sixth Sense must in microseconds: search the properties of all the known domain-specific concepts, and affirm that there are no sets of features in the input stream, which do not match one of the familiar set of features of the current domain-specific paradigm or model. Thus, the Sixth sense must certify that no new concepts or anomalies exist in the impulse stream. The rule is that the nonverbal mind cannot execute any script that is not closed. Anomalies put the brakes on or stop the execution of any script that is open. The nonverbal mind only allows action to continue when the input stream is closed or free of anomalies.

Anomalies play a complex set of roles and not all are negative. Kuhn (1962) focused solely upon the important negative role, "the brake role," of an anomaly.

If anomalies played only negative braking roles, then mammals would never have become self-organizing systems, and F3 would never have found her way through the matrix maze. The other part of the story, the positive side, is that anomalies are in the current context, bundles of unrecognized features and thus qualify as an anomaly. But this bundle of unrecognized features is a gold mine of facts. As nonverbal scientists, the rat or human, must then switch roles, put on their Sherlock Holmes hats and pipes, and begin reading the potential narrative that lays untold in the features of the anomaly.

The anomalous finches consisted of a complex set of features that supplied Darwin (1859) with clues he required to answer his scientific questions about their origin.

When rats are retracing in the maze on the causal side of an anomaly, they are probing nature for clues, which actually are potential knowledge. Nonverbal communication generates the nonverbal interaction that activates all these clues.

Kuhn (1962) recognized that the normal scientist simply fits pieces of some accepted puzzle into place, but he failed to recognize that the normal scientist in playing this role with the features of an anomaly is playing a critical role in decoding the mystery bound up in the features of each anomaly.

How Galileo read Anomalies

The nonverbal mammal is decidedly at a disadvantage in reading anomalies. Yet F3 only encounters four anomalies in the maze, which she reads very easily and quickly. By comparison, I have taken fifty years to master and learn to read the anomalous behavior of rats and mice in the matrix maze.

Consider the reasons why Galileo (1989) first discovered the anomalies of Aristotle's (1952) astronomy. First, there was his new telescope, which explains why he was the first to see the moons of Jupiter. However, his telescope was only part of the story.

Second, in order to prove that the objects he observed were the moons of Jupiter, he had to draw and record the position of each of the objects for several days in order to demonstrate that they orbited Jupiter.

Third, as an expert on Aristotle's model, he was all too well aware that Aristotle claimed that all heavenly objects orbited the earth daily. His discovery in other words was no accident. He knew what he was looking for.

F3, however, had no model to oppose, so she came to each anomaly unprepared and she had no lab book in which to keep records. Yet the rat's nonverbal mind with its one rational wisdom was able to deduce, quickly and efficiently, the implications of each of the evolutionary levels of the hierarchical model emerging as features within her nonverbal mind, which she converted into knowledge. As each new level emerged, F3 was able to deduce its implications and let these implications direct her behavior in the next exploratory cycle. In the second-grade stage of the model, she logically followed these implications too carefully and for nine times she repeatedly made the same error. Only on the tenth entry into the third row was she able to alter her behavior and turn right, but the fact that suddenly she was in the open and not confined within the blind units made quite an impression. After that right turn, she was finally able to run through the maze without making a mistake.

Anomalies play two roles, one negative, and one positive. Reading a relative anomaly involves extracting the nonverbal knowledge it holds nonverbally.

Revolutionary scientists play two roles; first destroying a paradigm, second creating a new replacement paradigm. Anomalies play a critical role in both situations. Reading anomalies is the key to the creation of a new replacement paradigm. (Kuhn, 1962)

The Internal Milieu of the mind: IMm

The genome supplies the body with the Homeostatic knowledge that enables all the organs of our body to regulate the Homeostatic functions they were designed to regulate.

However, the nonverbal mind comes with no instruction manual on how the solve the problems occurring in consciousness. Yet there are two kinds of problems, first those problems that involve injury or pain to the body, and secondly errors that are exclusively concerned with how a nonverbal mammal is to define a goal and discover the means to achieve that end within some domain specific consciousness. My concern is exclusively with this second kind of problem.

When I place a rat in an anomalous matrix maze, the rat must define its own goal and discover the knowledge required to achieve that goal. The nonverbal mind was designed to enable the rat to be an effective self-organizing system. The internal milieu of the mind must create in consciousness an awareness of the spatial structure of the matrix maze, the external milieu, in enough detail that will allow the rat to achieve its goal.

In the most important sense, homeostatic problems are simple compared to the problems of the external milieu. If the rat has no food, then the homeostatic nature of the physiology of the rat will make the rat hungry, which motivates the rat to search for food. However, when a rat is placed into the matrix maze, what motivation is there to alert the rat to what it should do in that domain?

Actually, the answer is simple. With the emergence of the nonverbal mind, each anomalous environment motivates each nonverbal mammal to activate the technology of the god-function, for there are surely anomalies up ahead that must be searched out and destroyed. Yet these anomalies cannot yet play Kuhn's (1962) specified role for anomalies, it is impossible to destroy a paradigm, which does not yet exist. What must be recognized here is that anomalies play an additional role. While they do not induce a homeostatic motivational state like hunger or thirst, they nevertheless induce a heterostatic motivational state called uncertainty, surprise, or anxiety. The nonverbal mind has a special physiological technology that these states turn on that was designed to take and induce the rat to emit the appropriate kinds of behavior that destroys anomalies. The nonverbal mind is endowed with a technology to know both the nature of the anomaly, and to know how to destroy the uncertainty induced by the anomaly. Anomalies induce the nonverbal mind to activate the knowledge creating structures of the mind, which create a coherent set of ideas. Ideas enable the rat to conceive a valid model of the maze, valid in the sense that the knowledge provided by the mind will empower the nonverbal mind to extrapolate and conceive a plan to run the maze without error.

The homeostatic mechanisms are genetically programmed to act in the most specific way to reduce each specific physiological need. This induces the rat to seek water and food, etc. These needs have specific objects as solutions and programs for solving their homeostatic problems programmed into the genome.

But how can the mind be genetically supplied with a specific solution for making a model of every conceivable maze, or every conceivable house, or ship, or field in which rats must find their living?

I propose after observing several hundred rats solve this particular problem of the matrix maze that the nonverbal mind is genetically programmed with a technology that will enable rats to model a maze or almost any possible domain. The heterostatic technology enables nonverbal mammals such as rats to conceive a coherent set of nonverbal images that enable the rats to see in their mind's eye the hierarchical structure of the maze, a model of the maze, which the mind can use to create a plan of action that will achieve its goal.

One interesting possibility, which must be examined further, is the role of the Place cells, which Moser has so remarkably recently discovered. (Moser & Moser, 1998)

The nonverbal mind was endowed at birth with a Heterostatic technology, which can enable the rat to solve problems of the external milieu, by creating a model in the internal milieu that is a valid representation of the physical features of the external milieu.

"The familiar world of objects, people, and events is a mental creation, an internal model we build that represents a selected part of the physical world outside. It is related to the physical world only as a model is related to the object modeled." (Kelley, 1968, P. 36)

Every anomaly says that the internal milieu, which induced the anomaly, was in effect invalid. The nonverbal mind has been endowed at birth not with the solution to any spatial problem but rather, has been endowed with a technology that can solve every problem by creating a model, which the mind can use to solve its problems.

Just as Cannon (1932) showed, the purpose of Homeostasis is to regulate the IMb, just so the purpose of Heterostasis is jointly to regulate/control both the EMm and the IMm. An error in the IMm simply means that there is a discrepancy between the internal and external milieus. The purpose of the nonverbal mind is to enable mammals to know or to be prepared to act.

The simplest species of life survived instinctively; their only source of knowledge was genetic. Most of these species died off after one of the great extinctions, and evolution began to design a new kind of adaptive organ, the nonverbal mind. There had to be a better way for life to get the knowledge required to survive.

The purpose of the nonverbal mind is to enable mammals to act not to talk. Evolution designed the nonverbal mind as a communication specialist, but again not to communicate verbally or socially but to communicate nonverbally with the external milieu. The purpose of nonverbal knowledge is to enable mammals to act, to interact with nature not to talk with other mammals.

My smartest rat F3 illustrates how this works. After she had explored a few rows of my maze, she had created a valid model of the maze; she knew the maze and she was then able to run the rest of the maze that she had never explored without error. She was able to achieve her goal free from errors.

Cannon (1932) showed Homeostatic species are born with their instinctive knowledge as solutions for regulating some one of hundreds of the EV's of the body, each of which has its own regulator.

However, mammals are born with a mind. They have a better way to know. The purpose of the nonverbal mind is to regulate the EV of the internal milieu of the mind. The mind is, in effect, one great regulator/controller, whereas the body had many homeostatic EVs and evolution has stored these homeostatic solutions in its gene pool. This instinctive knowledge relieves the mind from having to learn: it is born knowing.

However, the nonverbal mind, in effect, has only one EV. One of the profound differences between the physiology of the body and of the mind is that Homeostasis has one regulator for each EVb. A Homeostatic regulator is a relatively simple technology, and each regulator has only one purpose, to regulate only one EV continuously. (Cannon, 1932)

However, the mind works the other way around. The mind itself is one all-purpose regulator/controller and consciousness is the CPU or the units which Aristotle (1952) calls the common sense. Pure Homeostatic species do not require consciousness, but mammals are a new kind of life; they not only can but also must identify every error in the IMm, but even more remarkably, most rats enjoy solving problems.

The conscious function of the nonverbal mind is dominated by heterostasis, and is the one and only controller the mind has. However, the mind is a general-purpose controller and uncertainty turns on Heterostasis, which grabs or takes control of the nonverbal consciousness. The mind is a general-purpose control mechanism, which solves problems by making models of space. As Kelly showed, the most powerful way to solve navigation problems in the External milieu is by making a model of the domain.

With the evolution of the mind, which is designed to reduce uncertainty, we can now talk about the mind's needs in everyday terms; consciousness, ignorance, and knowledge enter the picture. Uncertainty is just a euphemism for ignorance.

The difference between Homeostasis and Heterostasis is that in the wisdom of the body each EV had its own regulator, which functions automatically without conscious awareness to reduce the error, and happens genetically.

However, when the human body temperature dropped below 98.6 degrees, below the set point, humans feel cold. Being cold is a physiological state that has a unique feeling in consciousness, so that a mammal may know it is cold as opposed to many other physiological states. What is unique about Heterostasis is that with the emergence of consciousness the possible solutions become almost infinite. How many ways can a mammal get warm when it is cold? Each culture in each domain can creatively solve that problem in anyone of an almost infinite number of domain-specific ways.

V-Aristotle's Twelve Station model of the Horizontal mind

In this book, I will propose a theory of the nonverbal mind that functions in two dimensions, the horizontal and vertical. In this chapter, I will focus exclusively upon the horizontal functions; in the third chapter, I will describe the vertical functions.

In this section, I will present a systematic overview of the horizontal structure and function of the nonverbal mind, the GNMM, which is a concept of mind so very different from all other theories, from Plato to Sperry. Although I have searched for an ideal way to introduce my theory, yet the best I can do is to start with the basic theory of Aristotle, the first person to have ever presented at least a serious fragment of a monistic theory of the horizontal mind; close enough to offer a starting point for my own theory.

An aside

Perhaps, my purpose and my pleasure in this book are to build upon the foundation of the men and women of science who preceded me. Aristotle (1952) proposed a model of human nature based upon experience and knowledge of matter. I will propose a 12 then 13-station model for explaining mammalian and then human nature. Aristotle insisted that Plato's model was wrong. The only concept of human nature, conceived as a SOS2, that can function in this world is one that knows how to obtain the knowledge it requires by interacting with Spinoza's pantheistic god. Only a model built upon the solid foundation of the material universe, one with an empirical nature, can survive.

The nature of Religion

As scientists, we all must take a fresh look at the nature of religion. There are at least two different roles for religion to play in the life of a mammal.

- First the verbal role of religion

The metaphor functions of religion. The human verbal concept of God/religion

In the human mammal, the nature and role of religion has been defined by metaphors:

The family metaphor, of course, dominates some religions with the metaphorical concept of God the father, Christ the only son, and Mary the mother.

The life and death metaphor is also a dominate theme in some religions. Many religions claim to have the power to get the soul into eternity.

The magic of religion is based on metaphors: the dirty souls, and spiritual washing, with the blood of Christ the ultimate Soap metaphor.

- Second, the nonverbal role of religion

The no metaphor religion: The nonverbal religions

Perhaps the most tragic consequence of human speech is the fact that humans lost touch with their nonverbal soul, and thus with the nonverbal god.

I am reluctant to put into words what most desperately needs to be said. The primary role of religion should be nonverbal. The life of every nonverbal SOS1 mammal is constantly in the hands of Spinoza's (nonverbal god. Our life, every second of our life, we put our life, into our knowledge; into the hands of the nonverbal god. It is impossible to outline or list all the nonverbal laws of the nonverbal god whose constancy every moment of our life depends upon.

"Give us this day our daily..." measure of gravity, light, heat, sound, Etc.

Aristotle, created the generic model upon which the nonverbal mammals depend in order to act rationally every second of their life.

However, I should begin by saying that my purpose in adopting an Aristotelian type model was that I required a model upon which I could hang some broad theoretical/conceptual elements, which provided the elements of a nonverbal mind conceived as a self-organizing system. It is impossible to sketch the structure and function of this invisible nonverbal mind without an overview of the structure and function of a self-organizing system. One perhaps not valid in a microanatomy sense, but rather a molar theory able to convey the essential functions a completely nonverbal mind should provide.

Today, too few scientists are trained to conceive a Gestalt of all the rational functions a nonverbal mind should provide to enable an individual nonverbal mammal to function as a self-organizing system.

In this book, I am proposing that rodents, rats and mice, are prototypical mammals whose nonverbal mind describes the basic structure upon which all mammal's nonverbal and even the human verbal mind are built.

The revolutionary nature of this data has forced me into regions, which since the beginning of human thought about human and pre-human nature have been conceived entirely outside of the proper domain of science. Science, I am arguing, has brokered over the last four hundred years a Faustian bargain with philosophy and religion and has partitioned the cosmos into two most unnatural domains, the sacred, and the secular. Unfortunately, my rat data has implications, which do not respect this classic distinction, which I perceive to be an unnatural division of labor.

The basic distinction is between verbal and nonverbal. My challenge is that this distinction is one with which most humans have had no experience, and which at first, will seem most unnatural.

The nonverbal mammals' bottom up perspective on life has a decided advantage over human's top down perspective. Their nonverbal mind gets them through their time on earth without having to fight the unnecessary fight with the being whose only ontological status had a verbal/virgin birth. The story of my life has had only one theme, fighting my way out of the bondage into which my culture's vivid verbal imagination forced upon me as my unfortunate birthright.

I may be the only human who feels compelled for some deep personal reasons to claim an alternate birthright to live and work as a scientist within a forbidden land which science long ago ceded to philosophy, religion, and the humanities.

The Aristotelian model allows me to go into places where no scientists have gone before, places that for too long have been a scientific no man's land.

Back to Aristotle

The discussion of my Aristotelian type model of a nonverbal self-organizing system will be conducted in two parts. First in terms of a 12-station assembly line, and then in terms of a 13 station assembly line, which is but the 12-station model, with one new station, A 12 + 1 model.

FIGURE0.1B

Structures: Station 1-12

How did the nonverbal mind enable mammals to become the first self-organizing form of life? In his model, Aristotle (1952) focused upon the input side of a theory, so I have restructured his model into twelve stations with the first seven on the input side and the last five on the output side; see FIGURE 0.1B.

While Aristotle (1952) was a student of Plato, he listened to what Plato taught, and then spent his life doing the opposite: exhaustingly cataloging the empirical forms of things. Aristotle liked matter as much as Plato despised it. He was as close to being a scientist as was possible in his day. However, he was more a naturalist, in the mold of Darwin, than an experimental scientist, which did not evolve for several thousand years.

Ontology: Monism vs. Dualism

While Plato (1952) conceived the mind as a spirit, which had no material structure to describe empirically, Aristotle (1952) proceeded carefully to describe a monistic model of the six senses, which provides a great empirical foundation: a place to start to evolve a material theory of the nonverbal mind.

The Generic mind

The Generic Nonverbal Mammalian Mind, GNMM, served all the mammals that ever lived in all kinds of the earth's domains or milieus. If we can finally propose and get accepted a material model for a universal material mind, then perhaps we can then finally banish the tacit spirit-minds that somehow supposedly magically appeared in the head of humans without supposedly any help from evolution or any other natural forces. Aristotle's (1952) model provides sufficient detail so that we can scientifically propose and discuss many of the basic structures and functions of such a material nonverbal mind.

The advantage of getting a scientific consensus over, at least, a rough functional outline of a fully material mind is that we will be able to turn our attention to defining all mental terms as real material things; terms that the linguistic experts historically could not, or would not define.

However, Aristotle's model allows me operationally to define the nature of language, meaning, consciousness, and rationality, etc. while carefully examining each of the twelve stations of the action-cycle of the nonverbal mind.

Each of the twelve stations in this model is a transducer that changes energy from one form to another. No physical thing enters the senses and then moves through the twelve stations unchanged. Light, for example, upon entering the eye is transduced into neural impulses, which in the next station change into sensation images, which next change into perceptual images, etc.

Aristotle's (1952) model is crude, but even crude is a great leap forward when all we have had for thousands of years was a formless spirit-model of mind.

I describe the nonverbal mind as essentially structured as a self-organizing system, or even more operationally as a 12-station assembly line. We next must assume that this model is a neuronal processing system, but what does the SOS construct? What kind of a factory is a nonverbal mind? The nonverbal mind is a complex piece of technology that enables mammals rationally to move according to a script or a narrative. In my modified model of Aristotle's Twelve Stations, I can propose the generic nature of an action cycle see FIGURE 0.1B.

FIGURE 0.1B is a modified block diagram of Aristotle's Twelve-Station model. Each station is represented by a block within the model. Each of the twelve stations is labeled from left to right. The discussion of these stations will be from Station 1 to Station 12.

Station 1

The first station, Station 1, shows a circle in the first position on the left side of the model and contains three objects, which are represented as a circle and two hexagons. One of these three figures could represent the organism itself, which is sharing the space with the two other objects say, a rose, and a girl.

Station 1 could represent any one of the domains of the planet earth, which mammals have ever occupied. However, in the language of the physiology of Heterostasis, it represents the external milieu of the mind, EMm.

What is mind? Is there a difference between matter and mind?

Plato's (1952) dualism conceives of the human body as a material being, with the mind a spirit being. Plato's concept of mind and all its parts are spirit beings, which have no material structure, which means making a model or a flow chart is impossible.

The objects of Station 1, for example, are objects made of matter, but one or more of them also have a mind. Generally, science accepts matter as real; the mind consists of a large system of neurons that produce mental states, which we experience as sensations and perceptions as qualities. Yet by way of contrast, we must carefully note that the objects of station 1 as physical objects consist of nothing but physical attributes, which all have quantitative properties.

A	B	C
Quantitative world	Qualitative world	Quantitative world

Quality vs. quantity differences

At this point in our discussion, historically, it was so much easier to talk fast and think slow by which I mean, it is easy to fail to make an important distinction and hope that no one notices. At this point, whether we like it or not several critical topics enter the picture, which can continue to remain either seen or unseen. I choose not to play the typical game and move so fast that no one notices we are skipping the most critical topics; I choose to make them seen.

What are the issues that have been passed over historically?

1- The problem of meaning
2- The problem of quality and quantity
3- The problem of metaphors

The scientific description of matter is relatively simpler than describing mental qualities. For example, sensations require a three dimensional -space description. Colors or sensations differ in ways that matter does not; what is the difference between yellow and red? They are not only different in quantity but also in quality; what is a quality difference? The physiological psychologists have shown that sensations require a three-dimensional vector; for example, hue, saturation, and brightness for color vision. In addition, they have the added feature of enabling the usage of a flood of additional words, particularly ones used by poets. The nonverbal mind plays a rather dramatic role by creating the nonverbal consciousness. The difference is that physics describes physical things in physical terms as differences in terms of quantitatively simple numbers. However, in order for the mind to describe the physical objects of the world in consciousness requires a new vocabulary. In fact, each of the senses requires a different vocabulary, a different kind of pre-wisdom qualities for describing the sensory-specific sensations unique for each of the five senses.

Metabolism

The difference between mental and biological metabolism is primarily that in biological digestion, objects enter the body at the mouth, but the girl of Station 1 does not enter the eye. Only the light waves reflected off the girl enters the eye.

In homeostasis, each object eaten enters the mouth and the matter of each food object is transformed as it goes through the process of digestion.

Yet in the nonverbal mind's digestive system, no material thing enters at one end or exits at the other. No material object ever enters or leaves any of the senses. Things, or objects, do not enter any station of the mind. Each object of the physical world either emits or reflects some form of physical energy which stimulates one of the 5 senses, but these energies last in the form they enter each sense only long enough to be transduced into a different form of physical energy, for example to cause one of the five senses to emit a neural impulse.

Station 2

The five arrows of Station 2 represent all the different kinds of physical energies such as light, each of which stimulate a different one of the five senses. Each real object shown in the domain of station 1 normally reflects or emits many different unique sets of physical energies, which each potentially could stimulate one of the five senses to start the action cycle of the mind. Collectively, all the physical energies are the nonverbal mind's food for rational thought. These physical energies are the essential ingredients for sustaining an empirical or sensory world, as well as the essential ingredients of a rational world. For it is these sets of physical energies which start the action-cycle of the nonverbal mind, these physical energies initiate some one of the sensory functions which enable the nonverbal mind to empirically recognize and know each object as an exemplar of some class. These physical energies are the second sets of material elements within the nonverbal communication cycle. These physical energies, each supply the rational mind with one unit of empirical knowledge. Collectively, all this empirical knowledge feeds the mental digestive system and provides all the knowledge the nonverbal mind has to process. This knowledge enables the mind to conceive concepts, forms, or cookie cutters, which allow the nonverbal mind to recognize each object and conceive an exemplar and eventually to act appropriately.

Station 3, the Five Senses

The column of five blocks in the diagram of station 3 represent Aristotle's (1952) famous five senses. Structurally, each one of the physical energies of step two may be an effective stimulus for some one of the five senses, such as light for the eyes. The eyes sense the properties of all of the frequencies of the electromagnetic spectrum for light. The ear senses the full range of the mechanical properties of the air; and so on for the nose, which senses the chemical properties of a domain, and the tongue etc. The five senses collectively supply the nonverbal mind with massive complex patterns of neural impulses, or precept patterns, which collectively may represent a unique syndrome of some class of objects within the external milieu of the mind, or EMm. The nonverbal mammalian rational mind must identify these complex precept patterns and give them ontological status as an exemplar of a unique real class or category of things.

Aristotle (1952) recognized that the mind consists of two kinds of senses; first, Station 3, which shows Aristotle's five senses, most of which are on the face or on or in the skin, which were the only ones Aristotle could observe, but today receptors have been described that are all over and in some of the organs of the body. Today many more special senses exist than the original five, but that does not change the general structural model Aristotle originally described. The purpose of the special senses, no matter how many there are, or where they may have been placed, is to monitor the External Milieu of the Mind, EMm, and to inform the mind's agent about everything that is going on in the space surrounding each of the senses in particular and the mind in general. Several of the senses provide knowledge that help localize place upon the skin or on the body. For example when a person hurts their finger, the senses leave no doubt about where the pain is located.

As noted above, there are two different versions of the external milieu: the external milieu of the body, EMb and the external milieu of the mind, EMm. Please note that generally, I will be referring to the latter milieu, EMm. The first difference is where the two external milieus start. The EMb, starts outside the body, while the EMm starts outside of the brain/mind. The most important difference between the two is that the body is included as the first element of the EMm. The EMm organizes hierarchically with the mind its inner most object, then next comes the body, etc. The compound mind of the human child, in other words, must first gain somatic control of the objects of all levels of the EMm, beginning with its own body. For humans and mammals, the body is the first anomalous object encountered that needs mastery.

Consciousness evolves, grows, or expands as the nonverbal mind successfully masters each new successive layer of the hierarchically organized levels of the EMm. The body and all the objects and events, which surround the body, are elements of the EMm. The rule is that every object within each of these nested layers needs mastery grade by grade from the inside out.

In order to understand the mind, I must now make explicit exactly what events do and do not happen at each of the Twelve Stations. Technically, each of the five senses are transducers, which are devices, which transform energy from one form to another. Light, as a part of the electromagnetic spectrum, consists of many frequencies, and the eye's receptors, the rods and the cones, transform these different frequencies and amplitudes of light into neural impulses. Light enters the eye and then enters the retina where the rods and cones within the retina transform the light into neural impulses that next move down the Optic nerves of Station 4.

What is an experience?

In this section, I will begin to explain the nature of the experience of an object like a rose. The physical rose of Station 1 reflects a particular wavelength of light, which is one of the several kinds of energies shown in Station 2. The light energies next enter the eye in Station 3, but the rose itself never enters the eye, the retina, or the optic nerve and never enters the brain/mind. So how can we experience in consciousness a rose, the object of Station 1 when all that enters the nonverbal mind are several kinds of energies from the rose, which enters one or more of the five senses, and begin to move through the twelve stages of the action cycle? Where does the "Rose which smells so sweet" develop into what we experience in consciousness? How does that sweet smelling thing get into our consciousness? In addition, of course, that sweet smelling thing also never enters our nose; only a bit of chemicals from the physical rose of Station 1 enters our nose.

Take another example, when Pinker (1999) is explaining the meaning of the word rose, he states that words have meaning and to illustrate he shows a picture of a rose, which he says was the meaning of the word rose. Does he mean that the ink rose entered the eye and proceeds into the brain, and this ink rose is what we experience? Alternatively, did he mean that the ink rose no more entered our brain than did the rose growing in a garden entered our brain? I cannot say which rose Pinker was referring to - the rose in the garden or the ink rose on the page! However, I suspect that he was referring to the rose of Station 1, that I have tried to show cannot enter the eye or any of the five senses. So what do we experience? What is the nature of the object we experience in consciousness, which we call the meaning of the word rose?

Station 4

Each of the arrows of station four represents one of the afferent nerve pathways that take the neural impulses created by one of the five senses into the next station. Each nerve pathway transmits neural impulses, which inform the mind of the physical properties of some attribute of the objects its sense receptors are monitoring. Each of the five senses is a transducer that transforms one kind of physical energy into neural impulses, these neural impulses are the raw material, which Kant (1952) called precepts or the mud, for which the mind nativistically furnishes a mold.

There are only two routes into the brain/mind, the blood vessels, and the afferent neural pathways. Neural impulses and blood vessels enter the brain, but even the blood never crosses the blood/brain barrier and enters the nerve cells.

Station 5: The Primary Projection Area, PPA

Each of the five sensory systems has its own PPA, each of which transforms neural impulses from one of the five senses. In FIGURE0.1B, note that the five elements of station 3 and station 5 are parallel. In station three, each of the five senses transduces one of the five sets of physical energies emitted or reflected from an object. The outputs of station three are neural impulses. The five senses represent a barrier or a transition zone between the physical world and the physiological world, or mental world. Physical energies, light, mechanical vibrations, etc. are three of the physical energies that enter the senses and transduce into neural impulses.

Station 5 is the next great transition zone. Neural impulses enter the fifth station, which is called the Primary Projection area, PPA. However, what leaves are sense-specific sensations. This station is a major factor in the function of the nonverbal mind. The neural impulses, which enter the PPAs, are physiological events, but they can't enter consciousness. However, the outputs of the PPA stations are sensations, which are the first products of what can be called mind. Neural impulses are not sensations. The mind must transform neural impulses into mental events before they can enter consciousness. Sensations are the first elements of the set of primary symbols of the mind. The sixth sense of the NV mind is a coloring book, a smelling book, and a tasting book, etc. The occipital lobe is the PPA for the eyes. If we are not color blind it generates just the proper color sensations given a particular set of neural impulses, to color the rose appropriately at which we are looking. Therefore, each species of rose may have a unique look, color, and smell that allows each connoisseur to recognize each exemplar of the Mr. Lincoln rose by its proper name.

The different kinds of sensations are each unique sets of primary sensational symbols. Every kind of technology creates a different kind of ontological stuff, and each kind of sensation consists of a unique set of symbols to code the various kinds of ontological stuff, which embody this unique place. I will come back to this in the discussion of meaning.

The ontological machine cannot do its job unless there are enough kinds or varieties of ontological mental differences to allow unique syndromes of attributes to be attributed to each image. Each image becomes recognized as an exemplar of a nonverbal named class.

Station 5: PPA, Symbols, Sensations

What are sensations and perceptions? I propose that sensations are the nonverbal mind's equivalent of alphabetical symbols. Just as we use letters of the alphabet to create a name for a verbal class of objects, just so the nonverbal mind creates unique words or nonverbal images out of the sensations contributed by each sense's PPA.

What is a verbal symbol? What is a nonverbal symbol? Every kind of wisdom must have a unique set of symbols and each kind of wisdom must have a unique domain in which to function. The nonverbal third wisdom of the generic mind has a remarkably rich set of symbols. The English alphabet has only 26 symbols, but the primary language of the nonverbal mind has thousands of sensational symbols out of which images can be constructed.

Musical symbols are only functional in the domain of music. The German alphabet or symbols are functional only in a German-speaking community. Therefore, each language generally has its own set of symbols. The nonverbal mind generates both the primary kinds of symbols, both sensations and perceptions. However, historically they have not been called symbols because nonverbal behavior has never been recognized for what it is, the primary language.

The wisdom of the nonverbal mind has its own representational system with sensations being the equivalent of symbols. Each kind of wisdom has to have a unique set of symbols, such as music (music symbols), dance (choreography symbols), math (numerical symbols); chemistry (chemical symbols), and so on. The fourth wisdom is the verbal mind. It requires alphabetic symbols, which each language community creates for itself.

Station 5 is a major component of the nonverbal mind, so it should not be surprising that theorists of different domains independently conceive ways to describe the entities, which are a function of this station. Two historical examples will be included; first the classical British Associationists Model system; and second, the current philosophers who have spent a considerable amount of time thinking about the function of this station. For example, the afferent nerve pathways of Station 4 take neural impulses into the brain and deposit them in the fifth station of the monistic mind.

Station 5 is responsible for some of the most dramatic transformations of any station of the cortex: symbols, sensations, and meaning. Station 5 generates the primary sensation symbols: colors, tastes, and smells, etc., all different types of sensations. It produces the primary sensation symbols of the mind: Red is a primary symbol, which under ambient light represents a particular wavelength of light; each musical note/sound is the symbol for some particular sound wavelength.

The nonverbal mind of all nonverbal mammals including rats, which are color blind, having no cones, has only one kind of symbol. The sensations and perceptions are generated by the primary mind. However, the rat's nonverbal mind is one of the first minds that evolved and has not evolved to the point where it has the fourth wisdom; where it can create higher order symbols. By contrast, one of the dramatic features of the human mind is its capacity to expand its heterostatic behavior into new domains, especially domains that were not spatial domains.

Pythagoras was a Heterostat who began to explore the world of sounds; he just happened to have the unique set of manual and mathematical skills, which required applying whole-number thinking to the analysis of the vibrations of strings, and before long, he had discovered the invariant relations between whole numbers and the first elements of the musical scale. (Burkert, 1972)

The primary sensations (red, blue, yellow) are the primary symbols, which represent the primary colors, the colors out of which the production all other colors happen.

All alphabets are sets of symbols made out of patterns of sensations (lights, smells, touches, tastes, and pressure senses) which are the primary symbols, and they are the primary language for all other symbols. The letters of the alphabet represent human symbols for verbal language.

The primary symbols can make symbols out of any sensations. The alphabetic symbols can make symbols out of any other sensations: any letter can be any color, etc. The PPA of Station 5 is the factory, which creates all sensations. All the sensations manufactured in the PPA then go to the Sixth Sense.

Station 5 creates the sensations for all of the five senses, which are the empirical foundation for the nonverbal mind. Each set of sensory-specific sensations are conceived in Station 5 as seen in the model. The PPA's do not turn water into wine, but they do create the first sets of mental qualities which seem almost as miraculous; they turn neural impulses into sensations, which can seem intoxicating under the proper conditions. Nothing in this world seems quite as sensational as a sensation.

The things of the physical world are best described by science as a quantity, but sensations seem different, as qualities rather than quantities. Quantities perhaps actually also seem strange, but they can best be described as consisting of something which can be parsed and counted but not directly experienced. For example, all of the elements of the color-spectrum, which consists of a large subset of colors, can emerge by mixing sets of primary "colors"/ sensations.

The mystery consists of this. What is consciousness? Why can these qualities created by the PPA be experienced? Therefore, it seems that Qualities are different from Quantities, and the latter, whether they are a function of matter or a state of matter.

Station 6

Station 6 consists of the nerve pathway from the PPA to Station 7 over which the five sets of sensory-specific sensations travel into the Sixth Sense of Station 7. Each of the five PPA's is a critical member of one of the five sensory systems and each performs a special function; each creates a unique set of sensory-specific sensations. This collective set of sensations created by the five PPA's form the palette of primary symbols, which can be used to create or to paint; say the image of a rose or an image which will be experienced as the exemplar of some class.

Station 7 -the Sixth Sense: The Common Sense, Association Area, AA

Once delivered to Station 7, the sensations are used to draw complex class-specific, domain specific, or concept-specific images which are experienced by the agent of consciousness as a perception; or as an exemplar of some class such as the image of the rose which the "experts" tell us is the meaning of the word rose. Their assumption contains many errors. First, it is assumed that the image we experience is the object of Station 1. Second, it assumes that sensations are objective properties of the world, of Station 1, rather than the subjective mental properties that first have their origin in the PPA's of station 5. Those that believe and teach that the human verbal behavior is a one-mind system have no idea where ideas originate; they still reject the possibility of the evolution of the mind just as much as do those that reject the idea of the evolution of the body. For those individuals that are dualists and believe the mind is a spirit, they obviously must reject the idea that a spirit mind evolved. However, those that are monists who believe only in matter then the concept of mental evolution is one with the idea of body evolution.

"Sensing involves the creation of qualities of experience based on neural response to forms and patterns of energy." (Kelley, 1968, P. 39)

Do not be confused by the big red 6 in the center of the block in Station 7, it refers to the fact that Station 7 is home to what is known as the Sixth Sense. In Stations 3 and 5 all the five blocks, which are labeled 1 to 5, refer to the fact that each of these parallel sets of five boxes represents one set of Aristotle's (1952) five sensory systems. The six in Station 7 is just the next sense after the first five.

All the sets of sensations created by the PPAs next enter the Sixth Sense, which Aristotle (1952) also calls the Common Sense, since it is the only sense, the only place within the brain/mind with access to all the afferent sensory knowledge; all the sets of sensations, which are all the primary nonverbal symbols created by the five senses. Additionally, Aristotle called it the Association Area; for as the Common Sense, it was the only place where all possible combinations of sensory qualities are available to paint a sensory image. An image can represent all the variety of physical properties possessed by any class of objects such as a rose. Again, I must belabor this point, objects of Station 1 have no sensory qualities; their only attributes are physical. The role of the nonverbal mind is to create a mental world populated with objects painted or created out of just the existing sensory and class qualities, to create a rose that smells so sweet or a sunset rich with color. The Sixth Sense is the only place where the nonverbal mind can paint the complex multiple sensational-perceptual images, which are the meaning of the word "rose." The image- based consciousness created by the Sixth Sense consists of concepts embedded within rationally selected sets of sensations. Why do people love a Mr. Lincoln rose, perhaps because it consists of the same set of reproducible sensations year after year!

The magic of the nonverbal mind is that nature has fashioned a neuronal technology that creates a nonverbal image-based consciousness that seems real, so real that its nonverbal scripts elicit behavior without hesitation. Rational nonverbal animals never doubt consciousness.

However, humans such as Plato were cursed to go through life with effectively half a mind. My problem with Plato is that I don't know how to characterize his mental problem. First, I feel it is fair to say that he effectively turned the world upside down for a large proportion of the population; he induced many people to wear Stratton glasses. Second, I feel that Plato in effect tied half his mind behind his back. As a nativistic philosopher, Plato was estranged from the physical world. Plato was trapped within a metaphor driven world of his own making. Plato was first, a strange mammal destined to walk the face of the earth without the direction of that half of the human brain/mind designed to know the nonverbal world. Plato was obsessed with the question of what was real, yet knew nothing of how to drive the nonverbal component of his mind, which was specifically designed to answer the kinds of questions he asked. Plato was a paradox, he insisted in asking verbal questions for nonverbal problems. The problem with Plato was that his life space was conceived as a philosophical cloud created almost exclusively of metaphors.

Animals are paradoxically blessed by not having verbal speech; they have no metaphoric being to contend with in consciousness. As a once upon a time Christian, I once was bedeviled with deep questions like will there be animals in heaven. I now have to laugh when I think how lucky animals are; they don't or can't worry about what's happening in all the metaphorical based spaces.

Physicians call this brain center the sixth sense or the Sensorium. When a patient is confused or disorganized, they say the sensorium is cloudy.

In Aristotle's (1952) day, there were limited numbers of functions that Aristotle was aware of for the Sixth Sense to execute. Today there are too many functions that have been identified which logically must be executed by such a place as the Common Sense, or Association Area, AA, or the Central Processing Unit, CPU of the nonverbal mind to name them all.

However, the first and the primary function, which Aristotle first noted was that that the sixth sense had the task of integrating all the data from the five special senses. Today we know as Aristotle could not have known that the sixth sense must process all the neural impulses, which all the special senses send in massive quantities into the brain. I call these raw complex sets of neural impulses "precepts" after Kant who distinguished between the precept and the concept. (Kant, 1952) However, all the neural impulses in Station 4 enter Station 5, the PPA where they undergo their first major transformation from neural impulses into sensations.

What enters the Association Area are all these sensations, which are the primary set of mental sensory images or sensory symbols, a massive palette of sensory images ready to be assembled into the shape of a class or category and become an exemplar, or in Piaget's (1952) terms, to be assimilated.

The CPU or the Sixth Sense

The Sixth Sense or its functions are the central components of the rational mind; in computer terms, it could be thought of as a super CPU. More than any other place within the mind, this sense performs the functions, which Plato insisted were the functions only a god could execute. Anatomists, thousands of years after Aristotle, found this place in the brain, which executed these functions and gave it Aristotle's name, the Association Area, AA.

Ideas or categories must be conceived here, for again, this is the first place where all the elements of sensory attributes of all the objects in the physical world come together to have their essential attributes identified and combined so that they or their source can be classified by a unique set of empirical sensations.

Aristotle's Sixth sense- The ontological machine

One of the critical nonverbal functions of the mind is to detect patterns, forms, or sets of syndromes of sensations, which have the pattern of attributes or features, which define a class thus verifying or justifying the creation of another exemplar of a concept. The mind must have a place like the association area to sort through and to identify those classes that have a match feature for feature for the sensations entering the mind, which collectively justify an assimilation of some input syndrome by one of the concepts previously created. Thus, the association area must authorize an exemplar-image for inclusion in the narrative or script playing in consciousness. Assimilation is an old concept in medicine and physiology, but Piaget first talked of psychological assimilation.

"Among the aggregate of external realities assimilated by the organism are some which are incorporated into the parts of physicochemical mechanisms. In the first case, there is physiological assimilation, whereas the second may be called psychological assimilation."
(Piaget, 1952, P. 43)

Aristotle's (1952) Sixth Sense, which more than any other organ of the brain, is the embodiment of the nonverbal mind, which evolved to detect these material features of the cosmos, just as the eye evolved to detect the features of light, the nonverbal sixth sense is class conscious; it is designed to detect classes, categories, or forms. The law today claims to be blind to class differences, but that is all contrary to nature. A mind blind to class is like an eye blind to light.

The Sixth Sense was designed to detect and constantly monitor coherent sets or schemes of classes, the momentary patterns of neural inputs, which represent the actions and the states of the objects or things of Station 1. The sixth sense is designed to detect each syndrome, which represents a class of things. However, the Sixth Sense can only experience, know, or detect the presence of things, beings made of matter. The Sixth Sense cannot detect the presence of spirits or any supposed being, which is not a thing or is a nothing!

By contrast, a spirit being, god or soul, etc., cannot be known, they can only be accepted by faith since spirits have no matter at all. They cannot excite any of the senses, nor empirically excite the nonverbal mind. A spirit god presents nothing physical for the senses to transduce into neural impulses and thus no data for the nonverbal mind to certify or for science to study. Spiritual beings have no physical attributes, which the sixth sense can recognize and know.

Real things, on the other hand, classes of beings which embody any or all the intrinsic universal attributes of the cosmos are the natural components of the diet of the Sixth Sense; a diet which evolved to digest all those real things, which are a class or causal parts of the cosmos. Then it will classify and certify them as real exemplars of some class. The nonverbal mind assembles an image for all classes of things for inclusion within the nonverbal consciousness. The agent or the "I" of consciousness depends upon the content of the nonverbal consciousness for a complete accounting of the events of the external milieu within which it is struggling to survive. The sixth sense is the producer, which assembles the final "take" of the current state of the real world.

The foundation for a self-organizing system

The nonverbal mind constructs a concept or category for each exemplar of each domain specific thing. This closed set of domain specific concepts defines all the real things that can be known in a particular niche, and all that can be said or thought, but even more importantly, all that can be done in any domain.

A closed model, one without uncertainty, defines an anomaly as any object that emerges that does not qualify as an exemplar of some model and thus is one that induces uncertainty. An anomaly is defined as a thing, object, or event for which the mind in the present domain has no concept, cannot recognize, and cannot assimilate.

The chief experimental goal of this book is to propose a theory of how the nonverbal mind constructs a hierarchically organized model for recognizing each class of domain-specific real object. Given such a closed model, the nonverbal mind will not allow any false exemplars to enter consciousness. The first thing the nonverbal mind does upon entering a new domain with anomalies is to create a set of concepts which will be used to construct a model for recognizing each class of real objects within the domain.

The "real" is operationally defined by the sixth sense of the nonverbal mind, not by any magical being that occupies a magical world, which it shares with the magical spirit god. The real is that which can be detected by the ontological machine. Simply put, the nonverbal mind was designed so that it cannot be fooled; only "things" can be recognized as empirical, as reproducible, and as real and can thus enter the nonverbal consciousness or script by which the nonverbal mind certifies what is happening in the real material world and thus causes the appropriate nonverbal behavior.

The mind's programs are not error free. Science has identified a sizable set of illusions, but these only occur under special conditions and probably few life forms have been lost because of them.

The ontological Machine: The scientific concept of the Real

It seems a simple thing for me to raise the question about how real things are to be certified as real, through reason or empirically. In this book, I see no reason why science should fail to raise this question as a straight empirical scientific question.

The most dramatic answer to this question I found in a movie about Rasputin. Queen Alexandra was interrogating Rasputin about his vision of the Virgin Mary. She asked him, "Where did you see the virgin, behind or before the eyes." Rasputin insisted it didn't matter; "they are both the same."

However, as I noted earlier one of the primary functions of the sixth sense is the certification function. Any image during the waking state that is seen before the eyes is real. Any object with the empirical muscle to qualify as material to be displayed in consciousness, before the eyes has achieved the maximum certification.

First, the Sixth Sense creates all the real classes or sets of things

The Sixth Sense functions, which create all forms or classes, can be called by many names. In terms of evolution, this sense is the mental Speciation machine. In reference to Plato, this sense can be called the God-Function for it does what Plato insisted only god could do, but the issue is straightforward. Plato was obsessed with the problem of what is real! Therefore, it is important that I keep pushing the various ideas of what is real. However, for Plato real beings were spiritual beings. In Aristotle's case, on the contrary, real things have empirical sensory properties as attributes. The ontological function of the nonverbal mind defines things as sets of exemplars or syndromes of sensory attributes.

The mediating role of an idea

It is impossible to conceive any function more central to the success of the nonverbal mind than an "Idea." Our task is to explain the nature, function, and origin of ideas. Historically, there have been various theories proposed which would eliminate the need for ideas.

- The Associationist claims that learning is dependent upon evolving an "association" between two things. (See Pavlov's model of learning. FIGURE0.2G).

- Hull (1943) conceived of learning as occurring between a Stimulus and a Response, or what he called a habit, an "$S^{H}R$"

In the case of the evolution of an idea, we must now think of the growth of an idea as occurring within the cortex, specifically in the sixth sense at the peak of the pyramid, or at that transition point between the afferent and the efferent sides of the nervous system.

Where do ideas come from?

My purpose is to explain how a rat, say F3 exploring the matrix maze, evolves the set of four ideas which will enable her to extrapolate and run the rest of the maze without error. Consciousness emerges at that point both in the mind and in behavior where a choice must be made, and where that choice can only be made if the mind evolves the essential new idea that reduces the particular uncertainty evoked at that point in the maze where the choice must be made.

The nonverbal mind has the prime responsibility for certifying reality.

Ontology is the study of being; what is real. Plato was obsessed by the question what is real?

Paradoxically the nonverbal mind is also rightly obsessed with ontological questions. One of primary function of the nonverbal mind is to construct a real consciousness that has no objects that are not real. The mind does not put exemplars into consciousness until they pass the mind ontological tests.

In order for Plato (1952) philosophically to satisfy his philosophical needs, to discover the nature of the real, he created a magical place full of equally magical beings who were the source and keepers of his magical eternal truth and the magical real.

Again, Aristotle (1952) took a different tack. Instead of designing a magical place, he conceived an almost magical machine, an ontological machine. Almost all of the answers to Plato's verbal questions about what is real can be found in the behavior of the nonverbal mind, in the role of the ontological machine, the sixth sense; if only he could have known, he had one. Nothing can be more critical than understanding the ontological role of the nonverbal mind.

Second: mental certifier

The ontological function must also create a rational detection or recognition mental functions, which can detect the presence of every class or set of sensory attributes, which match the respective sets of sensory properties created by the ontological function of the nonverbal mind as concepts for defining each class of thing.

Third, consciousness

One of the chief functions of the sixth sense of the nonverbal mind is the creation of the unitary display center, where all images are displays, and where the big picture must be put upon the purpose of the whole. In short, consciousness is always a script-driven space.

An ontological machine must next create a certification mental function, which determines which sets of attributes qualify as exemplars and therefore qualify to enter the nonverbal consciousness. Exemplars are certified as real by having their exemplars written into the image-based narrative, which is running in consciousness. On the practical side, the role of the ontological sense gets involved in real life in the predator and prey function. The goal of both the predator and prey is to avoid detection by the ontological detection function. For example, evolution created the ontological functions of the nonverbal mind to detect, classify, and initiate action toward every set of neural impulses, which enter the mind through the five senses that qualifies as empirically real. However, this recognition function of the ontological mind is both an advantage and disadvantage, as was famously noted, "Nature is bloody tooth to nail." Most animal life can be described as in a constant struggle for superiority between predators and prey, with the predators trying to avoid detection as long as possible as they stalk their prey, and the prey likewise trying to avoid detection from its predators. Yet the nonverbal mind is constantly doing its best to recognize everything that enters into the life space of an organism for both predator and prey.

The foundation for a self-organizing system

Mammals emerged after the great extinctions because a mind confers an additional survival value. Such a kind of being is called a self-organizing system. The chief new function of such a new rational system is its ability to construct a concept or category for each domain specific thing. This closed set of domain specific concepts defines all the real things that can be known in a particular niche, and all that can be said or thought, but even more importantly, all that can be done in any domain.

False exemplars: The ontological certification of consciousness

The prime role of the ontological function is to guarantee that every object that enters the nonverbal consciousness is real. The nonverbal mind will only include in consciousness an image of a class of objects whose defining attributes are sensory images. The real are a set of things established in the nonverbal mind by a set of empirical attributes, which enter the Sixth sense. The ontological function of the nonverbal mind does not permit a false exemplar to enter consciousness and cause the organism to act by responding to exemplars of objects that are not empirically real in the monistic sense.

One of the strange blessings of the nonverbal mammal's mind is to be blind to all beings not made of matter, to all spirits and verbal gods. The nonverbal mind believes only in objects that it can know.

Plato (1952) taught and untold millions of humans fell for his argument, that things made of matter are not real. Only spirits are real! Nonverbal mammals never fell for Plato. The nonverbal mind was immune to words and thus to all beings whose existence was entirely defined by words.

What a profound difference it would have made in human history if humans shared this power for all of Plato's spirits were beings without any sensory attributes. Each human should be envious of the nonverbal mammals that presumably don't or can't worry about gods or the afterlife. The anthropologists brag about the fact that humans are the only species that bury their dead. Nevertheless, from the monistic point of view, this ability to be blind to all spirits could just as well be considered a blessing.

The chief experimental goal of this book is to propose a theory of how the nonverbal mind constructs a hierarchically organized model for recognizing each class of domain-specific real objects. Given such a closed model, the nonverbal mind will not allow any false exemplars to enter consciousness. The first thing the nonverbal mind does upon entering a new domain with anomalies is to create a set of concepts, which will be used to construct a model for recognizing each class of real objects within the domain.

The nonverbal mind conceives closed sets of categories by interacting or probing a closed physical terrestrial space by the rational logic of Heterostasis.

We have now completed the discussion of the sixth sense and are ready to return to a discussion of the remaining stations of Aristotle's model.

Station 8 A and B

The summary nonverbal perceptual narratives or scripts displayed in the nonverbal consciousness are designed to induce nonverbal action in the nonverbal mammals, but in humans, additionally it can also induce speech. The signals leaving the sixth sense that call for nonverbal action leave through pathways 8A. Those that call for speech leave through 8B and are routed into Broca's verbal somatic area 9B for speech.

Originally, this model was intended to represent only the nonverbal stations of the mind, but to save time I have taken the liberty to add the verbal motor centers, Broca's area into this model. The two B pathways are only found in the human left hemisphere. I have put them here in order to make the point that motor centers in humans and the nonverbal animals differ in that only the human brain has a separate somatic motor speech area, Broca's area.

Station 9 Rolandic region

I stress that the first step in the evolution of the nonverbal mind was designed around rational somatic action. All the behavior of the nonverbal mammals was executed out of the Rolandic Region of the cortex. But in the second evolutionary step, with the emergence of verbal behavior, evolution was required to create a new somatic motor center, which was exclusively devoted to verbal behavior. This rational verbal somatic center was Broca's region. The difference was that in the first instance, the nonverbal act of communication engaged the primary Rolandic region, while in the second, the verbal act of communication also involved a somatic motor region, but this time: Broca's area. However, the words and rules used in the primary step involved a communication with nature while in the second step they involve a social communication with beings of their own human kind.

Station 9A and B

While primarily we drew Aristotle's model to explain the nature of the nonverbal mind, yet in station 9 we broke the rule and included one of the features that is unique to the human verbal mind. In humans, there are two somatic motor regions in the Rolandic motor cortex, but there is only one in the nonverbal mammals. The Rolandic region, 9A, is the central somatic motor control region of the cortex in both hemispheres in all mammals. In humans, the special somatic speech center, Broca's area, 9B, is located in the lower region of the left hemisphere.

Both of the somatic control centers are like keyboards, upon which the nonverbal mind can "play," and cause all the somatic muscles to act in concert to accomplish the goal the mind's plan produces. Station 9A allows mammals to do anything they can, but only humans have station B, which allows them to talk about what they have done or are planning to do.

Station 10A

The output of the Rolandic center is millions of coordinated efferent signals which next travel to the somatic muscles and induce all the somatic muscles to act in concert to do what the nonverbal mind intends. All actions are initiated as afferent signals sent to our somatic muscles to say smell the rose.

Station 10B

If the reader will review the outline of Aristotle's model, they will find that according to this model all the nonverbal class or conceptual images are created in Station 7, the Sixth or Common sense.

When humans talk about reality and perhaps say, "It's a fine day," the words with meaning describe the images created by the nonverbal consciousness in station 7. When the human mind in station 7 is experiencing a perception or an exemplar of a class, they are not experiencing events occurring in Station 1 as if they are occurring in station 1. All the events seen in the nonverbal consciousness are images assembled by the Sixth Sense by using all the sensory-images created from all the sensations made available moment by moment by the PPA centers of the five senses.

It is physically impossible to experience in the nonverbal consciousness in Station 7, the objects that reside in station or region 1 of Aristotle's model. The physical objects themselves never normally enter the brain, nor do the physical energies, which are the effective stimulus for the five senses ever enter the afferent nerve pathways. Of course, objects should never enter the eye! In case of an accident, objects can and have penetrated the receptors and even entered the brain, as with a gunshot. When we see an object, we assume it is a Station 1 object, but in fact, all we can experience is an exemplar of an object or class-type nonverbal image assembled by the sixth sense from all the sensation-type images. Normally the only things that enter the brain are neural impulses in Station 4 in Aristotle's model. One of several great illusions induced by the nonverbal mind is this illusion of the reality of the images created by the Sixth Sense, and moreover the illusion that each of these images is located outside the body within the EMm within it proper place in terrestrial space and time.

Verbal behavior in humans is not produced exclusively by the verbal mind. Verbal behavior is always the joint product of both the nonverbal and verbal mind. If we say, "That's a pretty dress," the nonverbal mind creates the primary sensory and categorical images of the dress, which are exported for sharing to the verbal mind, which constructs the verbal statement, which is a string of memes, about the dress. However, the image we see, which is the meaning of the words is a product of the nonverbal mind and is in its place in the nonverbal consciousness.

The nonverbal mind makes it possible for humans now to understand consciousness. Finally, we can understand how the two minds work together to produce consciousness. Now we can understand where the verbal mind originated, how the GNMM evolved millions of years before the evolution of humans and the human verbal mind. The verbal mind evolved in order that humans could export the narrative created by the nonverbal mind. This replaces the classic role of the verbal mind as the primary mind and conceives the verbal mind in a secondary role; it was designed to expand the power of the nonverbal mind: to enable mammals to share their actions with others.

Station 11

This station reminds us that there are two kinds of somatic behavior, nonverbal and verbal and that even human speech is but another kind of somatic behavior. Speech historically has been conceived as a magical function that is independent of matter. Particularly I have always been annoyed with the idea that gods could speak. In the beginning god said "let there be light." I am proposing that words are not magical, that words evolved, and that there was no speech in the universe until human speech evolved, which required the evolution of the somatic muscle control of the physical speech centers.

Station 11A refers to the set of somatic muscles, which move the body in all the kinds of niches occupied by all the nonverbal mammals: that all the nonverbal motor behavior originates in the Rolandic motor cortex. By contrast, Station 11B refers to the somatic muscles, which move the human speech center muscles and cause speech. These signals would originate in Broca's area.

Station 12A and B

Station 12A and B are the somatic responses of the respective muscles, which are the final somatic responses of station 12A or 12B, the final nonverbal and verbal behaviors.

Aristotle's Functions: Human Verbal language behavior as a compound function

Consciousness, the Language function

In the creation of a self-organizing system, the first challenge was to create a set of domain-specific ideas, which can create a domain-specific consciousness, which allows the nonverbal mind to direct the body to act out the particular scripts written by the nonverbal mind. The challenge is to create a set of exemplars so sensational that they create a consciousness, which causes the organism to act out the script, etc. as will be described below in the discussion of the "Magic Cockpit."

For some readers the point of my statement above may not be obvious, but the mind as a self-organizing system was not designed to create systems that would meet Plato's academic needs. Plato wanted truth to be universal, unchanging, and eternal, etc. However, the rat's nonverbal mind doesn't need all that refinement. The rat's mind was designed as a pragmatic self-organizing system; a domain specific model is only useful in the domain for which it was designed. Most human models are also particular. A road map of the state of Texas, for example, is only useful if one is traveling in Texas.

Plato (1952) required a spirit God, one that supplied Plato with perfect eternal ideas that never changed.

However, the nonverbal mammals require no spirit god to give them anything, as nonverbal pragmatists, as self-organizing systems; they have no need for knowledge other than that produced by their nonverbal mind. Furthermore, being nonverbal makes it impossible for them to share with others. All a nonverbal mammal needs in order to create its own knowledge is a hierarchically organized material universe, cosmos, or god; one that it can interact or communicate with to create the domain specific knowledge they require to act with purpose in a domain.

The magic cockpit and meaning

One of the nonverbal mind's best "magic" acts is its creation of meaning.

The nonverbal mind communicates with the agent of consciousness via images not words. The symbols of an image based communication system of the nonverbal mind are sensations and perceptions, not alphabetic symbols. Red, indeed all sensations are the first order of primary symbols of the nonverbal mind and create or provide the foundation of the superior nonverbal communication power of the third wisdom. The nonverbal rationality, created by the nonverbal mind by the Heterostatic interaction with the nonverbal world, is the first rational wisdom, but is the third wisdom, HHH. The fourth wisdom is the human verbal rationality, which is built upon a set of second order alphabetic symbols.

If, "A picture is worth a thousand words," how much more is an image-filled consciousness worth? The nonverbal mind keeps the agent of consciousness informed of all the objects sharing its life space by producing an image with all the objects of that domain in their proper spatial position relative to the organism's image of itself, and with the agent center-stage and fully in control.

The magic cockpit can only be understood within the context of the meaning machine and is most obvious if we try to imagine the nature of an image-based consciousness, which is the kind of display chosen as the primary output medium for the primary mind. The target of the nonverbal knowledge created by the primary mind is the end user of the images, which are displayed in space and time in the nonverbal consciousness. The "I" of the nonverbal consciousness is, as it were, the pilot of each living self-organizing mobile unit of mammalian life. The pilot of the magic cockpit occupies a body, which is a mobile unit that the nonverbal consciousness located at the center of a world, both physically and dynamically, for the pilot is in, or feels, if they are not depressed, as if they are in control of some limited region of space and time. Consider the nature of the cockpit. The pilot does not feel like a pilot in an airplane cockpit, as confined within some enclosed space, but rather it experiences itself not even as an image among images, but as one real live sentient being sharing with other real beings an extended terrestrial life-space.

I know, in one sense, as a scientist, that my image-based consciousness is but a dynamic simulation, a virtual display, occurring within the neurons of my brain. Yet my nonverbal mind in a way that seems magical at this stage of our understanding of the neuronal technology gets my consciousness out of my head, and somehow I experience myself as surrounded by beings, any one of which I can reach over and touch or smell or interact with in any number of ways. I experience myself as not in my head, or even in my body, but as somehow inhabiting my body, which I can control by thinking. All I have to do is think that I want to smell that red flower which in English is called a rose and my body automatically makes the set of appropriate moves, which ends when I smell that object which would smell the same independently of whatever meme every community agrees is its name. The meaning machine's image-creating-"magic" gives that flower its color, size, smell, etc., which are all embedded within the consciousness the nonverbal mind creates to communicate to the agent of the mobile unit of mammalian life. Additionally, the meaning machine provides the agent within consciousness a full knowledge of all the real exemplars within its field, which must be included within its plans, if it is to succeed.

All the nonverbal and verbal mammals have or share a Meaning Machine. The generic nonverbal mind evolved to enable nonverbal mammals to adapt to change, which forced evolution to bestow upon the nonverbal mammals the function that enables life to become a self-organizing system. The "god-function" endows life with the power to categorize, to adapt, and to change, by creating a full set of domain-specific concepts or ideas. It is this power, which grants or enables life the rationality to enter an anomalous domain and take charge.

Heterostasis, the wisdom of the nonverbal mind, enables neuronal systems to emit the rational elements of behavior to create operationally the forms that take the uncertainty out of all the anomalous objects in a domain. Empirical self-organizing knowledge, not revealed metaphors, allows the nonverbal mind to create quickly the conceptual forms, which give life the capacity to know and predict the behavior of every object with which it shares its life space. The meaning machine confers upon mammals the powers of primary communication; the rational power to know: nature, physical reality, the cosmos, or mother earth.

Nonverbal communication consists of several key functions, but chief among them was evolution's endowment within the nonverbal mind all the technical physiological apparatus required by a self-organizing system to function as an independent agent in almost any nonverbal niche, which exists upon the planet earth.

Born literate

Each nonverbal mind is born nonverbally literate; born with the capacity to read the book of nature, to read "Nonverbal." Mammals are born into a world for which they are not genetically programmed to survive; nor are they born as Plato (1952) imagined humans to be born, with a head full of eternal truths which some spirit god supplied at birth, but rather they are born with a nonverbal mind that is a blank tablet. Yet, the one-mind mammals had one saving talent; they were born literate; they are proficient at reading each page of the nonverbal book of life specifically written with a narrative designed to inform "whomever it may concern" with a local version of relevant features of the story of nature. Each spatial structure such as the matrix maze comes with nonverbal instructions, which describe nonverbally its hierarchically organized spatial structure and its emergent properties. The nonverbal mind is born literate; born knowing how to read the nonverbal books of nature, the nonverbal books in which the cosmos endowed potentially all the knowledge required to survive. The cosmos is nonverbal, and the nonverbal mammals are born nativistically endowed with a mind designed with a sensory system designed to read each page of the nonverbal book of creation.

The nonverbal mind loves to read "Nonverbal," the nonverbal mind is endowed with a new wisdom, Heterostasis, the "wisdom of the mind" that automatically knows how to interact with the physical parameters of a new anomalous domain. It knows how to communicate with it so that as it moves nonverbally forward, it is reading the book of nature and transforming those causal structural forms into nonverbal mental forms, into nonverbal ideas, into a nonverbal narrative, into a nonverbal script or into a model; one which encapsulates a formal model of the maze.

The nonverbal mind knows how to ask nonverbal terrestrial structures such as the maze how it is organized, how many parts it has, how these parts fit into a whole, and how many kinds of parts are there etc.? The nonverbal mind knows how to communicate nonverbally with these spatial nonverbal components of the universe, how effectively to communicate with nature and glean the knowledge the nonverbal mind requires to create a model of the maze.

The primary function of the nonverbal mind, the 90-10% function

Most language theorists, including Chomsky (1997) are stuck having to propose a nativistic theory of the origin of verbal behavior, for they know nothing about the nonverbal mind, which produces the image-based language, which provides all nonverbal mammals the nonverbal knowledge by which they survive. In terms of Aristotle's 13 station model, they are stuck, in effect, in the unenviable position of assuming that human verbal behavior is exclusively a function of only one station, the thirteenth station of the mind. Human language behavior however, is the joint function of both the 12 station nonverbal model of the mind and the 13th station of the verbal mind.

The primary language function is nonverbal

The nonverbal mind is the exclusive source of all nonverbal knowledge. All nonverbal actions originate and execute from the nonverbal mind, and so it has been for millions of years before human speech emerged. Yet what the classic linguist has tried to do was to explain language behavior as exclusively a function of one verbal mind.

In short, in their attempt to explain human speech, the one- mind theorists historically for an obvious reason have ignored the role of the nonverbal mind, which is responsible for 100 percent of nonverbal knowledge and all knowledge driven behavior. The purpose of this book is to call attention to and finally make science aware of the role of this primary mind, which the one-mind theorists don't just ignore, but of which they are unaware.

Nonverbal communication, which is exclusively a function of the primary mind, is the primary kind of knowledge and communication and supplies all the meanings of the words of verbal two-mind communication, and supplies all empirical narratives.

The verbal mind is not capable of executing on its own, a nonverbal action. The verbal mind can't control nonverbal behavior; it can't dance, can't walk, can't climb a tree, and can't play marbles. The classic one-mind theorists have always claimed that verbal behavior was a function of one mind. My goal is to correct this position and show that verbal behavior is a compound mind function. The verbal mind does not have its own independent sensory-motor system. The only function of the verbal mind is to export the nonverbal narratives created by nonverbal mind as memes.

Communication- image based and word based.

The primary narrative employed by the nonverbal mind is exclusively image based. Image-based communication is the only kind from which nonverbal mammals can obtain knowledge about their external milieu on-line and in real-time. This nonverbal image-based consciousness enables the nonverbal mammals to interact with objects and events of their current domain-specific niche.

These images, which come in two kinds, sensations and perceptions, are created by the Primary projection areas and the Sixth sense respectively and are elements of the nonverbal mind's nonverbal consciousness. The nonverbal mind is a speciation mechanism, which creates or evolves sets of coherent ideas, for example, in the case of Female 3 as she explores the maze.

There are two kinds of communication, the primary, and secondary. Verbal communication is the secondary kind; nonverbal communication is the primary kind. One of the central features of the primary kind is the image-based consciousness, which is the only way that the agent of the nonverbal consciousness, the "I" of consciousness receives the knowledge generated by the interactions of its nonverbal behavior with the objects of some niche. Under the proper circumstances, while the secondary or verbal kind of communication can be an exciting exchange of knowledge, verbal communication pales by comparison with the power of the image-based movie created by the nonverbal mind to excite and entertain. This nonverbal consciousness consists of all the exemplar images the Sixth Sense constructs at each moment of an action cycle. The nonverbal image-packed narrative by which the nonverbal mind informs the agent of consciousness serves as the primary communication created by the primary mind and exclusively informs and directs all the nonverbal behaviors of every kind of mammal that has ever lived for millions of years. Whereas, the agent of the verbal mind could be known as a “talker”, the agent of the nonverbal mind could be known as the” doer” or in this case, also as the producer and/or the artist.

What makes the nonverbal mind’s finest nonverbal 3D movies so exciting and action inducing is that it engages more of the senses than any Hollywood production possibly could or can; in fact, the nonverbal consciousness engages all the senses and all the emotions. The nonverbal mind, which is the producer of the primary consciousness, produces such a knowledge- packed movie that all mammals, humans, and pre-humans believe it to be real.

The implications of the concept of nonverbal communication

Watching the spontaneous behavior of F3, the reader should be able to understand the strange concept of nonverbal communication, and hopefully begin to understand the almost sacred nature of Heterostatic behavior, which was designed to activate the potential empirical truths stored within the nonverbal pages of the cosmos. The nonverbal mind was designed to self-organize, to activate the nonverbal truth latent within the spatial structure of the maze. The nonverbal mammals' nonverbal mind is bound intimately with the nonverbal nature of the cosmos. The nonverbal mind's six senses bind the nonverbal mammals to the cosmos. The power of this bond is what makes the nonverbal mind work as a self-organizing system. Because of this bond, the nonverbal mind’s somatic nervous system is then bonded to the cosmos. Because of this series of bonds a third bond exists, every somatic muscle action produces an appropriate feedback from the cosmos. On the anatomy side, we are not at any place on our body directly connected to the environment, but on the behavioral side, the behavior-feedback circuits are so direct and so powerful that each mammal behaviorally or functionally can be said to be one with its environment.

Sentiment aside, the nonverbal mind was designed to function so closely with the material feature of nature that the chief survival tool of the nonverbal mind is its rational power to know and to identify the invariant laws of the cosmos, i.e., to read the material god's conventions. The nonverbal god however, is forever potentially in direct communication with each mammal but not everyone is listening.

The nonverbal behavior, which F3 emits, enables her to glean a sample of all the physical energies emitted by the objects of station one, which are captured by the Five Senses and processed by the twelve stations of her mind. By repeated exposure to this invariant string of signals generated by running the maze, the nonverbal mind discovers how to know where it is within the 96 units of the hierarchically organized space at all times. It knows where it is, what's up ahead, what is behind, and knows its current position in the maze.

Plato (1952) argued that knowledge couldn't be derived by empirical contact with matter; that knowledge can only be created by reason. Plato could not have made a statement that was more misinformed if he tried. I am prepared to reverse his argument and say there is only one royal road to knowledge - through the senses by employing Heterostasis. Strangely, today I don't hesitate to say that knowledge, empirical knowledge, is everything Plato said it was not. Contrary to Plato, while the empirical representation of knowledge created by science is itself constantly a work in progress, yet the physical laws themselves are presumably an eternal truth, an invariant. The empirically derived laws of science are the best approximation to the truth, which hopefully consists of a large set of invariants. Empirical knowledge is an invariant, that which doesn't change during change. The law of gravity, the speed of light, and science are the repository, the library of the invariants of nature. Science assumes that the physical laws themselves are facts that don't change with change, thus, how we state and understand those laws are subject to change.

I must keep restating the concept of heterostasis. Nature designed the nonverbal mind to create a better way for life to adapt to change. The nonverbal mind is pragmatic, it was not designed to create the eternal kinds of laws Plato required. The nonverbal mind is content with knowledge that is good enough for it to achieve its domain-specific laws.

However, these big scientific laws generated by the hard sciences are not the only kinds of laws to which the nonverbal mind has access. Each rat as it runs the maze is constantly in contact with the physical reality of the maze and the nonverbal mind from its constant contact with the cosmos, soon creates a valid model of the maze that too is an invariant, and that enables the nonverbal mind to plan and act to achieve its goal without error.

These invariants are sacred; the laws of the cosmos big and small are invariants and it is sad that Plato and his followers still don't know the power in the empirical stream of feedback even in such a seemingly insignificant place like the matrix maze. The nonverbal laws of the nonverbal god come alive as nonverbal "words" in the interactions of the nonverbal mammal and the closed domain of say the Matrix maze. The nonverbal truth of the pantheistic god does not need to be written in Holy Scriptures. The laws of the material god do not need to be carved upon a tablet of stone. These truths are inherent within space/time; they are engraved upon, or within the physical objects of the cosmos.

Where does the law of the gravity dwell within the cosmos? God and god's law are inscribed within and upon the fabric of space/time.

Every school student has the profound right to know how the material god and their nonverbal mind communicate, to understand the concept of knowledge, and the difference between faith and knowledge.

What is an idea?

As noted in the Aristotelian twelve-station model, evolution designed the nonverbal mind with the five senses to read or communicate with the nonverbal features of the physical world, to create the particular domain- specific knowledge required to enable the nonverbal mind to self-organize in each domain.

The dualistic/spirit god, its spirit world, and all the spirit beings had their origin in some verbal metaphor. The spirit world's origin was more like a dream, which escaped from Plato's head into the human consciousness, a dream with no reality in this physical here-and-now world.

When discussing the dualistic mind, the spirit nature of reality, the spirit nature of god, and the spirit nature of mind are all simply three versions of the same metaphor. It's time now to expand this topic into one conceptual domain, the concept of truth.

What is truth?

The most difficult aspects of the new scientific concepts of mind to be discussed are the ontological differences, the differences between the Platonic dualistic set of concepts of spirit: god, mind, and ideas, etc. on the one hand, and the monistic concept of a material god, mind, ideas, etc. on the other hand.

For Plato (1952) a spirit god created all the forms of mental (spiritual) beings. Thus, one of the major challenges for a monistic theory of mind is to explain how all the former so-called god-given ideas had their origin in the nonverbal interaction of the nonverbal mind and some physical domain of the cosmos.

Paradoxically no humans, from philosophers to religious holy men, to Chomsky, have ever been able to explain or even to offer a natural theory of the origin of human speech. Now however, with a scientific awareness of the existence of some of the more important features of the nonverbal mind, the answers to some of these questions of the origin of all the former spirit beings should now have an answer.

The origin of human speech is at least a two-step process. First, the narrative expressed in human speech was the product of the conscious, waking nonverbal rational mind; second, this nonverbal narrative was transferred across the corpus callosum and transformed into memes. The nonverbal mind, however can only conceive material things. For example, a nonverbal mammal could never imagine a spiritual being. The nonverbal mind was designed to naturally execute or exercise its powers of rational selection i.e. mental evolution, but only by interacting with the material world.

On the other hand, dualists of each of the transcendental religions each have their own books, which contain the words of their special verbal god.

However, the nonverbal mind has no divine verbal book. The whole point of the emergence of the nonverbal mind was that evolution changed its focus and shifted from the design of instinctual, genomic-defined beings to the design of mammals as self-organizing systems, as relatively genomic-independent beings.

The primary function of the nonverbal mind is to transform the anomalies of each nonverbal terrestrial domain into the salient nonverbal knowledge required for each mammal to self-organize its own domain-specific nonverbal behavior.

The chief function of the nonverbal mind is to explore and read anomalous nonverbal regions of terrestrial space, which are effectively the nonverbal books in the vast library of the cosmos. They are effectively pages out of the nonverbal mind's nonverbal bible because the nonverbal nature of each domain embodies or encodes the nonverbal knowledge, which will inform the nonverbal mind of the critical nonverbal features, the parameters of each space. The essence of the nonverbal rational power of the NV mind is to create the primary nonverbal knowledge, which it gets from reading the nonverbal books of nature or the cosmos: knowledge, which is then transformed into a nonverbal model of the local domain, such as the maze. The NV mind turns the nonverbal parameters of space, which are written in the nonverbal books, such as those found in the Matrix maze, into NV sets of coherent hierarchically organized ideas or knowledge. This knowledge is in the form of images, which populates the nonverbal consciousness and allows individual mammals to create a nonverbal model of a space and then use that model to extrapolate, to conceive a plan, which can be executed so the mammal can achieve its goal without errors.

This ability to interrogate or communicate nonverbally with some sample nonverbal book from the cosmic library allows the NV mammal to read the nonverbal anomalous parameters of each local domain and turn that knowledge into a nonverbal narrative within the image based consciousness, which allows each mammal to extrapolate. The creative power of the rational nonverbal mind reduces to simply the power to convert the feedback of Heterostatic behavior into classes, categories, or forms that inform the mind.

With the emergence of the nonverbal mind, the cosmos began the first step in the last lap of the long developmental process of acquiring the power of speech. In the beginning, the cosmic god was nonverbal but with the emergence of the nonverbal mind, after billions of years the cosmos finally began to evolve the power to engage in nonverbal communication between the primary nonverbal mind mammals and the physical parameters of the empirical world.

The Aristotelian model of the mind discussed above shows how a physical mind with five senses could process the empirical physical properties of the cosmos into mental images or knowledge.

It should be noted that the cosmos consists of far more electromagnetic forms or bands of energy than the five senses were designed to process. Light for example is just a small band in the center of the electromagnetic spectrum. NASA today has identified the entire band of the electromagnetic spectrum and developed technology to monitor all of these bands. Still the logic of understanding the universe by empirically processing the stream of energies being emitted by the universe is no different from the technology evolution built into the mind for processing light.

However, with the emergence of the two-mind mammals, the cosmos now thanks to a completely new host of wisdoms, each of which expanded the range of consciousness, the range of topics with which the NV minds can know and communicate has expanded. This set of rational skills included the power of verbal speech and then the verbal power of reading and writing, and the power of numbers, the power of song, and on and on and on.

Kinds of truth

Nothing is more important in understanding the nature of the nonverbal mind than the concepts of truth and consequences.

So far, I have limited the discussion of ontology to a discussion of the two concepts of god, the spirit, and the material god. Of all the things a materialist can say about the nature of the material god or the cosmos, nothing is more important to keep in mind than the fact that the material god is nonverbal.

In the pantheistic context, mammalian evolution is an aspect of the evolution of god, and with each advance of the capacity of the nonverbal mammal's mind; the nonverbal god, in effect was acquiring the power of communication. With the evolution of the three-and then four wisdom minds, the cosmos took, in effect, another big step by acquiring the power of speech.

Truth and consequences

Next, consider the concepts of truth, and consequences. My task is to distinguish the nonverbal and the verbal side of truth with respect to the consequences of truth.

A pure verbal truth, one without nonverbal consequences is toothless and meaningless. A man's laws or a city's laws are no laws at all without one or more enforcers, without cops; the idea of truth is dependent upon the concept of there being a nonverbal accounting of responsibility, of some judgment, on some judgment day, and the sooner the better.

Can the "I" of consciousness always act upon the content of its nonverbal consciousness? The nonverbal truth of the cosmos is serious; how many times does fire have to burn one's hand before one believes in fire? How many times does one have to fall to believe in gravity? One of the profound differences between the NV and the verbal mind is, as I said above, that the NV mind is the certifier of reality. Plato was obsessed with the nature of reality. For the monistic NV mind, I must note that every nonmaterial or spirit being, which Plato (1952) believed was real could not be experienced by the nonverbal mind. The nonverbal mind cannot experience anything but material beings for the simple reason that all Plato's nonmaterial spirit beings are not real, they have no material nature, which can affect even one of the senses. All empirical knowledge begins with the five senses and then with the sixth sense interacting with the nonverbal parameters of some physical domain.

But the verbal mind's verbal wisdom can conceive of metaphor-based beings that are not certifiable by the nonverbal mind. In fact, all of the beings that Plato claimed were real were beings, which he proudly proclaimed could not be known empirically but only by reason. And so philosophers like Plato spend their life talking about beings that had their origin in some obtuse metaphor on the one side and perhaps on the other side, some family-driven metaphor. Such metaphor driven concepts like heavenly father, etc. were the steady diet for the verbal mind to believe but such food offered no protein, nothing that the nonverbal mind could empirically know.

Plato's verbal mind followers lived by faith in their reason, while the nonverbal mind lives by the knowledge of its senses. However, no one would ever know the difference between scientific truth and philosophical truth for thousands of years.

So, humans for thousands of years were left not even knowing how to know the nonverbal truth. Why, for example, should we not lie? Christianity says because all humans on the judgment day will stand before god who will judge the virtue of our lives, and decide how much pay-off that virtue has earned; how big a mansion our eternal home will be.

Pantheistic god

On the other hand, one of the supposed criticisms of the pantheistic god is that it cannot play any role in some final future judgment day in eternity. Indeed this argument or criticism is based upon on a grievous misunderstanding of the physical god's role in his/her physical world and in this place and time. So must we jump to the conclusion that the physical god does not judge, does not hold each being accountable, and will hold no judgment day?

On the contrary, the answer is that there are two concepts of an accounting; one physical and one spiritual but more importantly, we must distinguish between two concepts of judgment. First, consider the difference between the spirit and the material god's judgment; the material god's judgments occur here and now. They occur only in space and time. The material god is a traffic cop, judge, jury, and executioner that are always on the beat exactly when and where required; one that never sleeps is everywhere and gives immediate feedback or judgment for every infraction of its laws.

The action cycle

Nevertheless, the material god as an executioner exhibits one aspect to it sentences that seem to complicate the immediate feedback aspect of its judgments. In the case of violations of the law of gravity, if one walks off a cliff, the judgment takes effect immediately; only in the movie is the rabbit able to take several steps after walking off a cliff before it begins to serve its sentence.

The bitterest aspect of the way the material gods laws are manifest is that while the clock begins to count down immediately upon the violation of some law, yet the rate at which the sentence unfolds is most complex and may take days, years, or even centuries before the final sentence plays out depending upon the crime. For example how long after one begins smoking, does the damage to the lungs or the tongue become manifest. How long will global warming, due to human behavior, take to reveal itself, in melted glaciers', rise of sea levels, etc.? How long after exposure to atomic radiation until the symptoms are made manifest?

On the other hand, the spirit god supposedly, verbally records our behavior in a file upon which we will be judged but only on some single final future judgment day! What happens in the meantime, in the here and now?

The challenge of this book is for each of us to transform our thinking in order to understand the nature of the nonverbal mind, which for millions of years has directed the life of all the nonverbal mammals, which have only one kind of communication, the nonverbal kind. The fact is that this kind of truth employed by nonverbal communication is impossible or meaningless without immediate feedback.

The behaviorists got at least one behavioral law correct; the effectiveness of reinforcement, or feedback of material judgment is directly related to the swiftness of its delivery.

Perhaps because the two concepts of the nature of truth/judgment are so different, we have become so accustomed exclusively for so long to the verbal kind that we have lost our ability to think critically about the nonverbal concepts of truth and its functions.

I propose that the success of the creative power of the nonverbal mind is directly a function of the immediacy of the empirical feedback that is an integral element of the nature of the physical world--a feature embodied in nature. The power of the nonverbal mind to know the truth is a direct function of the perpetual and immediate feedback from the judgments of the material god. This core feature of the pantheistic god is that all the laws of nature are, in effect, active at all times and in every domain and can be operationally known by every temporal causal-effect relation enforced; the material god is known by its active empirical nature. The speed of the nonverbal mind to know the material god is a direct function of **the action cycle** time of each empirical domain-specific causal relation. The physical cosmos works because it leaves no doubt of its judgments of our actions. The cosmos insinuates itself immediately into every act; the material god judges every cause-effect relation providing immediate feedback into every truth-judgment relation; every cause-effect relation always bears the mark of time. The material cosmos-god can always be known only by the immediate consequences it builds into every causal relation.

Every mammal is born literate in the nonverbal language of the universe and each knows the swiftness of nature's judgments. How easy it is to die for the infraction of any one of a long list of nonverbal commandments. For example, gravity is one of these nonverbal commandments. Anyone only has to defy gravity, if there is doubt about the vigilant nature of the cosmic god, if there is doubt that gravity is real.

Did you ever see "road kill?" How many animals will have to be killed by cars before this consequence is no longer observed?

One of the evolutionary principles, which humans should learn to take seriously, is this: every organ bears witness to nature of the god in whose image it was designed. Just as the eye was created in the image light, etc., just so the emotions are created in the image of the environment in which they emerged. Fear only emerges when it confers some adaptive advantage.

For example, Darwin reported that the animals on the Galapagos Islands had no fear and proposed that this reflected the fact that these animals had no natural predators.

Why do we see road kill in so many species? Natural selection will, presumably, perhaps over hundreds of years evolve new strains of each type of the animals that are now the subject of road kill; ones that have an instinctual fear of cars or some aspect of cars, so that their species will no longer be subject to road kill.

There are not just ten of these cosmic laws nor are they written on stone. The material god's laws are always live, always on; they are inscribed into the material fabric of nature.

Another example, the "visual cliff" for children is necessary because the material god makes no exceptions to its laws for anyone or any age. Moreover, the whole foundation of the concept of nonverbal communication is the fact that the physical god gives immediate feedback for every cause-effect relation evoked by any and every somatic nonverbal action. Again, the nonverbal mind only has to ask, intentionally or not, its god a nonverbal question, and it will immediately be given a nonverbal answer on some time scale of the material gods choosing.

The difference between the nonverbal and the verbal mind is that nonverbal truth has an immediate nonverbal meaning, while verbal truth has only a verbal meaning. In the other words, it has a meaningless verbal meaning. The nonverbal god does not bother with books to record or enforce its nonverbal laws, but science is trying to remedy that problem. Our trouble is that the laws of the nonverbal cosmic god and those of the verbal tablet-of-stone writing god are the nature of the laws. The laws of the nonverbal god were not "given." Nonverbal truths are nonverbal; they are never given in verbal form; laws were never put into verbal forms until, for example, scientists such as Newton and Galileo stated the cosmic god's nonverbal laws in scientific mathematical and verbal forms.

By comparison, the laws of the Ten Commandments seem amateurish compared with the laws by which NASA planned the path of the orbits of the satellite GALILEO through the planets and around the sun to reach some final destination years later and millions of miles out in space.

Communication

So let's start over and examine the facts. Actually, the first fact is that only the physical god can create physical beings. Things are beings made of matter, only the physical cosmos exists in space and time. All things are bound together causally as real physical beings in the material universe, cosmos, or god. All physical beings are expressions of the nonverbal laws of the cosmos, and only a physical world has a physical nature, which can be objectively known, and described by nonverbal knowledge. More critically, it has a nature for which evolution created a special sense, the sixth sense that monitors, reads, and recognizes where all the relevant objects of any domain are within a life space by creating a model of the local objects and their properties of the physical world.

But allow me to jump ahead for a moment. The physical world provides immediate feedback, immediate closure on every causal nonverbal act, and this is the foundation for every aspect of nonverbal communication for every nonverbal mental event.

As the behaviorists say, "feedback" is only as effective as a function of the immediacy of the feedback.

Feedback or Consequences

The word feedback is a technical term and often not an item in most people's vocabulary. So let's talk about consequences. One of the most profound differences between Plato's spirit god and Spinoza's (1933) matter god is the time between an action and its consequence. Science studies matter, but not spirits because nature, which consists of matter, is characterized by immediate feedback or consequences. The cosmos generally responds or communicates immediately for our every somatic nonverbal action, or immediately to every cause, our mind emits. Nature directly supplies an instant effect or consequence, which Aristotle's (1952) Twelve-Station model then processes immediately. This immediate feedback is registered within the data stream, which reports the feedback for our actions. The Sixth Sense transforms all the class based sensory input directly into consciousness. This nonverbal content of the nonverbal consciousness is experienced as nonverbal reality. This nonverbal reality is time-tested, reproducible, and empirical; these kinds of nonverbal facts contained in the generic nonverbal consciousness drives our lives and should justify our every action.

For humans to base their actions upon a meaningless spirit-based words or meaningless ideas; words with no empirical content, which satisfy some philosophical reasoning, only illustrates once again that much of the human population has been wearing, as it were, Stratton Glasses for far too long. Most humans are perfectly adapted philosophically to this upside down world. The content of the nonverbal consciousness is for the nonverbal mammals not just the primary kind of meaning but the only actionable meaning. It is the content of this image-based consciousness, which provides the empirical truth. If we experience a red apple on the table before us, that red apple is an image-based exemplar of the physical object on the table but the red apple cannot enter into any of our senses. However, it provides a variety of physical energies, which enter one of our five senses, which start the nonverbal mind into action, and in only a matter of milliseconds the nonverbal mind will give its ontological judgment of what kind of an object is on the table before us.

In order to avoid some confusion, perhaps I should point out that the mouth is not one of the five senses, but rather has several different kinds of sense organs embedded within it, particularly embedded within the tongue. There are four kinds of taste buds in the tongue.

The face, it seems, was designed for the sake of eating; take for example, the behavior of putting an apple in the mouth. The eyes are located relatively high above the mouth and assist each individual to find food, and then focus upon an apple as it approaches the mouth to check it out as fit to be put in the mouth. Likewise, our nose is placed just above the mouth, so that the quality of the chemical properties of an object can be tested up to the last second before being put in our mouth. Even after we put an object in our mouth, the taste buds on our tongue can still check out the apple as we chew the apple and can decide whether to swallow the apple or not. There are also temperature receptors in the mouth and receptors that sense pain and sharp objects. All the facial receptors can be thought of as jointly organized to provide the mind with the knowledge necessary to make the critical decision: to swallow or not.

I used to ask my class why the senses were placed on the face the way they are. Why I would ask, was the nose placed where it was and aimed down not up. One student suggested that if our nose had been turned up we might drown during a rainstorm!

But the rule is that it is the various kinds of energies from the apple, but not the apple, are what stimulate our senses.

These sensory-perceptual judgments, which the nonverbal mind prepares for consciousness, are the only ways that the nonverbal mind communicates the consequences of the nonverbal physical action initiated by the nonverbal mind.

For example, if the object in our mouth has something sharp embedded within it, the only way we become of aware of that is by the sense organs in the mouth reporting that sharpness in consciousness. Consciousness reports the nature of the objects. Consciousness is the means by which nature or the cosmos talks to us.

The creation of consciousness

The creation of consciousness can be said to evolve in three steps:

- Function one- the Sixth Sense evolves, conceives, or constructs the forms or patterns of features, which define a coherent set of features, which define a concept, which define a class. The Sixth Sense is an association area that must first sort and then synthesize or associate all the sensations from the PPAs, which belong to each exemplar and next compose a sensational-image or an exemplar of a class of thing, an image of a girl or a rose. The sensation type images are the base level foundation for the nonverbal sensory symbols. They correspond to the alphabetic symbols, which compose a word. Each hue or color represents one of the wavelengths of light, but it also represents one of the physical attributes of the rose, which shares the external milieu with some organism. All of these sensory symbols must be combined into an image of an object class which has all the sensations as attributes; red, smells as sweet, size, weight, etc., which the senses testify are empirical elements of the current life space.
- Function two- each of the class images must be combined into a life space, with a nonverbal narrative which varies with physical events of each moment in time, which corresponds to the frames of a life space or in effect a mental movie.
- Function three- each of the frames must then be assembled into a series of frames, which generate a nonverbal movie or

> consciousness. The Sixth Sense is the master control center for the primary nonverbal language center where *all images are created and assembled* into an action movie called the nonverbal consciousness, and creates and certifies all actions, and all nonverbal meaning.

Kinds of ontological certification

The classic Platonic view of the rational mind has always been a talk centric view. However, the nonverbal mind is the court of last appeal with respect to ontology; it is the chief, the only classifier and certifier of reality. For the nonverbal mammals, the nonverbal mind classifies everything before it enters consciousness, but not everything enters consciousness. For example, each of what Kuhn (1962) calls an anomaly is a cluster of sensory events that the nonverbal mind doesn't recognize, hasn't classified, and cannot assimilate. Anomalies are classes for which the nonverbal mind has no matching concept. Anomalies are things that surprise, that cannot enter consciousness as an exemplar of some known class of object. While anomalies cannot enter consciousness as known classified event that does not mean they can have no effect upon consciousness.

On the contrary, those things that the mind cannot assimilate, these anomalies, have the most profound effect upon consciousness. As anomalies, they induce surprise, or they induce some degree of uncertainty depending upon the degree of surprise they induce. Other criterion will be identified in the next chapter, such as how intensely they diffusely desynchronize the frequency of the EEG signal throughout the cortex. The physiological nature of the signal frequency that is dynamically changing within the canopy of neuronal branches within our cortex is one of the most interesting aspects of the primary mind's function, which begs to be told.

Unidentified or unclassified objects

The nonverbal mind is the ontological authority; it says what is real and creates a concept to identify each exemplar of each real class. For Plato (1952), the real had to be eternal and spiritual not material, but for the nonverbal mind the real is any class of material object; any reproducible event, any reproducible syndrome of physical properties; any invariant. This is true even in human science. Ideally, the real is a set of, or a syndrome of physical features that strike the senses, which the sixth sense judges to be an exemplar of some class, an invariant that doesn't change during change.

Certified

Years ago in my graduate school days in Texas, one of the most advertised beers was called "Lone Star beer." Its advertisement said only, "It's certified." I always thought that was the funniest thing, certified how, that it's safe to drink, that it's beer.

But the nonverbal mind certifies each empirical object that is a candidate for reality. The nonverbal mind is the sole judge of what's real; what's safe upon which to act. If the image-based consciousness includes an image, an exemplar of some class of object, the individual mammal can be sure it's safe to act upon that image, it's certified.

The nonverbal waking consciousness only includes images in the nonverbal consciousness, which pass the ontological scrutiny of the sixth sense. The role of the nonverbal mind is to create a nonverbal consciousness, which can be the reliable basis for nonverbal mammal's action. The nonverbal mammals need what Plato wanted; they need to live in world they can trust, in which they can reliably act. The role of the sixth sense is to insure or to certify that any object included in the nonverbal consciousness is real.

To Avoid Surprise

In today's modern sense, Plato did not know how to know what is real.

The purpose of the nonverbal mind is to warn its agent of consciousness of the real state of the immediate world so that the agent can act appropriately. All the surviving mammals are rational beings that trust their nonverbal consciousness.

The point of this section is to discuss and explain that in the first step of the creation of verbal behavior, the nonverbal mind produces a nonverbal narrative, which is the primary meaning of a sentence. The nonverbal narrative is created before the second step where the verbal mind imposes its meme words and rules upon the narrative.

The second step, however, simply enables the organism to export and share its nonverbal narrative. The nonverbal narrative is the primary language function. Only in the second step do the verbal words and rules enter the picture, only in the second step does the verbal mind play a role in verbal communication.

Dreams

By contrast, when we are asleep and dreaming, the objects in our dreams are not real and are not certified as real. Evolution designed into the sleep circuits of our brain/mind, a mechanism that disconnects our dream consciousness from the somatic motor center; our somatic muscles are paralyzed. Part of the nonverbal mind's task is to insure that dream images are not certified. Evolution built in circuit breakers to insure that we don't act upon images that are not empirically real, that are not certified by waking consciousness.

Occasionally this fail-safe system doesn't work, humans "sleep walk" and can injure themselves while acting upon uncertified images.

The secret of rationality is the power to execute the god function; to create the forms, the concepts, which give meaning to the syndromes of neural impulses the five senses deliver to the sixth sense. This god-power is ontologically serious, its job is not only to create classes, but it also has the critical ontological responsibility to identify when a syndrome of precepts qualifies as an exemplar of a nonverbal concept. The power to certify when the physical world contains an object to which it is appropriate to respond means consciousness has a critical ontological responsibility, to certify when and which objects are real and are safe to include in our acts; this is a critical life and death function.

Suppose you are driving on a highway and an eighteen-wheeler crosses the center lane and is coming straight at you! What will you do; A, turn the car and try to avoid a collision or, B, go into an ontological debate with yourself about how empirical data cannot give knowledge. Since all you have is sensory data, and how in Plato's eyes, matter only gives appearances, etc., I could go on, but you have been dead for several second now.

What is real? Anything the nonverbal mind places in consciousness, before the eyes, is real.

In the compound mind human case, somehow, the nonverbal mind lost control of consciousness and the verbal mind enables and allows the one most harmful ontological failure to occur; it allows metaphors, all kinds of uncertified images to enter the waking consciousness. Unfortunately, how this section of the mind works is not a part of rodent domains and is not anything that I can experimentally study. But nothing is more serious for me than to show that somehow once the two stages of the human mind began working together, the nonverbal mind has no longer been in absolute ontological control as it is in the one-mind animals. More human troubles are caused by metaphors than I could ever estimate. Could the nonverbal mind of a rat ever fall for Santa Claus as I did or a verbal spirit god?

My problem in this book is to convince the human experts that the nonverbal image-based content of their consciousness that they experience is the product of their nonverbal mind, is the product of station 7 and certified by the sixth sense, the ontological certifying sense.

Faith vs. knowledge

Two of the most misunderstood words in English or probably in any language are the words faith and knowledge.

Again, Plato's (1952) concept of knowledge proves to have corrupted and dominated the West's concept of knowledge. Plato employed two geometric metaphors, the metaphors of the divided line and the slave in the cave to define his concept of knowledge. Plato, first, absolutely rejects empirical data and empirical argument as having any role or virtue in defining reality. The real is transcendental as is his first assumption, and the reality by this assumption can only be explained and verified rationally or with words. With the divided line metaphor, Plato is generally credited with having proven rationally that only eternal truth and eternal beings are real. For a philosopher, this so-called knowledge is knowledge only because it is eternal or transcendental in nature.

Mathematical proofs, however, only establish that a proposition is tautological; that it does not contradict itself.

In this book, the use of the word or the concept of knowledge is restricted to empirical truth. Knowledge statements are statements that only inform us about the nature of physical nature or the physical god. Only the cosmic god can be known, and all knowledge is derived from nonverbal behavior. Historically and theoretically, the West recognizes two gods: the verbal-god of faith and the physical-god of knowledge

Paradoxically or should I say, pathetically, the nonverbal mammals in a strange sense have an advantage over their human decedents; the fact that they are nonverbal or preverbal means they are naturally immune to metaphor-created ideas or beings, including all verbal created gods. I will come back this topic later.

Yet the other side of the coin is even more paradoxical. Humans have long believed they were the only species of life with the gift of communication, of verbal communication. Yet again, facts are stranger than fiction.

The nonverbal mammals invented and perfected communication, nonverbal communication, and here the paradox goes full turn. The nonverbal mammals emerged because nature required a better way to endow life with knowledge: its first solution was the so-called "nature" solution that endowed life only once with one big dose of knowledge: genetic knowledge at conception. But mammals emerged as a part of one of the great evolutionary experiments, to design a form of life that could self-organize, which means they were endowed with a mind by which they could extract knowledge directly from the domains of earth, the domain specific knowledge required them rationally to design the means to their domain specific ends.

Communication emerged so that life could communicate directly and nonverbally with the cosmos in order to achieve its earthly needs. Where Christ taught the verbal faith based believers of the verbal god to pray by saying "our father who art in heaven, give us our daily bread," this prayer consists of many metaphors and is a homeostatic prayer.

The nonverbal mammals communicate directly and nonverbally with the cosmos, but they come seeking only knowledge, the knowledge they require as SOSs. Humans in the meantime only pray the verbal god's, Lord's metaphor-based- homeostatic prayer. Meanwhile Plato has his followers all tied up with geometrical metaphors.

The role of the nonverbal mind is to produce nonverbal knowledge; knowledge is an idea empirically certified by the nonverbal mind. Nonverbal ideas are created by the nonverbal mind and certifiable as real, as having been approved by the Sixth Sense, and being exemplars of concepts. The origins were created by the nonverbal mind and based upon empirical data.

Faith, on the other hand, is acting on an idea that cannot be certified by the nonverbal mind. Faith is acting upon beliefs that are not and indeed cannot be empirically certified. Faith is acting upon ideas that have a verbal or metaphorical origin and are orphaned concepts that have no empirical parents. They begin life in the verbal world without certification in the nonverbal mind.

So, it follows, that ontology recognizes two worlds, two gods, two truths, two kinds of life, two religions, etc., the knowable world that can be known and the unknowable world, the world of faith. Presumably, there are an infinite number of possible faith-driven worlds but only one knowledge-driven world. There is and can be by definition only one "god of knowledge," the god of science. How many word-gods are possible? The simple answer to that question is that all word gods have their own books. All word gods are metaphor driven beings. All verbal gods' books are, themselves, elaborate metaphors, and a type of graven image of god.

However, there are an infinite numbers of words, books, or faith-based gods. How many faith based gods are there, how many could there be? How many metaphors of god are possible? These questions have no scientific answers.

The two time-based concepts of god: one in time, and another out of time

There can be only one god of space/time; only one god of all things; only one god of the cosmos; only one unitary god; only one fair god, only one of Spinoza's (1933) gods; only one nonverbal god, one emerging god, only one science god.

Most of the ideas that Plato (1952) idealized had their origin in his metaphor of the divided line. Plato took pride in the fact that all spirits by definition are meaningless; as empirically, they have no material being; as spirits they cannot be seen, smelled, or touched, etc. spirits have no empirical or nonverbal meaning. So they can never be certified by the nonverbal mind. They are not capable of producing an effect upon any of the five senses, and thus, cannot be detected by the sixth sense. What color is a spirit god? What does a spirit god look like? How can a spirit speak? Any image of a spirit god is a graven image, an image created as a metaphor by some human artist or sculptor, etc.

God

Most professional philosophers and even ordinary humans probably are never consciously aware of Plato's (1952) insistence that knowledge cannot be acquired empirically but only rationally by reason.

One of the hardest questions for me to get an answer to was a question that I didn't even know how to state in a philosophically correct form. For almost all my teenage years until my senior year in college, I was burdened by the question: how can we know god? Only in my senior year in college did I know how to phrase the question. Which god should we worship: the god of knowledge, or the god of faith? My problem was that I did not understand how to ask the proper ontological question. My problem was that I naively believed that there was only one kind of god. Once I understood the ontology of dualism the question suddenly had a simple answer; there are two kinds of gods, the material, and the spirit. The material god can be empirically known, but since the spirit god has no material nature it cannot stimulate any of the five senses so the sixth sense has nothing to certify as real. Even Plato understood this! He clearly says the spirit forms cannot be known empirically but only by reason.

I didn't understand this until my senior year in college. Immediately I withdrew from my bible major, dropped out of the ministerial society, and never attended church again. I was only able to find the answers to my questions because I finally met someone who understood the problem. I was finally helped by my Greek teacher's mentor who was visiting the campus. He helped me to understand the meaning of faith or the ontological significance of god being a spirit. That day was one of the great turning points in my life.

How for example, can each Christian know which kind of god their church is worshipping? How can you know which kind of god you are worshipping?

Actually, it's very simple and straightforward. If they ask you, "Do you believe in god?" Then your god is a spirit god. The real god can be known.

Atheism

Atheists unnecessarily allow dualists to define the term atheism. The classic dualists define an atheist as one who doesn't believe in god, meaning the spirit god.

However, monists should define an atheist as one who doesn't "know" god. It is interesting that by that concept everyone who has faith but no knowledge of god is an atheist. I will not allow anyone to call me an atheist. I don't understand how so many authors today can't wait to defend atheism. There seems to be a rush of authors willing to call themselves atheists, while it is still in style. I would bet the pope, if he could be bothered, would call me an atheist and that would please me, for then I could wait my turn to call him an atheist.

Words vs. Action

One strange belief that too many humans share is the belief that words command the physical world. The fact is that if you want to open a door you must do so nonverbally. You can command a door to open verbally forever and a day but the door will not open, but movie producers make a fortune creating worlds where words have powers, magical powers, worlds where words can do things scientifically impossible.

What is magic? Magic is invoking words that are believed to have physical effects.

Only in a Harry Potter world does magic work, words only have magical powers in storybooks. Authors nevertheless can become billionaires by creating a verbal narrative, which in a strange way supports or reinforces Plato's (1952) idea that words/ideas are real. Magic creates a world that would have made Plato's day, one in which words are real. Plato always insisted that the real world is verbal but the fact is that the real world is nonverbal.

In fact, the verbal world is always the product of the human compound system, but I would like to understand how evolution plus plasticity created a two-mind system where the nonverbal world is no longer in charge and thus no longer serves to certify ontologically what is real. Today in the verbal world, no one seems to understand how we all have a choice: to live a fact based life, or a metaphor based life.

Nevertheless, this nonverbal mind is the only source of ideas that are safe for the nonverbal mammals to act upon; only nonverbal ideas certified by the sixth sense of the nonverbal mind are real ideas and are represented in the nonverbal consciousness. The wise man is one who does not trust words or live by words but trusts and acts only on the ideas, which are created or certified by the nonverbal mind. In the human mind: knowledge is certified by the sixth sense of the nonverbal mind, while faith is the belief in concepts that are not certified or certifiable by the nonverbal mind.

Meaning: two sides

The discovery of the generic nonverbal mammalian mind finally makes it possible to answer many questions that no one, or almost no one, ever asked. Possibly the best example historically has been called the problem of meaning. Actually, there are two problems of meaning or two sides to the problem. However, those few scientists that have raised the problem of meaning have only focused upon the second half of the problem, the problem of verbal meaning. Pinker (1999), for example, in his book about words and rules does note that a word has a meaning and shows a picture of a rose.

There are two kinds of meaning: Nonverbal and Verbal.

Verbal meaning focuses upon the meaning of words, which is half the problem, the second or newest half. This half of the problem has and is on rare occasion discussed in textbooks. But, generally, it gets at best a perfunctory treatment.

One of the best-kept secrets is the primary meaning problem, the nonverbal meaning problem. There are several ways to state the difference between the two halves of the problem.

First, how does the nonverbal mind discover the meaning of the objects of the external world since these things never enter the mind?

Second, the other half of the problem of meaning is the words. How do words get associated with the physical object they represent?

Plato (1952) holds the record for stating this first half of the problem in his famous "Slave in the cave metaphor."

The Slave in the cave metaphor

Plato (1952) paradoxically was almost the only theorist who ever recognized or who almost recognized the nonverbal side of the meaning problem. His famous statement on this phenomenon was in his "Slave in the Cave" metaphor.

Plato asks us to imagine that we are slaves born chained inside a cave with our back to the cave entrance. Plato then asks us to consider this question, what could such slaves know about the objects and events happening in the outside world, outside the cave. What makes the metaphor interesting is that he then states that all the slaves can ever know about the world outside the cave must be inferred from the study of the shadows entering the cave cast by the objects that cross the cave's entrance. Plato allows the sun to shine by day and fires by night to furnish the shadows.

FIGURE0.1C
Mark's NV model

My son Mark drew his nonverbal understanding of Plato's (1952) metaphor, which is shown in FIGURE 0.1C. On the left, inside the cave, a stick-figure slave can be seen chained to the wall in the cave with his back to the cave opening. Outside the cave, he has drawn the sun with two human stick figures between the sun and the cave opening. Mark has drawn inside the cave on the far wall of the cave two shadows as two letters I's with the height of the letters representing the height of the shadows cast by the two men outside the cave. Notice Mark has tried to capture the slave's problem by drawing the shadow on the left smaller than the shadow for the man on the right. The dilemma for the slave is that the shadow on the left is shorter even though he is taller, but he is further from the sun.

The point is that the only things that can enter the cave are shadows. The slave chained to the wall of the cave obviously cannot leave the cave nor can anyone enter the cave. Then, how can, how much can, and what can a slave empirically know about the physical world with only shadows to inform him?

Mark's art represents the physical world problem as stated by Plato. Now what we need and what Mark did not try to draw was the mental world aspect of the physical world problem, the problem of the mind.

If we translate this metaphorical statement into modern terms, it suggests that Plato (1952) can be seen to have been remarkably insightful! However, the problem is that the mind aspect of the problem is my understanding of the physical world problem, not Plato's understanding. The serious question, which I cannot answer, is how much insight Plato had in the content of his metaphor. Let me call your attention to two advanced points: could Plato have seen these analogies?

In modern terms, Plato could have meant:

First, we are our brain,

Second, the brain is inside our head, or inside our skull. Plato's metaphor now is starting to make sense; the idea of a slave chained inside a boney cave could be our brain born inside our skull and a prisoner for life, which starts putting some kick into his metaphor.

Thirdly, I have said this before and will say it again, for strange as it seems nothing is more critical to understand than the nature of consciousness, but nothing ever enters our brain/mind but neural impulses. Plato (1952) was right, the only things that enter the mind are shadows, which we call neural impulses or neural events, and Kant (1952) called "precepts."

It must be just a coincidence but if my metaphor is even in the least bit a valid restatement of Plato's metaphor, then Plato's questions were brilliant and far before their time. My version of Plato's metaphor couldn't have captured one of the key problems any more insightfully.

Plato's metaphor allows me to make the critical point that the objects of Station 1 never enter the cave/brain. Therefore, all that we and Plato's slave can know about the external world is by the inferences drawn from the study of shadows or from the knowledge which the afferent neural pathways carry via of the shadows like neural impulses.

The first problem of meaning is a technical point that historically never gets the attention it deserves. The problem is that the direct awareness of any physical thing is impossible. All objects are outside the cave/brain and the slave is chained inside the cave. How does the slave ever get any knowledge of the things outside the cave, for neither objects nor the physical energies, which are the effective stimuli for the five senses ever enter or travel within any of the afferent nerve pathways.

Historically humans have chosen to taken one of two positions:

First, most people, and most textbooks ignore the problem all together or

Second, there are those that technically admit there are problems of meaning but then immediately ignore the problem.

Pinker's concern with the second aspect of the meaning problem

Pinker (1999) recognized the second aspect of the meaning problem with the lightest touch and then immediately went on to topics that he thought were more important.

Pinker says that verbal behavior consists of two components, words, and rules. However, he adds in passing that each word such as a rose has a meaning, but after drawing a picture of a rose which he says is the meaning of the word rose, he then fails to tell us either where meaning comes from or what function it serves! Words and rules then must be are all there is considering that's all he actually finds it necessary to discuss.

Conventional wisdom tacitly assumes that what Pinker (1999) calls "meaning," the image of the rose, is the physical object of station one. This idea is the only one possible within the classic Western one-mind concept of human nature and language. However, first we must understand that humans are a one-mind species, and then we can understand that they process sensory data in two stages, nonverbally then verbally. The primary nonverbal mind constructs image-based nonverbal exemplars of the physical world in Station 7 and then we have a science-based alternative for understanding the meaning of words. In Station 7, the sixth sense constructs an image-based exemplar of all the objects that share the life space with a subject. Moreover, this nonverbal consciousness does double duty. The verbal mind perhaps only modifies the nonverbal consciousness by transforming the narrative created by the nonverbal mind into memes.

The primary mind executes the most profound mental function; the evolutionary speciation function, which creates or evolves, de novo, the forms or concepts required to understand what happens in the external milieu of the mind in an anomalous domain, and thus enable a mammal to act; both rationally and with purpose, while understanding changing situations. By conceiving the forms, which clothe each image and establish its meaning, the magic of the nonverbal mind then goes one-step further, gets the agent of consciousness out into the world, and enables it to take control of its External Milieu of the mind, EMm.

The Impossible function

It is impossible for me to overstate the facts or their importance, but the primary role of the nonverbal mind is executing a function, which essentially all the philosophers and linguists for thousands of years have insisted was an impossible function or a nativistic function. By which, they mean that in their professional wisdom, there was no way for anything, any being made of matter, i.e. the mind to execute this function. Why, because the creation of the forms, concepts, molds, cookie cutters, and so forth were beyond the power of matter, i.e. the human mind to conceive. This speciation function was conceived as a function reserved for a god.

Yet, F3, a 90-day-old nonverbal mammal loved nothing better than to find a good anomalous domain where her nonverbal mind's main function, the impossible function, could get a good workout. The nonverbal mind's favorite sport is to execute the impossible function; to play god, by spontaneously exploring an anomalous domain such as the maze, and to create a set of coherent hierarchically organized forms to give meaning to each of the ninety-six vectors of the potential classes of objects within the 96 units of the maze.

The nonverbal mind uses its speciation skills to conceive the structure of the maze, which it then uses as a map to organize and execute a plan of behavior appropriate for the domain. The point is that the rat or any nonverbal mammal has, in completing this most complex act of speciation, acted rationally and completed the execution of a primary act of communication between the rat's nonverbal mind and the structure of the maze. This communication was rational for it was based upon a nonverbal image-based consciousness conceived by the rational nonverbal mind.

Purpose in biology: natural selection

One of the major themes which I should have been advancing more than I have, is the concept of purpose. One of the cardinal principles of biological or Darwinian evolution is that all the changes induced by biological evolution; all the new species that emerge are not planned by any being with a purpose. Rather as the complexity of the physical world evolves, creating new environments in which evolution can occur, then, new species can evolve, but only because a new set of environmental conditions were at work operationally defining natural selection.

Purpose in mammals; rational selection

The adaptive advantage afforded by the evolution of the nonverbal mind can only be understood by recognizing that mental speciation and biological speciation employ completely opposite principles. Biological evolution employs natural selection while mental evolution employs rational selection. The difference between the two is entirely a matter of purpose. In biological evolution, natural selection does not employ purpose; but in mental evolution, rational selection employs purpose.

For example, mental evolution:

The mind employing rational selection designed new species, which adapt by making models. The concept of purpose comes into play in process of designing models. The design of models is a four-step process.

A model is the embodiment of a representation of a set of variables selected for some purpose.

The design of a model temporally emerges in reverse order to the rule above.

- Step one: The selection of a purpose. Suppose we select for our purpose the creation of a road map of the state of Texas. This decision must be the first one made. If we haven't a clear purpose, we could spend or waste our time making a map of any or all the states.

- Step two: The selection of a set of variables. Since our purpose is to travel from Place A to Place B, we need a road map, which will contain two variables - cities and roads. How many kinds of roads are there, and how many kinds do we need on our map? How many kinds of cities should we include on the map?
- Step Three: The selection of the representation. Next, we must choose how we will distinguish the different kinds of roads, and the different kinds of cities. What color, what size dots for cities, and what width for the roads.
- Step four: The selection of the embodiment. In step four, we must select how to embody the map. What size paper, what kind of paper, or what kind of ink will we use? Where will it be published, etc.?

The difference between natural and rational selection can be reduced to one purpose. None of the above decisions can be made without a clear purpose in mind. And while the above decisions for each step may not entirely be made in the above order, yet this order works.

The payoffs, however, for the emergence of mind are many, yet the most obvious one is time; how long does it take evolution to create some goal? Natural selection may take millions of years, yet a professional mapmaker, for example, may create a map in days or less.

However, the big difference, which is so obvious it almost embarrassing to mention is that biological evolution can only do biological evolution! While, on the other hand, mental evolution can in principle create any physical thing, wet or dry beings, such as: cars, swimming pools, radios, space ships, clothes, tomatoes, radishes, bowling balls, computers and iPods, etc.

What is a communication?

All the rats in this set of experiments chose essentially the same purpose/goal for exploring the maze. Their self-organizing minds chose to discover, model, and then run through the maze by the shortest path. The fact that they were able to achieve this goal is the fulfillment of an act of communication where the rat's NV mind must interrogate or probe the matrix maze and induce the maze to reveal its causal structure to the rat nonverbally. The model of the maze the nonverbal mind constructs is the successful completion of a nonverbal act of communication. It was the completion of a purposefully somatic motor act, but this communication involves ideas but not words or memes. The primary act of communication is an exchange of nonverbal knowledge between mind and matter, or mind and nature, or the cosmos.

Since the rat is a one-mind mammal (twelve-station model), it has no power to export a verbal mind account of its factual behavior. Only a human NV/V mammal (thirteen-station model) has the power to take this second step and export a verbal account of what it has done nonverbally, and what it has learned about the maze. But this nonverbal private communication is the first step in the evolutionary emergence of verbal behavior. All revealed religions are based on a verbal book. All are based on metaphors

I would like to ask a provocative question: does a verbal communication that involves both words and grammar, but has no meaning constitute a valid verbal communication? Is meaning really a trivial aspect of communication?

Does a human verbal act of communication where the words are legal expressions of all the communities' rules, but have no meaning qualify as a legal act of verbal communication? Do meaningless words, which conform to all the rules of grammar, qualify as a communication? My question concerns whether or not Pinker is correct. Are the essential ingredients only words and rules, or are there three essential ingredients: words, rules, and meaning?

In my opinion, if a verbal construction has no meaning, yet all the words conform to all the rules of grammar, it does not qualify as a valid verbal communication. A sentence composed of a word or words without meaning is a meaningless sentence.

Meaning of the rose revisited

The rose, B, an image-based unit of consciousness is created in Station 7 by the sixth sense of the nonverbal mind.

What is the meaning of the word rose? In order to understand its meaning, we must understand and distinguish the separate role of the events in the twelve-station action cycle, the role of each of the physical objects: light, neural impulses, sensations, and images. Once we understand the roles of each then we can ask, which of these are the exemplars of consciousness?

What is the meaning of the word rose? It's time now to start applying Aristotle's model to evolve a theory of meaning. Is the physical rose of Station 1 the meaning of the word rose, C? Where in the brain does experience occur, where does the experience of a perception of a rose occur? Since the classic linguists conceive of the verbal behavior as the only function of the verbal mind, they have no option but to assume that the meaning of the word rose, C, is the object rose, A, of Station 1.

There are not two minds in the human brain, but verbal language behavior is processed in two steps. The human 13- station mind has two functions that are each processed on one of the two sides of the corpus callosum. This is called a division of labor. In other words, each of the two sides of the brain executes a different function, yet the two jointly cause not only the nonverbal consciousness, but also the human verbal consciousness. The former, the nonverbal consciousness, is the main event showing that the verbal mind is no independent mind, which duplicates all the functions of the nonverbal mind, but rather the verbal mind's functions pick up where the nonverbal mind's functions end. The verbal mind does not start from the senses and create a completely new independent consciousness, but it supplements the image-based consciousness created by the nonverbal mind, which it cleverly dubs with memes. The verbal mind's role is, in effect, like converting a foreign language movie into another language. Each mind is playing a different but essential part of the verbal language function.

My primary task is to describe the role of the twelve stations of the nonverbal mind, which has served all the nonverbal mammals that ever lived.

Consciousness, meaning, and the primary metaphor or the ABC of mind

A	B	C
the physical rose	**the rose of consciousness**	**the word rose**

What are consciousness, meaning, and a primary metaphor? For a change, let's begin at a new place with the concept of the primary metaphor. I have been negligent for not getting around to this topic sooner. My task is to distinguish two kinds of metaphors: the verbal and the nonverbal.

The verbal- believers have had thousands of years to consolidate their beliefs in spirits, the "no things" which occupy that nothing place called eternity. Hopefully by now I have successfully convinced the reader that these things don't enter the mind. All that enters the senses are energies, but they enter each sense only as far as they were designed to function. The stimulus for the eye is light, which enters the eye only deep enough to excite the rods and cones; there it is converted into neural impulses, which only enter into the brain where they hit the Blood Brain Barrier. But the point is this. There is a profound difference between the quantitative physical properties of light and the qualitative mental property of sensations and perceptions or even more simply, mental images.

Metaphors

How are the two modules of the human mind bonded during verbal speech?

Human speech is the joint product of a two-stage communication function: the verbal and the nonverbal or we could describe it as the joint product of the 12/13-stations functions. Meaning is not just something piecemeal that accompanies each meme. The consciousness, which human's experience, starts life as a nonverbal narrative that the nonverbal mind creates and is all that the nonverbal mammals experience. But in humans, the script of the nonverbal mind is transferred to the verbal mind, which dubs the nonverbal codes into memes of a second language. But the nonverbal mind is the primary mind because the conscious content of the verbal narrative is the pure uncut sensory-image narrative created by nonverbal mind. The nonverbal mind furnishes the images used by the verbal mind essentially uncut. When the linguistic experts parse their human consciousness, they are unaware of the nonverbal mind and unaware of its contribution to verbal speech. The best they can do is to talk about the meaning of a rose as if it were within a one- mind context. The human one-mind theorists have no choice but to parse the product of the two modules of the human minds as if it were the product of one mind, as if A, the rose growing in some garden, were the meaning of C. Sadly, the fact is that within their context there is no such thing as B, the nonverbal narrative created by the nonverbal mind for the use of the nonverbal mammals because in the theorists' mind the nonverbal mind doesn't exist. What is technically wrong with their oversight of the role of the nonverbal mind? Simply, the image of the rose consists of an image painted in the sixth sense, which incorporates every sensation into the image that was triggered by the physical energies, which stimulate the eye. The one- mind theorist, in effect, must suppose that the sensations created by the PPA of Station 5 are objective physical things. So, what then stimulates the eye? Does light or hue stimulate the eye?

Heterostasis is a theory of the technology of the non-nonverbal mind, which creates meaning, rationality, and consciousness and generates the nonverbal narrative, which induces the organism to act.

The two-stages of communication in the verbal mind

Why can't the verbal mind function as a stand-alone mind? Generally, it is impossible for the verbal mind to construct a sentence until after it has something to say that has sensory meaning. The primary mind controls the nonverbal somatic motor system, while the secondary mind controls the verbal somatic motor system. Only the primary mind can initiate action, or "do something," and only then can the verbal modules talk about what was done. Only the secondary mind can cause the verbal motor centers to utter sentences, but the secondary mind cannot talk about anything it has done for it can't do anything. So the two minds complement each other perfectly, the secondary mind talks about what the primary mind has done, is doing, or will do.

Which comes first, having something to say or a fully formed sentence? The verbal mind's function can't be executed until the nonverbal mind has first executed its nonverbal meaning-generating function. In other words, it is impossible to understand the function of the verbal mind without first understanding the function of the nonverbal primary mind.

Is the meaning of a word rose, the object in Station 1, or the object in Station 7? To take a very simple example, assume there is a compound function, which requires two successive changes. In step one: A is changed to B. In step two: B is changed to C. The primary or nonverbal mind is the first operator, which transforms A to B. The secondary or verbal mind is the second operator, which transforms B to C.

With respect to the problem of meaning: the word-rose is C. The image-rose is B but the physical-rose is A.

The primary or nonverbal mind in the right hemisphere is the first operator that transduces the stimuli from the physical rose, object A, into the image-rose, object B. The secondary or verbal mind in the left hemisphere is the second operator that transduces the image-rose, object B, into the word-rose or the meme-rose, object C.

The physical rose is an object, a thing that has a physical nature or an invariant set of physical properties. The primary mind executes the impossible function, the programming function, or the god-function. The primary mind has a nature, which was designed to detect the pattern of features of the physical nature of each class of things in Station 1. Each thing has a different set of attributes of a physical nature and different sets of invariant properties. Each individual "Mr. Lincoln" rose lasts only at most a few days and is gone just as Plato (1952) observed and complained. So according to Plato, each Mr. Lincoln rose is not real, it is but an appearance a transient not an eternal thing, but that's where Plato went wrong. He failed to note that year after year each Mr. Lincoln Rose, a physical thing, is a copy of a physical exemplar of the same physical cookie cutter, of the same physical form or idea. The attributes of the Mr. Lincoln rose are physical invariants; they are that which doesn't change during change from year to year. Each species is a species only so long as it reproduces its kind, generally year after year.

At the heart of the primary mind is the Sixth Sense, a machine that has the power:

- First, to detect the presence of almost any set or syndrome of physical invariants within the input stream to the Sixth Sense

- Second, to create an exemplar of a class of perceptual images that embodies all the invariant physical features, which distinguish every class, including the Mr. Lincoln rose.

- Third, the words Mr. Lincoln Rose, object C, calls to mind an image of the unique features of this particular rose, object B. This image, B, is one of the meanings of the rose.

- Fourth, the real physical Mr. Lincoln rose, object A, has a set of physical quantities but no sensory qualities. This set of physical attributes of physical energies can excite many or all of the senses. The images, object B, are created in the Sixth Sense in Station 7.

- Fifth, for the nonverbal mammals the image, B, of the Mr. Lincoln rose is the one and only meaning of the physical rose, A.

- Sixth, for humans there are two meanings of B; as verbal mammals, B is the meaning of the word, C. But humans are also nonverbal mammals, so B emerged in the nonverbal mammals to give meaning to A; therefore, B, which is the meaning of A, is also the meaning of C.

- Seven, B is the meaning of A, and B is the meaning of C. A is the flower that grows on a Mr. Lincoln rose plant. B is the product of the Sixth sense of the nonverbal mind, while C is the product of the verbal mind. A is a physical rose, B is a nonverbal image of a rose, and C is a verbal mind meme.

- Eight, A is an objective public thing with a set of physical attributes with sufficient variety to enable the nonverbal mind to identify a set of mental attributes, B, to represent all the classes of physical objects that must be distinguished. The main point is that A consists of a set of physical energies.

- Nine, B is a subjective private set of mental symbols, sensations, and perceptions, consisting of a set of mental symbols, which seem to have been conceived so they can blend into a coherent pattern, which can unite into an essentially infinite set of possible images, which can represent any imaginable empirical class of objects.

- Ten, the set of B events are the ground events out of which all higher symbols can be drawn.

What always amazes me is that most people have no intuitive feel for the need for consciousness, to understand the critical role that a nonverbal consciousness plays in the design first, of the consciousness of the nonverbal mind and then in the design of the consciousness of the verbal mind.

Let me state the problem one more time, in a different way.

The physical world is populated with objects of various sizes that consist of matter. In this context, at least two classes of physical interaction must be distinguished: object level and energy level interactions. Car wrecks are object type, which are energy intensive; but mental interactions are extremely low-level energy interactions. If I look at something, each of my five senses may glean the different kinds of sensory energies the examined object emits. In a car wreck, everyone involved if they survive are aware of the interaction. However, most objects are never aware of having been viewed. Physical interactions often involve dangerous amounts of energy, but mental interactions involve an exchange of such low levels of energy that they generally go unnoticed.

However, the significance of an interaction cannot be measured in terms of the quantity of energy exchanged.

Physical objects when they interact with one another physically can release a great deal of energy, and it is possible that one or both objects are damaged. Physical interactions between objects, in other words, are one thing, but a mental interaction is very different!

When a mammal probes a domain, generally, no object interactions are occurring. When the mind of a rat integrates a matrix maze, the five senses of the rodent simply glean the stimuli, which the object being explored has shed, and the object being sampled is never aware of having been probed.

My version of Aristotle's twelve-station model was designed to explain what happens in mental exchanges. Mental exchanges begin with one or all of the five senses absorbing one of the various bands of energy to which the five senses are sensitive. For example, the light reflecting off the girl's face enters the eye, which starts one of the twelve stage action cycles. The eye doesn't transduce a girl, the senses do not sense objects; they sense some one of the so called stimuli, small bits of physical energy, in this case light, which the eye was designed to transduce reflected off the girls face.

The girl as an object of Station 1 does not enter the eye or any one of the other five senses. Objects are invisible to the senses, but each object reflects untold photons of light, which can activate a rod or cone within the retina of the eye. The point is that the mind cannot experience a whole object but starts the mental cycle by experiencing possibly millions or billions of physical stimuli from each object, and each stimulus is processed individually until the sixth sense in the association area begins putting all the individual sensations together into a perception of the whole object. The nonverbal mind assembles all these sensations, which are primary mental symbols out of which an image of a girl is constructed. B is a mental image in which the nonverbal mind has put the object girl, Humpty Dumpty, back together again. The image of an object is assembled in the sixth sense or the association area by assembling all the sensations from all the senses, which had their origin in that object in Station 1.

What is the difference between the object girl in Station 1 and the perception of the girl in Station 7? What does the input side, the first six stations, of the nonverbal mind do? The nonverbal mind creates consciousness and awareness of the events of the external milieu of the mind, EMm. The events of Station 1 are physical objects, while the events of Station 7 are mental events. What is the difference between a physical and a mental object? First, the mental objects are representations of the physical events, which exist only in consciousness; they are not physical events in the station one sense. Rather, they are images of the objects of the physical or material world; they are, in effect, a set of a special kind of representational event, which give the physical events they represent a presence within consciousness. Each image in consciousness represents some physical object so perfectly that the mammal in whose mind the consciousness is occurring believes that the representation in his consciousness is the object in station one, which it represents.

Images can be manipulated in the mind as if they were physical objects and arranged into a model much like a map, which is so real that the "I" of consciousness can make plans of action based upon the sensational qualities, which the nonverbal mind believes are physical-like features of the model.

For example, the set of four ideas which form the structure of the model of the maze; A2, B3, C2, D8 are created in the nonverbal mind, but these ideas are the seeds of rationality, which can sprout once they are imported into a verbal mind and converted into memes. There they can escape the private world of the consciousness of an individual mammal and when they become elements of each new wisdom, they can assume mental roles that would be impossible for the physical object itself; they can be transduced into stories, pictures, songs, operas, and movies, etc.

A physical object apple, which Plato (1952) would say is not real because it will spoil or be eaten, however, once it is an image created by the mind can become a topic of discussion in one of Pinker's books, which provokes me into a long discourse. Who knows how long the original physical apple, once the mind has processed it, will continue to exist in the mental world-long after the physical world apple ceased to exist. Maybe Plato was insightful. Perhaps mental world beings are more real or survive longer than their physical counterparts survive.

How many books on Lincoln are written each year? Which one was mortal, which one immortal: the man or the idea of the man?

The nonverbal mind does something for which we don't have a word. In religion, they have the word incarnation, which means that words become flesh. But one of the primary roles of the primary nonverbal mind does exactly the opposite; it turns matter into ideas or words. My word for this is inideation.

I am still in awe of the sensations created in the PPA. The primary mind in this station has created a set of symbols, which all have a common denominator. The mind can combine all the properties that were an attribute of a physical object within Station 1, and that Picasso himself could not have transformed into a single image, yet all the sensations created by the PPA can be combined into a single image in consciousness.

The word Apple, which is C, is a set of English symbols, which are an abstract meme and don't intrinsically have any meaning. Yet, once we learn to associate the image of an apple then the word "Apple" will recall or reactivate in consciousness another image-exemplar of the original image conceived by the mind perhaps long ago. How the mind keeps images alive, as it were, is impressive but is the images stored within the mind themselves mortal or immortal?

Where do nouns come from?

A noun is a name for a class. Symbols provide a unique attribute for each feature of a class.

B puts a face, an image, upon either some part or some whole object of the physical world, i.e. .nature, cosmos, or god. But a spirit god has no material nature that can induce the nonverbal mind to create an image. The Jewish god forbade images but most religions create them. It seems it is most difficult to get the common person to believe in any being without an image. One of the critical roles of the primary mind is to create a graven nonverbal image for every physical thing. One of the classic problems of the concept of spirit-gods is that they have no image. One of the laws of the nonverbal mind is to supply everything real with an image.

But again, for the classic language theorists, like Pinker (1999) who has no knowledge of the nonverbal mind and believes that verbal language is a one-mind function, there is no other option but to believe that A is the meaning of C. Never knowing that speech is a two-station function the classical theorists have never been able, and are not even now able, to attribute to the nonverbal mind the creation of the meaning. The primary role of the nonverbal consciousness is first to provide the only consciousness the nonverbal mammals can ever have. But the role of the nonverbal mind is simply to prepare the raw nonverbal narrative passed over to the verbal stations, where words are added so that members of the verbal community can share and discuss the meaning of the experiences they are sharing.

Two kinds of languages: image based and word based.

There are two kinds of languages: word-based and image-based. The language of the nonverbal mind is the image-based language. The secondary language is a word-based language employed by the verbal mind. But human verbal language behavior is a two-station function even though humans are only consciously aware of the secondary language function, the word-based component, or kind. Humans, it seems have always believed they had only one mind, and that they used only one kind of language, the word-based type, which explains why my most difficult task is to persuade humans that they have a nonverbal mind for no one can initially grasp the nature of anything nonverbal.

In the new paradigm of human nature, human language behavior is for the first time recognized to be the joint product of two communication modules, the twelve station nonverbal function, and the 13th station verbal function. Each type has its own unique language with its own symbols, much as computer's behavior is a joint function of both machine code and source code. The human NV/V mammals speak both kinds of languages at the same time. Human language behavior is a joint function of both the nonverbal mind's image-based language, and of course the verbal mind's word-based language. The nonverbal mind's image-based language constantly supplies image-based knowledge in consciousness to the nonverbal mind, which later in the human mammal supplies the meaning of the verbal mind's word-based behavior.

The novelty of the idea of the verbal behavior being the product of two stations may make the idea difficult to assimilate for those who have long believed otherwise. But next, we must seriously consider the quality of nonverbal images, which comprise the nonverbal consciousness the nonverbal mind, supplied the nonverbal mammals. We must proceed slowly at first and consider one carefully chosen nonverbal mammal at a time, but after considering a fair range of mammals, the nonverbal mind has the capacity to create a near perfect image based consciousness within every conceivable domain. The record will show that the nonverbal mind is a general or universal technology that can conceive a superior consciousness of any conceivable type, even a bloodhound, a horse, or a bat. If we carefully search the mammalian kingdom for mammals with a superior consciousness, I suspect we would conclude that evolution obviously provided the human verbal mind with a foundation in the nonverbal mind. A foundation that can provide the human consciousness with a range of possible nonverbal images, which will tax the verbal mind to find the words to describe the range of states the nonverbal mind can assume.

The nonverbal mind is no second-class mind. Evolution provided humans presumably with a superior communication skill, but in designing the nonverbal mammals, evolution designed several if not many kinds of species with sensory capacities that no human could easily reduce to memes. Evolution did not shortchange either kind of mind; most kinds of mammals were given sensory capacities that essentially were as perfect as is physically possible.

What capacity does the verbal mind have to impose order on verbal behavior: does the verbal mind have the capacity to make humans a perfect self-organizing system with respect to verbal language behavior? For example, does the verbal mind get its grammar from the nonverbal mind or does the verbal mind have its own independent self-organizing capacity for words. Where do humans get their grammar, syntax, and pragmatic skills, from the nonverbal or the verbal mind?

Knowing and the two -chambered mind

How can the nonverbal mind create a consciousness we can trust? The most basic mental nonverbal function is the knowing function. What does it mean to know? Actually, conceptually the essence of knowing is most succinctly expressed by the Boy Scout motto, "be prepared," and thus not be surprised. Operationally, for example, when F3 finally understands the A2, B3, C2, D8 hierarchical structure of the maze, which she was able to infer in the third row of the maze, she turns around and runs the rest of the way through the next unexplored nine rows without making an error. How could she accomplish this "magic?" No magic was required; to know means to be prepared, to know means that there are no surprises waiting up ahead; all the anomalies have been digested by the mind. Knowledge inoculates the mind for some closed domain.

The nonverbal knowledge or narrative created by the nonverbal mind is not given to the individual user in book or in word, but each mammal must work through the hierarchical structure and grow or evolve a consciousness that can create all the knowledge it requires as image based knowledge. If each nonverbal mammal could magically obtain verbal technology, the image-based knowledge conceived in its nonverbal mind is ready to be transferred across the corpus callosum, as Sperry (1961) showed, to the verbal secondary mind where the nonverbal knowledge contained in the image-based consciousness could then be converted into memes. As each unit of this nonverbal image based consciousness acquires in the human compound mind system its meme counterpart, the image-based component of verbal behavior is acquiring an association, which makes it the meaning of the particular word, which is the meme.

Perhaps I am wrong but I conceive of words having two components, the image component, which is the meaning of its companion word and the word, which is its meme. But the meaning of words is a function of the nonverbal mind, with the primary mind, creating the primary image based meaning component, which is displayed as nonverbal images in consciousness, these images certainly are not memes. The verbal, or secondary mind function, furnishes the conventional sounds associated with alphabetic symbols, which are the means by which the verbal community's names are associated with each meme.

The verbal mind at work in a scientific mind can cause, of course, the nonverbal mind to expand its nonverbal awareness of presumably any object of the physical world. For example, the verbal mind is full of textbooks that force the mind to look more closely and carefully in order to see features, which normally would never be seen.

Petersons for example has a series of small pocket sized books that teach where to look to find the features, which are the critical defining feature for each species for all kinds of plants and animals. Science must train the eye to look where it is most important. I remember on my first trip to Florida years ago, trying to identify a water bird, which was standing in the water, and my guide pointed out to me that I was wasting my time; the critical features were the feet, which were under water at the time.

Female Three, F3, primary communication

Female Three was a master at the skill of primary communication:

First, by emitting somatic behaviors which rationally and physically interact with the nonverbal physical world of the Matrix Maze, which is the object of her nonverbal communication.

Second, in creating a working consciousness that allows the "I" of consciousness to take causal control of the salient objects of the domain.

In less than thirty minutes, F3's nonverbal mind created a valid model, which tells her what is real and where she is within the whole from which she can extrapolate the specific nonverbal knowledge, which enabled her mastery of the hierarchically organized, structured maze, and enabled her to move easily without error through the maze to achieve her goal.

F3 explored the matrix maze and evolved a valid model from which she could then extract the precise kind of nonverbal knowledge for extrapolating and then running the maze without errors: you learn a spatial form of rational behavior. Behavior, which continued to evolve for millions of years before evolution, was able to evolve in the Broca's region of the human motor cortex, a technology to generate the special kind of complex verbal sounds required for human speech.

F3 has the extraordinary capacity to self-organize and to use the knowledge her nonverbal mind evolves as she explores/runs the maze, but F3 was a one-wisdom species and had no capacity to share her knowledge of the maze with any other nonverbal mammal. All of her nonverbal behavior is accomplished without a verbal mind!

The nonverbal mind communicates nonverbally with the objects of the maze. The nature of this communication can be observed in the behavior of the rat seen in the display created by the Graph 53 program. Both the nonverbal mind and the nonverbal objects communicate by means of an exchange of knowledge via the causal nature of the physical world. The Sixth Sense of the nonverbal mind knows how to extract the causal nature of the physical world, which the mind requires to model and understand/predict/expect the spatial dimensions of the local domain.

Image-based consciousness

The nonverbal mind supplies its agent of consciousness knowledge about what is happening within its external milieu of the mind, EMm. By creating an image-based consciousness, a nonverbal or somewhat silent movie, which the agent of consciousness believes is not a product of its mind but rather it experiences the image-based objects as real and it, the agent, at the center of its IMm is in control of the events and everything that is happening according to the agent's wishes.

The image-based objects in consciousness are exemplars of real objects in station one. The meaning of the image of a girl is the physical object "girl" in Station 1.

The nonverbal images created by the nonverbal mind in consciousness are responsible for perhaps ninety-five percent of the content of the so-called verbal consciousness. However, it is my opinion that the verbal mind only supplies a meme translation of the nonverbal image categories.

Nonverbal meaning

My primary concern as a rat runner is not with the verbal behavior created by the verbal mind but rather with how the nonverbal mind creates the nonverbal image-based meanings. In order to understand this chain of events, one must understand how mental evolution functions in all Twelve Stations of Aristotle's model: how the objects of Station 1 induce neural impulses which progress into Station 2 and then through all Twelve Stations of the nonverbal mind.

In philosophical terms, my primary concern is especially with Kant's (1952) distinction between the concepts and precepts. The classical problem was the origin of ideas with Kant and all the continental philosophers insisting the concepts were the gift of god. My self-defined goal is to challenge their Nativism and show that the nonverbal mind cannot just create simple concepts but can create hierarchically organized sets of ideas. My purpose in presenting the Aristotle Twelve Station model is to show how the nonverbal mind creates sets of concepts; how the Sixth Sense is the station in the mind, which does what Kant and all his colleagues said, was impossible.

Kant's error: the origin of order

Kant (1952) was the first to discuss verbal behavior as a product of two different functions: one-part rational and one-part empirical. The rational element, the concept, was the form, which he took most literally, while the empirical element was the mud. Kant drew upon a brick maker's model, where the A component or the mold was literally a mold, a wooden box, into which the mud, the B component or empirical contribution was packed and left in the sun to dry.

Kant's (1952) contribution to our understanding of the nature of experience, then was his distinction between concepts, the A component, and precepts, the B component. Kant claimed the former were gifts of god, as had Plato, and which I refer to as the "the god function." With respect to concepts, Kant was a nativist--along with Plato and Chomsky! However, he was the first to recognize that neural impulses, which he considered precepts or mud, played an empirical role in the perceptual process.

Where Kant's mind went aground was in not being able to conceive an empirical role for the neural impulse to play in the creation of the forms or concepts.

My theory says that both the concepts and the precepts play a role in the perceptual process but two different roles.

With regard to the evolution of concepts, the nonverbal mind, specifically the sixth sense, conceives concepts, which recognize all the critical features of each object's sensory attributes and stores that pattern. So in the future, any syndrome of attributes can be examined and tested; any object that can match the original conceptual pattern is then recognized as an exemplar of that class. Each concept is a frozen conceptual context, which can be objectively used to determine whether any particular pattern of mud qualifies as a member of a class. But let me finally state this criticism as clearly as possible. Kant (1952) and historically all the other philosophers have insisted that the forms or concept were nativistic, that only god could conceive a form.

But my experiments showed the primary role of the nonverbal mind is to transform anomalies into forms or ideas. I keep taking pleasure is stating that even the rat's nonverbal mind excels at the executing of Plato's/Kant's "god-function."

But additionally, Kant did not recognize that concepts were hierarchically organized. After a bit of mud passes the test of level one, it will then be promoted to the next grade where it can then become a member of a two-dimensional vector; next, it must pass the test applied at the next level. And so the evolutionary growth can continue.

The maze is a four-dimensional hierarchical structure, and so the mud at each of the four levels must be given the next idea of the second level of a four-stage nested set of meanings. The question is how the events within the stages within the cortex furnish a context within which these successive sets of syndromes of neural precepts acquire a nested or hierarchically organized meaning.

This trick is achieved by the arousal system of the vertical brain. As a third generation pragmatist, my question is; how the meaning of nonverbal hierarchically organized concepts evolves within the primary nonverbal mind in order to create a nonverbal life space derived from empirical reality. My questions are first how the forms evolve, and second, how finally the neural impulses, the mud that has a four-level meaning forms concepts and precepts. How does the nonverbal mind impose rational nonverbal meaning upon the patterns of nested neural impulses entering the brain? How does the nonverbal mind execute the god-function; how does the nonverbal mind create a nested set of nonverbal ideas?

The role of the primary nonverbal mind is that of a meaning-creating machine; a self-organizing meaning generating machine that invents all the domain-specific classes or categories required to create a nonverbal narrative that provides the nonverbal images needed by organisms to achieve their nonverbal domain-specific ends.

The challenge is to understand how one-level meanings first emerge in the first grade, and then how they evolve or grow in the second grade until they are organized as a part of an N-dimensional hierarchically organized space. For example, how does each of the 100 units of the matrix maze acquire a separate or unique vector identity? What I had to recognize and now explain is how the meaning of each unit in this four dimensional space acquired a unique vector identity.

One of my big problems as a rat runner was to understand how the nonverbal mind of the rat creates a model of the maze simply by emitting a rational sequence of hierarchically organized somatic moves. Then to understand how this results in sets of nonverbal vectored spatial meanings emerging within the mind; one that is organized to function in a hierarchically structured space. What is the meaning of each particular unit of the 100-unit maze? What I had to recognize and explain is how the meaning of each unit in this case evolved a complex identity within a set of 100 nested units within a four dimensional vector in the nonverbal mind.

The functions of the primary mind of F3 can be described several ways:

- First, it is a rational speciation machine, a mental evolutionary machine, which must give meaning to every salient element or object of a domain. How she was able to identify it as an exemplar of a set of four nested classes, A2, B3, C2, D8, where A is direction with two levels, B is replication with three levels, C is row type with two levels, and D is eight units per row. Concepts are not transcendental gifts of god but are the products of the nonverbal mind.
- Second, each concept is a piece of a rational puzzle; each conceptual element must be conceived as an essential element of a coherent domain-specific set of elements; a closed conceptual system, which takes the uncertainty out of the domain so that the action of the organism can be planned within the context of a closed probability structure.
- Third, the primary mind is the first self-organizing rational machine. Plato imagined humans as being the first rational beings, yet he was never able to believe that the human mind was a self-organizing system; Plato (1952) assumed that the core conceptual system, the core ideals of the mind, were nativistic, the gift of a transcendental being. But the radical concept/paradigm of this book is that even a sixty-day-old female rat already has the nonverbal neuronal structure, which executes the "god-function" in its nonverbal mind up and running.

Verbal Meaning

Humans are not born talking; talking is a skill that is mastered in many stages, but an interesting fact that is too often left unsaid, is that humans learn to talk while they are nonverbal beings.

The cosmos was also born not talking. The physicists say that in the beginning the cosmos was a great fireball of pure raw energy. But slowly, very slowly, over billions of years, the cosmos evolved complexity and finally the no-mind forms or species of life appeared, and then one- mind mammals emerged, but still the cosmos was speechless. But silently, unbeknownst to today's science, the foundation for verbal speech had been evolving in the one-mind mammals for perhaps fifty million years. Specifically the nonverbal images, which had been created by the nonverbal mind and then assembled into nonverbal consciousness became for the NV/V mind mammal, the content of their verbal communication. When humans talk about the rose, the meaning of the word rose is the image created by the nonverbal mind during the one-mind mammal's primary communication with the objects of its nonverbal world.

Humans are carried away with their verbal skills. But the ability to talk and having something to say are two different skills. The nonverbal mind creates everything real empirically that can be said; the verbal mind only provides the ability to talk, the power to export the nonverbal narrative conceived by the NV mind. The primary acts of communication, driven by the nonverbal mind, create in its action cycles a nonverbal consciousness expressed in images of everything real that can be said with empirically driven words. All empirically derived words have empirical meaning expressed in the images created by the nonverbal mind.

Verbal metaphors are not real words they are words without sensory meaning. They could well be called false words, or lies, or myths, for they pose as real words with real images created by the nonverbal mind.

One point, which Sperry (1971) did make while discussing the relation between the two minds, was that the nonverbal mind was responsible for the first stage of verbal communication, the act of recognition. Sperry went on to explain that in the second stage whatever form the knowledge was in that came across the corpus callosum to the verbal mind was then transformed into memes.

I assume that the content of the communication of the nonverbal to the verbal mind is that the nonverbal mind simply supplies images originally created for display within the nonverbal consciousness where the "I" of consciousness reads them.

In the beginning, the cosmos was nonverbal. Speech did not emerge in the cosmos until the human mammals emerged, a process that required roughly over twelve billion years.

Verbal behavior evolves in two big steps; first, in the primary step, evolution designed the nonverbal mind, which confers upon life the almost magical skill, the power to, in effect, nonverbally communicate with the nonverbal nature, the cosmos, or the pantheistic god. The nonverbal mind communicates with nonverbal physical objects, such as the maze, in order to induce the structure of the maze to reveal its hierarchically organized structure. By means of physical communication, or Rolandic induced actions, the nonverbal mind probes the maze with nonverbal behavior and the feedback allows the nonverbal mind to create the conscious knowledge or the image-driven knowledge required to create a working conscious model of the life space in question, which enables the mind to achieve its domain specific goal.

From the nonverbal mammals to the emergence of the compound 13 station mind, verbal communication was still millions of years in the future, but right here and now, the rat's amazing rational nonverbal mind causes nonverbal rational things to happen, which are the topics of this book.

The birth of Consciousness and the Meaning Machine

There are many stages of nonverbal communication:

- First, the primary kind of communication is the nonverbal dialog between the "I," or the agent of consciousness of the nonverbal mind and the physical parameters of the maze. This nonverbal communication is how the nonverbal mind forces nature to reveal the parameters of the physical maze, which enables the nonverbal mind to create the knowledge, required for it to achieve its goal with a minimum of errors. Female Three illustrates how the nonverbal mind interrogates nature as she explores the maze, and creates the hierarchically organized knowledge required to create a valid model of the maze, which allows her mind to extrapolate, to create a plan to move through the maze without errors.
- Second, as magical as is the communication function of somatic behavior interacting with the maze, the crowning achievement of the nonverbal mind is consciousness or the design of the meaning machine, where the five senses of the nonverbal mind collects all the sensory data. Then the five senses deliver this total set or ocean of neural impulses into the Sixth Sense, which organizes it into categories, classes, molds, and forms, which enable the nonverbal mind to conceive an overview model or conceptual structures. These conceptual structures allow the nonverbal mind to conceive rational plans of behavior, which allow the organism to conceive the means to its end.
- Third, what is so great an achievement is the nonverbal mind's capacity to integrate all this knowledge or all these images in real time? The nonverbal mind then displays within consciousness a comprehensive overview of the objects and events that are happening within the terrestrial space of the maze, second by second for a lifetime.
- However, there is more. Then the nonverbal mind places the perception of the self as an agent in the center of consciousness. The nonverbal mind creates a model, a life space in consciousness so real that each individual agent of consciousness feels at the center of its world and can feel their somatic behavior causing the events that are a part of its plan to achieve some goal.
- The nonverbal image-based consciousness is how the nonverbal mind publishes the results of its real-time experiments. The nonverbal consciousness gives each mammal not only an active awareness of what is happening in the real material cosmic world, natures' world, but, and this is the revolutionary part: it enables each self-organizing system, as it were, to get in the driver's seat and take charge, to act and immediately experience the cosmos's response. Nonverbal communication, i.e. Heterostasis, is a form of communication, which is, in effect, a dynamic on-line real-time prayer. The nonverbal mind by acting in specific ways can ask any or every domain-specific response-specific question possible and in effect immediately receive nature's response, and thus each

nonverbal mammal can elicit any specific knowledge the individual requires to achieve any domain appropriate goal. "Ask (The nonverbal god any appropriate nonverbal question) and you shall receive." Ask, and instantly, nature responds by doing nothing if the actions were irrelevant or inappropriate. But when the nonverbal mind gets it right, as can be seen in the behavior of F3, then, everything happens the way a nonverbal mind was designed by evolution to expect. And so the nonverbal animal's science, or the formal human science, both employing experimental methodology get the same results; nature never fails to respond appropriately to the salient nonverbal inquiries of the nonverbal mind.

- Any species that employs experimental and control conditions can induce nature to reveal her laws. So, the generic nonverbal mammalian mind was endowed with the knowledge of how to induce nature to reveal the domain-specific knowledge it requires for mastery of the maze. First, the nonverbal mind creates a valid image-based model of the hierarchical structure of the target domain, then the test phase begins and F3 created a plan of how to complete her self-chosen goal. This self-organizing female rat, in executing her self-organized plan, is able to move through almost two-thirds of the maze, which she had never explored and pass through all those unexplored parts of the maze without error. The role of consciousness is to give self-organizing individuals immediate feedback on the validity of the model, the knowledge-organizing structure it created in the first stage, the exploratory stage.
- Seeing is not a passive function where reality determines the content of consciousness, but is a dynamic function that evolves or changes each time F3, for example replaces an anomaly with nonverbal knowledge, which the Sixth Sense should immediately pick up and use. The knowledge is used to expand her awareness of the hierarchical structure of the maze, which in turn changes her behavior. Now she no longer stops at the place of the old anomaly but runs forward in a new exploratory loop; past that place where, just moments before, she had been blocked from passage by the uncertainty or ignorance induced by the now late anomaly.
- What we see, and later what we talk about, is only a matter of describing the original nonverbal sensory/perceptual narrative, the images displayed in the nonverbal consciousness, which the primary nonverbal mind created to communicate with the agent of consciousness about the causal structure of the physical events in the nonverbal physical world.
- Words cannot describe the external physical world. The verbal mind can only talk employing two kinds of words: words with meaning or words that are metaphors. Words with meanings

enable the verbal mind to talk knowingly about the empirical world. But these words with meaning are not the product of the verbal mind, but are only meme translations of the nonverbal narratives exported from the nonverbal mind. Real talk employs the nonverbal images created within the nonverbal mind for the nonverbal consciousness by the meaning machine as it directs the nonverbal somatic behavior to probe the cosmic objects and force these real objects to reveal their causal structure, their nature, by interacting with each appropriately.

- Words that are metaphors never exist in the nonverbal world for several reasons. First words don't exist in the nonverbal world. But, suppose for a moment that they did, but then, they would never get by the sixth sense, which is the certifier of reality. The nonverbal sixth sense only includes within the waking consciousness objects that are real and are appropriate with which to communicate or to act. The nonverbal sixth sense only creates exemplars of concepts that that have empirical attributes. The nonverbal consciousness is a product of the nonverbal mind. Nonverbal concepts only have empirical attributes. A metaphor is a word whose attributes are not empirical. By definition, metaphors have no empirical meaning. For example, Santa Claus has no empirical meaning anymore than a spirit god has empirical meaning. Plato's love of god was entirely because his spirit god had no empirical meaning. Spirits by definition have no physical natures, no physical beings. Their purity is the fact that they are not contaminated by matter by their being free or outside of space and time, by being eternal and thus not having to suffer the temporary fate of matter. Plato (1952) hated matter because in his biased mind it was not real; it suffered change. For Plato real things, eternal things, don't change, which can't happen in anything composed of matter.

Chapter Three: Where did the parts out of which the nonverbal mind evolved come from?

What was the nonverbal mind made of? I propose that evolution designed the nonverbal mind out of three structures that were in place inside the skull of the pre-mammals; redefining these structures; modifying their functions, and thus creating a revolutionary organ, the nonverbal mind.

Historically theorists of various kinds, philosophers, psychologists, and physiologists, have suggested that the mind and the brain were two different things and that the mind was located somewhere within the brain. Descartes (1946) is most famous for locating the mind within the pineal gland, while Sperry (1961) located his two minds in the cortical hemispheres. However, the one localization fact that I take seriously is that the evolution of the mind and the evolution of mammals occurred in parallel. Or in other words, only mammals have a cortex and only mammals have a mind, or only species with cortexes have a mind. While the brain has been evolving for hundreds of millions of years, evolution did not begin the revolutionary transformation of the particular neuronal systems that transformed the brain into a mind until nature was forced to design a new family of animals, mammals. My position is that the brain was transformed into a mind with the evolution of the cortex, and that the cortex/mind was constructed of at least three parts or building blocks: a horizontal, vertical, and an emotional or chemical component.

The brain is a large neuronal system: the question is how the neurons of the brain were finally organized into the new revolutionary functional unit called the mind. In this chapter, I am proposing that one of the three large subsystems out of which the mind evolved was the sleep-waking system.

Sperry's (1961) idea that each hemisphere of the cortex is an independent mind is an error based in part upon the belief that the two cortical hemispheres are whole systems capable of controlling themselves. My thesis is that the mind had three parents whose descendants still actively define the structure and function of the mind. The functions of the mind are a product of the interaction of all three structures. The big questions are what these three structures are; second, where are these structures?

First, the horizontal concept of mind

The cortex, the outer layer of the brain, which covers the inner or core layer of the brain has long caught the imagination of theorists. While this horizontal layer of the brain consists of billions of neurons in humans, yet in rats the cortex or bark is only a thin layer covering the brain.

FIGURE0.1B

In Chapter 1, with Figure 0.1B, I began to describe and explain the systems of both the nonverbal and verbal minds as if all the functions were taking place horizontally, and from that two-dimensional perspective, the nonverbal mind system breaks down into twelve functions each occurring on one of the twelve-stations of the assembly line of Aristotle's model. However, the verbal human mind breaks down into thirteen functions each occurring in one of the thirteen-stations of my revised Aristotelian model.

In this section, I attempt to identify other preexisting components: the arousal circuits which evolution most cleverly incorporated into the nonverbal mind. I will describe several new roles that evolution created for these structures and functions that were originally simply arousal functions.

In the 1950's, the physiologists discovered the arousal circuit, the Ascending Reticular Activating System, the ARAS. This arousal circuit originates within the deep brain stem centered in a large set of Reticular cells. Reticular means that a collection of neurons is more or less a massive random network of cells, which then rise up through the brain vertically to induce sleep and waking rhythms within the neuronal circuits of the cortex. These reticular centers are the primary pacemaker of the mind.

This chapter is devoted to describing the sleep-waking arousal mechanism; one of the ancient mechanisms out of which the mind evolved that has functioned for millions of years as an arousal center.

Chapter 2 was devoted to describing the structure, function, and origin of the horizontal part of the mind. The biggest difficulty many humans have in understanding the structure and function of the mind is that they focus their attention exclusively upon the cortex; believing the mind is a function that takes place entirely within the cortex.

Pavlov (1960) was one of the first that made this error of thinking. Pavlov believed the physiology of the mind was exclusively a Transcortical function. Pavlov was just born too soon: in spite of his trying to cram the whole mind into the cortex, Gastaut (1957) showed, many years later after the vertical arousal circuits had been identified, that Pavlov's theory made several glorious errors. Gastaut showed that all the physiological forces, which Pavlov believed had their origin in the cortex, on the contrary could all be explained equally well by the phenomena that were uniquely a function of the arousal circuits.

Second: the Vertical component of mind

In this chapter, however, I will focus upon one of the original structures and functions out of which the vertical component of the mind evolved. Specifically, this vertical structure was designed and originally functioned as the arousal system, which was a core structure of the brain. The arousal system of the mind originally evolved to regulate the sleep and waking behavior of biological life. One of the most intriguing component elements of my theory concerns the way this arousal structure was reconceived or redesigned to become the pacemaker for the nonverbal mind as a whole.

FIGURE 0.2A

The best single structural description of the vertical arousal system is FIGURE 0.2A, which I call the Candelabra model.

This FIGURE makes the point that the mind is not just a horizontal structure, which creates human verbal behavior, but also has a vertical dimension, which first emerged as a Sleep-Waking, or arousal system, which created and maintained the Circadian rhythm cycle that is about 24 hours long. This arousal function in different species is partitioned into two phases, a sleep phase, and a waking phase, where the length of each stage varies with the species. In humans, a common division would be (8 +16 =24).

The primary function of the original structure was to move each mammal through its daily sleep-waking cycle. First, each mind must provide a proper time and place for several hours or several cycles of sleep each day. The planet earth had a roughly 24-hour daily orbit long before there was animal life, and this base rhythm continues yet today. My chief responsibility as a rat runner is to describe how a nonverbal mind manages the hours of the waking portion of the day.

The vertical mind evolved to supervise or implement a circadian rhythm that once the sleep cycle has received its allotted share of time each day, the vertical mind then turns its attention to its primary waking function.

The Origin and role of the two-phase human mind

The human mind must be thought of as a great factory, a great assembly line. Sperry's concept of the two minds recognized only two stations of the assembly line.

For example, according to Sperry (1961), in phase one the nonverbal mind in the right hemisphere nonverbally recognizes an object; then in phase two this knowledge is transferred across the corpus callosum into the left hemisphere's verbal mind where it is then transformed into memes.

Sperry seems to imply that a mind can be housed and functioned out of the cortex or actually out of one-half of the cortex, out of one hemisphere. Since there are two hemispheres, there are two minds. Alternatively, the brain is a great industrial building with a duplex in the cortex, which leaves us with the question; what does the rest of the brain do? Again, the brain is an iceberg with the two minds at the tip.

The idea that the mind is a part of the brain or is different from the brain creates more problems than solutions. I begin with the assumption that there is one mind in the whole brain. This assumption opens the door through which I want to travel in this chapter and begin to discuss what role some of the other parts play. The idea that the mind or minds are in the cortex has been the fashion for two long. It started with Pavlov (1960) and ended with Gastaut (1957), who I will review in this chapter.

In this chapter, I will propose a whole-brain theory of mind. In particular, I plan to put on the map of the mind, the sub cortical structures, and their functions in the nonverbal mammalian mind and thus extend the classical concept or the role of the mind. Humans have too long attributed all their rational behavior to the horizontal verbal mind and conversely have falsely denied that the nonverbal mammals had any rational abilities.

One big break with classical theory is my claim that rats have a nonverbal rational mind. The classical theory of mind demonstrates that the classical theorists had no concept or certainly, no valid concept of nonverbal rationality, which is the idea that mind, evolved and in the first stage emerged as a nonverbal mind for all prehuman mammals. The prehuman nonverbal rationality is the first to evolve, and as the nonverbal form of rationality, is the major kind of rationality, and the concept, which I must define and defend in this chapter.

The fact that the generic nonverbal mammalian mind is a nonverbal one has enabled human hubris to be blind to the fact that this mind has served the nonverbal mammals well for more than fifty million years. This nonverbal mind, however, is the first stage of our human two-stage mind and is not to be despised or ignored.

The origin of the mind

Some of the radical things I have discovered by studying the rat's behavior are:

- That the human verbal mind is a two-stage mind;
- That the rat's nonverbal rational mind, the generic nonverbal mammalian mind, GNMM, is the first phase of the human two-stage mind;
- That the GNMM had nothing to do with verbal speech or verbal rationality;
- First, the nonverbal mind evolved, in part, out of the horizontal structures of the brain.
- Second, the nonverbal mind evolved, in part, out of the vertical sleep-waking circuits;
- And finally, the human brain is home to only one mind, which fills the whole brain; the brain is not an apartment house for minds, there is not a mind in each hemisphere as Sperry claimed.

The iceberg metaphor

Historically the mind has always been conceived as an organ of the cortex, but then Sperry (1961) told us there were not one, but two minds in the cortex, one in each hemisphere. Sperry, however, was totally overtaken by the iceberg illusion; both of his minds occupied, as it were, only the tip of an iceberg.

My task is to redefine the structure and function of the mind, especially to describe the submerged portions of the iceberg. In chapter two, the focus was upon those functions within the horizontal or tip of the iceberg. In this chapter, the focus is upon the vertical control circuits within the vertical core of the mind, the submerged part of the iceberg.

My definition of mind

"Some mythmakers sought to sell the idea that human beings had neither the whole and unified brain described by Descartes, nor the half brain of Broca and Wernicke, but rather two brains, each devoted to its own tasks and operating essentially independently of the other." (Levy, 1985, P. 43-44)

Sperry's (1961) concept of mind was derived from the historical human fascination for verbal behavior. Thus, the two-minds were a one-function organ, an organ for talk.

Our pride and conceit has too long allowed us to not only ignore but also deny our animal ancestors, however, our understanding of the human genome is demonstrating that our genetic nature makes our body one with even the first and oldest form of life. However, with respect to the mind, our genome has not yet revealed exactly who our mental ancestors were. Yet it is already clear, that any species with the possible exception of the chimp that tries to muscle in on our mental verbal ancestry is not going to be welcome.

My thesis is that all mammals have the nonverbal first-phase mind; including, in particular, my ancestor and research associate, the rat.

In this chapter, I will discuss the physiological facts that the horizontal region of the brain is not suspended in midair and is not a spirit that can only function in the airlessness of eternity. How do spirits function?

I do not know how spirits function, but I know a bit about how the cortex functions, and that is the topic of this chapter. Where does the nonverbal mind come from? What parts of the brain came together to form the mind?

The first radical transformation in my thinking about the human mind was forced upon me when I realized that the nonverbal mammals all share the first stage mind, the nonverbal mind. It took me a while to catch the contradiction: they have a mind, yet they can't talk! Even Plato and Chomsky ought to find that fact inconvenient. So now, we must start over. For thousands of years we have been fed the idea, the myth, that our mind gave us our verbal skills! What does a mind do that isn't verbal? Ok, it obviously does nonverbal things! But what are nonverbal functions? It is very simple, watch any nonverbal animal, and everything they do is nonverbal! But you may soon realize that the nonverbal mammals do almost everything we humans do except talk. So what did the verbal mind add to our behavioral repertoire, nothing but talking.

Nonverbal mammals have the GNMM, and both of their hemispheres are nonverbal! The nonverbal mammals are nonverbal not for lack of something to say, but because the nonverbal functions of the nonverbal mind, evolved long ago before evolution designed the verbal motor centers, i.e. Broca's area.

Calling the human right hemisphere the nonverbal mind is not useful, for that name should be reserved for a brain with two hemispheres that both have no verbal capacity and have no Broca's area. What we need to know is how the human right hemisphere differs from the right (and left) hemisphere of rats; how, when, and why the human mind evolved the thirteenth station of the assembly line. This whole topic is confounded by the fact that the brain has plasticity, which means that functions, which are located in one hemisphere, can be implemented at different places because of injury or surgery, etc.

In a most profound sense nothing has changed; the nonverbal mind first emerged as a product of two brain components, one of which had probably been designed to regulate the sleep-waking cycle but now was recreated as it assumed its role in the newest creation by evolution, the nonverbal mind. The first great change was the design of the nonverbal mind.

Then the next, the second great change, was the design of the verbal mind.

First, what obviously changed in this second step was the emergence of the human species. What we don't know is how the thirteenth assembly line station emerged and what other changes were made at the same time in the design of the verbal mind. Given the plasticity of the human mind, there may be no way to know how much rearrangement was made in the brain as a whole, or how much was made just in the cortex in order to accommodate the evolution of Broca's area. What we need to know is how the two hemispheres came to differ; again, plasticity was the wild card.

My chief defense for my theory that there is only one mind in the human brain is that the singular vertical arousal pathways are shared by both hemispheres.

Sperry was rightly concerned with making as much out of his discovery as possible, but he overlooked the fact that both hemispheres have to function in a coordinated fashion in order for a human to talk. **The human brain is home to only one complete mind, the only one with a full thirteen-station assembly line**. This thirteen-station mind, however, still functions with the support of only one vertical component. In order for this thirteen station mind to work with all thirteen stations, twelve parts nonverbal and one part verbal, all these 13 assembly line components must be aroused in a coordinated fashion by this one unified set of vertical arousal pathways. There are only one set of these arousal pathways to go around.

The vertical component of the mind serves as a very fancy clutch, which enables mammals to shift gears rapidly as we change tasks at various time scales many times a day.

Extending Bernard: two sets of milieus: one for the body and one for the mind.

The set of Homeostatic Milieus

Claude Bernard (1957) introduced into physiology the concept of two milieus, the internal, and the external milieus of the body, IMb, EMb. Bernard's concepts became an integral component of classic Homeostatic Physiology. For Bernard the IMb was the body water, which is the general internal environment in which all the body cells, including the brain cells, live and which have to be regulated.

The set of Heterostatic Milieus

In this section, I discuss the origin of the mind by extending Bernard's (1957) concept of the Internal Milieu of the body, IMb, into a comparable distinction for the mind. Just as the body has an internal and external milieu, IMb, and EMb, just so the mind has an internal and external milieu of the mind, IMm and EMm. The difference is that while Homeostasis regulates the IMb, Heterostasis controls/regulates the IMm.

The body has many Essential variables, EVs, which must be regulated and each has its own set point and regulator. The body temperature is one EV of the body and has a set point of 98.6 degrees Fahrenheit.

By contrast, there is only one essential variable of the nonverbal mind, which is critical in regulating the IMm, which is a measure of the arousal level as discussed in the Yerkes Dodson Law (2004), Y-D law. The Y-D law described below represents the momentary level or quantity of ignorance in the EMm.

The Electrophysiological dependent variable: The S-D Function

The arousal level, I must admit is a most complex concept, for there are too many ways it can be measured. In this chapter, I will explain the concept of the arousal in terms of the EEG. The vertical mind is a complex pacemaker, which regulates the EEG's, which can be described in terms of frequency and amplitude. The EEG reflects or reveals the arousal level, which moves upon the Synchronization-Desynchronization level, S-D.

Technically, the IMm, on the one hand, has only one essential variable, the set or arousal level of the mind as discussed in the Yerkes-Dodson law (2004). Technically, this is correct, but on the other hand, this position doesn't tell the full story. The arousal level of the IMm is a function of errors in the EMm of every model that is functional at any time.

The behavioral dependent variables

In book two, the experimental section of this book, one of my first responsibilities was to create and describe the set of three unique dependent variables that I invented for measuring the Heterostatic behavior of the NV mind.

There are three of these special variables. However, one of these dependent variable scores is more revealing of what going on within the mind, the BFE (Between First Entry). The BFE typically measures retracing. Moreover, the strength of every anomaly can be measured several ways: First, by the amount of retracing as measured by the BFE score; or Second, by the Arousal level which various autonomic functions could measure, for example the heart rate or some EEG variable. A full discussion of these new variables will be found in book two.

The Error in the internal milieu of the mind

In developing the concept of the third wisdom, the heterostatic mechanism, there are elements of the Homeostatic function that can be preserved, yet on the other hand, there are several critical changes.

It could be said that the primary task of the nonverbal mind is to control the Essential Variable, EV, of the Internal Milieu of the mind, IMm. Also, it could be said that's its task is to control the External Milieu of the mind, EMm, but the truth is that the control is in both directions. The control of the IMm is the means to the control of the EMm and vice versa.

FIGURE0.2B the Yerkes-Dodson law, the Y-D law

The emergence of the nonverbal mind in mammals puts on the face of the earth, a new family of beings, mammals, which were the first self-organizing systems. Instead of being given all the knowledge in their genome they would ever need to survive, the task of the nonverbal mind is to create the relevant knowledge from the structure available within any domain. Forgive me if I make this point in excess, but nothing is more important than understanding that the nonverbal mind did not evolve in order for humans to talk to each other but rather that mammals could be prepared for any nonverbal contingency hidden within any environment. What makes this task formidable is that mammals were designed to survive in a variety of environments or to deal with change. If a mammal finds itself in a strange anomalous domain, the anomalies of the new domain turn on the Heterostatic technology of the mind.

Figure0.2B shows a colored version of the Y-D law, with grey identifying the region where the arousal level of the organism, is too low. The Green Zone is the ideal level of arousal, while in the Red Zone the organism is too aroused. This law will be discussed extensively in later chapters. If the arousal level is normal, in the B zone of the Y-D Law, however, when an error occurs, the arousal level enters the C-zone of the Y-D Law (2004). Any error in the C-zone turns on the creative function of the mind, which was designed to destroy anomalies and convert them into knowledge. (Female three, F3 in the first experiment, illustrates how this works and how efficient mammals are in making a valid working model of an originally anomalous domain.

The arousal functions of the vertical mind are dynamic ones. The vertical mind has two functions: first, to maintain the arousal level of the IMm at some optimal point, and second to be the Boy Scout mechanism of the mind. With the onset of the conditioned stimulus (CS) at Time 1 (T1), the Boy Scout mechanism must begin to activate those centers, which must be prepared to process the unconditioned stimulus (US) at Time 2 (T2). By activate, I mean that the CS must get those cortical centers prepared, activated, or ready to spring into action at the first instance of the neural impulses arriving in the cortex via the classical pathways that require processing in the next frame of the script that is playing in consciousness.

Bernard IMm

Historically physiology has been exclusively focused upon that part of our nature that is a function of our body, as opposed to a function of our mind. However, Claude Bernard's (1957) work has two sides: first, on the negative side, it is a good illustration of the bias in physiology discussed above, but on the positive side, his concepts offer a good way to solve the problem induced by inserting the mind into physiology. Bernard's concept of the internal milieu offers an easy way to expand our concept of human nature by putting both sides of our nature, mind, and body, on equal footing, a way to create a level conceptual playing field for both our body and our mind. However, Bernard's problem was that he failed to include the concept of the mind in his thinking about milieus.

Candelabra model

In this section, the structure of the arousal system, which is one of the three brain/mental structures of which the mind was constructed will be described using a model, which is a variation on the work of Gastaut that I call the Candelabra model simply because it reminds me of a candelabra.

FIGUREo.2A

FIGURE o.2A in one sense gives us a complete vertical overview of the two component models of the nonverbal mind.

First, the rectangular box across the top is a model showing the cortex from a side view. This box is one view of the horizontal model of the mind, the Aristotelian model, which was discussed in chapter two. Second, this figure nevertheless primarily shows the structure of the vertical components of the nonverbal mind. Third, at the bottom of this model one medium sized box labeled "the receptors" is shown, which is Station 3 in Aristotle's model.

This figure shows those parts of the mind that were not represented in Figure o.1B and gives a different perspective of the receptor systems. It shows the afferent nerves leaving the five senses but then drops them there. This figure shows that the afferent pathways are picked up again once they get inside the brain. FIGURE o.2A shows part of the story left out of the Aristotelian model in FIGURE o.1B. The question is what happens to the afferent nerves once they enter the brain. In FIGURE o.2A, these pathways are called the classical pathways, CP, and are labeled Pathway # 1, which is a bundle of ascending fibers consisting of all the afferent nerves. The Classical Pathways, CP, goes straight up through the core of the brain directly into the cortex and then splits to make connections with their respective Primary Projection areas labeled station 5 in FIGURE o.1B.

FIGURE o.2A shows more than how the afferent nerves get from the receptors through the brain into the cortex. This picture shows the revolutionary new structures and their new functions, which are major collections of elements, which convert the brain into a mind. The new features are the four collateral pathways that leave the CP and become the arousal system for the mind; vertical circuits which prepare the cortical cellular matrix to receive the particular content of the CP, which is about to enter into the cortex. These four arousal pathways were designed to get the cortical centers prepared for the neural inputs, which contain the knowledge, which will sustain the narrative playing within consciousness.

The four collaterals take off from the CP at various levels on the way through the Reticular pathways, and into the brain's arousal system. They can be thought of as complex spray guns that control the synaptic input circuits of the cortical blanket by spraying or releasing just those particular Neurochemical neurotransmitters, which each pathway that the plans of action generated by the nonverbal mind needs.

Parenthetically, I have no chapter devoted to the third component of the mind, the chemical component because the chemicals are generated in a distributive way throughout the brain but especially the cortex at too many local synaptic sites. The above paragraph makes one of the few references to the chemical functions, which are the third critical parent of the mind.

The five vertical pathways consist of two kinds of pathways, Extrinsic and Intrinsic. The Classical pathways consist of a set of afferent nerve pathways, the extrinsic pathways, which have their origin outside the brain in the five senses. The other four pathways are intrinsic in origin, since their origin is within one of two levels of the reticular pathways within the brain. The Candelabra model shows how the structures of this set of four pathways collectively form the vertical arousal system of the nonverbal mind.

"The intrinsic portions of the thalamus project to those sectors of the cerebral mantle usually referred to as "association cortex."(Pribram, 1960, P. 1324)

Technically this set of four intrinsic pathways, from the Reticular structures to the cortex, formally represented the arousal structures of the mind. Only one of the five pathways of FIGURE 0.2A, channel one, the Classical pathways, CP, is shown leaving the receptors. This extrinsic central pathway collectively represents all that the afferent nerves bring into the cortex - the sensory-specific neural codes from the five senses.

The four collaterals

The features that make this figure interesting and important are the four pathways that branch off the CP at one of two levels on the way into the brain. The first two collaterals branch off at the level of the Brain Stem Reticular Formation, ARAS, and then the next two branch off at the level of the Thalamic Reticular Formation, TRF, which is a bit higher up on the Classical Pathways or Pathway #1 in FIGURE 0.2A.

Magoun was well aware of the role of the ascending reticular influences upon the cortex in 1963.

"It now seems clear that the ascending reticular influences upon the cerebral hemispheres are importantly concerned with initiating and modifying such states as arousal and wakefulness, as well as orienting and attention."(Magoun, 1963, P. 74)

These four pathways control the Synchronization-Desynchronization, S-D, continuum of the cortical networks. The point of controlling the S-D continuum is to get the cortical networks prepared to process the raw data, Kant's (1952) precepts, which the Classical pathways supply. These arousal pathways prepare the NV mind to process the pragmatic domain-specific, purpose specific potential knowledge that the nonverbal mind needs to acquire and store within its model of the current domain. Knowledge prevents surprise! Ignorance allows the cortical cells to be surprised by the input from Pathway #1. Ignorance makes events anomalous. Knowledge destroys anomalous events. Yet on the positive side, the arousal system uses knowledge to prepare the cortical matrix for processing the knowledge regenerated by the self-organizing mind.

The whole reason for evolution designing mammals was to create a better way for life to obtain/create knowledge; to go beyond the simple forms of life bound or limited to just genomic knowledge. The vertical arousal system transforms the static knowledge stored in the model of a domain into dynamic knowledge that gets the cortical center prepared to receive and process the signal using the knowledge that predicts what precepts are about to arrive within the cortex. The function or the vertical arousal component of the mind is to implement the creation of knowledge. The essence of knowledge is expressed in the Boy Scout motto "be prepared". The arousal circuit has an impossible job; to prepare and implement in real time, the cortical centers for all the salient knowledge, the knowledge that the self-organizing system of the nonverbal mind can predict is about to enter the cortex over pathway #1. I repeat myself, the nonverbal mammalian mind evolved to create or grow domain specific empirical knowledge structures.

The classical physiological theorists have for too long focused exclusively upon the cortex as the only tissue of which the mind was constructed. At best, the cortex is but a third or half of the mind.

The thalamic arousal function

There are also the focal arousal circuits, Pathways, 2+, and 2-, which produce local arousal or focal excitatory or inhibitory arousal effects upon any region or set of regions of the cortex. Focal means these circuits have less centrality. These four vertical circuits were designed to function together. The diffuse, the D's, and the Focal, the F's arousal circuits were designed to create complex patterns of arousal and to focus either broadly or narrowly upon the afferent centers of the mind to collectively regulate and control the arousal levels of the cortex. They are capable of implementing within the IMm those arousal levels and patterns, which are appropriate to process the knowledge which the self-organizing machinery of the mind predicts will be required to process the signals, which soon will be arriving within the cortex.

The five vertical pathways:

First, pathway # 1 is the Classical pathway, CP. This pathway is a bundle of all the afferent nerves that are aimed, as it were, at the horizontal component of the mind or at least are aimed at those parts of the cortex, which function as the screens (the PPAs).

The next four vertical pathways are the arousal pathways, per se. All of these pathways leave the CP via the collaterals that branch off the CP at various levels within the brain.

Pathway 3 + and 3-

Pathway # 3, I call D+ and D- leaves the Classical Pathways deep in the brain, at the level of the Brain Stem Reticular Formation, BSRF. This pathway has centrality, which means it is connected to and thus can regulate the arousal level of every neuron in the cortex; its effect upon the cortex is diffuse. Diffuse communications, D+ or D-, are exemplified by a fire alarm system that delivers a critical simple message to the widest possible population in the shortest amount of time. This simple diffuse signal carries only a little critical knowledge that everyone needs immediately about some unexpected event that has suddenly appeared. This D function is the cause of "surprise" behavior, or Pavlov's (1960) Orienting Response, OR.

This is same channel that in Cannon's (1932) Homeostatic illustration of the Fight or Flight reflex activates the body to be prepared for some homeostatic emergency. However, with Heterostasis, an anomaly is not a bear but ignorance, yet not being prepared is a threat no less dangerous. In general, Heterostatic threats are anomalies, which announce that there is some surprising object in the EMm, which the mind cannot assimilate. An anomaly activates this D+ channel, which effectively activates, or desynchronizes the entire cortical mantle; every cell in the cortex is aroused. D+ has the most important communication responsibility, to alert the whole community of neurons of the discovery of some probable most important unidentified lethal object or event.

This signal gets its name, D, from its effect that is Diffuse. The arousal system of the brain has two diffuse pathways: D+ and D- and in evolutionary terms, these pathways were probably the first to evolve. They regulate the Electrophysiological behavior within the horizontal nerve net of the cortex. One channel D+ wakes up the brain centers of the mind, by blocking all the alpha waves and instilling the cortex with high frequency low voltage Beta waves. On the other hand, the D- pathways evolved to put the cortical networks of cells into sleep cycles smothered in delta waves.

Pathway 2+ and 2 -

The two Pathways labeled 2 + and 2- leave the CP at the level of the TRF. This pathway has a limited spatial ability to alter the EEG frequency. The arousal role of these two pathways are focal as opposed to diffuse, and are therefore called F+ and F- which means they do not have the centrality that characterized the two D pathways. These pathways move local networks of cells in the direction of sleep, F-, or waking, F+.

The revolutionary components of this model, Pathways 2 and 3 were only discovered in the 1950's. These arousal circuits can be seen in FIGURE 0.2A

Pavlov

Pavlov (1960) was the first person to expose me to the idea almost 60 years ago that sleep was a force, like a knife within the mind. The verbal version of Pavlov's ideas that intrigued me was his phrase that "internal inhibition was local sleep." I am still in awe of the paradoxical role that neurotransmitters play in their new roles in the nonverbal mind that first evolved to put the cortex into states of sleep or waking.

The evolution of the Reticular Formation

Plato (1952) could not have been more wrong: even the nonverbal mind of rats is more than capable of executing his so-called "god-function," and of course, to make matters even worse for Plato, the Generic Nonverbal Mammalian Mind is made of matter. The Physiological transformation that endowed mammals with the makings of a nonverbal mind was in part, a consequence of the division or the evolution of the reticular formation in two stages:

BSRF

In the first stage of the evolution, the Brain Stem Reticular Formation, BSRF, emerged. The original pathway 3 + and 3- components seen in FIGURE0.2A could induce the cortical neuron masses to shift, in mass, between the two ends of the arousal stick, the ARAS Plus and ARAS Minus. The ARAS+ woke the cortex up, while the second put mammals to sleep.

TRF

In the second stage of evolution, the Thalamic Reticular Formation (the TRF) emerged. The emergence of the Focal, F+, and F- channels revolutionized the physiology of the cortex. Whereas the BSRF was a pacemaker which acted upon the organism as a whole, this new TRF level of arousal pathways created a whole host of new pacemaker arousal functions, but again all these new functions would be focal pace-making effects. All we must now do is to understand the difference between diffuse and focal pacemaker functions.

The functions of the TRF

In mammals, the four-arousal circuits of FIGURE 0.2A organize the cortical matrix. The question is how the evolution of the arousal system could have advanced the evolution of the nonverbal mind, specifically the consequences of the emergence of the TRF. First, what role could the BSRF play in the absence of the TRF and the cortex; second, what difference could the TRF and the cortex make?

The evolution of the cortex transformed circuits that had evolved to execute Function A to execute Function B. The BSRF presumably evolved to execute Function A, but the evolution of the TRF enabled the brain to execute Function B.

Function A was the sleep waking function, which is mediated by the pathway 3+ and 3-, which became the major pacemaker of the brain, and metaphorically regulates the brightness of the diffuse flame, the S-D (Synchronization-Desynchronization) continuum of consciousness. The BSRF is a holistic or diffuse function, which controls the flame of the brain or controls the mean brain temperature, or arousal level.

Function B, on the other hand, is a secondary pacemaker with an effect that can best be conceived of as functioning as a focal flame, as a cutting flame or KNIFE that parses the shapes of things within consciousness.

The TRF: the Set or learning function

The TRF circuits main function is to focally activate just those specific cortical centers of the nonverbal mind that are required to process the incoming salient knowledge on its way up the classical pathways and soon to enter the cortex.

The TRF arousal circuits enable the nonverbal mind to self-organize, to use these circuits to be prepared for the salient incoming knowledge. The four vertical circuits arouse the cortex in several different ways. They modulate the EEG and thus control the distribution of arousal within the cortex, both the level and the location of the arousal.

I will return to this topic below in the section on Pavlovian conditioning where I will show how the TRF gets the brain SET during learning.

The BSRF functions

The Sleep-Waking functions

The BSRF arousal circuit contains the principle arousal function, which was perhaps the original and the most ancient function. The original Reticular function was a diffuse type of arousal that had an identical arousal or enflaming function upon every cell of the cortex i.e. had centrality. The excitatory half of this function, which originated in the Brain Stem Reticular Formation is expressed in Pathway 3+ and produces uniform electrophysiological effects upon all the neurons of the cortex. This function was probably originally designed to simply wake an individual up.

The Y-D Function

Later this same pathway began to play a larger role than supplying a waking function in the nonverbal mind. The arousal function can be thought of as a Heterostatic EV of the Internal Milieu of the Mind, IMm, namely keeping the mind within its optimum arousal level.

External Inhibition function

Additionally, it activates the Pavlovian external inhibition reflex, which is the master reflex of the mind. This master reflex is capable of interrupting instantly and absolutely any and every other reflex that might be functioning in the cortex at the first sign of danger. (Pavlov, 1960)

Kuhn's Function

This BSRF arousal circuit, at least the excitatory or 3+ circuit, plays the central arousal function role, namely Kuhn's (1962) anomaly trigger.

The Yerkes-Dodson, Y-D law

The vertical component of the mind plays a primary role in all functions of the nonverbal mind; not a secondary or supporting role, but rather a role while not the same character as those roles of the horizontal stations, yet one no less vital. The vertical component of the mind does not constitute yet another station of the assembly line rather it energizes, implements, or enables all the other functions. It organizes the function of each station of the horizontal mind so that each station can execute its particular function or role quickly and efficiently.

In fact, one of the most famous laws in the history of psychology, the Yerkes-Dodson law, recognizes that the vertical arousal system of the mind plays a primary role, and the universal nature of this role determines the efficiency of every function of the mind. The arousal system, as I have noted several times, is one of the three founding elements of the mind. The Y-D law puts the spotlight on the nature and importance of the arousal system and states that the effective rate at which every mental function is executed by the nonverbal mind is a direct function of the arousal system component of the mind. (Yerkes and Dodson, 2004)

FIGURE 0.2B

The Yerkes-Dodson law can be seen in FIGURE 0.2B. The two experimental variables, which form its two axes, are behavioral efficiency on the ordinate and the arousal level on the abscissa. The law states that there is a curvilinear relation, an inverted u-function between performance efficiency and arousal level: that performance is a function of one and only one variable, everything else being equal the arousal level is imposed upon all 13 of the horizontal stations of the mind by the vertical arousal system.

The curvilinear curve of this law is divided into three sections: A, B, and C. where

In section A, the gray zone, the organism is not sufficiently aroused to perform at its best.

In section C, the red zone, the organism is too aroused to perform at its best.

In section B, the green zone, the arousal level is just right.

Heterostasis is the wisdom of the mind and was designed or programmed to keep the EV of the mind in region B of the arousal curve. This knowledge, which makes Heterostasis work normally, works automatically, and if nothing else we should all learn to keep our arousal or signal level at B. (Hanouch and Vitouch, 2004)

With regard to performance efficiency, region B, the most important point on the curve is the high point on the curve, which instinctively kept Female 3, F3, moving in the right direction in the maze. F3 and all mammals must learn if they are to stay alive and well to maintain their arousal level in region B.

Arousal level B is the set point for arousal on the Synchronization-Desynchronization, S-D curve and identifies the optimal level toward which the organism should strive. The sweet point, B on the curve, shows that the performance efficiency is at a max when the subject is somewhere in the middle of their arousal range, not too little and not too much. One of the main roles of Heterostasis is to keep the arousal level at the optimal point of arousal, B, which is the set point of the mind. When an organism is bored, its arousal level falls and in order to regain its optimal level of performance it must seek and find some arousing event, which will restore its arousal level to optimum. In the maze, when the rat is too aroused by an anomaly it will stop exploring in order for its arousal level to drop so that its performance level will once again become optimal.

The Y-D law puts the spotlight on arousal and makes arousal the key measure of the performance efficiency rate at which the nonverbal mind is performing.

The nonverbal mind is a self-organizing system and the Y-D law is possibly the best measure of how efficient an individual is at self-organizing. (Yerkes and Dodson, 2004)

One of the paradoxical central features of Heterostasis is the role of the anomaly in the function of the nonverbal mind. All mammals have a rational mind whose chief phenomenon is its dual needs: first, the need for uncertainty inducing anomalies as food for thought, and second, the need for the causal knowledge created during retracing to destroy anomalies. Heterostasis doesn't make any sense until one understands this dual role of an anomaly; these seek and destroy behaviors of the mind.

The nonverbal mind feeds upon the uncertainty induced by anomalies. When it has no anomaly induced uncertainty with which to deal, it has nothing to eat, but once it has discovered an anomaly it retraces until it creates the knowledge required to digest or destroy the uncertainty created by the anomaly.

Mammals are a set of species born without knowledge but with a nonverbal mind that knows how to create knowledge, how to transform anomalies into knowledge. Within any closed domain like the maze, the nonverbal mind may soon exhaust the supply of anomalies; so the mind needs a new domain to conquer and another: to seek, to destroy, to discover, etc.

Heterostasis, arousal level, and performance

Strangely, as noted before, the nonverbal mind makes mammals into self-organizing systems: the prime function of the nonverbal mind during the waking state is to regulate the arousal level of the mind, to keep the arousal level at B, the high point on the curvilinear function. Thus, the nonverbal mind is regulating or controlling the organism's performance efficiency. The question becomes how the nonverbal mind regulates its arousal level.

When the mind encounters a relative anomaly it may become too aroused which causes it to enter level C on the Y-D law. The organism stops exploring and begins retracing, which is the first phase of mental evolution. In this phase, the organism seeks the causes of the last anomaly; the primary mind probes or investigates/interrogates the region before the anomaly, which will create causal knowledge that will destroy the uncertainty induced by the anomaly. All too soon, the organism will begin to get bored and so the cycle begins again. One of the tragic examples of a two-low arousal level is in driving. How many accidents and deaths are there each day due to drivers getting drowsy and then falling asleep at the wheel?

Applying FIGURE 0.2B, let's imagine that for the nonverbal mind the inverted U curve itself is consciousness with three speed zones: A, B, and C with zone B the ideal speed zone for maintaining an optimal consciousness performance level.

The chief role of the vertical component of the mind is the management of the circadian rhythm. Since I am a rat runner, my primary interest is in the 16-hour waking component of the curve. The Yerkes-Dodson law can be thought of as a waking law, which the mind must observe, a speed law as it were, a consciousness speed limit.

The Speedometer on consciousness

Each normal mammal is a Heterostatic being. Each day, the human circadian rhythm breaks up into two phases, an X-hour home/sleep component, and a X-hour away/work component. The value of X varies with age, etc. The Y-D Law makes us aware how important the sleep cycle is; the mind must be stocked during each sleep cycle with sufficient neuronal supplies for a full duration of the away-work phase. Metaphorically, we are cars or trucks that once we leave home or leave the station each day we must make a 16-hour run between servicing. Each day we must pace ourselves, in our IMm, so that we can sustain consciousness at some kind of optimal work/play state for 16 hours.

As a teacher, I always tried to remember Pestalozzi's (1859) instructions for teachers. In the ideal classroom, a teacher should feed the class with the quantity of knowledge each day that sustains each student's consciousness within Zone B. Zone B is the Goldilocks zone. Students should find in the classroom just the ideal quantity of knowledge each day, but not too much knowledge to scare or overwhelm their consciousness. Also, Zone B should not contain too little knowledge to bore the consciousness; Zone B should contain just the right amount of knowledge each day to maintain each student's consciousness at the optimal level of performance, zone B.

Habituation

The four ages of stimuli

Evolution designed the arousal circuits of FIGURE 0.2A with a set of five circuits each with a different set of critical features, yet collectively the set of five circuits had the requisite versatility to keep the stations of the horizontal component of the mind at the optimal point of arousal. It had enough arousal so each mammal could execute its plan for achieving its goal in a most efficient manner.

The five channels of FIGURE 0.2A are not just a hodgepodge of channels but are a brilliant set of coordinated pathways designed to execute the functions, which make the nonverbal mind do all the impossible things that down through the years philosophers from Plato (1952) on have insisted the human mind couldn't do. One of the chief arguments for a spirit god was the insistence of philosophers and theologians that there were things a physical mind couldn't do, that only a god could do. For example, Plato insisted the mind was only a half-rational machine.

These five channels are able to execute innumerable functions including the impossible function. The philosophers and even modern linguists insist that the mind had a nativistic origin or that the mind did not evolve but was a gift of god and executes functions that no machine can compute.

Obviously, they never dreamed of these five pathways, which rise up through the core of the brain to create the master arousal control center in the nonverbal mind, which creates a confluence of several rivers of knowledge from the afferent nerves and five master arousal circuits to control and process them. This configuration of the master control center enabled the nonverbal mind to transform the massive raw material from the five senses into nonverbal images within consciousness of the objects and events occurring in the IMm, which induces the body to act. This confluence of machines and knowledge enables the mind to create an active consciousness, which can direct the life of any mammal in any domain to achieve any purpose the domain can support.

In order to illustrate how the technology of the vertical pathways work, I will describe five stages of functions which Pavlov (1960) called Habituation. See FIGURE0.2E. Habituation is but one of the 10 sets of mental functions, which Pavlov identified but time prohibits me from describing all of them. Pavlov, in spite of being born too soon, (he never heard of the arousal structures of the mind), was able to explain all the basic mental operations that make the mind work in his physiological theory of mind.

FIGURE0.2E

Gastaut (1957) describes Pavlov's (1960) Habituation procedure as a passive function which any stimuli automatically goes through simply by its being repeated a number of times. Gastaut identifies four ages or stages of habituation which each stimulus passes through. This process creates four kinds of stimuli, which are each important events in themselves and each of which can be used in various procedures. As stimuli age or habituate, they change properties.

His four stimuli ages are:

1-Novel, N
2-Absolute, A
3-Indifferent, I
4-Old, O

Habituation is a passive stimulus aging process and all five channels of FIGURE 0.2E play a different role at each step of mental aging. In FIGURE 0.2E, these arousal channels are not labeled in the four sections of the diagram. Habituation is a process where a stimulus is simply repeated; it is not paired with any other stimulus nor does anything change in the environment during the Habituation process. This series shows that a novel stimulus, N, which initially is absolutely exciting and arousing, eventually becomes old, O, simply as a function of repeated use. In other words, this illustrates what Pavlov (1960) meant by his expression "extinction is inevitable" where extinction in this case literally means an excitatory kind of death.

I have included three quotes from Pavlov below, all of which make the same point about inhibition, yet are slightly different. What nature did in the design of the brain was to take what was an originally an element of a sleep-waking mechanism and created many new uses for this original force.

"One of the most important of these properties is that under the influence of conditioned stimulus they pass, sooner or later, into inhibition." (Pavlov, 1960, P. 234)

"...since all our experiments abound with observations showing that internal inhibition invariably passes into sleep unless special precautions are taken." (Pavlov, 1960, P. 252)

"It is thus clear that the administration of the unconditioned stimulus results in an inhibitory state of those cortical elements on which the conditioned reflex depends- these cortical cells becoming temporarily unresponsive to their normal exciting stimulus." (Pavlov, 1960, P. 243)

FIGURE 0.2E describes what happens at each step of the Habituation procedure. Each of the four stages of stimuli is shown in this figure.

Step one habituation: the N state.

Step one in the upper left hand corner of the Figure shows what happens when a novel stimulus or an anomaly is presented for the first time. Pathway D+ is the home for what Pavlov (1960) called the master reflex, external inhibition, which is the most sensitive and dominant of the three-arousal centers. A novel stimulus, an event presented for the first time, causes a massive activation in the entire cortex by way of the diffuse force, which I call D+. This set of pathways probably first evolved to manage the sleep-waking function of the organism. Pavlov was the first to recognize that the basic mental functions were originally sleep waking functions.

In this figure, Gastaut (1957) shows above each of the four divisions a rectangle, which represents a region of the cortex. Above each cortical region, he shows a sine-wave function, which illustrates how the EEG looks in each lobe as a function of age. In other words, he shows how age alters the arousal function, which affects the lobes of the brain. In these four sets of figures, only two EEG sleep-waking frequencies are employed D+ and D-. The D+ arousal force desynchronizes the cortex, which means the EEG above each of the three rectangles show a high frequency low amplitude wave. In the extreme case D+ or at this age or stage of habituation this N-stimulus overexcites the arousal level of the cortex or it diffusely desynchronizes the cortical EEG by moving the arousal level into Level C of the Y-D law.

The D+ arousal signal causes the cortical neuron to change to a too aroused state. The figure shows that the stimulus when in the N state elicits a relatively high frequency low amplitude EEG signal across the entire cortex, which is characterized as a state of diffuse activation or desynchronization.

One of Pavlov's (1960) most important mental forces was internal inhibition, which he recognized had originated as "local sleep" which I call K-.

Step two: the A State.

In the second step of habituation, we must compare the two elements of FIGURE 0.2E, N and A. In N note that the N stimulus event activated the BSRF, which has the maximum quantity of centrality and sent out massive amounts of arousal stimuli, D+, to all four lobes of the cortex as required by its centrality properties. Note however, in the N state that for some reason Gastaut only included three of the four lobes in this illustration.

However after N has been presented a few times, the stimulus ages until in a massive quantity of internal inhibition had built up within the BSRF center deep in the particular region of the brain that was active for N. This buildup of inhibition releases the thalamic reticular formation, TRF, and now the TRF activates focally just that portion of the cortex, which is the Primary Projection Area, PPA, for the A stimulus. Note that two of the three sections of the cortex are in a state of synchronization or sleep. Now the TRF component or the arousal system, which creates an F+ arousal signal, activates only one of the three portions of the cortex shown. In A, this figure shows that the BSRF has rings of concentric circles around it that represent the fact that during the first stage of aging inhibition accumulated as the N stimulus was presented by itself; signaling that internal inhibition had accumulated while the N event was repeatedly presented.

In the first section of FIGURE 0.2E, note that each of the collateral pathways was initially in an excitatory state, but every time the N event was presented the active region accumulated a bit of internal inhibition which is represented as rings of inhibition or localized sleep which Pavlov (1960) considered as two uses of the same arousal Neurochemical.

Step three: the I State.

In this third section of FIGURE 0.2E, look at the figure and see if you can understand on your own what happened next. Note that in this section of FIGURE 0.2E both the BSRF and the TRF have accumulated a massive quantity of internal inhibition and subsequently have lost their capacity to desynchronize regions of cortical neurons.

What the reader must recognize, even though it is nowhere stated, is that a novel stimulus has an automatic capacity to desynchronize the entire set of cortical neurons. This activation effect wakes the organism up, but once inhibition has accumulated on one or both of the reticular activation centers then a novel stimulus, which was once a wake-inducing stimulus, an anomaly, then has the opposite effect, the synchronized signals are inhibitory in their effects, which is to say they induce sleep.

The high frequency low amplitude stimuli are events that arouse, awake, surprise, or desynchronize the cortex. While the same stimulus, depending upon how many times it has been exposed, may initially be an anomaly, it will eventually become an "Old" stimulus that puts an organism to sleep. In a most interesting way, this effect proves the fact that heterostasis is a physiological force that feeds upon novelty and destroys anomalies. Yet, it is imperative to note that the nonverbal mind requires a constant supply of anomalies for it digests anomalies almost as fast as they can be discovered. F3, I must keep reminding the reader, created a model of the matrix maze, which means her mind destroyed all the anomalies in less than thirty minutes.

Mammals feed upon anomalies; yet, they require anomalies to stay awake and keep their nonverbal mind usefully employed.

Step Four: the O State.

In the fourth section of FIGURE 0.2E, the effect of habituation has run its course. The stimulus that was initially an N has now been habituated through each level of the reticular formations and once the habituation "took," the habituated stimulus is called an old stimulus. That means that the habituated stimulus is now an indifferent event. Once an event has been fully habituated then every channel now fails to respond to it in any way. It is now appropriately called "an indifferent stimulus." This expression is somewhat misleading for technically the arousal centers, which have accumulated internal inhibition, are indifferent to the habituated stimulus not the other way around.

Why is this? Note that internal inhibition has built up on each of the three-arousal pathways, simply as a function of age, which prevents any of these channels from responding any longer to this stimulus.

The Match-mismatch mechanism

The one mind theory proposes there are not two minds but only one mind in the brain, which was probably constructed of parts that had originally evolved for other purposes. In this chapter, I describe one of these ancient parts, which in its new role as an element of the mind evolved into the vertical arousal system of the mind and show its role in the mind. I will also show how it interacts with the horizontal component of the mind.

In this section, I will introduce the concept of a Match Mismatch Mechanism (MMM). This mechanism most effectively shows how the two components of the mind, the horizontal and vertical, interact. This mechanism compares two products of the mind; the Observed, O, and the Predicted, P; the former is computed by the horizontal mind, while the latter is computed by the vertical mind. But I am getting ahead of myself.

FIGURE 0.2F

The concept of "getting set" is as old in psychology as it is in sports and has played a central role, especially in Piaget's (1968) concept of assimilation. However, in this chapter my purpose is to go at least one step beyond Piaget and discuss how the structure of the vertical arousal system gets the cortical matrix set or how the vertical mind gets the horizontal mind set. The second function of the vertical mind discussed above is to get the cortical centers prepared, activated, or ready, all of these words refer of course, to the concept of set.

Hilgard and Bower simply review what is so obvious about conceptual assimilation, but my purpose in this section is to take the concept of assimilation one-step further.

"An adjustive process is needed to fit external reality into an existing structure (assimilation), and to modify that structure while this is taking place (accommodation)."(Hilgard and Bower, 1975, P. 323)

FIGURE 0.2F shows a model of the Match Mismatch Mechanism, which plays the role that embodies the essence of rationality. The match-mismatch mechanism determines which patterns of inputs are to be or not to be assimilated and thus determines which enter consciousness ready for a response. All of these options ultimately refer back to what is real, which was Plato's passion.

In FIGURE 0.2F, on the left is shown a "Thing," T, in some X-domain. This figure begins at the same place where FIGURE 0.3 began in Aristotle's model. Next, there is a Receptor, R, one of the five senses as in Aristotle's model, which sends a signal, which in this context will be labeled Observed, O, into a place, which executes a new function, the comparator function, which is but another name for Aristotle's (1952) sixth sense. What is actually different in FIGURE 0.2F from FIGURE 0.1B is that this figure involves both horizontal and vertical circuits, while Aristotle's model was purely a horizontal model that can do nothing by itself. The nonverbal mind, in order to function, requires the interaction of at least three kinds of cellular or modular components.

In this section, the focus is upon how the vertical and horizontal components of the nonverbal mind interact to make behavior possible. The match-mismatch mechanism is one example of how this interaction works.

The vertical arousal function in FIGURE 0.2F is labeled the modeler, and it provides the mind with a prediction, P, which is compared with the observed, O, produced by the cortical horizontal model. The comparator compares O with P and renders one of two verdicts, match, or mismatch: if the verdict is a match then the O event is certified as real, is recognized, inserted into consciousness, and it is safe to respond. However, if the verdict is a mismatch then the O object or event is classified as an unknown object and thus it is not safe to respond.

The essential elements of rational behavior: P and O

Predicted, P, Heterostasis, and the nature of nonverbal rationality

The concept of the predicted, P, conceptually carries the full weight of the concept of nonverbal rationality. Historically rationality has been conceived as a verbal function, but I propose that rationality first evolved as a nonverbal mind function and was the first line of perceptual defense for the nonverbal mammals.

Nonverbal rationality embodied within Heterostasis is the *sine qua non* of the rational mind. Evolution, as I have stated several times, of the nonverbal mind began to evolve when the great extinctions killed off most of the one-niche forms of life. With the evolution of mammals, nature, in effect, found a better way to supplement life with the knowledge required to survive other than the pure genomic one. Rational behavior, Heterostasis, was the solution nature adopted; mammals were the species of life that were the benefactors of the evolution of this revolutionary new rational technology.

What does nonverbal Rationality mean? What kind of technology makes a mind rational? What kind of behavior does a rational mind exhibit? The Match-Mismatch Mechanism consists of a big piece of the rational technology out of which the mind was constructed. Yet, it is possible to point much more specifically to the function, which is the heart of nonverbal rationality. As noted above, the concept of the predicted, P, conceptually caries the whole weight of the concept of nonverbal rationality. The principle, which embodies rationality, is the overall function of a match-mismatch mechanism; Heterostasis is the embodiment of rationality. Heterostatic behavior exhibits two kinds of responses: explore and retrace, which are responses to the two verdicts of the comparator, match and mismatch. Once again, Heterostasis embodies nonverbal rationality.

Mammals are rational beings not because they talk rationally but because they act rationally. The rat's behavior in the anomalous hierarchically organized maze is a manifestation of nonverbal rationality. The rationality rule as Ashby (1960) proposed: "move on a match, stick on a mismatch."

In the matrix maze, however, I prefer two different words: "explore on a match, retrace on a mismatch."

The comparator gets its name because it is the place or station of the nonverbal mind where O and P are compared. FIGURE 0.2F shows that the comparator has two output options, Match or Mismatch, M or M-. The figure shows that the judgment, Match, leads to the rational mind emitting a normal response, or certifies that it is safe to continue with the domain specific script that is playing in consciousness.

However, this figure also shows that upon the judgment of Mismatch the only option the rational mind can emit for an anomaly is to do as Kuhn (1962) described. The human scientist must respond by constructing a new model or paradigm or by modifying the existing one.

Nonverbal Rationality is simply the wisdom to make exploring a function of knowledge or to pick up where I left off in Aristotle's model; a rational mammal should only act or respond by exploring exemplars that are certified.

There are only two ways to behave or respond, rationally or randomly. Later I will describe Female 3's behavior, which operationally defines rational behavior, while one of the many random rat's behavior will be described and compared with F3's behavior.

Observed, O, behavior as rational behavior

The five senses provide the sixth sense with an Observed state, O, in the EMm, while the vertical mind supplies the sixth sense with a predicted state, P. In this figure, the sixth sense plays a new comparator role, which is, of course, to compare O with P and to render a verdict of either Match or Mismatch.

A Match means that the O stimulus has been assimilated using Piaget's term, while a Mismatch means that the O stimulus has not and cannot be assimilated, and is an exemplar of one of Kuhn's (1962) terms, an anomaly.

Note that each of the two verdicts has a different consequence. First, a match means that assimilation has occurred and thus it is safe for the organism to act appropriately to the stimulus. A match justifies exploring as a normal response. Second, a mismatch means the model or paradigm is in error, wholly or in part.

A rational organism

A rational organism is one that emits normal or learned responses that in this context means, explore but only to a known or certified object, an object for which the comparator renders a verdict of "match".

The organism cannot respond knowingly, safely, or rationally to a mismatch because the O object is an anomaly. An anomaly is an unknown input pattern for which the nonverbal mind cannot generate, rationally or knowingly, an appropriate response. However, one appropriate response can be made to an anomaly, to activate that part of the mind that executes the retrace function. An anomaly means the model or paradigm responsible for the anomaly is invalid and it is time to go back to square one or for a rat in the maze, it means, "retrace" and begin the search for the cause of the anomalous effect.

In the maze, there are two responses the mind can induce in the somatic muscular system, either explore or retrace. A rational organism can only respond to a match by exploring, likewise, a mismatch must induce the organism to retrace.

To come straight to the point, the four-arousal pathways of the vertical mind are responsible for generating the "Predicted", P, which is compared with the "Observed", O, which is supplied by Pathway #1 or the classical pathways.

Kelly in the quote below uses his expertise in the science of manual and automatic control to point out just how critical the concept of the predicted, P, is in control theory. Prediction is but another name for extrapolation. Female three's behavior showed that when her model of the maze was valid, it enabled her to run two-thirds of the maze for the first time without error. In effect, her model was allowing her to predict every feature of those unexplored units of the maze and make the correct turn at every one of the several choice points the first time through the maze. Once the nonverbal mind conceives a valid model of the maze then a rational organism can move through those previously unexplored units of the maze with absolute confidence. A valid model predicts what to do at every choice point. Predictions that are made based on valid knowledge can't be wrong. The house can only win in a gambling casino when it can guarantee not all its gambling devices can be predicted. Valid predictions take all the uncertainty, all the chance out of a choice situation.

The nonverbal mind creates valid models, which destroy all the uncertainty in the situation so that no guessing is required, and instead a rational organism is one that bases its actions or its behavior upon empirical knowledge (actually empirical knowledge is the only kind of knowledge there is).

"The primary difference between the experienced operator of a complex control system and the novice is in the ability to predict."(Kelley, 1968, P. 42)

A match mismatch mechanism provides one of the few vital rational machine functions. For example, let's go back to some Chapter one topics. Consider the ontological function with two halves. The first half is what I have shown was historically the sacred function, the god-function: the conceiving of ideas or Plato's (1952) ideals. Only God, the obsequious theorists proposed, or the great programmer in the sky could execute the creative function of conceiving a concept, idea, law, or model. The second half of the ontological function is defined by the Match Mismatch Mechanism, MMM. This function in chapter one was called the Certification function. For a rational nonverbal mammal, the MMM renders one of two verdicts, a match, in effect, certifies each observed empirical event entering the sixth sense as a valid exemplar of some known domain- specific concept. This certification validates each syndrome of sensations as safe to act upon as "real," which was Plato's compulsion; but of course, nothing empirical could ever be certified as real according to Plato (1952). Everything made of matter was for him only "an appearance." Little did Plato know that every rat and possibly every mammal had a material mind with a Match Mismatch Mechanism, MMM, that could certify every sensory pattern in real time as it entered the sensorium of the mind as either a match or a mismatch, as known or recognizable, or as unknown or unrecognizable.

When certified as a match, a nonverbal image of the class enters consciousness as an exemplar of its class and takes its place in consciousness precisely within the place the thing represented occupies in the EMm. The certified exemplar within the IMm seems so real that no mammal, not even the human authorities, has ever suspected that images in consciousness are not the physical things in the EMm or of station 1 in Aristotle's model. In other words, the rational mind has been successful. Consciousness is the ultimate trick. The mind constructs models of a virtual reality precisely to cause, to induce, or to trick an organism to act.

But in this chapter, one of my ancillary tasks is to make the reader realize that the mind is not solely in the cortex, but also, there are not two minds in the two human hemispheres. The mind is in the brain and my task is to argue logically that one of the many other regions of the brain, in particular, the arousal structures play a critical role in the certification function, which is implemented by the Match Mismatch function.

Within the cortex, two forces have been at work: first, the division of labor, and second, plasticity. Together they created the human verbal mind out of the nonverbal mind and it is likely that many stations of the nonverbal mind were relocated. Perhaps, they acquired a different set of functions in the human two differentiated- hemispheres for reasons that are not necessarily obvious. Yet the point is that it really does not matter why the particular functions were redistributed between the two hemispheres. Whatever the reasons were, the localization of function that occurred still does not alter the fact that the cortex's role in both hemispheres is limited to the conception of O, while the vertical mind's role is generating P and is just as vital for establishing rationality, as is the cortical function. The MMM is partly in the cortex and partly in the ancient sub cortical regions of the brain. Evolution must have been forced to restructure the organs of the nonverbal mind in the construction of the human mind, for there was only a finite volume of space in the human brain cavity. There was no extra space: evolution did not have the luxury of creating a second, or an additional mind, a spare mind. I propose that one mind fills the brain; there are no "room for rent" signs placed around inside the brain. On the contrary, the facts seem to be that evolution, or actually, plasticity, at least in humans, had to move structures around in the cortex to maximize the use of the limited space available within the skull.

My evidence for the duality of the function of the MMM is actually the most dramatic property of the mismatch half of the MMM. The verdict of mismatch initiates Pavlov's (1960) External Inhibition or more commonly, what is called the startle response, or what Kuhn (1960) calls an anomaly. Yet a mismatch puts the mind in a particular state where it is impossible for it to execute any normal response from the Rolandic motor strip. A mismatch puts the cortex into the C region of arousal in the Y-D law. In this state, a rational organism must retrace and reexamine all the conceptual elements of the model, which allowed the anomaly to occur. (Hanouch and Vitouch, 2004)

Hierarchical learning

The GNMM was designed to create the critical domain-specific knowledge required for the resolution of some goal. In the traditional language of psychology, the mind is an advanced kind of growth machine. As a self-organizing system, the mind can only create knowledge by destroying the relative anomalies that induced the uncertainty in the first place.

One of the remarkable features of the nonverbal mind is its ability to self-organize in anomalous hierarchically organized domains. This last feature puts a special requirement upon the arousal system, which controls the activation level of the horizontal component of the mind.

Let us assume that the matrix maze used in this series of experiments is hierarchically organized with four parameters or is a four dimensional object within the EMm; an A2, B3, C3, D8 hierarchically organized object. The challenge for me is to explain how the rational nonverbal mind conceives a model that destroys or rids the mind of each of the four regions of ignorance or uncertainty induced by the maze.

There are two tasks here. First, I must explain how the nonverbal mind by interacting with the maze conceives perhaps an abstract model of the physical maze in its EMm. Then, the mind must translate this abstract code into a virtual reality called consciousness, which seems so real that there is never any reason for the organism to question whether it should act. If a perception is experienced "before the eyes" then the object of that perception is certified. Mammals are a set of species that "took over all the habitual niches of the earth," because they were endowed with a nonverbal mind that can model any spatial domain or thing, well almost anything. That means their mind can make models that can take the surprise out of all these domains or that their mind can predict everything that can happen in each of these domains so that it is impossible to be surprised. A prey cannot be caught without being surprised. Knowledge is a mammal's most precious possession, which is another way to say that a nonverbal mind that can create knowledge deserves to be recognized as executing a function that the classical theorist claimed only a god could do.

The nonverbal mind confers upon mammals the third wisdom, the wisdom of the mind, HHH, which includes Homeostasis, Hedonism, and Heterostasis.

This does not include any of those wisdoms, which enrich and bedevil the human mind.

The task of the self-organizing system of the nonverbal mind is to create knowledge and to make mammals rational, to enable them to act only when justified by empirical knowledge. The mind regulates knowledge so that the mind can act only when the difference between the internal and external milieu is zero, so that there is no ignorance or uncertainty within the IMm. By carefully exploring the maze, the nonverbal mind discovers one at a time each relative anomaly, which is the source of ignorance or uncertainty. Each of these relative anomalies turns on the negative feedback waking circuits, which are designed to rid the mind of uncertainty by retracing in the maze to create a model or knowledge of each continuum one at a time, and thus rids the mind of the differences between the internal and the external milieu. As soon as the mind has discovered the cause of each of the relatively anomalous objects or dimensions and can predict where each object or dimension is in the maze then the mind knows the maze, and the mind has destroyed all the ignorance/errors that originally existed between the two milieus.

Conditioning: A Pavlovian type Human Experiment

It is time to begin to illustrate some of the consequences of the evolution of the emergence of the TRF or of the focal pathways 2+ and 2- seen in the Candelabra model. The first of the revolutionary consequences is how this particular change enabled mammals to become self-organizing systems. The emergence of the TRF intrinsic pathways made possible the most important phenomenon, which in different contexts has gone by several names, two of the best-known are "Set," or "be prepared."

First, perhaps we should ask, what is the essential mental skill that converts the traditional Nature species into one-mind Nurture species? The no-mind or Nature-species' survival was exclusively based upon the genomic knowledge they received at birth. After the "great extinctions," nature designed a better way. The solution nature adopted was to create a new species of life, mammals that were given a new organ, a nonverbal mind, which could interrogate nature and conceive the domain specific knowledge they required in real time.

So, let's consider an experiment by Pavlov.

FIGURE 0.2G

Pavlov's theory of conditioning

FIGURE 0.2G is a model of how Pavlov's (1960) theory of acquisition works. He calls this the Stimulus Substitution theory of Classical Conditioning. This figure uses an Aristotelian type model of the cortex.

Compare FIGURE0.1B and FIGURE0.2G

In fact, it is interesting and instructive to compare FIGURE0.2G, Pavlov's model, and FIGURE0.1B, Aristotle's model. Pavlov lived over two thousand years later than Aristotle did, yet Pavlov's model of the mind was almost identical to

Aristotle. Incidentally, Pavlov did not receive a Nobel Prize for his work on conditioning, but for his work on digestion.

In his original conditioning experiments with dogs, Pavlov paired a bell, the CS, with food, the US. Each pairing was a trial. At first, the dog only salivated to the US, not the CS (BELL). However, after a few pairings of the two stimuli, the dog began to salivate to the CS, as if it were the US.

According to Pavlov (1960), stimulus substitution had occurred; in effect, a short circuit had developed between the two stimuli: the dog could no longer distinguish between the CS and the US. The dog now responds to the bell as if it were food. Alternatively, or more specifically, Pavlov's exact prediction was that the dog could no longer distinguish the two stimuli. Pavlov's argument was that if the dog cannot tell the difference between the two stimuli then it should make the same response to both. The CS is now equal to the US so the CR = UR (conditioned response=unconditioned response).

In this instance, Pavlov (1960) is displaying his great metaphorical driven imagination. He imagined that a short circuit gradually developed in the association area between the two stimuli, the CS and the US. See FIGURE 0.2G, which indicates that a short circuit developed by showing that concentric circles had developed around both stimuli. The idea was that when the two sets of concentric circles overlap a short circuit would have developed, which is represented, as path segment Z. Path Z would have effectively established a connection, literally a short circuit, between the CS and the US that made the two stimuli identical. This was what Pavlov meant by stimulus substitution.

In Figure 0.2G across the bottom, the two pathways taken by each of Pavlov's two stimuli are shown:

The Unconditioned Stimulus, US, takes path A, B, C, and D, while:

The Conditioned Stimulus, CS, takes path X, Y, Z, C, and D. Where Z, which is shown in the figure, represents the place within the brain where the short circuit evolves. What did Pavlov (1960) mean that learning is due to stimulus substitution? Did he mean this literally? Of course he did, and for good reason, it works; the CR = UR, what more could you want?

Gastaut: A Human variation on Pavlov

Gastaut (1957) replicated all of Pavlov's (1960) original experiments on dogs upon humans. Of course, in order to replicate Pavlov's conditioning experiment, he had to choose a different pair of stimuli and responses. With dogs, Pavlov had actually done surgery and externalized a salivary gland on one side so he could easily collect and measure the volume and onset of the salivary response; both timing and volume were his dependent variables.

So exactly, what pair of stimuli did Gastaut chose for his human experiments? Gastaut chose a Light and a Bell. His question was what would happen if a bell, an Indifferent, I, stimulus, was paired with Light, an Absolute stimulus, A, as defined in FIGURE0.2E. Of course, the light is not going to make a human salivate. So what is the logic behind the choice of these two stimuli?

In FIGURE 0.1B, it was noted that every one of the five sensory systems has its own Primary Projection Area, PPA, and that the specific stimuli which are the effective stimulus for each of the five senses, are an Absolute stimulus for its own PPA. For example, an auditory stimulus is an A stimulus for the auditory PPA in the temporal lobe, while a visual stimulus is an A Stimulus for the occipital lobe, which is the PPA for vision. The rule is that each absolute stimulus, as shown in FIGURE 0.2E activates or desynchronizes only its own PPA. Every stimulus, when it is in the A state, activates or blocks only its Target PPA as seen in FIGURE 0.2E.

In a human experiment with a light as the A stimulus the light's target would be the occipital lobe, the light as an A stimulus would block the occipital lobe.

Now what can we expect from the bell, which on trial one must have been converted into an I stimulus or what Pavlov (1960) called the CS? First, the bell as an indifferent, I, stimulus means the two reticular structures (of pathways 3 and 4) are indifferent to it since their centers are surrounded by inhibition (See frame three in FIGURE 0.2E). Why would Pavlov require, in this case, that the bell be an I stimulus? Because, and this is important, in order to do a Pavlovian acquisition experiment in the first step an A stimulus, which will serve as the CS or as Pavlov liked to say, the "to be conditioned stimulus," must first be converted into an Indifferent, or I stimulus so that initially it has no conditioning effect. However, the light, which is an A stimulus is fine as is. These terms were defined in FIGURE0.2E.

On trial one, the bell must be an "I," or an indifferent stimulus as defined in FIGURE 0.2E, while the light must be an A stimulus. The remarkable causal effect Gastaut (1957) discovered was that after a few pairings the onset of the bell would start to desynchronize the occipital lobe, and so as Pavlov's theory of Stimulus Substitution predicted what had been the Indifferent stimulus had the same effect as the A stimulus; the CR was now equal to the UR.

But think about this! What lobe should a bell desynchronize? The temporal lobe is the PPA for the auditory system, but only when the bell is an A stimulus. So what business could an I- bell, have in desynchronizing the Occipital lobe? This effect is not normal, right. Have we created a crazy bell? Why should a bell block the occipital lobe, which is the light's business!

Why should this occur? Why should a bell desynchronize the Occipital lobe?

Get Set, Go, or be prepared

The answer is obvious or even simple if one remembers the purpose of learning; namely, that learning is a matter of being prepared. So what is the role of the bell in this acquisition testing procedure? The results are that the role of the CS bell is to get the occipital lobe set, or prepared for the light, so that the occipital lobe will instantly be able to begin to process the light. In order for the occipital lobe to be ready to process the light, it must be desynchronized, expecting or "set" for the light. Learning, once again, is a matter of being "set," or being prepared. So how and why does the I-bell desynchronize the occipital lobe?

The concept of set has a long history, in sports, as in science. In sport's races, all the athletes line up and get set for the race by assuming a position that will allow them to make the optimal start at the sound of the gun. The standard referee says, "on your mark, get set, go." In the nonverbal mind, the CS serves as the "on your mark" signal.

Why then do I challenge Sperry's claim that there is a mind in each hemisphere, that each human has two minds. I must insist that the horizontal component of the mind is not a standalone mind. Why did evolution create the mind out of both a horizontal and vertical component? My point is that the horizontal mind cannot function without the pacemaker function that only the vertical component of the mind can supply. One of the major advantages that emerged generally due to the evolution of the nonverbal mind, but specifically due to evolution of the TRF, is that the cortical processing centers probably cannot perform at all when deprived of the energizing and/or pace making effects of the arousal pathways. Without the arousal energy of the vertical arousal pathways, the cortical centers would be frozen or locked in sleep.

The Indifferent bell after being paired with an Absolute light now acts like the Absolute light stimulus, by activating specifically only the Occipital lobe, which gets the Occipital lobe set. The TRF serves as a specific pacemaker by activating the Occipital lobe. Once the I bell has been paired with the A light, the bell becomes functionally identical with the A light. The bell now can activate the TRF to alert the Occipital lobe.

If we stick with our match-mismatch example, the horizontal region of the mind can only respond to syndromes of sensations for which it has been put on alert by the predicted, P signal. Specifically, only the focal TRF level of the arousal circuits can generate and deliver the arousal energy when and where it is required.

The collection of five vertical pathways evolved as a coherent set of pacemakers. They have been tested for over fifty million years, throughout the full time mammals have lived upon the face of the earth.

These four intrinsic vertical circuits are the pacemakers of the mind. They have the critical technology; they can induce a bell to predict the light.

The match-mismatch mechanism

In the match-mismatch mechanism, I noted that the role of the vertical component of the brain was to predict, P, what would next enter the cortex via the CP, which is the O. The purpose of the nonverbal mind is to enable mammals to survive in a world of change. The point of Heterostasis is to take the change out of change. The sooner the nonverbal mind can conceive a valid model of an anomalous domain the sooner the effect of change will be negated. In a causal world, all the changes introduced by "change," are predictable events, and so the role of the nonverbal mind is to identify and model the causal parameters of the new domain. As soon as the mind can model the "change", embodied say in the matrix maze, then the new model created by the nonverbal mind will have become the new norm or the new paradigm, and the mind will then have reduced what were anomalous events to scripts, which transform the strange to the familiar. As soon as the mind can model a domain and can predict every event of this new domain, then this new world will have become redundant. Heterostatic organisms are genetically programmed to require enough novelty to maintain their level of arousal within the B range of the Y-D law and will have to leave the security of this now "known" world and explore until they can find a domain with a new plate full of "change." The nonverbal mind is addicted to change, and no domain can hold a mind without a quantity of change sufficient to entice the mind to stay a while (every married couple take note) until, yet another model can be created, which once again models and destroy the anomalies, which induced the mind to stray. Mammals, thanks to their nonverbal mind, are restless souls, which cannot stand still. Heterostatic behavior is a base-two pattern of behavior that cycles as every base-two system must with the inner-most rhythm incessantly cycling between its two base values, Explore, E, and Retrace, R. Heterostatic behavior is a start-gradient form of behavior that retraces only long enough to discover the causal structure of a domain sufficient empirically to justify the beginning of a new exploratory cycle.

The design of Environmental Hierarchical Place/Base Systems

FIGUREs 0.2H, I, J, and K

How does the nonverbal mind conceive an abstract nonverbal set of Kantian forms of the maze from which a concrete set of nonverbal images can be constructed to create the exemplars of consciousness? Where does order come from?

THE Place-Base, P-B Problem: a pure Place one-base structure

What is nonverbal knowledge? How can nonverbal knowledge be encoded and stored? Is nonverbal knowledge an abstraction that can be stored off site in a book or journal, or is nonverbal knowledge domain specific, purpose specific and concrete? The latter I presume. The rat's nonverbal knowledge of the maze is stored as a structure with a dual function; first, it must embody a virtual model of the object represented in long-term memory. Then at the other end of the continuum, the nonverbal mind must create in consciousness a virtual world that allows an organism to rehearse its behavior in the privacy of thought and only issue a manifest copy of the planned narrative after the individual has satisfied itself that its behavior will achieve the desired consequence. In this section, I will present some ideas about how an organ, the nonverbal mind, evolved that can engage in these kinds of rational behavior.

Lilienthal

I must attribute in some degree, whatever insights I may have in these matters to my father's insistence that I read David Lilienthal's small book on the TVA when I was a teenager. Lilienthal discussed how the TVA constructed a model of the Tennessee River's drainage regions, which would then allow the TVA experimentally to compute the drainage time for x inches of water released over some particular watershed. They then could measure the rate at which any given quantity of water moved down stream, with or without a dam in place, at various projected dam sites. The TVA's model of the Tennessee valley was, of course, physical, yet it was a scale model never the less. (Lilienthal, 1944) The point was that their behavior was rational as I define the term, for they did not act, i.e., build a new dam, until they knew how it would affect the flow of water moving down stream.

Two concepts of rationality must be kept in mind: verbal rationality and nonverbal rationality. The classic assumption that human uniqueness was entirely attributable to their verbal rationality is still very much alive and well; yet one of the main themes of this book is that nonverbal rationality is the primary kind.

The problem with the classic verbal conception of rationality is that it had its origin in the days when no one had heard of the nonverbal mind and nonverbal rationality. The nonverbal families of phenomenon are a package of mental properties that are attributable to mammals; humans are mammals, but still the idea of the nonverbal is meaningless until, one at a time, readers understand the nature of nonverbal behavior and nonverbal ideas.

Pragmatism has long had as its goal the advancement of the concept of behavior independent of words. But the first two generations of pragmatists were still unaware of several meanings of the nonverbal. Presumably, the nonverbal was what they were after all this time.

My proposal for the role of the nonverbal mind supposes that the nonverbal mind can construct a virtual model of any spatial domain and can empirically generate a Prediction for the consequence of any plan of behavior, and thus empirically and rationally rehearse various scripts and determine the consequences of any that they desire.

FEMALE THREE

For a rat example, I propose that Female 3 conceived a virtual spatial model of the maze with a rational estimate of the empirical dimensions of the hierarchically organized maze. This virtual model of the 96 cellular structures of the maze embodied all the D2, R3, RT2, C8 features faithfully in their place. Female three, in her mind, could then rehearse her moves through this virtual model with complete safety. Only after she had proven to her satisfaction that her plan worked would she emit an overt set of moves, which made manifest her plan.

My question here is a most practical one. What kind of mental structure could the rat's mind construct, which could permanently store a model of the virtual world of the maze? Also, one that can be called up on occasion as needed. I presume that the rat's mind can construct a covert model; much like what the TVA constructed and what I constructed as a human scientist as I struggled to understand her behavior in the maze. The rat is a nonverbal scientist, whose nonverbal behavior is most similar to my human nonverbal scientific behavior.

I suppose that the rat's nonverbal mind and my nonverbal mind both roughly employ the same generic nonverbal mammalian mind, GNMM, and that both minds could create a virtual model of the maze. The first difference is that as a human, I am many times larger than the rat and thus the rat and I experience the maze from very different perspectives; and then secondly as a human, I can export my model employing the resources of the human verbal wisdom.

A mixed base conceptual system

In this section, I am attempting to describe some of the conceptual models I employed to conceive the maze. I propose that such models are a mixed base conceptual system. My purpose now is to describe how a mixed base system is designed, created, or implemented. However, a mixed base system is an advanced compound system, so I will begin by describing one or more, simpler single-base systems, in particular a Base two and a Base three model. Afterwards I will then discuss a mixed base model.

How to construct Place/Base Rhythms: Base 2, 3, and mixed-base

FIGURE 0.2H: a base-two system

FIGURE 0.2H

FIGURE 0.2H shows a base 2 design; a 2^{5} design; an A2, B2, C2, D2, E2 model. This figure illustrates how the structure or how the construction of a Base-2 to the fifth system works. However, this illustration only shows half of this design; only sixteen of the thirty-two rows, a completed system requires. What is missing? Can you complete this illustration?

In this section of this book, I am claiming that even the rat's nonverbal mind can execute those functions, which Plato claimed only a god could do. Moreover, I am ready to propose a theory to explain how evolution created a nonverbal mind in rats, in which these god functions are the primary sources of the adaptive powers of a rational nonverbal mind.

Kant (1952) was the first to recognize that experience was function of the joint product of nature and nurture, with the mind providing the forms or concepts, while experience or the senses provided the precepts, or mud. By analogy, he argued that the forms gave shape or meaning to the precepts or mud. Yet, Kant was nativistic with respect to the origin of the forms, and an empiricist with respect to the origin of the precepts or mud. My break with Kant was because I propose that the forms as well as the mud had an experimental or experiential origin.

Instead, I am proposing a theory of where the hierarchical forms for giving meaning to both space and time came from. How can a rational mind create a set of forms for conceiving or parsing space and time? I am proposing that an experimental rational organ, a nonverbal mind for creating hierarchically organized forms for parsing space and time first emerged in the earliest mammals such as rodents.

I am proposing to reveal the secrets of the rhythm master, the mastermind, the rhythm-compounding component of the mind, the one that conceives the forms and creates sets of Place/base systems.

Moreover, I am intrigued with how evolution brought together within the mammalian brain at least three different tissues or organs whose interactions became the nonverbal mind. In particular, I propose that evolution took the vertical arousal circuits which were probably originally designed to supervise the circadian rhythm and transformed them into a technology which encodes information into rational Place/Base systems, where the element of the place/base systems are nested or stacked rhythms.

Choosing the place

The nonverbal machine mind designs models by conceiving and assembling the column elements of Place/base constructs from right to left. The first place, or column, on the right is the inner-most or minor element of the system, and each new element is added by moving to the left one column or place at a time as each next higher-order Place is added to the design. The left-most variable or place is the outer most variable of a design, and is referred to as the major variable.

The variable on the far right is the minor variable and paradoxically it cycles the fastest. This scheme then conceives of the left most variable as the major variable and that it cycles the slowest.

Choosing the base: the kinds of bases

The set of bases can be chosen two ways:

The nonverbal mind creates models by designing sets of vectors, which provide a unique place, which identifies or names all or each of the cells of the matrix with unique vector names. Each individual cell's numerical name within a matrix must have a unique place or nonverbal name in the hierarchical design. Each cell's numerical name must be unique to associate it with one and only one level of each of the parameters of the matrix. The two kinds of place/base systems between which the nonverbal mind can choose are a pure one-base system, or a mixed base system.

FIGUREs 0.2H and I, illustrate pure one place-base systems, which means that the system always employs the same base at each Place.

FIGURE 0.2J, illustrates a mixed place/base system, which is the only kind that can be used to understand the four experiments of this book.

What roles do Place/base systems play in the nonverbal mind?

What do these Place/Base systems have to do with rat behavior? What do these systems have to do with nonverbal rational behavior? The Place system illustrated in FIGURE 0.2H is not the only kind with which evolution endowed the nonverbal mind.

One of the most impressive features of the nonverbal mind is its ability to self-organize, which must not be conceived to mean that mammals directly impose order upon their milieu. Nothing is that simple or easy. Rather we must slow down and understand the several steps in the mental process, which initiate action. For example, in the first step our mind must get organized to conceive the order we wish to impose upon the EMm. In the first step, the requisite order must emerge in our mind, second, this order then must be imposed upon the data or precepts entering the mind, and third, which in turn will impose order upon behavior; and then fourth or last, behavior imposes order upon our external milieu.

A base-two system uses two levels of symbols or numbers to create a nested series of rhythms at every place in the system.

Every place in a series of places is written or filled from right to left.

In FIGURE 0.2H a five places system is shown; see columns A through E that are labeled across the top of the page.

The next row shows that each place or each column has a base value of two. My purpose now must be to explain in words and deeds the rules for stacking a series of rhythms.

STEP 1: Writing column or Place A

Starting in the column furthest to the right, write in each cell one of the two levels of the base two symbols. Write 1 or 2 down the page starting at the top, and proceeding down to the bottom of the page. For this first column or place write, by taking turns between the two base symbols, 1 and 2, with absolute resolve. Additionally, I find it useful to mark the end of each rhythm, after every cycle with say a half-inch line beneath each second number, after each symbol two. The line-marks can be easily seen in FIGURE 0.2H. By the way, count how many cycles we got to a page of our illustration for place E. What did you count? The simplest way is to count how many half-inch lines were drawn for the Place E rhythm. There are eight of these lines, which mean that we drew eight full two-cycle bases for the E Place set of rhythms.

Step 2: Writing column or Place D.

In the next step, the column D place will be written, and now things get interesting, so I must state two general rules. The first general rule states that the next column to the left of the last place or column for which a rhythm was filled in is the next higher order, or major column or place.

The right-most column is always the lower order or the minor variable and as such, its rhythm proceeds with the two symbols alternating without interruption. The rule says the cycling of the inner most columns of a base two-symbol system cycle incessantly, at the highest frequency.

But for the second column, the nesting of the rhythms gets more complicated, yet the rule for creating the rhythm for this second column is the same rule that will be used for creating and stacking the rhythms for each of the successive higher order, or major set of columns or places.

The higher order places can be thought of as counting the lower order rhythm. Each next higher order rhythm counts the lower order rhythm. The complexity, the thing to be remembered, is that the rule says each level or each base of the higher order rhythm counts one cycle of the lower order rhythm. For example, since a base-two system has two levels or two base values: 1 or 2, then each level of the higher order cycle, both the 1 and 2, each count or covers one full cycle of the lower order rhythm. Look at the symbols in column D, 1, 1, 2, 2/1, 1,

2, 2/ etc. (each slash represents an end of a cycle.) The higher order place can be seen to have the values: 1, 1 that covers the 1, 2 cycle of the first column etc.

So in this base-two example, each base symbol of a higher place counts a full cycle of the lower place, or each cycle of a higher order place covers two cycles of the lower order place. In this example, for the variable in the second place, one cycle of the first D rhythm will cover two cycles of the E place rhythm. How many lines were drawn for the first column? How many lines did you draw for the second column, etc.? How many lines will be required for place C? How many for place B? How many for the A place?

How can you calculate the total number of elements in a base/place system? The number of elements is calculated by multiplying the base values times the place values together, or is the product of the bases of each place. For example, the total for the three Figures 0.2H, I, and J would be calculated as follows:

Figure 0.2H - 2*2*2*2*2 = 32 elements
Figure 0.2I - 3*3*3 = 27 elements
Figure 0.2J - 2*3*2*8 =96 elements

Nonverbal Mind technology

One of the central scientific challenges for me as a rat runner was to discover what skills or what technology was built into the mammalian mind. My task, in effect, required that I learn to read the rat's mind.

Single base VS mixed base designs

The nonverbal mind is genetically programmed to process the sensory input arising up the CP from the external milieu of the mind, EMm, within the context of the conceptual framework online, which will be some base/place scheme.

Choosing the base: the kinds of bases

The base for any parameter or place is simply the number of elements in the cycle. What is the base of D in the maze design: A2, B3, C2, and D8? The base for D is obviously eight but why, because the number of units in each row is eight. So why does A = two? Why does B = three? The base is but another name for the number of seasons in a year, or the days in a week, or seconds in a minute, etc. There are two directions in the maze: out and back so A=2 when A = Direction.

Anova talk

First, all the data in these experiments were statistically analyzed using ANOVA designs, and all the complex data sets were described using only mixed base data systems. In order to enter data into these designs for processing, it was usually only necessary to inform the computer first, of the parameters of the design and then, secondly, to enter each data point into the computer in its place within a series of scores dictated by the vector names. My computer had programs, which could read and recognize or sequence each data point by virtue of its vector name. These unique mixed place base names are required to create a unique place for each of the data points within the complex mixed base metric system, which had been designed for this domain specific place.

RULE ONE

Rule one: The Place-Base form-creating system of the nonverbal mind is a rule based system that biological evolution built into the nonverbal mind.

RULE TWO:

Every place in a place-base system is an X or a holding slot into which some data point can be slipped and assimilated as an exemplar of some place within the mixed base emerging system. In order for the nonverbal mind to model each spatial structure, each structure must be assigned a set of mixed places and bases. The two elements of a modeling system are places, and bases. A Modeling Machine within the nonverbal mind must create or assemble a set or series of Place-Base elements into a coherent structure or model. On the other hand, more simply, one of the roles of the Primary mind is to create a Kantian set of forms into which each data point has a unique slot, or cell, or a home place. (Kant, 1952)

A base-three system

Figure 0.2I

This place-base system was left for the reader to complete. The reader eventually should recognize that the nonverbal mind is the primary mind, and thus there are times when the nonverbal mind should be exercised. As John Dewey (1916) said, "We learn by doing."

A Mixed Base Structure

FIGURE 0.2J

In science however, a mixed base/place system is far more useful than a one base system, not only in the everyday real world, but also especially within the scientific context such as in the experiments used in this book.

What distinguishes mixed base systems is that they provide, in my case, the means for creating the type of conceptual system actually employed in any scientific community that used statistics such as ANOVA, but mixed base symbols also are used, to my delight, by the nonverbal mammalian mind even by rats in their everyday life.

The nonverbal mind is endowed with the technology of Heterostasis, which moves the rat through the maze while employing only two moves, explore and retrace. Heterostasis is the wisdom of the mind just as Homeostasis is the wisdom of the body. Heterostasis, however, induces a rat to emit behavior, which will make a hierarchical model while encoding the information or the knowledge about the hierarchical structure within a multiple place/base combinatorial type code.

I assume that the creative technology of the nonverbal mind was designed to encode the knowledge created or gleaned by the interaction of the individual's overt behavior within the maze into a mixed Place/base structure.

Based on my observation of rat behavior, rats have the mental power to conceive the maze as a mixed base system. For example: Both the rat and I conceived the maze structure as a: D2, R3, RT2, C8 hierarchical structure; with D2 as Directions two, R3 as Replications three, RT2 as row types-two, and C8 as Cells 8. This is a mixed base conceptual structure since it employs three base values: 2, 3, and 8.

FIGURE 0.2J

FIGURE0.2J shows how a mixed base matrix is operationally defined or constructed.

My training for analyzing behavior using ANOVA designs always employed a mixed base system, and as noted above and to my great surprise, rats employ this type of design.

Transformations: their nature

Evolution transformed the organs, which were originally designed for regulating circadian rhythms to function to create and monitor other type rhythms employed in rational behavior. The new generalized functions of the arousal system of the NV Mind did not replace the original sleep-waking functions, but supplemented them.

Transformation # one: T#1 the emergence of the Place System

But I digress: the purpose of this section is to describe and explain the remarkable achievement revealed in the emergence of the particular rational capacity of the nonverbal mind to employ Place/base systems, which provided another rational nonverbal element, or another function, which extends the adaptive powers of the nonverbal mind. Eventually these new rational functions at the human level extend human behavior into new formal domains of both counting and combinatorial, but at the level of the earliest mammals such as a rat, the studies of this book show that these place systems' functions emerged to enable the rational behavior of the rat to model hierarchically organized spatial structures.

Chomsky and all the verbal language experts are at a loss to explain where verbal behavior originated. I propose that verbal behavior grew out of nonverbal behavior that verbal skills evolved out of nonverbal skills but first, the nonverbal role of the place system function must be described. Perhaps a large part of the problem is that too many imagine a great impassible gulf between the nonverbal and verbal minds.

In its first life, the arousal circuits regulated the behavior supporting the circadian rhythm, but in its second life the vertical mind in conjunction with the horizontal mind broadened its function and became the Shepherd for the creation of all rhythm-driven forms. What happened was the emergence of a new kind of transformation. With the evolution of the vertical component of the mind, evolution broke new ground and produced a new structure/function that became the foundation for rational nonverbal behavior, but also provided the foundation for all future developments in mathematics, music, and science. Rodent exploratory play behavior paradoxically employs the new rational skill conferred upon mammals by this binding of these two old structures into the organ of the third wisdom of the nonverbal mind.

Why use mixed base systems?

Heraclitus's real: Two concepts of "change"

The words, "never change," seemed to Plato to have a simple, straightforward meaning. I suspect that Plato was incapable of changing his mind. Yet, Plato for some reason never grappled with Heraclitus's concept of the real. Heraclitus is famous for noting that it is impossible "to step in the same stream twice." Heraclitus's concept of the real was considerably more complex than was Plato's concept. Heraclitus accepted Plato's rule that "everything changes," but then Heraclitus takes another step and says, "Except the way things change never changes." The water in the stream was always changing, but the stream endured nevertheless. Heraclitus seemed intuitively to have understood that the stream was an invariant. (Freeman, 1970)

Heraclitus however, lived thousands of years before science was discovered, but science is based upon the idea that every scientific fact and every scientific law must be invariants. The essential nature of every scientific fact is that it is reproducible. In fact, scientific results are generally not accepted until they have been replicated by independent research, until the same results have been found by different researchers in different places and time.

However, it's time to make the point that both Science and Plato seek the real, "that which doesn't change" but science defined "that which does not change" after Heraclitus not after Plato.

The nonverbal mind was conceived as a technology, which had a deep understanding of the concept of an invariant, which again is defined as that which does not change during change.

The key to answering Plato's question "What is real Things or Spirits" ultimately leads to the question; what is change?

Plato (1952) defined Spirits as a set of beings that by definition do not change.

Science on the other hand defines things as a set of beings that empirically do not change the way they change. Things change, but they do not change the way they change.

Which kind of constancy is more interesting, Plato's or Heraclitus's?

Plato's world would be boring, to say the least, no new cars, or radios but more simply there could be no news, no anomalies, no science, etc. Boring, boring, and boring! Plato would be the philosopher/king of a utopian world that can best be described with one word - BORING!

Again, I beat about the bush.

What is the mind? What feature of the cosmos required evolution to design the nonverbal mind, the very feature that Plato most despised!

There would be no need or no place for a mind in Plato's ideal world. The mind evolved to enable life to deal with and to survive in the world that Plato despised, a world of change.

What is the point of this difference

Let us get back to the nonverbal world, which is the theme of this book. Back to the world of the nonverbal: the nonverbal mind, the nonverbal mammals, the nonverbal language, the nonverbal symbols, the nonverbal god, the nonverbal consciousness, the nonverbal narrative, etc. For the moment, let us focus upon nonverbal science, the nonverbal real and nonverbal knowledge.

Two worlds: the world of time and space, and the world of eternity

The nonverbal mind has no problem with invariants. In fact, the arousal system, which supplied the vertical component of the nonverbal mind, got its start as an instinctive mechanism for circadian rhythms, which is the prototypical invariant. Every day is a day, in spite of each day having its particular name and what about years, or months, or seasons, etc.

Invariants are the most obvious elements of time and of space. See the example below.

The natural mixed base system employed in real world domains are illustrated below: Consider the hierarchal dimension of time.

Consider the several examples below of how humans measure or represent time. Each measure of time is a concept or a variable but more specifically each conceives of some unit of time as a cycle. What is a year? A year is a feature of time that is generally measured employing the base twelve, or each year consists of twelve months.

How many Type bases are used in the West to represent time?

What is the relation between these seven cycles? Roughly, they form the hierarchical structure by which the Western world describes time.

In other words, Time is described as a Mixed Base Hierarchical Structure: (Y12, MI30 or MII4, W7, D24, H60, and MN60.)

12 Months in year; Y12:
30 Days in month; MI30: or
4 weeks in a month; MII4:
7 Days in week; W7:
24 Hours in day; D24:
60 Minutes in hour; H60:
60 Seconds in minute; MN 60:

Stacked rhythms

This chapter is about that part of the brain, which in its first life was designed to impose a circadian rhythm upon biological life. The original technology was designed to coordinate life with the rhythms, major and minor, of our planet. Humans are said to have first evolved in parts of Africa such as in the Great Rift Valley where geographical anomalies were the driving force for change, and forced humans to evolve bigger brains and at a faster pace in order to master the CHANGE that characterized their homeland.

The mammalian mind was designed for change, was designed to master change. What better way to master change than to internalize change. The master organ, the nonverbal mind became the rhythm master by branching out in several directions. Obviously, the way to master anomalous rhythms was to internalize them, to predict them, to know them, and to be prepared for them. Mammals lived in the midst of a real world that they could not escape; that it did no good to disavow, there were no ivory towers where they could retreat and denounce "bad matter."

The question, which should have focused the course of human evolution, was how evolution could design a new kind of being: one designed to not only deal with, but to thrive on change. In this book, I claim to have found an answer to this question; namely, that evolution designed a new kind of organ, a rational organ. However, there are several themes in this book.

First, the nonverbal mind emerged by evolution causing a marriage between three systems which originally may have served independent functions. Nonverbal rationality emerged as evolution designed a technology for managing rhythms or for internalizing, modeling, or more concretely predicting rhythms. The nonverbal mind evolved as a function of biological life's need for predicting or factoring both time and space. However, this book reports four experiments which only studied rodent's rational nonverbal ability to factor and/or predict the properties of space.

The evolution of Heterostasis rationally enabled individual mammals to model the parameters of their niches just for fun. The remarkable trait of the nonverbal mind is its ability to reduce the parameters of the matrix to rhythms. Every parameter/idea is a rhythm!

The cycles/rhythms of life: of space and time

Humans obviously have real reasons to take pride in their verbal ability; however, I propose that paradoxically they have as many or even more reasons to take pride in their nonverbal ability. For thousands of years humans have a taken a perverse pride in their superiority over the other species of life even the mammals which Darwinian or biological evolution showed almost certainly were our ancestors.

However, now is the time to take biological evolution one-step further and show how our mind evolved in two stages with respect two languages and when the verbal language model emerged. It is now clear that our mind not just our body was the product of biological evolution: that our nonverbal mind evolved by natural selection. Or more to the point, it is time now to differentiate myself from the traditionalists. The nonverbal mind of mammals is not a spirit, a gift from some verbal god but evolved, as did every other organ of the mammalian body.

Plato's Mind: Nativism vs. matter

The basic idea, which is assumed here, is not limited to philosophers, but is why the universal assumption especially in the transcendental religions. The most obvious difference between Plato (1952) and the Christians is that Plato has his god put ideals or truths directly in the head of the philosophers at birth, while Christians have their god put its truths in a book. The god of the Christian version, at least the Protestant version, requires that followers be literate, while Philosophers, in principle had god's truth already stored in their head, they don't even have to read.

The origin of the functions of the nonverbal mind does not require the assumption of a set of metaphorical beings: a god who is a heavenly father with a son who was born without sin and with a whole host of heavenly beings. All of Plato's perfect beings are perfect primarily by definition; his ideals are perfect eternal beings because he said they were. Moreover, if you accept his assumption that because they are spirit beings who have no material being, then these spirit beings are perfect eternal beings, simply because Plato (1952) asserted that they are free of the corrupting influence of matter. Plato called himself a rationalist and would have been offended if he were asked for empirical proof of any of his assumptions.

The rat's mind: empiricism and matter

I propose: First that all mammals including humans share a generic nonverbal mammalian mind, which is the GNMM and is the primary mind and houses twelve stations of the thirteen-station mind.

Second, the human verbal system was not the only system, which brought to fruition physiological systems that have been evolving for millions of years.

In this chapter, I have probed even deeper into the origin of our supposedly unique human wisdoms or skills, and I am now proposing additionally that those mathematical skills, which also have been such an additional source of human pride, are no less dependent upon the nonverbal mind than is our verbal nature.

My rhythm, cycle, or loop bias

I must confess my bias for cycles. In these next few paragraphs, I will explain and describe my bias for loops.

Historically the behaviorist chose trials as the central methodological element of their testing procedure because presumably they could imagine no other option. However, intuitively I felt the concept of a trial was unnatural. The biological forms that evolved upon the earth adapted to its environmental cycles: years, seasons, months, and days. For this and many other reasons, I began my research over fifty years ago by assuming that cycles not trials was the natural unit of behavior.

The behaviorist defined a trial as a one-way trip through a maze from the start box to the goal box. If an experimental protocol required a rat to run several trials a day, the experimenter was required to take the rat from the goal box and carry it back to the start box by hand. This seemed to me most unnatural. In the real world, when an individual rat leaves its nest or den for some reason there are no experimenters waiting to carry it back at the completion of its task. It is more natural to assume that leaving the nest begins a cycle, returning to the nest completes a cycle; each individual is responsible for both legs of the trip out and back, for getting to some goal and getting back home again. In the real world, rats or any mammal for that matter that goes out and is not able to find its way back home simply would not have survived. Going out without any thought of how to return home would not be a successful survival strategy.

The Loop Testing Procedure, LTP, operationally inserts loops or rhythms into this set of experiments and assumes that: Life is commuting, and commuting is a cycle. Every day consists of one or more such cycles, for rats as for humans, a day consists of cycles: whether going to school, shopping, work, a movie, dinner, etc. every unit of normal behavior occurs as a cycle or loop. One of the most basic assumptions of the nature of mind is that the vertical component of the primary NV/mind emerged out of this circadian rhythm inducing pacemaker, which was designed to incorporate this basic feature of the planet earth. This neuronal circuitry of the brain/mind was designed in the image of the circadian rhythm nature of the planet earth. Its success is most profoundly reflected in the sleep-waking cycle.

Thus, I chose to study rodent behavior within the context of the concept of cycles or loops, the LTP, rather than the context of trials, the TTP. I recognized that music; rhythms, cycles, loops, phases, and even time and space itself can all be reduced to complex hierarchical patterns of frequencies, amplitudes, and phases. In the experimental chapters, I will begin to describe how the Heterostatic behavior spontaneously emitted by the primary mind is hierarchically organized in growths, runs, loops, waves or cycles. I will show further that the primary mind, which emits hierarchically organized behavior is itself organized by waves or circadian rhythms, which pervade both the body and the brain, especially the EEG rhythms.

Hierarchical structures

A hierarchical structure can be visualized as analogous to a set of nested bowls or mixing bowls.

For example, my human mind conceives of the maze as an A2, B3, C2, D8 hierarchically organized conceptual system--the four ideas can be or must be thought of as nested or stacked with D as the bottom bowl and with A as the top bowl. The metaphor works better if you turn the bowl upside down. Then D is on the bottom; C covers D; B covers C; and A is on top.

However, in the mental context I must describe technically, what is meant by nesting. In FIGURE 0.2J by convention these letters: A2, B3, C2, D8 are nested, and by convention A is the outer most loop and D is the inner most letter or bowl. By convention D8 would be the inner most loop and thus it would cycle the fastest and constantly, while A cycles the slowest.

Note in FIGURE 0.2J the base values of the D columns that consists of a sets of the numbers 1-8, which relentlessly cycle as a Hericilitian rhythm should. Such a rhythm changes but never changes the way it changes. As your attention moves left, the variables move slower.

FIGURE 0.2J

The mixed base model-The Heraclitus model

Heraclitus was a real-world theorist who is most famous for noting, "It is impossible to step in the same stream twice." (Freeman, 1970) The stream was an invariant that never changed being a stream and yet the water in the stream was constantly changing, and so it is in the Base-two system in FIGURE 0.2H. Each of the five columns of the figures is a stream of numbers that are constantly changing and yet the way they change never changes.

Does the Rational nonverbal mind only employ a single-base or a mixed-base system?

Which kind of Place-Base system did Female 3 use to create her model of the maze?

Can the nonverbal mind employ both? Based on the data of my four experiments, it is certain that rats can and do employ mixed base conceptual systems. For example, Female 3, F3, employed a mixed base system to model the matrix maze.

Nonverbal Mind reading

A mixed base system allows the nonverbal mind to choose any base at each new level: the nonverbal mind was constructed so that its technology allows it, in the extreme case, to employ a different base at each place or of course, to repeat any base any number of times.

The Mixed place-base Maze

The maze is a mixed Place-Base system, or a real-world system. In the illustration, the system is an A2, B3, C2, D8 design with bases: two--2's, one 3 and one 8.

The place system for this data in the figure is incomplete. It includes only 40 of 96 units. The reader should complete the figure on his or her own.

The sixth sense is the particular organ or place within the assembly line of the nonverbal mind where the nonverbal mind creates sets of nonverbal ideas, which captures the rhythms or the formal structure of each spatial domain. The purpose of the nonverbal mind is to create a map, stacked rhythms, or models of the rhythms of space, which consists of knowledge, which embodies or internalizes the forms or rhythms of space, which the nonverbal mind can then use to plan its behavior. The nonverbal mind creates the domain-specific knowledge, which informs the nonverbal behavior of all species of mammals.

The nonverbal mind must be capable of capturing the domain specific forms of any and every domain of the cosmos occupied by some species of mammal. The nonverbal mind's chief function is to capture an image of the formal structure of space as a set of coherent ideas, a set of knowledge, a set of rules, or a set of stacked rhythms or laws about the invariant structure of the space of any domain of the cosmos. This knowledge is then transformed into a virtual world, which functions as what Kurt Lewin (1943) called "a Life Space."

I assume that the nonverbal mind is in the image of the cosmos; or consists of a universal Pacemaker that can create any rhythm by which I mean it can impose any rhythm upon the EEG rhythms of the cortex. Furthermore, it can conceive or recreate the hierarchically organized intrinsic rhythms of a domain, by creating an idea or a form that simulates the rhythm of each variable.

The NV/mind is not always successful at conceiving an orthogonal design for any space; first because not all spaces physically embody a coherent design. My only prediction is that the NV/mind will attempt to conceive an ideal coherent form on a space, but, of course, it cannot conceive a conceptual space that is better than the space where it is playing.

How to construct Counting cycles

FIGURE0.2K

FIGURE0.2K illustrates a different kind of cycling structure than the three we have examined so far. The differences between FIGUREs 0.2J and FIGURE0.2K may not look obvious, but actually and historically, they are profoundly different.

There are two kinds of cycling systems, first combinatorial and second counting systems. The first three examples FIGURES0.2H, I, and J were combinatorial systems; while, this example FIGURE0.2K is the first and only example of a counting system. The difference between the two is too simple, so that at first it would be natural to discount it.

The difference, as you may have noticed by now, is the presence of the zero as an element of the base of each Place of this system. The Arabic numerals are famous for introducing the zero. In creating a combinatorial system, a zero does not make any difference but as for counting, it can't be done without a zero.

Look at each place and notice that at each place a zero is always the first element of each base at every place.

The figure is a pure base two cycling system with four places.

The counting rule

Before I describe the rule, we should note the two rows of numbers above the line, but under the alphabetic names for each of the four columns, A-D. In the first row above the line, the numbers labeled are the place values, which are the successive powers of the respective base, in this case, the powers of two. In the next or second row above the line, all the numbers are twos, which means that this counting system is an example of how to count by twos.

The counting rule states that the count for each row of the system is the sum of the products of the base value times the place value for each place for each row. How this rule works is shown on the right side of the cycling system. For example, consider the fifth row, the one with a count of four. In this row, there is only place with a base value other than a zero, place three from the right, which has a base value of one. All the other places have a base value of zero. What is the count for the fourth row? With the base value of one and with the place value of four, then the sum of the product for this row is (0 * 8) + (1 * 4) + (0 * 2) + (0 * 1) = 4

On the right side of FIGURE0.2K in red is a table that shows how by using this set of stacked rhythms, it is possible to count to fifteen row by row from the top down. Each row shows how to compute the count for that row.

These four examples illustrate one of the essential properties of the cosmos

First, these figures are examples of how cycling structures are constructed and work. These structures are also examples of hierarchically organized structures, but they are down to earth rational constructions, and Plato would be at a loss in comprehending them, while Heraclitus should experience the full measure of their power and the beauty of a construction of such symbols and rules. They are examples of an invariant, rules that describe the behavior of physical systems that do not change during change.

Only the human mind with the fourth wisdom could verbally describe these behaviors, but as strange as it seems, the nonverbal mammals can encounter these nonverbal invariants in the physical world, as evidenced by the behavior of Female 3.

The six properties of the three/five vertical channels

Why are there five arousal pathways shown in FIGURE 0.2A and only three arousal pathways shown in FIGURE 0.2L? Figure 0.2A is correct there are five arousal pathways, but there are only three kinds, D, F, and V for the CP. FIGURE 0.2L was designed to list the set of six technical properties of the three kinds of arousal channels ignoring the differences due to the plus and minus signs.

FIGURE 0.2L

In this Figure, there are only three kinds of pathways, types, 1, 2, and 3, or D, F, and V, if we ignore the differences due to the plus and minus signs, where the plus represents an excitatory channel, while the minus sign represent inhibition or sleep. In FIGURE 0.2A, which shows the four intrinsic arousal channels, the two minus channels are shown on the left, while the positive excitatory channels are shown on the right.

FIGURE 0.2L was designed to show the set of six properties of the three kinds of arousal channels including both the extrinsic and the intrinsic channels. The Plus and Minus channels of the same type have the same set of six features.

The meaning of Observed, O and Predicted, P, O-P

The knowledge in this figure requires some serious thinking to understand, yet the concept of the "observed" and "predicted" which we discussed in the match-mismatch system, and which at the time perhaps seemed to have a simple enough meaning. However, in this FIGURE 0.2L we must go beyond any obvious verbal meaning and in this figure begin to understand how evolution operationally defined these complex terms. These sets of six features of the three kinds of arousal systems require our attention.

The Three channels compared

Pathway one: In FIGURE0.2L column 1 on the left states that the pathway #1, or the classical pathway, CP, which is a bundle of all the afferent nerves has the following set of properties: It is the fastest pathway, or has the fastest latency, which reflects the fact that it has the shortest length, with the fewest synapses. The fourth row shows that the CP has the most specific cortical electrical effect. Actually, in the old "Ma Bell" telephone circuits of my childhood there were two kinds of lines, the party line, and the private line. In the former, the party line, several people shared a line, everyone on the line could hear everything that was said, and some listened to everyone's conversations. But if you were rich, you could get a private line. Anyway, all the afferent nerve lines from the five senses to the neuron in the PPA were private lines, or as it says in row four, the connection was specific.

The next two rows show that the electrical connections from the cones in the eye to the neurons in the Occipital lobe were not only private but were directly to the cell body not to the Dendrites. The monitoring of these communications requires a special kind of recording called an Evoked Potential, EP.

Pathway 3 on the other hand has the opposite set of features. It has the slowest latency, which reflects the fact that it is the longest pathway and has the most synapses.

The communication in this vertical pathway is via F, the collateral that branches off the CP, but then terminates upon the dendrites in the cortex. These signals in other words affect arousal or specifically they terminate upon the dendrites and may alter the EEG.

Note that Pathway # 3 has a diffuse effect upon the cortex. Diffuse means that it affects every dendrite in the cortex. This pathway is a great Party line but deliberately so and for good reason. This line has centrality and can immediately send an emergency signal to every neuron or to every dendrite. This line is the equivalent to a fire alarm in a community or of a civil defense alarm, etc.

Pathway two's features are all intermediate in value to channels one and three.

In our discussion of the match-mismatch system seen in FIGURE 0.2F, we learned that of the two elements of the mind; the horizontal and vertical components each contributed one of the two different critical conceptual elements that are compared in the comparator: O is contributed by the horizontal component, while P was the gift of the vertical.

So the horizontal component of the mind gets two kinds of inputs from these five pathways. First, they get their afferent nerve knowledge, which is headed to the PPA's, which are station 5 in Aristotle's model; secondly, they receive a complex pattern of arousal energy from the four pathways, which are the Pacemakers for the nonverbal mind.

The critical question is how the slow channels can predict the content of the fast channel. It would seem that it should work the other way around! Indeed, why should the vertical arousal channels all be slower than Pathway #1?

Discussion of the significance of these properties

We must keep in mind the idea that to know is to be prepared! Our technical issue now is how the four slower arousal channels can get the cortical center prepared for the data of the classical pathways, CP, when the CP is the fastest channel!

This first obvious question requires that we consider the significance of the differences between the features listed above; let's especially consider the time or temporal aspect of the differences. Consider the neuronal knowledge within the CP at any moment, the potential knowledge, which might be thought of as say the mail that is being delivered to the horizontal mind at time one, T1.

So why should the four arousal channels, D+, D-, F+, and F- all be slower than the CP? Even if they were as fast as the CP, they still could not race ahead and activate the appropriate PPA that will process the US's. So we have a time problem.

The "Get Set" time problem

The nonverbal mind emerged in order that mammals would have a better way to adapt to time: to get set, to get ready for some event at T2.

This problem goes by, of course, too many names. However, Lashley, proposed what I think is an ideal name, the serial-order problem.

Two times must be considered in understanding learning, Time 1, T1, and Time 2, T2. The CS is presented at Time one, T1, while the US is presented at Time two, T2. The role of the CS at T1 is to get the cortical PPA for the US, which will be presented at T2 set, activated, or prepared.

Trials	T1	T2
	CS	US

My goal now is to explain how the CS events at T1 can get the cortex prepared, set, or aroused, for the US events at T2.

If I may perhaps oversimplify my case, the central issue being discussed in this chapter is what the difference is between the nonverbal mind as a one-dimensional system, as only a horizontal system or as a two-dimensional function involving both a horizontal and a vertical component of the mind. The answer to the first question what is learning almost answers the second question below. Question one: does learning happen at one instance, at one time or something that happens at two times, at T1 then T2.

In the most abstract sense, learning is acquiring the knowledge that allows the organism to be prepared. Or more specifically, for F3 learning is more a function of the vertical than the horizontal component of brain/mind. The ultimate manifestation of "being prepared" is extrapolation. The ultimate proof of knowing is extrapolation, which enables an individual to explore a large proportion of an anomalous maze without making an error, errorless exploring, performing without error is the most certain proof of knowing.

Now I must show that this kind of behavior is a function of the Vertical arousal system of the mind. Performing without error requires that a valid model of the maze has been stored in memory, and by accessing this knowledge, the organism can be prepared to make a correct response at an intersection where it has never been before. Knowledge enables the arousal system to get the organism prepared to make the correct choice at each choice point. Knowing is being prepared. In any domain, to know means to be prepared. That may sound impossible, but that is what science is all about. Knowledge makes prediction possible, which enables the knower to be prepared.

Recall the Match-mismatch system

The match-mismatch mechanism only allows a syndrome of features to become an exemplar of a class to enter consciousness when it is a valid concept. Real things according to mind theory are predictable events.

In order for the mind to assimilate an Observed event, O, from the CP, the mind must have been forewarned, or O must have been Predicted, P. The mind can only assimilate those events that have been predicted. The rational mind is a self-organizing system. It is not born knowing but is born knowing how to know. A nonverbal mind that knows how to acquire knowledge must know how to arouse those cortical centers, which must process the signal. The arousal system bears the responsibility of preventing surprise: to forewarn every region of the mind that must participate in the assimilation of the afferent knowledge that is on its way into the cortex. The slow channel arouses or issues the warning signal, which must allow the cortical centers of the PPA enough time to get prepared

For example in order to catch the bus at T2, we must start getting ready to meet the bus at some earlier time T1. One of the most difficult tasks of growing up in the everyday world is the fact that the T1-T2 interval must be empirically determined. However, in Classical conditioning the T1-T2 interval is only a matter of .5 second. Yet for catching a bus, the interval may be thirty or more minutes, or hours if we have to travel to the terminal. In real life, catching a bus requires that we set our alarm for T1, at some time which will give us time to do all those things which must be done before we can catch the bus.

The role of the vertical arousal system originally evolved to adapt the organism for sleeping and waking; a function that occupied say eight hours out of a twenty-four hour day. But when the sun goes down mammals, such as gorillas, have to be ready; they had to have made their bed and have gotten ready to get into bed. There are no lights, candles, or flashlights, etc. Being prepared generally means a different thing in each context, but being prepared is a major consideration of every function, both sleeping and waking.

Begin with the fact that "Knowing is a matter of being prepared."

The two problems:

FIGURES 0.2A, 0.2E, and 0.2I all show some aspect of the vertical mind, which operationally defines the T1-T2 interval. It is important to understand how nature through evolution designed a neuronal technology, which can impose, in effect, any conceivable T1-T2 interval. In terms of classical conditioning, T1 is the CS and T2 is the US.

FIGURE0.2A, The candelabra model shows five pathways converging upon the cortex. These channels can be divided several important ways.

First: let's consider the 4-1 split. Channel 1 is the Classical Pathways, CP, which is a bundle consisting of all the afferent nerves, which bring all the sensory data into the cortex from the five senses.

Second, the other four channels are reticular channels, which are elements of the vertical arousal system. I like metaphors, and I like to think of these 4-1 vertical channels as a kind of racetrack.

Rule one: T1 can happen at any time.

Rule two: T2 happens .5 seconds after T1.

Rule three: In order for an organism to respond most quickly to T2, the PPA for T2 must be SET or ready to process T2. The PPA for T2 must not be surprised by T2.

FIGURE0.2E

Problem one; the four pathways 2+, 2-, 3+, 3- are all slower than # 1, so how can they predict any P#1, T1 events? Answer: they can't predict T1 of P# 1, but they could predict T2 events of P # 1!

What is the time of P# 1 event, T1 or T2?

What is the source of P? How many P's are there? Is there P #1 prediction, in addition to sixth sense P? FIGURE 0.2F shows that that the arousal or vertical components of the mind have a Modeler.

The vertical modeler

The vertical mind component has one feature, which I have ignored until now namely, that the four collateral pathways are collateral or intrinsic pathways. The four intrinsic pathways have only one place they get their knowledge from, the CP or P# 1!

So the extrinsic knowledge of P#1 must be understood as Knowledge at Time 1 with respect to the intrinsic knowledge of the four intrinsic arousal channels. I can only conclude that the T1 intrinsic knowledge of the channels 2 and 3 must be destined to predict knowledge at T2, or the extrinsic knowledge, which is in the next car in the convoy!

The Firecracker model: The Pacemaker Function

In chapter two, the horizontal model was broken up somewhat arbitrarily into 12 functions each of which was executed at one of 12 stations. There is a lot going on in the horizontal component of the mind, but what I did not show was that all the computing occurring in the twelve channels is constantly interacting with the vertical component of mind. The nature of these interactions will be discussed later in this chapter. However, for a moment consider this problem:

how does the vertical mind get the cortical cellular networks prepared at each of the twelve stations?

The Synchronization-Desynchronization continuum or the S-D function

The cortical matrix is a network of neuronal cells with a thick canopy of dendrites that are always active, always oscillating and firing. In the Firecracker model, I will show that these cortical networks are controlled by pacemakers that can control their timing or where they are on the Synchronization-Desynchronization continuum, S-D continuum.

First, what is the pacemaker function that makes an organ a mind? In some respects, the mind is like the heart in that pacemakers control the functions of both. The pacemakers of the mind, however, are many orders of magnitude more complex that those of the heart.

The famous electrophysiological response of the heart, created by a pacemaker is called the PQRST complex, while the corresponding pacemaker function for the mind is the EEG record. The most obvious difference between the two is that the normal PQRST complex is a constant, while the normal EEG has a thousand faces. Thus, the arousal system, which became the vertical component of the mind and the Pacemaker of the mind are massively more complex compared to the Sino-atrial node, the primary pacemaker of the heart.

The primary task of the arousal structure of the mind is to control the EEG rhythms seen in the cortex, which is a massive neuronal network that varies along the Synchronization-Desynchronization continuum or the S-D function.

What is the nature of a pacemaker function? Examine a general model of a pacemaker called the firecracker model designed to assist the reader in thinking about the nature of these cortical Pacemaker functions or rhythms.

A Model of Partitioning

Assume we have:

1000 firecrackers

Every pacemaker-induced rhythm has a sine wave function, which can be described by two variables or dimensions: FREQUENCY and AMPLITUDE.

FREQUENCY

2: Each firecracker fires with a fixed frequency once a second or has a frequency of 1 Hertz, 1 Hz

AMPLITUDE

3. Each firecracker fires with some measure of energy (for fun let's say, 1 horsepower, HP) when it fires.

Then the arousal or pacemaker circuit can induce this system of firecrackers to take any one of the three following states:

A. the Synchronized state

A. The arousal pacemaker circuit can synchronize all the 1000 firecrackers, so that all firecrackers fire at the same time, which would produce an EEG signal of one cycle per second, 1 CPS with amplitude of 1000 HP

B. Desynchronized state

B. The arousal circuit on the other hand, could also desynchronize all the 1000 firecrackers (the D+-function), so that all 1000 firecrackers fire at a different time but always with the same, one HP. When the cortex is in a desynchronized state, it exhibits a high frequency low amplitude signal, i.e. a frequency of 1000 cps with one horsepower amplitude.

C. Threshold

One of the central properties of neurons or dendrite networks is that they have a firing threshold; that is they require a certain critical total amount of electrical or chemical charge in order to fire. In the candelabra model described earlier, I propose that the cortical neurons all can receive a portion of the total charge they required to fire, their firing threshold from some combination of these five vertical pathways that all converge upon each of the cortical neurons or their dendrites. Each of these five pathways normally contributes some portion of the threshold quantity of energy they require to fire. This set of vertical arousal circuits can move the threshold of any set of neurons thus making the network easier or harder to fire.

Keep in mind that the quantity of knowledge any human possesses may well, at least in the human mammal with a verbal mind, be measured by how much they have to say. Yet in the nonverbal mammals knowledge is nonverbal, thus is more simply conceived as "being prepared" to act or to move. Knowledge enables the nonverbal mind to get that somatic element of the body that is required to execute some action to be prepared, to be ready, not to be surprised, nor caught off guard.

The nonverbal mammals in the real world do not have some official who says, "On your mark, get set, go." Yet the vertical mind in effect has the capacity to raise the threshold on those critical somatic regions of the Rolandic areas of the mind so that each mammal that has a model for its domain is always ready to execute appropriate behavior. The vertical arousal system of the non-verbal mind gets the organism ready and set for some particular domain specific stimulus, which says, "GO."

The 1000 firecracker model can be thought of as a model for all the dendrites in any nerve net, in any group of cells in any mammal's cortex.

The Synchronization-Desynchronization, S-D function

Unfortunately, the vocabulary the various kinds of Physiologists use consists of a rather long list of technical memes over which I have no control. I designed the firecracker model to enable me to began to acquaint the reader with a technical expression for the word arousal level employed based on the EEG.

The candelabra model with five circuits is a model of the vertical mind, which controls the arousal function of the horizontal component of the Nonverbal mind. These arousal circuits are the same ones that control the arousal levels discussed in the Yerkes-Dodson law. The optimal arousal level is some grand mean that brings all the critical action circuits up to just below the critical threshold so that the last final bit of energy, which causes the appropriate group of cells to fire is the small quantity of energy from some sensory input. At the intermediate level of arousal B, the next actor who will act will be ready to go given the proper final increment of energy by some final cue.

Partitioning

Home partitioning

The concept of Partitioning plays a major historical role in understanding how the massively large system of neurons within the cortex functions. Partitioning is a technical term employed by two very different theorists Ashby (1960) and Pribram (1960). Partitioning however is a word, or rather a metaphor drawn from architecture, where partitioning refers to the question A. A partition is a non-weight-bearing wall: Why have rooms in a house? How many rooms should be used or are required in a house? Frank Lloyd Wright was famous for his Glass House with no rooms. However, in home construction a partition refers to a non-weight bearing wall. Theoretically, a partition is a temporary part of a house that can be put up or taken down with no consequence for the overall integrity of the house that is removing a partition will not make the roof fall down.

The billions of neurons within the cortex can be partitioned, into particular areas or regions, where some function of the whole can be computed. The EEG frequency and amplitude gives us a rough estimate of how the cortex has been partitioned. In the sections below, I will try to give the reader some idea of how this function works.

Brain or cortical partitioning

The concept of partitioning plays a major role in how we think about the functions of the billions of neurons within the cortex. Two assumptions: first, the cortex must compute more than one function at a time, and second, the billions of neurons must be partitioned into as many rooms, or computing systems as are required by the task, assuming each function must have its own computing room or partition. The general rule is that the sleep-waking continuum is primarily a function of partitions. A rough rule answers this question; how many partitions are required in each stage of sleep-arousal? Again, roughly assume that at least one room is required for each function the computer is executing. In sleep with respect to the somatic motor system, we are in bed and generally motionless. So let's say that X functions are being executing.

In the B-zone of the Y-D law, let's assume that five X function must always be executed for the subject to prevent boredom.

However, when the subject, too aroused by the question, is in the Red zone the demand on the CPU may be excessive, say 10x or 15 x. With my discussion of the S-D function, I will try to make this clearer.

The S-D function continued

The maintenance of the optimal level of arousal is one way to understand the role of the Synchronization-Desynchronization, the S-D function.

What does an arousal circuit do? What does it control? With respect to the nonverbal mind in particular, the role of an arousal circuit is to control the S-D function within the cortical network of neurons within the bark that surrounds the core of the brain.

An arousal system must possess pacemaker circuits with different degrees of centrality, which instantly and simultaneously can shift the cortical frequency of cells up or down the circadian/arousal continuum. The arousal pacemaker can move the cortical EEG anywhere along the S-D continuum from sleep to waking.

The arousal system is an S-D function control structure, which was designed so that its circuits have direct access to the cortical neuron, which allows it to control the place on the cycle frequency at which each cell fires. It must have centrality, which means that one cell deep within the brain stem has the capacity to control or desynchronize all the rhythms of all the cells in the cortex. From the cells deep in the BSRF, there must be spot where the S-D function begins. Beginning at that point, there exists a great fan circuit with ribs extending out to each of the cortical neurons. This central circuit enables the BSRF region to act like a pacemaker for all the neurons of the cortex and controls the rhythm of the whole mass of cortical neurons. The cortical side of the fan must intersect with each synapse of each neuron. Effectively every cortical neuron should be capable of receiving the same dose at the same time of some neuronal chemical synaptic activator or inhibitor: from either the D+ signal or the D-signal.

This D signal must be capable of administering a uniform dose of some activating hormone, which could have an instant excitatory or inhibitory effect. The response to any dose of hormone can cause some interesting effects due to the curvilinear shape of the arousal curve. These effects were discussed in the Y-D arousal function section of this chapter.

The arousal function as a S-D function

The S-D continuum has one property that is not obvious. My firecracker model was designed to make the reader aware of this dimension.

1. What is the frequency when all the cells are synchronized? If all 1000 firecrackers fire at the same time, then the EEG will show only one firing a second with amplitude of 1000 HP.

A 1 Hz frequency and with 1000 HP amplitude = 1000HP per interval.

2. What is the frequency when they are all desynchronized? No two fire at the same time

A 1000 Hz frequency and with a 1HP amplitude =1000 HP

The Partitioning model

The numbers reflect a critical relationship, which I call the Partitioning model. In the Partitioning model:

A * B = C where A is frequency, B is amplitude, and C is a constant. An illustration of this model is show below.

A		B		C
1	*	1000	=	1000
2	*	500	=	1000
4	*	250	=	1000
250	*	4	=	1000
500	*	2	=	1000
1000	*	1	=	1000

Let's assume that A represents the number of partitions the billons of cortical neurons are divided into and B represents the average size, or the number of neurons comprising each partition.

So what is a partition? If we adopt for the sake of illustration a computer model, say a windows phenomenon, the most practical question is: what is the ideal average Partition size and number for any program. In windows, assume that on some computer there is C quantity of memory available for the windows program. The question is what is the ideal number or size of partitions where each partition enables the mind to run one program or window.

The curvilinear relations of the Yerkes-Dodson law simply reflect the Partitioning law or model. The B region of the Y-D law is in the mid-range of the Partitioning continuum. The A zone of the Y-D curve represents the sleeping end of the curve and is the point on the curve where in the extreme case there is only one or some too few partitions; where the size of the partition is too large, where essentially too many of the neurons are all in the same partition. The C region of the Y-D curve, on the other hand, represents the other end of the curve where theoretically there would be one partition for each neuron, or each neuron would have its own partition and one neuron would then obviously be too few neurons to run a program. (Yerkes and Dodson, 2004) On this end of the continuum, there are too many partitions each with too few neurons to run a program.

The question that should be asked is how many neurons ideally are required to run or support a program or window, or what is the ideal size partition required for region B?

This partitioning model raises a serious question even though the values of the numbers in my little model are off by many orders of magnitude. Why is the ideal arousal level some B state? The numbers I chose to illustrate the Partitioning relationships above are numbers from the two ends of the continuum.

When the model is synchronized, it will output a low frequency high amplitude signal. This is a sleep or A region of the continuum.

But, when it is desynchronized, it will output a High frequency low amplitude signal. This is a too waking or C zone signal.

Or when the cortex is ideally aroused and is in the B zone of the Y-D law, the EEG signal will be at some intermediate arousal state. (Yerkes and Dodson, 2004)

Please note the two most important words used in the two examples above: Synchronized and Desynchronized. The arousal system controls the S-D function.

The purpose of this section is to make the reader aware that the nonverbal mind is a system consisting of at least two parts: the cortical mantle, the horizontal dimension of the brain; and the vertical component, the arousal system. These latter circuits are the control centers or pacemakers, which regulate the S-D function of the horizontal mind. In order to understand the mind, it is essential that the function of each of the two pacemakers or vertical pathways must be distinguished and understood.

This point may seem trivial, but actually, it is an important physiological fact. The arousal system causes the EEG frequency to change along one dimension, the timing continuum, which causes the EEG frequency to change across the whole range of the arousal dimension of the Y-D law; from sleeping on the left to the too aroused state on the right; from A to C.

This Firecracker model works just as well in other contexts. Perhaps, I should not have mixed my metaphors and instead of employing the assumption of one horsepower for each second of the record, I should have stuck with a spatial model. For example, now assume that a builder is given 1000 one-foot square pieces of plywood. Now ask how a large building could be divided, how much floor space will be required for each computing room.

Once again, the Y-D law applies. Three size rooms could be constructed. (Yerkes and Dodson, 2004)

In the A zone, the builder could partition the building with a few very large rooms, excessively large.

In the C zone, on the other hand, the builder could partition the building into too many small rooms.

Or in the B zone, the builder could partition the building into just the right number of intermediate size rooms.

The pacemaker function again

The arousal system with a set of four carefully configured channels can serve as a pacemaker to partition the cortex at any moment with just the proper number and size of partitions, rooms, or cells required by the task.

What can the EEG reveal, what is occurring in the brain/mind? Historically, Psychology has enjoyed many schools of thought that claimed to have conceived some technology that has mastered some form of mind reading. Phrenology is one such ambitious program. However, while the EEG does not allow science to read the exact content of the mind of an individual, yet the EEG does allow science to read or distinguish certain critical features of the mind.

Again, I can only fall back upon the Y-D law. Now, hopefully the reader will be better able to understand what state of mind an individual is in by noting simply whether an individual is in Zone A, B, or C.

The power of an anomaly

One of the most ingenious consequences of the evolution of the Reticular Formation is that one of the most critical and interesting new functions of the nonverbal mind is the evolution of the Match-Mismatch system. A mismatch is the response of the Match-Mismatch mechanism to an anomaly. In science even, a mismatch is a kiss of death for any function being executed in IMm.

Two kinds of anomalies: logical anomalies and unexpected anomalies

Pavlov(1960) best described the unexpected discovery of an anomaly, which he called External Inhibition, EI, which is the master reflex for it has the power instantly and absolutely to interrupt every learned response.

What I find so interesting is the emergence of a new role for an extreme A-state inducing event. What makes an anomaly physiologically work? The answer is that a mismatch triggers an intense D+ arousal response to be levied or administered upon the cortex, which dramatically shifts the partitioning of all the neurons of the cortical mantel from a B state of partitioning to an extreme C state of partitioning. This means that all the right number of correctly sized partitions is instantly destroyed and is replaced with an excessively large number of too many too small partitions. Why do all the CR's (conditioned responses) go? They instantly and absolutely all lose their room!

The normal and abnormal EEG patterns

The mind consists of at least two major subsystems, the horizontal and vertical. The Horizontal blanket of synaptic connections can function two ways: independent or dependent of the vertical pacemaker circuits, but a normal brain's EEG pattern is always a joint function of both the horizontal and the vertical components of the mind.

The normal EEG rhythm observed from the scalp, I must take some extra time here and carefully make a technical point, are always a joint function of both the H and V components of the mind. In the normal case, the H-component is dependent upon the V-component for the pace making or arousal energy component to drive the cortex.

What is strange however is that abnormal EEG patterns, those seen in Epilepsy, are patterns caused by abnormal pacemakers, such as dying cells within the cortex that may have been injured by some disease or accident of many possible kinds. These abnormal collections of cells within the cortex can usurp the normal pace making functions, normally supplied by the V-component of the mind. These damaged and dying cortical cells can take control of the cortex and induce abnormal EEG activity that is the sign of Epilepsy. My point is that in a normal mind, the cortex depends upon the vertical centers for the required pacemaker signals, but in an abnormal situation, the cortex can function independently of the normal vertical pacemakers. In a sick mind, the cortex can act independent of the V-pacemakers. In a sick brain, generally a large collection of dying cortical cells can take control of the brain and make the horizontal brain act independent of the V-pacemakers. The abnormal cortical cells develop an abnormal rhythm of their own, which can become so powerful that it can recruit other cortical cells and as independent pace makers; they may recruit other surrounding neurons and induce epilepsy. Epilepsy shows in an unpleasant way how important and vital the arousal energies supplied by the V-pathways are for the normal functioning of a healthy mind.

Normal and abnormal pacemakers

Fireworks displays

Abnormal pacemakers create complex signal patterns in the EEG, which might be likened to a Fourth of July fireworks display. Professionals, as it were, come prepared with many sets of bundled firecrackers (neurons) where each has a unique set of timing circuits, which causes a different spectacular aerial display as it rises in the sky. Each kind of aerial display has a unique set of timing mechanisms built into each package of firecrackers at the plant, which fire all the sets of neurons over time as it rises in the sky.

However, if the pacemaker control circuits fail then all the firecrackers may become synchronized and release all their energy at the same time in the air or in the brain the energy may be released into the motor system at the same time. What do you think would happen? Various kinds of epilepsy are possible. Epilepsy in effect illustrates our point that the horizontal neurons of the cortex cannot be left to act independently of the vertical circuits. If we did not have the vertical arousal centers we would all be epileptics or incapable of acting in any organized way. The organized behavior exhibited in the cortex is a manifestation of the organization imposed upon the cortex by the arousal centers. Sperry's (1961) argument is that there are two minds, one in each of the hemispheres and each mind (or hemisphere) is responsible for its own organization. I am arguing on the contrary that the organization of the mind has its origin not in the cortex but sub cortically.

Obviously, the dendrites, which cause each EEG pattern occurring in the cortex, cannot be left alone to do their own thing.

I must say again; the vertical arousal circuits are the control circuits of the mind and are designed to control the partitioning and timing of the massively large system of oscillating groups of dendrites in the cortex.

Define Active

When is the cortex most active? The answer depends upon what is meant by active. If by active, you mean generating the most energy or HP out of the system, then the answer is the energy output of the network is always the same day or night. In our model, whether it is synchronized or desynchronized, the system always generates 1000 HP per unit time. Every entity, cell, or dendrite, in my model cycles once a second so the system is always generating 1000HP per second, and I assume that in the real brain something similar is true.

The central feature of the mind, the Essential Variable of mind, EVm, is the S-D state of the cortex, which is the state of the collective nerve network. The state of the mind, which is the EVm, varies along the Synchronization-Desynchronization continuum, the S-D continuum, with sleep at one end and waking at the other. The vertical mind was designed to regulate this S-D continuum. In other words, the synchronization-desynchronization organization is not intrinsic to the cortex, but rather the pacemaker of the mind is found deep within the sub cortical structures of the brain.

It is important to understand the nature of synchronization and desynchronization and the role this continuum plays in every act of the mind and body.

The discovery that rats and probably all mammals have a nonverbal rational mind, in effect, demands a new paradigm be constructed not only for animals, but also of human nature. Why, because the facts are almost embarrassing, the facts make it clear that even the rat's nonverbal mind executes functions, which philosophers and even modern "experts" on human language behavior, but especially the transcendental religions, have long claimed were beyond the power of even the human mind.

Summary of contribution of the vertical mind

I can only conclude that evolution's design of the nonverbal mind, as a technology for transforming life into a self-organizing system, was a complete success. Rats, a prototypical mammal, for example, are still around and follow us everywhere on land and sea. The best example however, was Female three who opened my eyes in several ways. First, clearly her nonverbal mind did not evolve for talking, but rather to endow F3 and all mammals with the rational power to communicate with Spinoza's (1933) cosmic god; thus, discovering some of the nonverbal spatial secrets of the nonverbal cosmos.

Moreover, evolution did not design the mind to transform some specific species in some specific niche at some specific time into some independent self-organizing system. On the contrary, evolution designed the nonverbal mind well, as an organ that could serve all mammals, each in its own home domain, in essentially any niche of the planet earth and perhaps all of the niches of the cosmos.

The only possible sour note that has emerged in the long line of nonverbal mammals is the evolution of the human mammal that was given a new revised model of the ancient nonverbal mind, a talking mind. The verbal mind somehow broke the nonverbal mind free of its absolute dependence on Spinoza's empirical world. (Spinoza, 1933) The human curse was that the verbal skills were so powerful that they shifted their allegiance from the earth god to a word or book god. The human curse was that with the emergence of verbal behavior humans lost their contact with and their dependence upon the empirical wisdom of the nonverbal mind.

Chapter 4 the Nature and Origin of the Generic Nonverbal Mammalian Mind

In Chapter 1, the topic was the theories of the origin of mind.

In Chapter 2, the topic was the structure and function of the Horizontal component of the nonverbal mind.

In Chapter 3, the topic was the structure and function of the vertical component of the nonverbal mind.

In Chapter 4, the topic will be the structure and function of the environmental context in which the nonverbal mind emerged, for the nature of the nonverbal mind was very much a function of the environmental conditions, which gave it birth. The properties of the nonverbal mind are many and complex, but perhaps the most succinct summary of the properties of the nonverbal mind is the idea that it is a self-organizing system.

I am suggesting that if we are to understand the nature of the nonverbal mind that we must additionally understand the conditions by which natural selection caused the nonverbal mind to evolve and emerge as a self-organizing system.

The human compound mind evolved in two steps

First, I describe the structure and function of the NV mind of rats and mice or rodent behavior conceived as a type one self-organizing system, SOS1. What environmental conditions were required for a SOS_1 to emerge? If we start with the obvious, nature doesn't give a species eyes if they live in an environment without light; my working assumption is that nature would not have designed mammals with a nonverbal mind as SOS if there were no environment where a SOS gave life an adaptive advantage.

Second, I describe the structure of the human mind: how the human NV/V mind built upon and extended the NV foundation built and perfected by the preverbal mammals. This revised human NV/V mind emerged as self-organizing system type two, SOS_2.

Evolution designed both the NV and the NV/V mammals to function as self-organizing systems, SOS_1 and SOS_2, as not the ultimate forms of life, but as the latest and most powerful species yet to have emerged. Evolution is not finished with the mind.

Why and when did the NV SOS1 and NV/V SOS2 mind emerge?

Nature made mammals or SOS_1 so that they could create the variety of knowledge required surviving in a world of change by nonverbal interaction with the cosmos. Nature designed mammals as self-organizing systems that only required one thing to survive in a world of change: knowledge, actual local knowledge, or detailed pragmatic knowledge of that portion of the cosmos, which they occupy so that a spatial map or models of these domains could be designed. An SOS1 requires one thing from the cosmos or the nonverbal god; access to the parameters of the local domains so that the knowledge needed to achieve some self-designed goal could be created by interacting or interrogating the domain in question.

The focus of this book is upon SOS1, first because I only have animal data, but secondly, the differences between SOS1 and SOS2, at least to me as a scientist, are scary. I personally have no idea how to explain these differences. Moreover, these differences are not trivial. One of these major differences, which had such a profound effect on human behavior, is the emergence in humans of metaphorical behavior. In science, metaphors are a priceless source of hypotheses. We have all heard the stories of Einstein's hypothetical experiments, which was his using metaphors scientifically, such as riding a beam of light. In literature, metaphors make a book. In religion, metaphors make a religion. In philosophy, metaphors enabled Plato to define truth, reality, etc. that effectively turned the world upside down.

The SOS1 mammals, I propose, base their life, their actions, and their ideas on empirically-driven knowledge, and while they can extrapolate and perform some remarkable mental gymnastics, yet compared with the metaphor driven behaviors of humans, animals' behavior is plodding and relatively unimaginative.

The gift of verbal behavior came with a great price. I sometimes feel that the price humans have paid for the gift of metaphors was so great that someone could argue the price was too high; it cost more than it was worth. In the next section, I will distinguish two concepts of god; one concept of god is empirically driven, while the other concept is exclusively metaphorically driven.

Two gods

From the dualists' framework, humans live in two worlds. Their bodies are matter and are subject to all the laws Flesh is obligated to, while their mind as spirit is free of matter but is subject to the laws of the spirit world.

While as a scientist, I am obliged to discuss the classic theories of mind, yet I am at a loss, to address many critical issues, which never arose in the context of spirit world theory. If there are spirit world laws then presumably, we can assume there are spirit world kinds of causality, such an assumption would be necessary to guarantee that the observance of a law would insure its stated consequence.

A dualist must then assume, it seems to me, that there are two kinds of gods and two kinds of laws:

- Physical laws, which are the manifestation of a physical or cosmic god are expressed in space and time, which science empirically and experimentally studies. As far as their bodies, they are made of flesh and are mortal; they are inexorably bound to the physical world, and with all manner of homeostatic body functions, and with all those functions, which space and time bestow.

- Secondly, spiritual laws, which are the manifestation of a spirit god, and which are expressed outside time and space, in eternity, and which, would escape the scrutiny of a material Aristotelian kind of mind. And so each individual dualist must choose which laws to focus upon-- the laws of the cosmic god of time and space or the laws of the spirit gods of eternity. As far as there are beings which are immortal spirits: the human mind must be concerned that it is inexorably bound to eternity; to that eternal duration after life when it must be prepared to "live" eternally in another spirit world.

At birth, humans, as tabula rasas, must obtain two kinds of "information." They need knowledge about the physical world and knowledge about the world of eternity.

Historically, the human information needs are two, and rationally there are two gods, which serve as the source of one of these two kinds of knowledge.

However, all the above is irrelevant, for the topic of this book is the knowledge requirement of the nonverbal mammals, who no one claims, to the best of my knowledge is in possession of a spirit soul or mind.

So since the nonverbal mind is a self-organizing system 1, SOS,1 we must ask, where and how does the neuronal mind get the domain specific empirical knowledge it requires to achieve its goals.

Perhaps it's overkill, but let's state- up front- that the nonverbal mind as a SOS1 does whatever it does independent of any spirit god. In fact, the facts here fit together nicely; the nonverbal mind communicates with the nonverbal god to conceive the nonverbal causal conceptual structures it requires to achieve its SOS1 goals and this has been going on for perhaps fifty million years.

Suddenly, within a relatively few years as evolutionary time goes, Broca's area evolves. Suddenly human kind has forsaken the cosmic god as its exclusive source of knowledge and empirical knowledge at that. Suddenly, humankind is now obliged to get its spiritual substance from "divinely revealed" books. Suddenly, all humankind has not completely lost its mind, but has lost the primary mental function of the nonverbal mind. Suddenly Plato (1952) announces this primary nonverbal mental function, which has been the right arm of all the SOS1 mammals is declared a function reserved for the gods.

So what is the point of this book? What is the high point of this book? Suddenly, we are allowed to see F3's nonverbal mind execute this supposedly dead function. F3's execution of this lost god function hopefully will encourage many humans to recognize that if a 90 day old rat can execute the god function, in spite of Plato's assurance that it is impossible, then they should have, at least, the mental power of a 90 day old rat.

FIGURE 0.3A

FIGURE0.3A shows the two competing concepts of the cosmos, the physical world and the spirit world or the secular and the sacred world. The spirit world is the home or the abode of the verbal god, while the physical world is home to **Spinoza's** (1933) nonverbal god, **the pantheistic god.**

For the verbal mammals like humans, a verbal god's book or a sacred text can be conceived to be the source of the all the verbal information they require to live and to die.

However, obviously the sacred books of the verbal god cannot provide the NV mammals the knowledge they require to self-organize. While theoretically the words of the sacred texts of the verbal gods might logically supply all the verbal information humans require, yet the same logic does not work for a nonverbal mammal. If a verbal mammal requires a verbal god as a source of verbal information, then a nonverbal mammal should require a nonverbal god as a source of nonverbal knowledge. Is there such a god?

Yes, the physical world is home to **Spinoza's** nonverbal god, **the pantheistic god.**

There are several features of the two gods: Dualism is the first we shall consider.

DUALISM

The dualistic concept of the cosmos, mind, and god

The dualistic god is a pure spirit, a being without any physical nature, and thus without any physical attributes that can be (empirically) known by the nonverbal mind.

What role can the spirit god play in science? None because a spirit god has no physical attributes with which the nonverbal, empirical mind can interact. It is impossible to know a spirit being because it has no physical nature by which the five senses could know of its existence or its nature. Thus, it can only be worshipped through faith.

Historically, Western culture has an elaborate and fanciful conception of the proper relation between humans and the verbal spirit god, but this is all a part of religion not science. Western culture, not surprisingly, says nothing about the proper relation between the nonverbal mammals and god.

The religious aspect of the topic of the human-god relation is necessitated by the myth that the human mind or soul is a spirit; thus, human contact with the spirit god is required for humans to clean up and "save" their soul and to assure its place in eternity. In short, the religious justification for much of the human-god interaction is based on the human interest in the "next" world; not for any interest in this world, nor is it this-world directed, but is based on the human's need to get their souls into heaven.

Some religions make this world's ethics a requirement for entry into the next world.

Generally, in the transcendental religions, the human soul or mind is conceived as passive, which means that a verbal spirit god gives humans a soul and everything the soul requires including both hardware and software. The human soul or spirit is understood as having its origin out of this world, as a transcendental being or reincarnated being. In the classic myth, the origin of Ideas was transcendental; the verbal God is the lawgiver, the ethics-giver, the fact-giver, etc. God gives humankind laws, truths, ideas, and ethics, etc. Plato's (1952) Ideas can be understood as having their origin transcendentally, as being the gift of a verbal spirit god.

The Western concept of the human mind is conceived as a spirit and a language function. Human nature is conceived as passive; as receiving truth, data, laws, or ideas from above: in a book, a tablet of stone, or inserted directly into the soul at birth.

FIGURE0.3A shows the verbal vs. the nonverbal world, which characterizes two ways of thinking about animal and human nature: First, the classic paradigm and second, a new paradigm. The latter, I suggest is based upon the data of the experiments in Book 2.

Dualism

On the left of the figure, the dualistic position is depicted as a top-down theory. The top- down theorist's claim that Human nature is dualistic: part matter and part spirit. Spirits are immaterial beings without matter and are out-of-this-world beings: gods, spirits, ideas, etc.

The classical Judeo-Christian origin story has a spirit god verbally creating the heaven and the earth: "let there be light."

The passive model

Similarly, the dualists conceive of human nature as passive. The classic passive concept of human nature is supported by some of the most famous metaphors, the sheep, and the blank slate metaphor.

The sheep metaphor

In the 23rd Psalm, one of David's most famous and beautiful poems, David sings and describes himself using a double metaphor, first as a sheep with "The Lord is my shepherd; I shall not want. He maketh me to lie down in green pastures: he leadeth me beside the still waters," (Psalms 23: 1-3) and then as a cup with: "My cup runneth over...." (Psalms 23: 5)

Yet as sheep or cup, David conceives of himself as depending upon god the good shepherd for supplying his every biological need. In the end, he concludes that the good Shepherd supplies his every need in abundance; indeed, his cup runneth over. David found the passive life satisfactory.

The Tabula Rasa metaphor

John Locke (1914) bested David in a scientific metaphor, by conceiving of the mind as a blank tablet upon which nature inscribes text, which is the foundation for empiricism. Man's empirical mind passively records what god writes upon his tablet. David's cup metaphor is close to Locke's tablet, yet whether humans are a blank slate or an empty cup is a superficial difference; the point in both cases is that only god can provide or supply our human biological needs. In the words of one of my childhood favorite hymns, "God will take care of you, through every day, o'er all the way."

The logic by which human nature has been defined or conceived, over the ages both philosophically from Plato to religiously from the poetry of David, has tragically been a big step backward. What is human nature: Are humans passive beings like sheep, or hopeless helpless beings such as wretches, or are they SOS_1 beings that have been updated to SOS_2 beings? One of the tragic consequences of the emergence of the power of metaphorical behavior is that by using metaphors humans continue to conceive themselves as not wisdom IIII beings but still as Wisdom I beings. Wisdom I is the wisdom of the body and first defined by Cannon (1932). This first wisdom conceives of the body physiology as being explained by Homeostasis, which is a physiology entirely defined by the genome.

What I am trying to show in this book is that with the nonverbal mind, mammals became SOS1, and then the human compound mind, in addition to the mammalian mind, made humans even more powerfully adaptive organisms, as SOS_2 beings.

Homeostasis: A passive mechanism

Homeostatic technology works automatically whether we are asleep or awake and generally outside of consciousness. Homeostatic species are pure nature beings whose nature is entirely defined by their genome. Homeostats cannot abide change. Evolution designed Homeostatic beings for closed domains.

On the other hand, Heterostatic species were designed as self-organized beings who cannot abide constancy; nature designed Heterostats for open nonverbal domains.

Yet, all the physiological organs of both homeostatic and Heterostatic species are homeostatic in nature; in fact, all the neuronal cells which are a part of the nonverbal mind of mammals, which enable or endow mammals with the rational power of self-organization, are homeostatic functions. One of the facts, which even Plato (1952), in his three-soul model, was tacitly or intuitively aware was that evolution in designing more complex forms of life, conceived of life as; in my theory, multi- wisdom beings, or in Plato's theory, as three-soul beings in the case of humans. The nature of the higher forms of life incorporates the older forms. If the reader will review Chapter 1, they will discover that I depicted each new life form as requiring an additional symbol to describe its nature. For example, the first life form was characterized as H, then the next as HH, and then HHH, etc. The Heterostatic forms of life are still employing Homeostatic mechanisms, etc.

The traditional top down concept of god conceives of human nature as passive, which in effect is oblivious to the power of Heterostasis, the wisdom of the nonverbal mind. The passive model of human nature is embedded within the dualistic concept of human nature. While our bodies are made of matter, thus each exemplar is like a flower that, as it were, lasts only a season.

This demeaning concept of human nature ignores all the empirical facts that have been accumulating for the last two thousand years.

The monistic concept of god and the monistic concept of human nature paint an infinitely more optimistic picture than the top down model. In the next section, I will outline a profoundly different theory of the nature and origin of the human nature.

MONISM

The monistic concept of the cosmos, mind, and god

In Darwinian monistic terms, evolution is a function of natural selection, so the theory of evolution compels us to understand the local conditions, which caused or selected the emergence of the nonverbal mind. For example, if a fish has an eye, we can assume that it was not a cavefish. In other words, we can make a rule that eyes evolve in an environment where there was white-light, or that the kind of eyes that evolve is a function of the kind of ambient light that is a property of an environment.

My question is what kind of a world, what kind or environmental conditions selected the features of the nonverbal mind? To make this as simple as possible, let us identify two opposite kinds of adaptive technologies and ask what environmental conditions would cause them to evolve?

For the two technologies, let's study two of the adaptive systems discussed in Chapter 1 that were called the closed and open systems, which are also names for the two models of FIGURE0.3A.

Two causal environments: the open and closed

One of the central themes of this book, which has been addressed several times, is that the emergence of the NV mind was in part a response to the five great extinctions. After most of the simpler closed species became extinct, nature, in effect, went for broke, sought, and found a better way to design a form of life to deal with change. Instead of creating another **closed or nativistic** species, one genetically designed to live in a closed domain, nature took a different tack and designed a revolutionary kind of species, an open system, one designed to thrive on change; one independent, rational, and in short a self- organizing system of life forms, i.e. Mammals.

This book has been designed to acquaint the reader with mammals, which up to now were given their name because the female of the species has mammary glands. Historically, human conceit advanced the idea that they alone had a mind. However, based on the behavior of the rats and mice of this study, it is now time to take a new look at mammals. Mammals' evolutionary successes have too long been attributed erroneously to the fact that the females of a species have breasts, not their minds, for their capacity to create milk, not knowledge, not ideas.

If we would understand mammals, then we must understand:

First, how mind creates ideas; and secondly, how ideas enable humans to deal with change. It's time to take a serious look at the family of mammals designed not just to live in and survive in a world of change not just to tolerate change, but rather a species of life that was born capable of being both the author and agent of change.

The monistic concept of mammalian nature

The study of rat behavior in this book forced me to conclude that rats, indeed all mammals, had a nonverbal mind, which functioned as a SOS_1. So why does science need to include a concept of a cosmic god into their concept of the nonverbal mind?

As long as Science is fixated upon god as a spirit then the concept of god has no place in science. However, there are important facts that historically have been left out of the picture. So, let's get back to basics. I am a rat runner, and no one to my knowledge has ever proposed that rats have a mind or soul. My first point which is the purpose of this book is, however to make just such a claim, but of course, when I propose that rats have a mind, I am not proposing that their mind is spirit. Nor am I proposing a theory that there will be rats in heaven. So for humans who don't like rats, in my best scientific judgment, I would not hesitate to suggest that heaven would be just as free of rat souls, as human souls, and all other possible kinds of souls.

In this book, I am proposing a theory of mammalian nature that is soulless and spiritless, that is monistic all the way.

God in science

Many may take offense at the quantity of god-talk to be found in this book.

However, the paradox is that my intent is not to promote the verbal spirit god, but rather to expose those places in science where the verbal spirit god is hiding, and there are many of them. My intent is to expose and rid science of the closet Christians who invent words to mask their advocacy of the verbal spirit god. My task is to expose the kind of god-talk, which is not compatible with science. There are two kinds of gods: the spiritual and the physical. Two features of the physical cosmos/god, in particular, must be noted: hierarchical evolution and emergence.

- **The Cosmos/God evolves**

  First, and foremost scientists generally laugh at or pity religious holdouts who don't accept Darwin's theory of evolution, yet one of the hiding places for the verbal god is in the work of Chomsky and Plato who also do not accept the concept of evolution, Chomsky, in effect, hides the spirit god behind the concept of Nativism!

- **The cosmos is hierarchically organized**

  The first interesting and most important of these features of the physical cosmos is the idea that God is hierarchically organized or nested.

- **The cosmos has an emergent character.**

  The cosmic god has a physical nature
  The cosmos has purpose.
  The cosmic god has potential.

The universe, in the most important sense, has purpose. The purpose of the cosmos is to realize its potential. The cosmic god differs from the spirit god in that the former has a nature, while the latter has none. The purpose of the physical god is to realize or express its endemic nature.

The simplest definition of god is to be found in the dictionary, where god is defined as "the cause of everything."

The mind is a physiological organ designed to be a SOS1, and at the same time, it is designed to be rational by rigorously employing the rule that action will only be taken in a closed domain where the cause-effect structure is known and has been modeled. A Rational mind only acts upon the knowledge embodied within a closed causal physical structure.

- **God is the cause of all things, and everything is caused.**
- **God is nonverbal and can only be known empirically.**
- **All nonverbal communication with god begins by establishing the conceptual or class identity of things.**

The existence of such a physiological organ as the nonverbal mind has already assumed control of not only the evolution of ideas but has already even begun to replicate the biological evolutionary process. Moreover, we must mention that it alone has the power to cause mental evolution, which changes the whole picture of what inferences can be made about the nature of the physical universe, or the cosmos.

The nature and function of nonverbal communication

In this chapter, I will start to describe a revolutionary new concept of human nature as an independent SOS_2 conceived as having a new and different relation to the cosmos. The image of the helpless sinner, the Wretch, has too long been the dominant image as the Christian religious attitude, the obsequious model of humankind's relationship with god.

The emergence of a nonverbal mind raises many new strange problems, all of which revolve around the fact that the nonverbal mammals had and have a mind that could not communicate verbally in conventional historical terms.

By conventional logic, a mind that cannot communicate verbally is not a mind.

However, there are several critical questions to be answered!

Communication

- What kind of communication is a nonverbal one?
- What kinds of things could a mind do if not communicate?
- What is the relation between the nonverbal NV mind and the NV/V mind?

Historically humans have assumed that verbal communication was the only kind of communication, and defined it as a social exchange of verbal information between humans employing some one of many language conventions.

The discovery that mammals have a nonverbal mind raises all kinds of vexing problems. For starters, let's begin with a big one: What is nonverbal communication? Can animals communicate nonverbally and with whom, with one another, with humans, with nature, or with god?

Many researchers who have studied animal communication have understood communication conventionally as between animals and humans. That really does not work in the case of rats. Let's redefine our question: Who does an SOS_1 rat communicate with to acquire the knowledge it requires to achieve its self-defined goal in the matrix maze?

The nonverbal mammals obviously cannot turn to books for answers, neither sacred nor secular. The nonverbal rats running the matrix maze get their nonverbal knowledge about the hierarchical structure of the maze directly from the maze itself and from the laboratory environment of the maze.

As a rat runner, I am obliged to ask where a self-organizing NV mammal gets the specific knowledge its rational mind requires to self-organize in all the changing circumstances life on earth affords.

Western civilization recognizes or claims there are two great repositories of information/knowledge: one sacred and one secular: the Bible and Nature. Unfortunately, they generally don't distinguish the utility of the two.

Our question is where do the nonverbal forms of mammalian life find or provoke the truths or the reality they require to act?

What is the role of the nonverbal mind in the cosmos? It is impossible to answer this question without first considering the companion question: what is the role of the verbal mind in the cosmos. We first must remember that the Human verbal mind is actually a compound mind, a NV/V mind.

"Biological Evolution has no purpose"

The Darwinian theorists who are strict constructionists hold most seriously the position that biological evolution reveals no purpose behind the forces of biological evolution. The genetic adaptive mechanism may in the short term, say in wet years; favor those genes that promote environmental survival in wet years. Yet, the dry year genes are still in there and are not being effectively expressed at this stage of the wet-dry year environmental cycle; yet they are just waiting their turn. Evolution does not favor wet year over dry year genes, it has no purpose, and whatever works is just fine.

Mental evolution has purpose

Mental evolution however, has the power to replace, in principle, all the powers of biological evolution. There are at least two kinds of evolution, biological and mental, with the former functioning as described by Darwin as natural selection.

However, mental evolution is a function of a completely different set of assumptions, but the main difference is between **natural selection** and **rational selection**.

But even more importantly, what I will propose is that the cosmos for which the mind was designed to know, is a much more interesting and complex being than Darwin was prepared to accept.

Mental evolution trumps biological evolution

There are many arguments for and against the concept of purpose in biological evolution, but one outweighs all the rest. The emergence of the nonverbal mind changed everything; except, no one as yet seems to have noticed. The evolutionary functions of the nonverbal mind take priority over biological evolution. This becomes especially true with the emergence of the NV/V mind. First, the nature of the cosmic god as revealed in the mental evolution supplants whatever might have been revealed by biological evolution.

- For example, in every case of biological evolution that science has examined the doctrinaire Darwinians are convinced that purpose has never been observed.
- However, in every case of mental evolution that science has examined, I am convinced that purpose is the central component.

The mind creates knowledge by encoding a model of the unknown domain in question, yet this statement is not complete. Every nonverbal act is always partial, every possible nonverbal act is but one of many kinds of ways of acting with regard to every object.

All the steps of model making are secondary to the purpose of the organism. Every model and every concept of every model is, as William James (1950s) says, partial. Which particular perspective or slant on the object forcing a concept is always just one of the many possible? Without purpose, action is impossible.

So Darwinian or biological evolution may proceed without a purpose, but the same is not true for mental evolution.

Example

F3 made this point most clear, but all the other animals made the same point. When she was inserted in the anomalous Matrix maze, without being deprived homeostatically, it didn't matter:

- For first, the anomalies of the maze immediately activated those portions of the mind, which exercise the god function, which seeks and destroys anomalies.
- Secondly and simultaneously, the anomalies of the maze activated the rat's primary rational need to understand the maze, which means to be able to run through the space out and back without error.
- Moreover, punishment forces the rat to reveal its other purposes.

If biological evolution is purpose free then mental evolution is purpose rich.

Finally, one of the most important lessons to be learned from these experiments is that biological evolution represents the growth process at the first stage of life. At the next stage of growth, mental growth replaces biological growth and takes precedence over biological evolution.

Open vs. Closed Species

Closed Species

Evolution's first strategy for defining life employed homeostatic wisdoms in fixed milieus using genomic technology to define species. However, this first strategy essentially failed because nature loosed upon the earth catastrophic events, which produced "the five great extinctions."

Secondly, this set of rare but terribly destructive acts of nature seemed destined to defeat the first forms of life designed by evolution. In fact, with this first set of hurdles, nature came very close to losing all the biological forms of life with which the earth had been seeded.

The first classes of life forms were designed to live in the Static cosmos. However, over billions of years the cosmos proved to be anything but static with a devastating consequence for those forms of life designed for a static domain.

Open Species

In contrast, a rat's nonverbal mind was not designed to get a rat into heaven but was designed to enable a mammal to function as a self-organizing system here and now. In this physical world, god is the totality of all that is physical. This pantheistic god has a physical nature in which the nonverbal mind emerged. The Aristotelian model of the mind, in turn, was designed to embody this generic mammalian mind and its environment, the physical cosmos.

The Evolutionary model of the physical world, as conceived by Darwin, is a partial model of this world. Darwin was only interested in a limited aspect of this world; the part required for supporting a theory of biological evolution. Darwin's model of evolution was not a comprehensive theory of evolution for it was so conceived that only one aspect of evolution was described, the natural selection component of evolutionary theory.

Closed and Open Loops

The design of a closed loop species makes two assumptions:

Only when the future can be known should a closed species be designed. Understanding the function of the mind requires first an understanding of the difference between two forms of life: open and closed. It is important to consider this theoretical question. Could the closed species have evolved naturally or were they designed? If designed, what kind of designer rationally and ethically could employ the closed type design? I propose that only a god, who from its place in eternity can presumably see the future, only if a god knows that an environment is forever closed, would it be safe and prudent to design a species with a fixed/closed program, and thus immune to the great natural calamities which cause the extinctions of closed species of life. Nature should only rationally design a species with a fixed or closed nature if it has foreknowledge that the respective environment will forever be closed. A "nature" species is a closed genome-designed being.

Our thought question is what would a closed species be like?

- First, they assume with respect to the outside that no change will come from the outside;
- Second, however, on the inside, within Bernard's internal milieu, it assumes that a reliable regulator exists for each essential variable, EV, and thus no change can come from the inside. For a closed loop species, the assumption is that nothing can happen that was not planned for in the design of the species.

Darwin

Darwin (1859) challenged this top-down theory by presenting scientific data to support a bottom-up theory of the origin of species, in his book, *The Origin of Species*. However, Darwin's theory of biological evolution does not explain the origin of the mind, ideas, or behavior.

So Darwin left the door wide open for the verbal-god religions, since his theory of evolution does not show that mind could have evolved by natural selection, and thus he allowed the basic ambiguity which creates the ground upon which religion thrives. As long as religion can objectively make the claim that the mind or soul could be a spirit, that humans are dualistic, then they have some kind of ground upon which to stand.

In the United States, since the law distinguishes between sacred and secular, the dualists do not have a problem; they can grant that each human's body is secular, while they can also claim that the mind is sacred.

In this book, I present a theory of mental evolution, which explains the origin of mental species or classes, not biological species.

Bottom up Beings

The bottom up design style

When the future cannot be known, or if the known future is open, then a rational god would create open species. A good god would not create closed systems for an open space. Thus, the great extinctions show that no rational top-down god designed all of the species that went extinct during the great extinctions. If the designer god knew the ultimate fate of its creation and still designed all the closed species and then sat back and watched as all its creation was destroyed: call that God sadistic.

The only rational option for designing a form of life when the future cannot be known is to design a species with an open mind for an open niche, an active self-organizing system.

The emerging universe: A Pantheistic god assumption

The role of the NV/V mind in the cosmos

My theme is that the compound nature of the NV/V mind should enable the human mind to play a unique role in the service of the nonverbal cosmos. This theme turns the normal god-man relationship upside down. The standard question has been what can god do for man? To paraphrase President Kennedy, the issue is "Ask not what your god can do for you, but rather, what can you do for your god?"

The Active Nonverbal Self-organizing Mind

In my theory, transcendental or out-of-this-world beings are rejected, *a priori*. They simply are not real and do not exist, and at the level of the nonverbal no one, to the best of my knowledge, has ever claimed that rats had a soul, mind, or spirit. The nonverbal mind, of say a rat, evolved to endow the nonverbal mammals with the set of nonverbal rational skills required to be independent agents capable of evolving their own domain specific knowledge, their own pragmatic survival skills, and their own sets of forms, ideas, or ideals.

In this section, I describe my first challenge as a rat runner: to design a test environment, an External milieu, which would spontaneously induce rodents to initiate Heterostatic behavior: to create rational models of space, to create coherent sets of domain-specific ideas. This task took me twenty-five years to accomplish. Some of the major conceptual phases through which the evolution of the maze progressed are listed in Book 2.

A new way of thinking is required to understand the nonverbal mind, yet there is no way that the concept of the rat's nonverbal mind can be understood within the passive historical spiritual context.

In the secular model of FIGURE0.3A, the right half proposes a new active role for the nonverbal mind within the cosmos. The secular nonverbal mind is a form of life designed to feed on uncertainty, on the unknown; designed to digest, to map, or to model the Cosmos, the unknown, great or small. Mammals could be called the Ptolemaic species. Ptolemy was the first to put a coordinate system on maps, which allows anyone reading one of his maps to know where they were with respect to the entire known world.

The nonverbal mind of some species still has sensory systems, which assist in long-range navigation, which science does not yet understand. Whether rats have such a system, I cannot say. However, for the short-range task, the nonverbal mind creates models or maps as a mainstay of its navigation strategy. The nonverbal mind employs these models to map, plan, and execute all the kinds of nonverbal behavior of which each species is capable, yet the NV mammals, even parents, do not have the luxury of deciding, whether or not to share their NV knowledge with the next generation. Conversely, this does not mean that the knowledge of one generation is not inherited by the next. We humans demean the nonverbal forms of communication; in the nonverbal world, each generation can and must inherit by observing their parents' behavior. Is it really so different in the verbal world? When will we learn that "Actions speak louder than words?"

Heterostasis

Heterostasis is a new kind of physiological function that emerges for the first time with the emergence of the mammalian mind. While all kinds of body cells germ, soma, and neurons are homeostatic in nature, yet the whole mammal, which consists of only these three types of cells, has a new emergent property, Heterostasis.

The Hierarchical Nature of the Cosmos

EMERGENT functions

Emergent functions are one of the most important properties of the cosmos. An emergent function is one in which the whole has a property which was not in the parts. For example, the wetness of water is the most famous emergent property. Water is H_2O. While both Hydrogen and Oxygen are gases, yet water is not a gas and has the property of wetness, which neither H nor O possesses. Where did the wetness of water come from? It is an emergent property of the new whole created by combining hydrogen and oxygen into H_2O.

The cosmos is said to have an emergent nature. The emergence of water was a prerequisite for nature to take the next step: the emergence of O_2 biological life forms, which in turn set the stage for the emergence of the nonverbal mind, which set the stage for the emergence of the verbal mind, etc.

The Emergence of Mind: The Emergent Properties of Mind

The mind as an organ has many emergent properties, most notably consciousness and speech. The nonverbal mind conferred heterostasis upon the cosmos, yet none of the cells of which it is composed has that property. Heterostasis is an interesting instance of a whole, which is different from and more than the sum of its parts.

The classical dualistic paradigm of human nature has for far too long attributed all the emergent features of human nature to its verbal function. There is only one thing wrong with this position, all the rational elements of human nature were attributed to the verbal function only because that was the only known difference to which the function could be attributed. The error with this logic was that they missed the emergence of one major organ, the emergence of the nonverbal mind. Evolution must be credited with the design of two revolutionary stages in the emergence of this organ: First, the emergence of the nonverbal model of the mind and second, the emergence of the compound NV/V mind. Both of these organs were emergent; both were stages of the evolution of the mind, or wholes, which were more than, or different from, the sum of their parts.

The emergent properties, which were the product of each of the two organs, were in effect confounded in the classical paradigm which credited all the revolutionary new human potentials to what the classical theorist believed was the only new organ of humans, namely the verbal mind.

One challenge is to distinguish the emergent features of the nonverbal mind from the new emergent features of the compound mind.

My present understanding of how the emergent features of the two minds must be scientifically attributed is as follows:

The emergence of the nonverbal mind revolutionized the nature of all life on earth, specifically the new life forms, i.e. mammals became:

1. The first self-organizing systems, SOS_1
2. The first rational beings
3. The first species with the power of nonverbal communication, the first-stage type of communication (with nature);
4. The first conscious beings

The emergent features of the compound NV/V mind.

1. The first species with the power of verbal communication, the first two-stage type of communication: first with nature; secondly, with others

The Scientific open concept of truth

The nonverbal communication functions

The nonverbal god, the silent potential of the cosmos, requires NV/V beings like humans capable of knowing god and participating with god several ways:

- First, to give voice to god and to publish god's nonverbal laws as verbal laws for all humans to share

The paradoxical fact should be that the role of the NV/V human mind is to write an open empirical book for the nonverbal god as a gift of science for all humanity: But:

- Secondly to implement the non-biological potential of God; the non-biological technology
- Third; to implement biological potential in all those ways through genetic evolution could not:
- The fourth potential of mammals with a rational Compound mind is to expand god into domains where rationality/wisdom have never gone before; this one could be conceived as fulfilling Adam Smith's (1993) vision of the role of division of labor in generating the wealth of nations.

The operator function

The concept of a team of two beings, the nonverbal god, G, and a NV/V human, G-H, raises several interesting possibilities. Both need each other in several different ways. First, a nonverbal god, the cosmic god, is a resource with infinite potential, which is an inexhaustible resource for the human Heterostat's compound mind. A verbal human, on the other hand, is in need of such a resourceful silent partner. Both need each other, the silent god needs a voice, but even more it needs a voice with the rational nonverbal communication skills that can extract from the nonverbal god, the universal truths that all humankind and/or all the nations of the world need to know. A compound NV/V human is a perfect partner for the silent god, to speak for god. The God of all things is an infinite resource, infinitely expanding, infinitely rich in potential. The material God is an infinite potential but can only become known and fully kinetic by having an army of verbal beings willing to devote their lives to turning god's potential both into reality and/or into a narrative. All rational wisdoms are paths strewn with different symbols, by which the unique energies of each domain can be energized, enabled to emerge, and by which each human can have some part in making god emerge; or making god be all that it can be.

For over fifty million years the nonverbal mammal probed and nonverbally knew the cosmos, yet all that knowledge was lost. What will human science know about the cosmos after having studied the cosmos for a comparable number of millions of years?

Two kind of gods: verbal-nonverbal; closed-open

All the sets of closed-book-transcendental-gods have long been essentially out of the communication business: their books were closed, whether they have nothing to say or nothing left they want or need to say doesn't really matter--"if it was good for Paul and Silas, it is good enough for me."

But an open, physical, nonverbal god with an infinite potential should be the ultimate life-long challenging conceptualizing partner for every human Heterostat.

The third wisdom

The third wisdom, the nonverbal wisdom, enabled or required each nonverbal mammal independently to conceive its own hierarchically organized conceptual system, to create its own model for each of its niches. Yet without the fourth wisdom, or verbal behavior, the nonverbal mammals could not share the nonverbal knowledge created by their third wisdom.

Only with the evolution of humans were mammals for the first time, not only able to conceive nonverbal solutions for anomalous spatial domains and thereby effectively become nonverbal self-organizing beings, but then with the emergence of the thirteenth station of the assembly line, the human mammal could export, share, or communicate the knowledge acquired by the nonverbal mind.

The Emerging, open, changing cosmos

However, there are yet other sets of hurdles that are, perhaps, even more threatening. Theoretically, the cosmos has at least two possible extreme natures: the first called the "static" Cosmos, and the second is called the "emerging" Cosmos.

The open dynamic cosmos

This book is about the first set of rational life forms, the nonverbal mammals, which were designed to live in an Emerging Cosmos, one expanding or changing effectively in one direction but in ways that Darwinian evolution could not anticipate. In the end, if Life were to survive in the emerging cosmos then a new form of life had to be designed. Mammals, I propose, were the solution. Mammals were designed to survive in an emerging cosmos, humans especially so.

Nature's big gamble was to design mammals as self-organizing systems, which means that the mammalian mind as a rational mind was designed to deal with and cause some kinds of emergent events, with the insistent change of the emerging cosmos. The mammalian brain has, however, become a master of this Emerging Cosmos (especially the human species), and its rational powers have been honed and perfected to the point that presumably any new potential emerging class of objects or events the organism will soon have become adapted.

The Open-classes of challenging events all initially have the same property; they are all strange, all anomalous events are the sign of change or symptoms of the emerging cosmos. Yet more personally, they invalidate the paradigm that enabled them.

The chief property of the generic nonverbal mammalian mind is Heterostasis, which is the behavior of seeking and destroying anomalies.

These Open- classes of events are usually, at least for the short term, bad news for almost everyone because they suddenly thrust upon life the worst that the universe has to offer or at least the greatest challenges the universe can offer. The Open-class type events suddenly and insistently thrusts anomalies on the nonverbal mind. Yet anomalies are the class of events for which the nonverbal mind was designed.

Is speech a communication or a programming functions?

What is language behavior? Is human verbal behavior the only instance of language behavior?

What is meant by a rat's nonverbal behavior? Is it a nonverbal language function?

According to Sperry (1971), human speech is a two-stage behavior; first, the right hemisphere creates a nonverbal narrative or script, which is exported first to the verbal side of the mind; secondly, it is exported to the external milieu as public vocal behavior.

First, the export function

The most benign way to understand the second stage of human speech is to conceive it as a way to export or share what were originally nonverbal narratives.

Second, speech as a programming language

The most revolutionary way to understand nonverbal speech is to conceive it as a nonverbal programming language. Within a computer system metaphor, we should begin revising the simple empirical concepts of John Locke (1914). Instead of thinking of human nature as a Tabula Rasa, the nonverbal mind is a self-organizing system.

However, I propose that human verbal speech was an idea whose time did not come until the nonverbal mind had been perfecting nonverbal speech for over fifty million years, and finally, the nonverbal communication of the nonverbal mind enabled the verbal mind to evolve in the next stage.

Speech, the ultimate threat to life on earth

The ultimate threat for human life on earth comes from our failure to understand our own nature; the worst evolutionary advance in the history of the earth and for the earth was the emergence of the two-stage human mind: the nonverbal/ verbal mind.

In the book of the Christian verbal god, one of the many gifts of the verbal god to human kind was the earth, and too many followers of this verbal god understand this to mean they own the earth, which is theirs to do with, as they desire.

The cosmic god model, on the other hand, conceives of man as the partner of god, effectively, the voice of god, and man is beginning to tell the world that his god is sick or even "in pain."

The verbal god's followers need to ask, can their god take care of them if they destroy the capacity of the earth to support life?

Our greatest danger is our ignorance of the invisible organ, the nonverbal mind, for it is this first stage mind that was designed to know god. The top down theorist, who worships the verbal god, is unaware that he has the power to know the cosmic god. Furthermore, the theorist feels no responsibility for the care of the earth god.

Consider the above assertion in terms of a computer model.

The computer model

Computers consist of two components: hardware and software, where the software functions are more than a communication. The proper meaning of software is in terms of programming functions.

The evolution of life programs

Assume that life evolved in several programming stages.

Stage one programming must be understood as genetic programming, where the first simple forms of life were programmed by their genomes.

Evolution by natural selection was and still is the first and primary kind of programming. This stage of programming is not only slow but suffers from many handicaps.

Stage 2 programming is the topic of this book. Evolution designed mammals with a nonverbal mind in order that they could function as self-organizing or self-programming systems. This new set of mental skills created a new family of species, mammals, which could adapt to change by writing their own new software programs. One of the intriguing new properties of the mammalian mind is how the self-organizing or self-programming function of the nonverbal mind works. This self-programming physiological behavioral technology I call Heterostasis, which has been discussed several times in this book. What is new is the idea that Heterostasis is a programming language, in fact the primary post-genetic language. Heterostasis enables each mammal to self-organize or program itself by the primary nonverbal forms of language. The nonverbal language emerging from the Rolandic cortex evolved to enable behavior to be rationally controlled and was the first and for millions of years the primary programming language.

Who programs the mind

The nonverbal mind programs itself by the sequence of behaviors it emits while interacting with its environment though nonverbal Heterostatic behavior or while emitting spontaneous exploratory-play behavior.

What is intriguing about this new style of programming becomes plain when I ask, who is the programmer in nonverbal programming? Is the nonverbal mind programming itself, or is the environment programming the nonverbal mind? Programming at this stage is a joint function of the mammalian-nature, M-N interaction during exploratory play.

With Heterostasis conceived as a programming function, what is the mind's goal when emitting Heterostatic play behavior? The two-stage or joint programming function between the mind and nature makes communication even more complicated. Peirce's concepts become more meaningful when applied at this level than at the verbal level.

What is the meaning of words? Peirce (1878) concluded that at the human level the meaning of words were actions. However, for this third generation pragmatist, at the nonverbal level, the new deeper meaning proposes that the meaning of images is action.

At the level of the nonverbal mind, programming was clearly an empirical or experimental function. Yet, at the verbal level for the first time, a programmer can program a computer where neither the computer nor the environment was involved in the programming. While I did not raise the issue earlier, the most important new aspect of human behavior comes into play at this new stage of programming. The question is one of autonomy, or what is politically called freedom as in Democracy.

When I take a rat out of its home cage and drop it into the matrix maze, at this first stage of the experiment, the rat is still most obviously my prisoner. Yet once I release her into the maze, at that point in time and space, I confer upon her a limited numbers of degrees of freedom. The rat is "free" to both choose its goal and to choose how it spends its allotted moves and time in the maze.

When a human child is born into a family, how much freedom are they granted by the family? At the verbal level, the communication between the parent and child is not purely playful but rather must be understood as a series of steps in the serious business of transferring all kinds of kin programming between generations.

Viruses

The evolution of the human two-stage nonverbal/verbal mind was and is paradoxically, in many ways, a step back relative to the freedom of a rat. At the level of rats, the parents in the absence of any verbal language have no real means to program their young.

While theoretically as a thought experiment, it would be possible to drop both a human and a rodent offspring of equivalent ages into a proportionally sized maze both could be equally free of their parents as soon as we drop them in the maze. Yet human verbal language allows various institutions to subvert the human child's self-organizing natural rights, while the rodent parent has no comparable ability to influence their offspring. In terms of personal freedom, the human mammals pay a great price for its acquisition. The fact is that humans lost a great degree of personal freedom with the acquisition of language. In fact, most humans are never given an opportunity to discover their own natural talent to be self-organizing. But of course, the blessing of the family and the larger community, one could argue, fully compensates for any loss of freedom that might have occurred.

However, in principle, any program that usurps the individual right to be independent, free, or self-organizing is and should be called a virus. A virus is any program that attempts to take control or personal freedom away from its legitimate owners.

Again, let me remind the reader that the nonverbal mammals paradoxically are lucky by not having a two-stage verbal mind; they are naturally immune to viruses.

Our utter lack of understanding of the nature of the nonverbal mind and the power of self-organization conferred upon each individual makes the verbal human fail to understand the nature and origin of knowledge and of freedom, and thus an understanding of the threat provided by faith-based thinking is severe relative to our lack of understanding of fact-based thinking.

The challenge of religion

Historically the open-class of events, events that demand change, have been so scary that science has been sadly obsequious before the Princes of the world of faith, even those whose power has long since peaked. The Princes of faith have the advantage over the practitioners of science, for they can instantly pretend they have solved any problem with words, gestures, smoke, idols, incantations, and all sorts of metaphorical magical words. Unfortunately, those committed to fact-based images with meaning have no comparable ability to offer instant verbal solutions to problems not yet amenable to science.

Science seems not yet prepared to start the factual battle, which is long overdue over the nature of human nature, the nature of the cosmos, the war to save the earth, etc. The Open- class events have the tragic consequence of inducing uncertainty, which the faith-driven believers are in most respects immune to or in denial of. They long ago immunized themselves to empirical knowledge, Plato (1952), I am prepared to say, started it all with his poisoned metaphors, which made all those who opposed his animosity towards the empirical world remain silent.

The open-class of events, the role of anomalies

Religion has long been Goliath; unfortunately, science has never produced its David. Yet, it has been a long time since the pope and/or the inquisition burnt their last scientist at the stake. Even Galileo, who among his other acts of defiance made fun of the pope, was only put under house arrest!

Science, for historical nonscientific reasons, grants religion the high ground. Science has no plans to stand against Goliath. Science does not yet ask, much less demand, a level playing field.

One part of the problem is that scientists are so busy doing science they have no time to understand or to get involved in the big picture or the big issues of their day. Most scientists do not recognize religious faith as a problem for science or the cosmos. Scientists may not think about gods or spirits for they find no need to worship a book god, but then they do not understand their possible roles in the support of Spinoza's (1933) Pantheistic god.

What scientist last called a Christian an atheist? What human, who partners with Spinoza's nonverbal god, last threatened the pope with excommunication? Who last called the pope an unbeliever in the real god, the earth god?

What would a level playing field mean? Why have those that refuse to live by faith, abandoned only the Book God. Why have they also abandoned the cosmic or the pantheistic god? Why do those that have abandoned the verbal god call themselves atheists? Why have they also abandoned the god of the cosmos? No follower of the earth god is an atheist! From Spinoza's perspective, all followers of book gods are atheists!

If freedom of religion was objectively observed then those that were followers of any god should be allowed to call all other nonbelievers in their god an atheist.

At least the scientists that have abandoned faith have not abandoned knowledge! On the other hand, tragically the spirit god's followers long ago abandoned knowledge not faith.

What is human nature? What rights should each nonverbal mammal have? What rights should a woman have, as Mary Wollstonecraft (1976) was the first to ask!

I am no atheist, yet I no longer worship a spirit or book god, and it has been a while since I believed in Santa Claus because I now collaborate full time with the Emerging cosmic god to study its nature, and to assist in the writing the god of nature's open book.

Plato and the dualists claim there are two worlds, the sense world, and the nonsense world. Plato, with his metaphor of the divided line justified to his mind the superiority of the nonsense world. Why do so many humans worship the nonsense god, a spirit god, the ruler of the nonsense world, a spirit world; a god that cannot be known by the senses; that can only be acknowledged by an act of faith.

Models of Mind

The known cosmos

Even scientists still insist that all they do is to record data, where data means "given."

However alternatively, the mind is active, and thus what science records is not given but rather "taken." TAKEN is a John Dewey (1916) word. The nonverbal mammals were designed to be self-organizing, to be active rather than passive; they were designed to create coherent sets of concrete forms through which they communicate with nature. The Heterostatic technology, which is built into the nonverbal mind of all mammals, enables mammals to execute the creative function, which "Nativistic" philosophers have for long enough said, "only a god could do". Mammals' successes can all be attributed to their having been given the technology, which enables their mind to compute the "hard" functions.

Mammals have a nonverbal mind, which has the technology of the third wisdom, designed to execute the hard-functions, the so-called god-functions.

Plato (1952) proposed that humans were helpless without god who executed the hard rational functions, which were beyond the talents of humans and slipped them the answers either at birth, or in some book, or on a tablet of stone, etc. However, Plato's god is a pure verbal being and all of its truths are in verbal form. What exactly would or could a pure verbal sentence one without meaning say?

However, for the nonverbal mammals with a nonverbal mind, their ability to self-organize is a function of the fact that the physical, this world god, the god of time and space, and the primary mammalian mind are both nonverbal.

Science should begin to explain the power of its mind. The nonverbal mind "talks" to the nonverbal nature or god, only nonverbally. Nonverbal communication is the primary kind. Verbal communication is a secondary kind.

The matrix maze triggers what Cannon (1932) called the "fight or flight" reflex but now with an entirely new meaning. The new purpose, the new reflex, is not fight or flight, but explore, "What was that" with the sense that triggers a deep rational epistemological curiosity; the nonverbal mind begins immediately to create a model that will reduce or destroy the uncertainty induced by the anomalous "bear."

Most rats respond positively to the anomalous events of the maze and almost immediately begin to explore the maze- space with a self-imposed purpose that dictates that each of the 96 spaces be conceived as either salient or not to their purpose.

The nonverbal mind of the nonverbal mammals enables mammals truly to be self-organizing beings. Their nonverbal mind cannot communicate with or be dependent upon a verbal god. The nonverbal mammals have survived independently of all verbal "book" gods. The nonverbal mammals live by communicating with the nonverbal god.

Where do all the nonverbal self-organizing mammals get their concepts, ideas, or truths? All the nonverbal self- organizing mammals survive because each of their nonverbal minds is capable of executing the first half of the rational functions, the hard half, conceiving of Ideas and laws entirely through nonverbal communication.

Ashby

Ashby (1960) made the first important rational distinction between regulation and control. Homeostasis can be understood in terms of the technology or the science of regulation. Homeostatic technology employs only servomechanisms, each of which has only to keep one of the many EV's of each mammal at their set points. Homeostasis is the product of the science of regulation.

However, Heterostasis is the science of control. The first technology designed to function as a control was designed by Charles Kelley (1968). One of Ashby's (1960) most impressive insights was recognizing that homeostasis was not capable of explaining the mind, and that a new technology, which he could not even imagine, was somehow evading science.

Car navigation systems

Even today, our car navigation systems typically embody a modern equivalent of Plato's distinction between two component functions: the god and the human function.

The man function

Our car Navigation systems consist of a technology that includes only what Plato (1952) insisted was the man function. Each year, the makers of my car's navigation system come out with a new DVD, which has a model of the whole United States. Our current navigation systems work only with the second half of the navigation function, the easy half, the man half of the control functions.

Our navigation systems' man-functions execute several component functions. First, space satellites download Global-positioning knowledge identifying each place upon the globe, so that our navigation systems can compute: first, where it is at all times; second, the driver must supply a destination; third, The DVD must supply a map of all the intervening space.

By knowing these things, where a destination is and the knowledge of the intervening geography, a navigation system can compute the route to a destination. Navigation is a matter of extracting the knowledge embedded in a relevant model.

The god function

The company that sells the navigation system executes, as it were, the god function and revises its DVD model each year. Each new DVD provides a copy of the work of some "god surrogate," an updated model of the whole United States: every street, every house, every place of business, etc. Each New Year's DVD corrects all the out of date knowledge because of the changes made each year.

The computer model

Today our home and business computers observe Plato's (1952) distinction between the G and M functions. The basic facts about our computers are that they execute programs, which they do not and cannot write. Only human programmers execute the god function. Additionally, human programmers conceive the model, which our car's navigation systems execute. However, Science fiction writers have long envisioned the day when computers can program themselves, and become, as it were, the equivalent of the nonverbal mind.

Computer Metaphor A

Computer programmers are to computers what myth attributes are to verbal gods. Computers are perfectly obedient, they only do what their human programmers tell them to do; humans as sheep are also obedient, only possibly less so.

Plato conceived of humans as, in effect, computers. A good Christian only executes Scriptural scripts. God is the great programmer in the sky. Again, let me remind the reader of Plato's consistency. Neither computers nor humans can execute the god-function. What is new here is that the god-function cannot be conceived as the programming function in computers.

Computer metaphor B: Charles Kelley's Zieboltz controller

Charles Kelley (1968), a human engineer working for the navy, designed the first computer capable of the execution the "control" functions, which he named the Zieboltz controller for US atomic submarines. I have never been able to discover how, or if, any relation exists between the car navigation systems and the military systems. Kelly's Zieboltz controller did for the Navy what our navigation systems do for our cars. Kelley's controller almost automated the entire control process, except again humans had to supply it with a relevant domain specific model with information of the sea-lanes, the reefs, the weather, the geography etc.

What is a mammal: rats, humans, computers, Zieboltz controller, etc.

Plato's (1952) concept of human nature was and always has been wrong; mammals, even one with as thin a cortical layer as a rat, can execute the god function. No mammal, not even a rat, is as incompetent as Plato insisted all humans were.

A few years ago, the DVD supplied with my car's navigation system did not include the ten square miles surrounding my home, and I drove to and from my home hundreds of times over several years, and each time the computer picked me up only after I got into the "known" region described upon its DVD. However, what always annoyed me was that no matter how many times I went from the known into the unknown region around my home and back, my car's navigation system never learned a thing! Each time I would get mad when I realized that Plato was correct; my car's navigation system never was able to self-correct or self-organize. My car's navigation system was only the master of what Plato called the man function. Still today, this navigation system cannot execute the hard side of the navigation or the mapping function.

How hard can it be, even a rat, such as F3 was able to conceive a map of the matrix maze in less than thirty minutes with only a few round trips through only about a third of the space of the matrix maze?

Even the nonverbal mind of a 90-day-old rat, however, in computer terms can write its own programs. Rats are born with the neuronal technology required to execute the god function.

What is the function of the nonverbal mind?

Therefore, I am obliged to ask the following questions: what is a nonverbal idea, a nonverbal word, a nonverbal narrative, a nonverbal cause, a nonverbal god, etc.?

Just as the eye was designed in the image of light and the ear in the image of sound, just so the nonverbal mind was designed in the image of the nonverbal cosmos.

For example, what is the nature of the eye? There are two kinds of sensory cells the rods and the cones. The rods are sensitive to the amplitude of light and the cones are sensitive to the wavelength of light. Combined, these two cell-types extract essentially all the knowledge required to describe light and to extract all the knowledge contained in light.

Just as wings must have been designed in the image of air and fins in the image of water, just so the mind must have been designed in the image of the nonverbal cosmos.

As strange as it may seem, the first role of science should be to capture the rational causal nature of the nonverbal cosmos, or the nature of "mother nature," or the nature of the cosmic god, or to put it most plainly: to know god, everything that is, is god.

Long before the advent of science, Western humans conceived of themselves as one-mind beings, or verbal beings, viewing the cosmos within the framework of one of the three books or bibles of the three transcendental religions, which all teach that god gives us "truth" in these three "bibles." This bible truth is "given."

Some of the eastern Religions, such as Daoism have conceived the cosmos not as an enemy, but as a great river of life, within which we all must find our "way."

Two worlds

The nonverbal and verbal minds differ in one strange way; each naturally fosters a profoundly different concept of the world and god. One of the strange concepts that may take a while to accommodate is that the cosmic god is nonverbal! But of course, the nonverbal mammal's nonverbal mind has no verbal skills, yet it served the nonverbal mammals well for over 50 million years of change. Column A summarizes the cosmos, as it is conceived by the nonverbal mind.

Thanks to Plato, the verbal mind brings humans a possible ontological reality that turns the cosmos as conceived by the nonverbal mammal upside down- See Column B.

A	B
SCIENCE	PLATO
SENSE WORLD	NONSENSE WORLD
material world	transcendental world
secular world	sacred world
god of time/space	god of eternity
god of matter	god of nothing (no things)
emerging cosmos	static cosmos
nonverbal	verbal
the knowledge-based world	the faith-based world

FIGURE 0.3A

FIGURE0.3A shows an abbreviated version of these two models of THE WORLDS of human nature AND two corresponding concepts of god.

In many ways, Plato (1952) was able to turn the world upside down only because he was at least asking vital questions, questions for which Plato's rationalism had no answer. Hopefully for the first time, this book will answer some of Plato's questions.

Plato wanted, more than anything, to know what was real. Plato was much like Descartes (1946), who thousands of years later claimed he had doubted everything, save that he doubted! Plato (1952) never played the doubting game. He never doubted the reality of all things, of matter itself, or of the cosmos itself: Plato never doubted the existence of matter. Plato's problem was that he doubted only the empirical value of physical things. Unfortunately, for all of western civilization, Plato concluded that matter, the world of things, had no redeeming empirical qualities for matter epistemologically was not real.

Neither Plato nor Descartes understood how empiricism worked. They had no understanding of how, at least in the nonverbal world, anything that entered consciousness had to have passed the most stringent ontological tests that nature could design. They had no way to understand that anything the nonverbal sixth sense authorized to enter consciousness was as certain or as real as anything can ever be known to be.

Closed truth

Plato (1952) disillusioned with the empirical world, which can be known through Aristotle's six-senses, conceived an ideal metaphor-driven world, a nonsense world, one outside of space and time. Plato conjured up his own version of a completely new world, where only his ideal beings, all nonsense beings, or nothings, met his highest spiritual standards. The real had to be nothing or outside of time and space dwelling in eternity and thus immune to change. Plato's ideal world, of course, he attributed to a nonsense god ruling a nonsense world, full of nonsense beings, spirits, created logically by appealing to a geometrical metaphor, ruled by his nonsense laws; a god which was a Santa Claus for adults.

Within this conceptual framework of the verbal god, historically, humans have happily allowed themselves to be demeaned too many ways to count. For example, as a child almost every Sunday my church sang “Amazing Grace" and celebrated their status as "wretches," or to celebrate being a sheep, or using a more modern metaphor being a computer. What does any owner expect of their computer; only to execute the programs it is given. I will never forget my father made his expectation of me clear by quoting First Samuel, "Behold to obey is better than sacrifice; to hearken than the fat of rams”. (I Samuel, 15:22) First, I was born into a top down world, children in my day were taught to honor/obey their parents, who, in turn, had been taught to honor/obey their parents, etc. to live within a top down chain of command. Why, because Plato and all the monotheistic religions and philosophy have long agreed that only one god can execute the god function. Only the god of faith or the god of grace, amazing grace, can do it!

Open truth

Meanwhile, the nonverbal mind of all mammals has been meeting the pragmatic challenge; the mammalian world requires only that real things pass the empirical tests. They are only required to prove themselves real by entering the nonverbal consciousness.

The mammalian nonverbal mind requires only a pragmatic source not a Platonic eternal source of truth. The nonverbal mind owned by the nonverbal mammals is NONVERBAL, and thus, they only require access to the empirical truth of the physical cosmos, the nonverbal god, and the god of all “things.”

The nonverbal god of the cosmos, the silent cause of all things, serves all mammals well. The nonverbal mammals were created as rational self-organizing nonverbal beings. Then at the next stage, the human mammal was conceived as a two-stage self-organizing being, one with several possible wisdoms. Humans were designed with a two-stage mind; first, a nonverbal mind then secondly, a new feature was added at the last minute, the verbal component.

Yet humans have paid a high price for this new verbal feature, which was added to their nonverbal mind. The problem started when they as mammals had only a nonverbal mind, which was originally designed only knowing how to know. Yet from the beginning, this one original passion was potentially positive. Aristotle (1952) first recognized "that all men are born with a need to know." Aristotle however, never imagined that all the NV mammals were designed as SOS1 and had as much need to know or even more than humans had.

Originally, Science has shown that the cosmos emerged as a nonverbal ball of energy with an infinite potential, and we in turn, billions of years later, emerged with a two- stage mind, which nature endowed with perhaps an infinite rational potential. The nonverbal mammals with their nonverbal mind had perfected a nonverbal rational organ but admittedly an incomplete rational organ; yet it was the nonverbal mammals, it should be noted, not humans, who first emerged in the image of the nonverbal god.

Spinoza's (1933) pantheistic god is neither out of this world, not outside of anything in this world. Everything is a part of the cosmic god, and science has much yet to learn/know about the cosmos. Perhaps the strangest aspect of his theory to internalize is his idea that each thing upon the face of the earth is an element of god. Consider the "plight" of humans with their rational compound mind against the backdrop of a cosmos with a vast unexpressed potential. Humankind was born with the rational capacity to participate in this emergence. Humans were born privileged; they were the first species born with verbal speech with rational powers, which far exceeded the power of the nonverbal god. What is special about human nature is that our skills, in one important sense, exceed the power of the verbal god. It is important to recognize the significance of the fact that the cosmic god is nonverbal. Science should be in an enviable position; humans alone have the rational empirical power not just to know the nonverbal god of the physical/sense cosmos, but they also know how to export and publish that knowledge for all humans to access. With the emergence of humans, there was a life form with the rational potential to match the potential of the nonverbal god.

Another advantage the nonverbal mammal had over the verbal mammal is that its lack of speech made it immune to the social pressure from many sources but particularly from religion. Yet on the other hand, once humans obtained the powers of speech, they must have immediately begun to suffer group pressure.

So the nonverbal mammals and the human mammals each went their own way. The NV mammals were able to live a self-directed life, but humans were forced to choose the dominant metaphor to live by. If humans finally discover they have a NV mind designed to be self-organizing, perhaps they will begin to think more for themselves. Humans have a rational burden that the NV mammals don't have; tragically, humans now have to learn how to know who they are, and they have new options for living their lives; especially they now have the responsibility to define rationally their place within the cosmos.

The cosmic nonverbal god is an infinite source of uncertainty, the ideal food for thought for the nonverbal mind. The nonverbal cosmos has enough uncertainty for all humans who choose to participate with it as partners. The nonverbal god has room for each fully rational being to participate in acquiring and writing, the book of science, an open book about the nature of all real things. Each human, theoretically, should be capable of writing an original book about some domain specific feature of god. The role of humankind is to make the nonverbal god known, to give voice to the nonverbal god.

The Truth-Sharing Mammalians

The rat, as the prototypical mammal, has forced me to consider another and wider point of view, namely that all of the thousands of different species of mammals have heretofore as nonverbal beings been unable to share their knowledge with even their own kind. In others words, the nonverbal mammals could not have gotten their "SURVIVAL-TRUTH" from a wordbook, divine or otherwise. The nonverbal animals cannot process words, either the words of man or the words of god, if there were such a god.

So where do these nonverbal mammals go to get the truth they live by? Why did this nonverbal mind emerge perhaps fifty million years before the human verbal mind emerged? This nonverbal mind was designed to enable mammals, both nonverbal and verbal, to self-organize, to create their own salient domain-specific, purpose-specific, species-specific life-space. Who will finally tell humans they were not designed to be sheep nor wretches? They were created with a compound mind designed to deal with change, and they have the ability and the right to change.

The NV-mind mammals must create their requisite nonverbal knowledge directly from the cosmos by nonverbally probing or interacting with local spatial hierarchically organized domains. Evolution did not give the nonverbal mammals all their knowledge in their genome or in books, etc., but rather gave them the power to know empirically the nonverbal nature of their local domains. So what is the nature of the generic nonverbal mammalian mind? This is the same as asking, "What are the properties of the cosmos."

By the generic mammalian mind, I refer to the invariant structure shared by all mammals independent of all the niches to which they are adapted, and of which I assume the mind of rats is a prototype. This generic nonverbal mammalian mind must be understood as having been designed in the image of the nonverbal cosmos, in order that this nonverbal mind should be able to capture an image of the invariant structure of the cosmos as different combinations of it are expressed in each domain of the cosmos.

The role of the nonverbal mind in the second chapter was abstractly described in my version of Aristotle's model. The sixth sense, which was described as executing the god-functions, must transform precepts into images whose function in consciousness is to induce or trick the organism to act, to believe that the subjective images in consciousness are not exemplars of physical classes of things in the real world, but are those real things themselves. The nonverbal pragmatist only asks, what is real--here and now? What any rat or any mammal with a first phase nonverbal mind cares about or is even capable of understanding the concept of Eternity? Yet every rational nonverbal mammal is vitally concerned with the objective "real" world" of space and time; the world that is independent of whatever they believe, but nature designed the Aristotelian ontological machine with the technology to know the domain-specific nature of whatever domain in which it was born. Thus, nature designed consciousness to reveal to each mammal, the objects, and events, which can be empirically known. Evolution designed the nonverbal mind with the sensory/motor skills to glean all the anomalous sensory data occurring at any time and place. The nonverbal mammals cannot survive in a world where they doubt the exemplars that enter the nonverbal consciousness. The sixth sense was designed for empirically driving the nonverbal mind; its control is absolute and comes with the technology like the Match-Mismatch mechanism capable of certifying the empirical validity of virtual images before they enter consciousness.

The nonverbal mind was designed to feed on anomalies. The nonverbal mind was designed to give its primary attention to sensory events that induce doubt; yet the nonverbal mind was designed to guide mammals through troubled waters, through bad times, and through times of change. The ultimate paradox is that exemplars are nutritionally deficient; they cannot induce mental growth. Think about this fact, only an anomaly can induce a new idea or a new set of ideas or as Kuhn (1962) would say; only an anomaly could destroy an old paradigm and make way for a new paradigm, i.e. growth.

The sixth sense is the particular organ or place within the assembly line of the nonverbal mind where the nonverbal mind creates sets of nonverbal ideas, which capture the formal spatial structure of each unique spatial domain. The purpose of the nonverbal mind is to create a domain-specific map or model of space, which consists of pragmatic knowledge, which embodies not a universal model of the forms of space but rather to create a domain-specific pragmatic partial model of space, which the nonverbal mind can then use to plan its behavior. The nonverbal mind of each mammalian species creates the domain-specific knowledge, which is capable of informing the concrete behavior required by each unique species in each of their unique niches.

The nonverbal mind must be capable of capturing the domain specific forms of any and every domain of the cosmos occupied by some species of mammal and use these forms to recognize the particulars of every niche-specific item. The nonverbal mind's chief function is to conceive the forms that can create an image of the formal structure of space as a set of coherent ideas, a set of facts, a set of rules, or a set of laws about the invariant structure of the space of any domain of the cosmos. This knowledge is then transformed into a virtual world of consciousness, which functions as what Lewin (1951) called "a Life Space," but one which Lewin probably conceived as a human life space. Lewin probably never imagined that a rat runner would propose that rats had a mind that created a life space for their pleasure, one unique perhaps for each mammalian species. Yet every species has a nonverbal mind, which has two components. First, all mammalian species have or share the design of the core structure, the A- function of the GNMM; second, the species -specific nonverbal mind must have what we call the B-function, unique set of senses, unique for its niche in the case of a bat, a whale, or a bloodhound. Additionally, the sixth sense of each species must have had its sixth sense adjusted for the composition of the particular combination of five senses with which they were endowed.

The challenge of the matrix maze for the rat's nonverbal mind is to understand or to model the formal structure of the maze in order to adapt the organism to perform successfully in the maze. The nonverbal mind's task is to "take on," capture, or embody those parameters of the cosmos manifest within its current domain.

Knowledge

The pragmatic challenge presented to the nonverbal mind by the maze is to model the complex hierarchically organized forms embodied within the structural properties of the maze space, the A-function, which then can be transformed into a set of ideas by the nonverbal mind, which must then direct the behavior of any mammal, the B-Function. The nonverbal mind conceives the form of the maze by creating a formal model, a causal conceptual structure that captures the form of the space, the A-Function.

Creating an Idea

Every factorial design, particularly those involving a mixed base design, must consist of a set of coherent ideas; each idea represents one element of the hierarchical forms embodied in matter. The pragmatic mind works with the sensory data generated by the organism's sensory systems then the Sixth Sense must transform the sensory data into a set of coherent nonverbal ideas. The function of a hierarchically organized set of ideas is to capture the form or nature of a set of causal features of a hierarchically organized mixed-base domain-space. Each element of a factorial design conceived by the rat's nonverbal mind must capture one component of physical nature as an idea, and as each rat in the matrix maze evolved a model of the maze, sooner than seemed possible, a valid pragmatic model or paradigm of the maze emerges. The rat's nonverbal mind cannot fake the number of factors or the kinds of factors they conceive in consciousness. My mind designed the matrix maze as an A2, B3, C2, D8 mixed base design, and the Twelve Station empirical model must conceive a model of the matrix maze with the same number and the same kinds of parameters in its model of the maze.

So what is a nonverbal idea, which is most powerfully expressed as images in a nonverbal image-driven consciousness?

Of the four factors which my mind conceives the formal structure of the matrix each of these four factors captures one of the distinct physical features of the domain-specific structure of the maze. Likewise, the rat's nonverbal mind must conceive a set of nonverbal ideas, which collectively captures each of the forms of the local manifestation of the cosmos. Each ideational element of the formal structure of the maze captures one of the features of the cosmos embedded within the spatial structure of the maze. If the forms conceived by the mind are a valid image of the formal structure of the maze then the nonverbal mind can act "knowing" the space. The purpose of the nonverbal mind is to capture, as a causal conceptual system, the form of the maze. By knowing, by creating a formal-conceptual image of the structure of space, the rat is able to plan the successful means to its end in that space.

The Emergence of the God-Function

One of my purposes in this book is to show how the technology with which evolution endowed the nonverbal mind enables the nonverbal mind to defy effectively everything that the experts have said about the human mind for thousands of years.

- Watson (1978), the father of behaviorism, denied that humans even have a mind.
- Sperry (1961), on the other hand, says humans have two minds, one in each hemisphere.
- Plato (1952) said the mind was an immortal soul, which could pick up or perhaps exchange ideas between reincarnations.
- Pinker (1999) says language consists of words and rules, whether a mind is included, I can't say.
- Plato (1952) created most of the myths that have become famous over the centuries. Plato's most famous/infamous metaphor was the geometrical metaphor of the divided line. No scientists, who generally settle disputes with empirical facts, have ever challenged Plato, who proudly proclaims his dualistic or spirit-based theory is devoid of facts!
- However, Plato slipped up! He defined verbal rationality as consisting of two functions.
    - First, the hard-functions, or the god-functions, which only god can execute
    - Second, the easy-functions, or the human functions, which only humanity can execute

- One of the most grievous conceits, which Plato has gotten by with for thousands of years, is his distinction between knowledge and belief. I dealt with this issue in my discussion of the divided line, in the section on Plato's theory of mind in Chapter 2.

Yet the rat's behavior has implications for all manners of topics, to take just one example: I never set out to solve Plato's (1952) problem, yet serendipity was on my side. It almost seemed decreed that the answer to Plato's problem, to understand the nature of human rationality, was paradoxically to be found in the behavior of rats spontaneously running the matrix maze! Unfortunately, thousands of years ago when Plato discovered it was possible to think about and perhaps even understand the nature of human rationality, the technology or the tools required to solve his problem did not exist. Indeed, the requisite technology would not come together for another two thousand years for the first time, in our time and in my lab! Moreover, I had not yet discovered Plato's problem, when I assembled the requisite technology.

The A and B elements of a class

Plato's (1952) problem can be conceived narrowly or broadly. In the narrow view, his problem was simply a question about the nature and origin of ideas, concepts, or forms. From his perspective, Plato wanted to understand the rational function, which he imagined was executed by the verbal mind of god. Plato's purpose was to understand how human verbal rationality works; what enables the human spirit verbally to act rationally? Plato had no way of knowing or even suspecting that there was a nonverbal mind, which he shared with all the nonverbal mammals, which was executing the functions which he attributed to god.

Yet in spite of his empirical blindness, Plato had some important insights, somehow he knew:

- that there were classes of things,
- that our knowledge of those classes was based upon the distinction between the two components of each class
- **The A component**, the concept, form or idea
- **The B component**, the exemplars of those classes, which Plato knew, the B component, and the physical rose we can hold in our hand was an exemplar and would soon be gone. Each particular rose was a temporary physical object that had taken on the form of the rose, the A Component, but that temporary things were just an "appearance," a too-soon-to-be-gone exemplar.
- Exemplars were copies of something real, which I call component A, the idea of the rose, which was eternal, perfect, and never changed.
- Each B was an attempt of the physical world to replicate A, but B was always an imperfect copy of A. B was a copy in matter, but matter can never perfectly copy A, a spirit.

Plato wanted to know and to understand A; the component that was permanent, real, eternal, i.e. Plato wanted to know A not B. His theorizing was concerned with one thing, what is A, where is A, where did A come from, etc. Plato, however, believed that while B was made of matter, A, was too good too perfect for matter to copy and so spirit beings or gods, ideas, forms, etc. out of logical necessity were born.

Plato's idealization of ideas forced him to create perhaps the most fanciful and fateful metaphor of all time, the geometrical metaphor of the divided line, which I have previously discussed.

Plato believed that human rationality had a Nativistic not an empirical origin, which is still the dominant perspective among linguistic theorists even today. For example, Noam Chomsky (1997) still believes in Nativism. In other words, the A component was too good, too fine, too perfect, to be explained as a function of matter. A, on the other hand, was beyond the power of the human mind to know. For all practical purposes, all we can say is that A happens, unfortunately, we will never understand how god works.

As a monist and an experimental scientist, I reject the idea that the verbal mind did not evolve but magically emerged in humans. I propose that the verbal mind had its origin in the evolution of the pre-human mammalian species. In order to understand the origin of human verbal behavior, it is essential that we first understand how the one-mind mammals evolved from the no-mind animals and then how the 13-station human species emerged from one-mind mammals.

However, in the broader view, Plato's (1952) too-narrow view of the problem was but a symptom of his larger problem. Plato's large-view problem was not that he just could not imagine a natural or empirical origin of ideals, See FIGURE0.1A, but his Nativism was possessed of a profoundly deprecating view of human nature. It was not that Plato simply could not imagine ideas having an empirical origin, rather he could not imagine how anything good, much less anything as perfect as ideals as having an empirical origin in human nature. Plato held Nature in such contempt that in his view nothing could be gained by its study, i.e. empiricism. Plato's Nativism/Rationalism had a profoundly negative origin. Plato's Nativism and Rationalism expressed his profound contempt for their opposite, empiricism.

Two schools

Today, Plato's contempt for nature is still alive and well in our schools and in our religions, but especially in those who claims to be linguists, the scientific study of linguistics. In spite of Sperry and his followers' brilliant experimental research on his concept of the two minds, yet there are those who still ignore Sperry.

Pinker

For example, Steven Pinker ignores Sperry's research and procedures. Pinker's science reminds me of Watson (1978), the father of Behaviorism, who claimed there was no mind. Watson imagined that behavior could be understood in terms of Pavlovian technology. Watson was famous for his Little Albert study.

Pinker (1999), on the other hand claims that (Human) language behavior can be understood simply in terms of words and rules, presumably no mind, verbal or nonverbal, is required. While Pinker studies human's verbal behavior in term of words and rules, I study the rat's nonverbal behavior by observing its nonverbal behavior in the maze. Lest there be some confusion, let me state the obvious. The first difference between the two minds, the verbal, and the nonverbal, is that in the domain of the nonverbal mind **there are neither words nor rules**. In the domain of the nonverbal mind, ideas replace words and causal concepts replace rules. After all, words and rules are memes or conventions. Each rat running the maze is alone in the maze; they have no dictionaries and no colleagues, much less a committee to which they can refer any ambiguity. The nonverbal mind is designed to work alone. Down here in the nonverbal mind, things don't work as the linguists think they do up there in the verbal mind, or perhaps I should say that since the nonverbal mind started maybe fifty million years ago, things in the nonverbal mind don't yet work, as they will in the future with the emergence of the human verbal mind. Back here, the nonverbal mind is concerned with the basics, ideas and causality; words and rules are still millions of years in the future. Frankly, there is for the moment no need and no place for words. Words are arbitrary beings called memes. In the beginning, when the nonverbal mind was emerging, the first responsibility of the nonverbal mind was to conceive ideas, which in this first stage are designed to provide within consciousness "stand in" images or exemplars, one "stand in" image for each class of things active in the narrative of consciousness.

This book is not about any of the hypothetical magical minds described in sacred and secular books for hundreds of years: gods, angels, devils, etc. However, writing about the differences between the classic concept of the verbal mind and the concept of the nonverbal mind has been unnecessarily difficult because the verbal-mind community has polluted and/or defiled almost all the words required to define and discuss the NV Mind.

In a nutshell

This book addresses the most central mental or rational problem: how the nonverbal mind of the nonverbal mammals like rats evolves sets of nonverbal ideas, which reduce the nonverbal uncertainty induced by nonverbal anomalies in a hierarchically organized spatial maze, the matrix maze.

Nonverbal Ideas, not verbal information, reduce the nonverbal uncertainty within the Horizontal-Vertical H-V Aristotelian type model of the mind during exploratory-play behavior, which I call Heterostasis.

Heterostatic exploratory-play behavior is an overt pattern of behavior, which executes the A-function with a series of only three moves, which enables a scientific experimenter to, in effect, read the rats mind. In other words, the experimenter can infer what physiological and psychological functions are operating at each stage of the creative ideational cycle.

In short, this book asks and answers how a material organ, the nonverbal mind/brain, conceives ideas. One place this topic is addressed in a discussion of the behavior of Female 3 in Book 2, Chapter 1.

Plato the origin of ideas

The central element of Plato's (1952) discussion of mind is the concept of a concept. At first, Plato had some remarkable insights. He knew about classes/concepts A and exemplars B as noted above, but then at some point in space and time, Plato's theory mutated and became a virus that for thousands of years has dominated and infected the discussion of the nature of the ideas, or forms up to the present time.

Aristotle's empirical model affords the only productive technology for studying the behavior of the rat.

Nativism: the rationalist's problem and an empiricist's solution

My running feud with Plato concerns the nature of human nature, specifically human rationality. Plato (1952), as mentioned above, insists that rationality comes in two halves: the easy and the hard half. Put more plainly, Plato holds that humans are only half- rational. Humans are capable of executing only the simplest half of rationality, the human-function, while the creative half, the god function is only done by god.

The thesis of Chapter 4 is that the creative half of rationality, the creation of the concepts, A, forms, molds, cookie-cutters, etc. are not so ideal as to require some master spirit artisan, some verbal god, to design, to conceive, and to distribute them as Plato assumed. On the contrary, this technology is an essential element of the nonverbal mind and rationally executed by every nonverbal mammal as a first step on every occasion requiring self-organizing behavior. The Nonverbal Mind has the requisite empirical god-function technology included. My task is, as ever, to read the rat's mind by studying the complex patterns of behavior emitted by its nonverbal mind while it is executing the creative rational function, A, while exploring the matrix maze.

However, in Chapter 5, I will put other kinds of complex behaviors upon the stand, so they too can help reveal the rational nature of the nonverbal mind.

Technically, my specific challenge in this chapter is to provide for the mind what Darwin provided for the body. Just as Darwin provided a theory of natural selection which is a general theory that explains biological speciation, my challenge is to provide a general theory of rational selection, which explains mental speciation.

Rational selection is a theory of what nonverbal ideas are and where they come from. What is an idea? Are they the gift of god, or are they the product of the core technology that makes the NV Mind rational? How does the rational technology work? How can a rational technology conceive ideas? What technology enables the nonverbal mind of a rat to do what Plato's followers still insist is beyond the powers of the human mind?

My entree into this master topic was my discovery that rat's have a generic nonverbal mammalian mind.

Plato and Chomsky have insisted that the language function was limited to a mammalian verbal species, if not only to god, then perhaps only to god and man. However, my data takes me far beyond the dogma of these two men. My data shows that rats are masters at conceiving hierarchically organized models of complex spatial mazes, which suggests that possibly all mammals share the nonverbal mind. What is an idea? What is a hierarchically organized coherent set of ideas?

The role of metaphorical behavior

How metaphorical behavior manipulates concepts and exemplars: A and B.

More about the origin of ideas later but for a moment, I want to look at another aspect of ideas, the nature and the role concepts play while humans are using metaphors; we will note how concepts evolve and change in the most interesting ways. This aspect of conceptual behavior obviously must be studied at the human, not the animal level. Concepts are a dynamic aspect of rational verbal behavior.

In this book, we are going to challenge **Plato's Nativism** a little bit at a time. Even if Plato were correct, and humans couldn't create new concepts from the ground up, yet there is still a world of twisting and turning that can be done to and with concepts while humans are thinking with metaphors. Specifically our topic will be to examine the historically important concepts of Accommodation and Assimilation.

Piaget: Accommodation and Assimilation: Domain and range

The word assimilation has a long history. For example, it was used in Roman times to refer to digestion, as when, we say we assimilate our food. Originally, assimilation referred to a physiological function but Piaget made it a psychological function, and the concept became a more complex and interesting process.

Our purpose is to add several more words to our nonverbal vocabulary: Assimilation and accommodation, and Domain A and Range B.

The words domain and range originate in the context of Kings and Queens, who rule over their Domain, which has some size or range. When two kings engage in war, the losing king loses all or some of his domain; the range of his domain changes size. Therefore, when we talk metaphorically, each word that is being used metaphorically has a domain, which can change both nature and size.

The domain is the place over which a concept reigns and gets it identity from the features of the concept, from which each class of known objects gets it identity. The concept identifies and defines each particular causal feature of the class. This set of causal features acts as a filter on the input side of the mind and selects each syndrome of sensations in the input stream, which matches the concept. The concept creates the domain of a class. Each match is an act of assimilation. Each match allows a previously unknown pattern to enter consciousness and to begin to participate in the narrative running in consciousness.

A concept defines a domain of a certain size or with a certain range. What I want to discuss now is how metaphors first, cause the concept of a class, or the domain of a class to be modified, and thus secondly, how metaphors modify the range of a domain, to either increase or decrease the range.

The invariant

Every idea establishes its own domain by creating a set of causal features, which identify the invariant set of properties, which justify the sixth sense including its exemplar in consciousness. Every idea must have a unique set of features, which define the domain, which is defined by its concept.

Heraclitus

Heraclitus is famous for saying, "you can't step in the same stream twice, which seems to have different meaning for different scholars. Some take it to mean that everything consists of a large number of elements, which are constantly changing.

However, I understand this statement to be one of the first references to the concept of the invariant, which is defined as, "that which doesn't change during change." In other words, while the water keeps changing, nevertheless there is something- the stream- that never changes, which is an invariant.

Yet, we need not work so hard to make our point; for the most basic facts, which sustain our lives are the laws of nature, which are invariants-for example, gravity.

Each feature of the concepts of every domain plays a role in determining the size of the range, or the number of individuals within the population specified by the domain.

Snow example

Let's analyze a simple metaphor from Julian Jaynes (1976), "Snow is like a blanket." Imagine we live in the north, and a friend from the south is visiting us. The weatherman says it's going to snow tonight. So, your friend asks, "What is snow," and you say. "Snow is like a blanket." You assume in your statement that your friend knows what a blanket is, that your friend has a concept of a blanket and are aware that a blanket is something to cover yourself with in bed when it is cold.

Talking, using metaphors, actually requires that complicated mental functions be evoked by both parties. When we say, "snow is like a blanket," in particular, two mental processes must come into play, which we must learn to distinguish. A concept has two properties: domain, A, and range, B, and we must learn to distinguish the domain and range components that are evoked when we are talking using metaphors. When we say snow is like a blanket, what must your friend's mind do, in order to understand the metaphor? First, you assume that your friend knows that the domain of blankets, A, is conceived as something you can fold and put in your cedar chest in the summertime, that you can wash, that you can put on your bed, and that is usually made out of wool, cotton, or some other manmade material.

Borrowing

When you say, "snow is like a blanket," whether you recognize it or not, you are requiring that your friend make changes in their concept of blankets, A. Actually, you are requiring them to change both the domain, A, and the range, B, of their concept of blankets. In order to understand the metaphor, they must be able upon looking out in the morning and see that the earth is covered with snow, which covers the earth like a blanket. This mental function extends the range, B, of the domain of the blankets' concept. Previously the range of blankets, B, was limited; blankets only covered your bed now a blanket covers the earth. Then, additionally your friend's mind's eye must be able to change their domain or their concept, A, of what kind of material forms a blanket. Of course, the earth's snow blanket is made out of something quite different then a manmade material. Yet, your friend must now think of snow as having the same function, if snow is like a blanket then snow must be thought of as capable of protecting the earth and keeping it warm. Up north in the arctic, there are animals called lemmings, which live under the snow all winter long, so when the temperature is forty or fifty below zero, down underneath the snow at the level of the ground the temperature may still be right at freezing. So the lemmings can live there under the snow, like a blanket, and be quite snuggly and do very well.

Inverse relation

One technical question that we will note, concerns how the number of exemplars, varies with the number of features employed by a class, or a concept.

This most critical relation is a function of the law of inverse relations, which states there is an inverse relation between the number of features of a concept and the number of exemplars it will extract from a population, which define the range of a domain. A concept during concept formation training evolves from concrete to abstract.

When a concept is first developing, it is said to be concrete, all the features of the first training object will have become features of the class. In reality, the features of a concept belong in two classes: figure and noise. In a concrete concept, all the signal and noise features will be perceived as causal features of the class.

At the highest level of maturity stage, when the concept is Abstract, the concept will contain only signal features, and no noise features.

The law can be described as:

A= K/B Where A is number of features, B is number of exemplars and K is the size or the number of individuals in the population.

A=20
B=5
K=100

Therefore, in a **concrete** situation, where there are many features, if the number of features is 20, then only five exemplars are possible in the population. When there are twenty features, two are relevant features and 18 are noise.

A=2
B=50
K=100

However, in an **abstract** situation, when the number of features is small, say only two, then there are 50 exemplars in the population. Now, all the noise features have been removed leaving only the two identifying features so more of the population qualifies as an exemplar.

When we employ a metaphor in our conversation, we are asking a lot of our listener's mind. In our example, the listener must change both the domain and the range of his/her concept of a blanket. In changing the range, the listener has to have gone from the blanket being something used in the house to something that has a much larger range, it can cover not just your body but also you can cover the earth with it.

Brown and Blodgett

In Chapter 5, I will show that there are many ways to change or alter the domain and the range of a concept.

1. We will illustrate this by teaching an animal to master an abstract concept of Triangularity.

2. The Brown and Blodgett study defined accommodation in a larger sense than did Piaget. The B & B procedure is important for they taught us how to carry this rational function even further while training Sam and Miss Sam for space.

Word-corrupting metaphor

One of the major difficulties is that Western civilization in the course of the last several thousand years has been infected with a series of metaphors/myths, many, if not most of which, were started by Plato that most humans take as truth. Plato's (1952) metaphor of the divided line discussed in Chapter 2 has been the most harmful. Moreover, these myths have managed to corrupt almost all the ideas and words a scientific theory of human nature should use by embedding them within the mythical foundation of our culture. A theorist has two choices; they can create a new scientific based theory of human nature defining all new ideas or words, or they can attempt to redefine the old words, but the latter is almost an impossible task. Sometimes redefining old word is counterproductive, for example, the words information, and grammar.

Usurped words

For thousands of years, the verbal mind theorists have been developing and spreading their myth that the verbal mind is self-sufficient and thus have been taking the credit for all the invisible rational functions executed by the nonverbal mind. In the process of selling their myths, they have defined and defiled all the words whose functions rightfully were created by the nonverbal mind. A large part of my task is restoring these words to their original nonverbal meaning. However, in order to talk about the nonverbal mind, these words somehow must be retrieved and rehabilitated. The usurped words, which must be salvaged and refurbished, are Rationality, intelligence, information, communication, Knowledge, god, Mind, atheists, homeostasis, uncertainty, and ideas.

1-Rationality

The first word, I will attempt to recover is the word RATIONALITY.

First, the word rational is too important to allow Plato's (1952) historic abuse of the term to stand. For example, Plato's philosophical use of the word rational and my scientific use of the word rational have almost nothing in common.

Plato called himself a rationalist, and he called the chief function of the human verbal mind rationality. Plato, in other words, was only aware of one of the several forms of rationality; the one I claim to be the secondary or verbal form of rationality.

By contrast, my concern is with describing the primary nonverbal form of rationality, the form employed by the rat's nonverbal mind. Obviously, Plato's verbal concept of rationality and the rat's nonverbal concept are very different. Plato was only aware of the human verbal form of rationality, while my task is to describe the primary nonverbal forms of rationality employed by the prehuman mammals for millions of years.

The Mental Periodic Table

The most profound difference is that my use of the term rationality refers to the power of the nonverbal mind to conceive concept, or forms, which is the primary creative nonverbal function. Concepts are the mental equivalent of the elements of the chemist's periodic table. Just as every chemical compound consists of a whole created by some combination of the elements of the Periodic table, just so every: model, act, narrative, or script consists of a whole created by some combination of conceptual elements, which are relative elements of some domain-specific environment. While the rat's nonverbal mind was perhaps the first rational system that emerged in the course of biological evolution, yet it was already a system capable of exercising the most powerful nonverbal rational functions.

Additionally, Plato (1952) insisted that the human mind was not even fully rational. In fact, he argued that the human mind was only half-rational; that it could not conceive those ideas, forms, or concepts, those functions, which in Plato's view were reserved for the gods. Plato was a strange genius, while on the one hand, he flattered humans by claiming they were the only life form that had a mind and that this verbal mind was the only rational organ in the physical universe, yet at the same time Plato claimed that the human mind was only half-rational. As a rationalist, Plato (1952) conceived of rationality as a pure verbal function, a kind of function that philosophers in their Ivory Towers could and did spend most of their life enjoying.

However, for my part, I am a pragmatist and a rat runner. I conceive of rationality as a pure nonverbal function that rats as prototypical mammals have enjoyed for millions of years. Long before there were philosophers, nature endowed rats with nonverbal rationality, which they employed as the central ingredient in their mental arsenal that made them the first Self-Organizing Systems, SOS1, upon the face of the earth. I conceive of the nonverbal form of Rationality as the only form of rationality that can be employed to empirically evolve or derive sets of forms by directly interacting with the physical nature of the cosmos, which Plato so despised.

Again, Plato (1952) conceived verbal rationality as an exclusively human function, a function in which armchair philosophers deduced truths from the nativistic ideals, which god magically stored in the mind at birth!

Yet, nonverbal rationality, as I define the term, refers to a creative empirical nonverbal mental function long enjoyed by all mammals. In fact, my concept of nonverbal rationality refers to the central mental function, which armed and endowed the nonverbal mammals as self-organizing-systems and which they employed to take possession of most of the niches of the earth.

In the four experiments described in Book 2, I finally achieve the empirical and experimental kind of data, which allows me to challenge Plato's concept of verbal rationality, both sides of his theory, the hard, and the easy sides.

For example, the rat's maze behavior in the Essential Path, EP and the Blind Path, BP of the maze are rational but in different ways. Both rational types of behavior are particularly adapted for that part of the maze for which it was designed. Both types make it not just empirically possible but imperative to challenge Plato. Plato's rationality has always been conceived as an ivory tower function, a mental function that rightfully could only occur for certain, in the heads of Philosophers, even when they were locked away in an ivory tower. By contrast, the pragmatic concept of the nature and origin of ideas, as described in this book, are such as could be expected of rats engaged in spontaneous exploratory-play in a matrix maze in some domain-specific environment: no ivory tower.

I am a rat runner

I am a rat runner. Perhaps it was no accident the requisite fix for the classical concept of the human mind and the nature of verbal rationality had to start where the system was broken. Humans dutifully ensconced in their mental beds, sleepwalking through life, do not stand a chance of repairing and respecting their nature until they can first wake up and overcome their overt contempt for nature, other mammals, and thus by implication, their contempt for themselves.

Most humans are oblivious to the monumental differences between what we think we are as a species and the facts. I am not suggesting that all humans are failing to keep up with the scientific facts, but rather the supposed scientific facts we are fed today about human nature are myths. I am also not suggesting that there are any deliberate conspiracies to suppress the scientific facts.

There are several sets of facts about human nature, which are grossly in error, some of which unfortunately falls into my scientific purview.

- First, the first major source of error about human nature concerns the nature of the human mind. Myths abound about the nature, origin, structure, and function of the human mind.
- For the moment, I will list only one of these myths, that humans are the only species with a mind, that the human verbal mind is the only mind. The purpose of this book is to announce to the world that all mammals have a mind but not the compound human mind but the precursor of the compound mind, the generic nonverbal mammalian mind, GNMM.
- The first implication, strangely, is that for the first time humans scientifically finally have reasons to acknowledge not just their primate ancestors but also their mammalian ancestors. Humans share the GNMM with all mammals.
- Second, the second major source of error about human nature concerns the taxonomic question: what is a mammal?

E.O.Wilson

E.O. Wilson (1998) observes, "The story of mammals is all about milk."

E.O. Wilson and the taxonomists have classified mammals as mammals because the females of the species have mammillary glands.

The alternative should be to call them mindmals, recognizing that the most revolutionary physiological organ of the family is the nonverbal mind.

I propose a new theory of mammalian/human nature based upon the empirical foundation established by Aristotle (1952), Plato's student, over two thousand years ago. I propose that the generic nonverbal rational mind of the rat, the first rational nonverbal mind to evolve, emerged as an empirical based twelve-station model, which later, in the human mammal, evolved as a thirteen-station model. I propose the radical hypothesis that the long demeaned human mind first proposed by Plato (1952), is simply a newer model, an updated version, of the twelve station generic nonverbal mammalian mind, a more advanced, updated nonverbal mind. Humans, since the time of Alfred Sloan and General Motors in the 1920s, have had the luxury of enjoying yearly models of their cars. Nature, on the other hand employing natural selection, simply may require thousands of years to produce a new model of the nonverbal mind of the nonverbal mammals.

If humans are ever to repair their own bad opinion of themselves, perhaps there is no better place to begin than taking the first step and repairing their bad opinion of rats. This repair can only begin by acknowledging the nonverbal rationality of rats, the prototypical mammals, authors, owners, and operators of the first nonverbal mind, which was the precursor of the human compound NV/V mind. Humans are mammals, but the whole thrust of this series of experiments is that rats are a prototypical mammal; they are exemplars of the kinds of animals within which human rationality emerged. Nonverbal rationality is the foundation out of which, or upon which, the secondary verbal human form of rationality emerged.

While Plato may not have started the Western denigration of animal nature, he did formalize and rationalize it and advanced its survival for all these years. Yet at the same time, Plato in effect, slammed shut the only real avenue for answering the big question. Empirically, where do the classes, the forms, the ideals within which NV rationality moves come from? Paradoxically, the verbal elite in the West, in effect, both philosophically and religiously, oppose the notion of "rational selection," just as the religious individuals opposed Darwin's (1859) concept of "natural selection." This opposition was manifest in the size of the gaps, humans imagined to exist at critical spots in the theory of evolution: gaps, which had the effect of decreasing the possibility of finding an evolutionary solution to Plato's problem effectively to zero. These two critical gaps were between god and man and between man and animals.

The rat's story

However, I am wasting time; let us listen to what the rat's story, as told by their nonverbal behavior, has to say about the rational nonverbal foundation for mammalian nature, which is the generic foundation of human nature. For with each new section in this series of experiments, we accumulate more and more of the facts upon which a new empirical or data-driven theory of mammalian rational nature can arise; one that could allow humans finally to evolve a scientific foundation upon which a revolutionary empirical conception of the origin of their two-stage rational mind can be erected.

Humans have only one mind, yet the human mammals evolved a unique set of verbal organs, located in the left hemisphere, Broca, and Wernicke's areas, which allowed them to evolve rational skills way beyond the classic rational adaptive powers achieved by the nonverbal mammalian mind.

For example, early man always attributed the creation of the universe to some god but then Darwin (1859) showed that creation was caused by nature or the nonverbal god. All forms of religion objected to this as demeaning god and failed to consider that religions are all too often, a zero sum game, generally played between a verbal god and humankind. Early man had perhaps good reasons to assume that god had all the power and they had none. However, as modern humankind recognizes their own significant role in directing the evolution of the cosmos, of recognizing the amazing nonverbal rational power in their nature, they should feel less and less helpless: for what strengthens the human side of the equation, weakens the arguments for the verbal god.

For a moment, consider the good news. If even a 60-day old rat with only the thinnest possible layer of cortical tissue has these god-like rational functions, imagine what rational powers the human mind must have! Humans, now for the first time, have an opportunity to understand how the nonverbal mammalian mind conceives, on demand, a closed hierarchically organized conceptual system and that should reduce human feelings of helplessness. The good news is that all mammals were designed to be Self-Organizing Systems, SOSs, with full rights to employ their rational nonverbal mind.

Self-organizing systems must create *de novo* their own domain specific knowledge, and my rat's behavior proves it is not just a master at this task, but even more interestingly, it enjoys the work.

Yet SOS systems require special kinds of motivational systems. For example, what conditions are required to enable or induce a rat to activate those structures of the nonverbal mind that generate self-organizing behavior?

On the other hand, what conditions will be required to induce humans to take charge of their nonverbal mind and begin to implement their empirical power to self-organize, to begin to evolve a scientific understanding of the structure of the universe, including an understanding of their own mind?

Thomas Kuhn gave us an example of how this happened in astronomy. It was one part Galileo, one part the telescope, and one part the moons of Jupiter.

Yet our question is what it will take to get humans to accept the idea they have a nonverbal mind and that all humans have the nature/god given rights to self-organize, the rights to do fact-based knowing and self-organizing. The confluence, in Galileo's (1989) case, of these critical events changed forever the way humanity was able to study and understand the place of the planet earth within the cosmos, but what is next?

What will be required for humans to have the kind of insight Galileo induced about the nature of the cosmos about their own nature? What is required to induce inert, but potentially self-organizing species, finally to begin to self-organize their behavior and to take charge of their life, and to abandon their metaphor-driven myths and crutches?

Where do nonverbal self-organizing systems go to create the specific data-driven knowledge they require to master each domain-specific problem. The nonverbal mammals take Female 3, for example, have no need for a verbal god, a heavenly god, or a book god. Female three created all the domain- specific knowledge her self-organizing system rationally required for her to find her way efficiently through the 96-unit maze. The nonverbal mind interacts nonverbally with those domain-specific parts of the cosmos where the questions are! The interesting point to recognize is that in the physical matrix maze, the questions and answers, the causes and effects were bundled together within the physical structure of the wooden maze.

The nonverbal mind was designed "knowing-how" nonverbally to know; knowing-how a nonverbal self-organizing system must rationally behave to provoke, prod, or elicit from the nonverbal cosmos, nature, world, or god the domain specific knowledge required to obtain all the nonverbal knowledge needed to fulfill its domain-specific mental construction's shopping list.

The rat has no need for Platonic ideals. The nonverbal mind does not require eternal truths, only domain-specific truths. For example, in traveling across Texas, the nonverbal rat would only require a Texas map, which provides the nonverbal mind with the domain specific knowledge to go from S to G, Start to Goal within Texas. Furthermore, the nonverbal mind requires only this year's Texas highway system knowledge; today's knowledge is good enough thank you! A nonverbal mammal is a pragmatist without knowing it.

Heterostasis enters the story

This is where Heterostasis enters the story. In this chapter, thanks to the behavioral telescope, the matrix maze, and the rats' primary mind, a new set of possible rational worlds are opening up. It is now possible for humankind to shed centuries of beloved fairy tales and evolve an empirical, scientific, or fact-based narrative about their own rational nature. To move inch by inch, step by step, closer and closer toward: first, understanding the rational nonverbal functions of their nonverbal mind and its place in the nonverbal cosmos; and secondly, understanding the rational functions of the human compound nonverbal/verbal mind.

Pragmatism

I propose that human verbal behavior is a product of a two-stage rational function. My task has been to devise a set of experimental test conditions where the nature of the primary nonverbal mind can be seen and objectively studied. My purpose has been to explain the role of the primary nonverbal mind in the evolution of the nonverbal forms, or ideals, which enable a NV mammal to act rationally. Strangely, the nonverbal mind achieves nonverbal rationality by destroying anomalies.

Plato (1952) proposed an impossible concept of the verbal mind; he conceived of ideals as transcendental beings without having any empirical idea of how this set of spirit beings created spirit "ideals"; nor how these eternal spirit ideals got to earth and stowed away in our head. Plato insisted that his ideas and ideals were not the product of matter, and that in his theorizing he will not be constrained by empirical facts. On the contrary, he insisted that his verbal mind be allowed to create its personal concept of an ideal world, populated with his personal concepts of a set of perfect eternal beings consisting of no matter, so they would be free of the corrupted nature of matter. Plato believed that things made of matter could not be known for their empirical nature, and always exhibited only "appearances," not their real nature. Only a Platonic conceived set of Ideal beings, spirit beings, perfect eternal beings; beings, which never change, could thus always present first, the true eternal nature of the universe and second, the true nature of human kind. Plato had the highest standard. His theory must satisfy his concept of the real, as sets of perfect ontological spirit beings.

Throughout this book, I have been shifting back and forth between two ontological models and two epistemological models. Above, we just discussed Plato's out-of-this-world concept of what is real. Now consider the opposite point of view, starting the position of the first original philosopher, Charles Sanders Peirce.

Peirce's Pragmatism

One of the first-generation philosophical pragmatists, Peirce (1878) started pulling Plato's ideals out of the heavens by proposing a revolutionary very set of American ideas: that ideas, words, and actions were real and inseparable. Peirce planted the concept that ideas had an earthly origin. In fact, Peirce proposed that actions and only physical actions define real ideas! The meaning of an idea was defined by the action it generated. Pierce was not a Platonic philosopher.

Peirce (1878) was aware that often ideas and actions were divorced, but he went so far as to propose that real ideas were indistinguishable, and inseparable from actions. There are two kinds of ideas and two kinds of doubt, real doubt, and feigned doubt. The two types of ideas were those defined by actions and those divorced of action; this was bold stuff in its day. My favorite example is from Peirce's critique of Descartes' (1946) theory of doubt.

Gravity is an example of a thing that is real. Peirce (1878), one of the founders of Pragmatism, was not impressed by Descartes' concept of doubt. Descartes claimed he could and had doubted everything. Peirce called Descartes' (1946) doubt "feigned doubt," not real doubt. Real doubt was nonverbal, not verbal. You may successfully doubt that gravity is real by playing Descartes' philosophical verbal game; first verbally say, "I doubt gravity" and perhaps you can talk yourself into believing that there is no gravity as Descartes claimed he had done. Next time, however play the game nonverbally. This time, however, do not doubt gravity verbally but nonverbally, may I suggest by "walking off a cliff."

As a third generation pragmatist, however, I am prepared to go where the first generation pragmatist could not or would not go. I propose that their action concepts of words were halfhearted or even upside down. They proposed, what was for that time a bold position, that actions define words, but Peirce was still living within a world of words, a world defined by words. By contrast, in my research, I have moved into a nonverbal world where I am concerned with how the generic nonverbal mammalian mind works. I propose that nonverbal actions are rational, that the actions of the nonverbal mammals are rational. In other words, I propose an extreme form of pragmatism, which Peirce could never have imagined. I propose that rationality first emerged in the nonverbal mammalian mind and that primary rationality for millions of years was independent of verbal rationality.

This first statement of the problem of nonverbal Pragmatism or the nonverbal problem of meaning is a big problem that cannot be settled on one occasion. An overall theory of meaning must explain the meaning of words, nonverbal actions, ideas, verbal actions, and things.

Chomsky's Universal Grammar, UG, assumption

One of Chomsky's (1979) most disturbing contributions was his concept of a universal grammar. This concept was provoked by his observation that children around the world in every language community acquired language much too fast and more easily than learning could explain. It was as if they were born with an instinct for language learning. This instinct enabled them to master features of language behavior to which they had not even been exposed. Chomsky personally developed no evolutionary explanation for why language behavior around the world was mastered so easily, yet he insisted that he could only assume that every human was born with knowledge or a universal grammar.

Chomsky's position that human language skills are nativistic is remarkably like Plato's nativistic assumption.

"For reasons that are not entirely clear, Chomsky for many years (1976, 1980a, 1986) claimed that human linguistic ability could not have evolved by means of Darwinian processes. Chomsky's position perhaps derives from his belief that language is a faculty that is completely independent of other aspects of human behavior. In earlier versions of his theory, Chomsky proposed that humans possessed an innate, species-specific "language-acquisition device" that enabled children to acquire the rules of grammar. Later formulations of his theories proposed an innate Universal Grammar that contained specific information concerning the possible syntax of all human languages." (Lieberman, 2000, P. 127)

The Universal Grammar UG assumption

Chomsky's (1979) concept of UG reflects or reveals the context within which he develops his theory of the control of verbal behavior. Chomsky's statement of the problem of grammar has its origin in the fact that Chomsky frames the question of verbal behavior or verbal grammar as unique in at least two respects.

- First, the problem of grammar is unique to the human species.
- Second, the control of human speech behavior is a unique problem in the context of human behavior in the sense that no other form of human behavior requires the same kind or degree of somatic control.

In fact, Chomsky conceives of human speech behavior as so unique and so superior to all other kinds of human behavior that he goes so far as to insist that it could not have evolved! Why is this? The complexity of human speech behavior exhibits, to Chomsky's way of thinking, the ultimate degree of motor skill level; let us say as quality Z. By contrast, Chomsky's problem is that no other form of human behavior has a difficulty greater than say M. If human language behavior had a difficulty of only N, then perhaps we could argue that N had spontaneously evolved out of the M pattern of behavior. Nevertheless, according to Chomsky human speech is a verbal skill so far superior to any other form of animal or human behavior that it could not have been attained by an evolutionary mechanism.

Lashley-The Universal Invariant, UI, assumption

In this section, my task is to revise Chomsky's (1979) concept of the existence of a universal grammar, UG, by starting at an empirical place, which Chomsky never knew, a place introduced to science for the first time with this book. My challenge to Chomsky, however, is not mounted one on one as me against Chomsky. My apparent boldness has its origin in my knowledge that I have two major theorists on my side: Frank Beach and Karl Lashley, the former, one of the greatest Comparative Psychologists and the latter one of the greatest physiological Psychologists. With the help of these two theorists, I will reframe Chomsky's concept of the UG and its origin.

The only way to solve Chomsky's problem is to start at a different place, one that allows us to see human speech behavior within a different context. My take on his problem is based upon my ability to bring to the table two kinds of arguments that Chomsky did not or could not consider.

First, it should come as no surprise to anyone that I see the problem of the organization of human speech behavior in the context of my works with the nonverbal mind, thus my point of view or my respect for the verbal component of human speech behavior begins with far less awe than does Chomsky. If Chomsky sees human speech as the ultimate grade of complexity possible in somatic motor control, say level Z; I, on the other hand, would be willing to go no further than level N, if even that far. My argument goes as follows.

- First, in step one, the nonverbal mammalian mind for perhaps fifty million years before human speech emerged advanced the technology of the nonverbal speech mechanisms. Then in step two, I propose that the human verbal component of speech evolved out of this physiological technology, which evolved for nonverbal speech behavior.

- **Second,** my theorizing begins with the assumption that all the organs of every mammalian life form evolved from lower forms. In science, there is no place for a nativistic assumption. In particular, I assume that the generic nonverbal mind of mammals had over fifty million years to perfect itself as a SOS. Yet the facts keep getting stranger and stranger. The primary function of the GNMM was to endow mammals with a primary nonverbal language, which is the primary language organ in the human brain. With no knowledge of the existence of this nonverbal mechanism, Chomsky has no choice but to conceive the verbal mechanism as a standalone language function, with the Z level or quality of function as discussed above.

The Evolution of the cortex in mammals drove the evolution of two kinds of communication behavioral skills.

- First the evolution of the nonverbal communication skills
- Second, the evolution of the nonverbal/verbal language mechanisms of the human mind, which is a joint function of the integration of the two communication skills

Beach, first, then Lashley

Before we consider Lashley's arguments, we should first review the works of Frank Beach. An essential part of the argument requires an understanding of the role of the evolution of the cortex for the evolution of both the size and the complexity of the mind. Beach makes the case for this aspect of the argument.

Frank Beach: the Mammalian sexual evolution

Frank Beach (1947), a comparative Psychologist, describes how the evolution of the mammalian cortex revolutionized mammalian sexual behavior. He showed that the mammalian mind at the beginning played no role in sexual behavior.

Frank Beach provided the critical scientific data needed to support the idea that the nonverbal mind evolved in parallel with the evolution of the cortex. Beach (1947) described the link between the evolution of sexual behavior and the evolution of the cortex. He showed that mammalian sexual behavior was a function of two new types of neuronal structures, which controlled sexual behavior: the arousal and consummatory mechanisms. In the lower pre-mammalian forms of life, sexual behavior was genetically defined or closed. However, the evolution of the neocortex revolutionized the nature of mammalian sexual life in ways that the classical theories of human nature find difficult to appreciate.

As the mammalian cortex evolved, both components of the sexual behavior systems, the arousal and consummatory systems, systematically evolved into a series of revolutionary new forms, but generally, they were transformed from a closed to an open form of behavior. Beach (1947) showed that as biological evolution expanded the cortex the sexual behavior gradually became open; first, the arousal type of sexual behavior became open, and then later, the consummatory type became open. Until finally in humans, both the arousal and consummatory functions are open. The critical element of Beach's (1947) research was that he showed that only as the cortex became involved in mediating sexual behavior did the instinctive closed sexual functions, which had dominated the pre-mammalian species, lose their absolute control over behavior. Only with the emergence of the cortex was mammalian life able to evolve both an arousal and a consummatory mechanism, which extended the range of sexual behavior. There are several clinical books, which catalog the varieties of Human sexual behavior such as *Abnormal Sexual Behavior* by Louis London. Even in a casual reading of these texts, it is almost inevitable that the reader will comment evolution went too far in opening up the human sexual mechanisms.

The open mammals-Nature/nurture

Evolution designed mammals because there had to be a better way to design life than by natural selection, by pure natural selection, i.e.by nature without any nurture. Psychologists have long talked about nature and nurture without ever confronting head on the contribution of the evolution of the mind, neither the nonverbal nor the verbal. The nonverbal mind is the organ that confers nurture upon mammals. This book is about how the nurture organ got started, how it works. Etc. The "Better way" was for evolution slowly to grow the nonverbal mind in mammals.

Why did mammals evolve? Nature needed a "better way" to get life the knowledge required to be an effective SOS1.

- First: instead of giving static-life-forms, static facts, i.e. to continue to make nativistic forms more and more nativistic, Nature chose to gamble and give life the rational nonverbal power to create on the spot in real time the domain specific knowledge required; the pragmatic knowledge needed for the successful completion of some task.

- The evolution of the arousal mechanism enabled life to be sexually aroused by new and different kinds of nonsexual stimulation. How can increasing the variety of arousal stimuli enhance the sex-life of mammals? The answer in one word is "fashion." The books on human sexual behavior answer this question. In fact, there would be no such books without a very active arousal mechanism.

- The evolution of the consummatory mechanism freed the somatic motor system from the bondage of the traditional way to engage in sex. Again, how many ways are there to engage in sex? Theoretically, the same degree of variety could be genetically hard-wired or programmed, but with a nonverbal mind, none of these varieties of somatic motor behavior requires a single additional neuron or gene.

Presumably, each neuron added to the cortex is a good investment in somatic variety in almost every possible domain of life, new styles of clothes, new fashions, new homes, new weapons, new transportation, new words, new religions, new sex, etc. as each neuron invested in the mind gets a lifetime return on some one or more kinds of freedom.

Beach's (1947) comparative studies of the evolution of mammalian sexual behavior showed how biological evolution slowly put itself out of business; how biological evolution slowly turned the keys over more and more as the latest mammalian species-specific genes emerge. Perhaps this point requires further development. In the lower species, all the nature-species, the genome made all the decisions. Under what condition will rats engage in sex? When the genome is in charge all that is required is that two adult rats of the opposite sex be put together in the same cage and for the female to be in estrus. No introductions are required, no flowers, and no courtship. At the level of rats, there are only genetically arranged marriages; the two rats have no say about any aspect of their sexual behavior.

With who, how, and when should sex happen? In the lower creatures, these questions are genetically answered; to have sex is genetically determined. Yet the male rat's arousal mechanism is already functional, and it can learn to use nonsexual cues to find a female.

- Perhaps it is not obvious, but sexual freedom, like all kinds of freedom, first means to be free of something. What does it mean to say that mammals gained their freedom in both the arousal and consummatory mechanisms? It simply means that **the genes stopped making some decisions and turned those decisions over to new cortical control circuits**! Freedom in humans means that, in both the arousal and consummatory mechanisms, the genes changed their design strategy; instead of designing closed structures, they designed open ones, which could rationally make their own species- specific and domain-specific decisions.

Monkeys, Chimps, and Humans

Beach (1947) showed that change emerged:

- First in the arousal mechanism
- Second in the consummatory mechanism

With respect to the sexes, change began in the male of the species and then in the female, but the quantity of the cortex was always the most important driving force. Beach also traced the evolution through a series of mammalian species: rats, monkeys, chimps, and humans.

Paradoxically, Beach showed how wrong Plato was. While Plato despised empiricism and matter, Beach showed that sexual freedom is attained by having more matter, more brain cells in the cortex, rather than less.

It would be useful if someone had made a comparable study for the evolution of the mind, but no one has. Yet Beach's research makes the point most clearly, the emergence of the mind was not some miraculous sudden event.

Human nature slowly emerged in the progressive advances in the general mammalian line with the central physiological advance being the quantity of the cortex with which each new species was endowed. The evolution of the quantity of the cortex set mammals free to become, not theoretically what nature required but rather a species designed to deal with change. What happened may seem a bit strange, but nature overdid or outdid itself. Heterostasis is not a mechanism designed just to deal passively with change, but nature designed a life force, which requires change, that can seek and even induce change. The evolution of the cortex enabled humans to be sexually excited by more kinds of stimulation and to engage in sexual behavior in more kinds of ways.

Lashley

Second, I must with a great deal of awe present the fact amassed by the master physiological psychologist, who in the Hixon Symposium of 1951, essentially answered Chomsky's problem years before Chomsky ever raised it. In fact, Lashley brought together a larger set of facts and illustrative problems that went even further in arguing for a new kind of UG (Universal Grammar) than Chomsky ever did.

Linguistic definitions

First, some linguistic definitions are needed. Human speech has been intensely studied for over a hundred years and the first product of that study has been a formal division of the study of language behavior into several parts: Grammar, Syntax, Pragmatics, etc. However, for all practical purposes, two divisions have been identified. One division can be successfully studied and another one for which little or no progress has been made. The successful division includes grammar and syntax. The unsuccessful sides include all the rest.

The problems of syntax and grammar study the rules for constructing sentences. The starting place for the rules is a major part of the theory, for the fact is that language is a powerful example of hierarchically organized behavior. How do letters form words, words form sentences, sentences form paragraphs, etc... We can then throw in clauses, phrases, adjectives, adverbs, etc. What are the rules for constructing a whole from a set of parts? What makes the science of linguistics interesting and difficult is that every language has its own rules.

If we choose to study linguistics and further we choose to study that part which science has mastered, then we can do as Pinker (1999) did and choose to study just the easy part, the parts consisting of just words and rules.

However, the point of this section is straightforward. The easy parts of the rational study of language behavior focus upon the problems, which Lashley (1951) called the "serial order problem." What are the rules for imposing order upon the elements of our hierarchically organized language: where do the various parts of speech fit into a sentence such as where do clauses go etc.?

Serial order Phenomenon: the human case

Lashley, in 1951, discusses the problem of serial ordering at the most abstract level, but Lashley was careful to insure that there was no doubt about the wide range of subtopics which he included under the topic of serial ordering.

All mammalian behavior, thanks to the evolution of the cortex, is capable of being rational. Where rational means the organization of the behavior, every kind of behavior, is organized as a product of the mind not the gene. At the most abstract level, the behavior consists of some temporal pattern or organization imposed upon some parts of a whole; several examples will be given below.

Playing a musical instrument like the piano is a matter of playing or striking some pattern of the 88 keys over time according to a score written by a composer. Playing a musical instrument is a matter of imposing some serial order upon the playing of the keys. The music is a product of the order with which the keys are depressed over time, i.e. serially.

Talking is also a serial order problem. Talking is a matter of moving the tongue, the mouth, and the diaphragm in complex temporal patterns.

Walking is a matter of moving the legs in a pattern, a temporal order, or pattern over time.

Typing is a matter of hitting the keys on the keyboard in an order that will reproduce what the boss said.

For Lashley (1951), human speech behavior was not a unique problem. Lashley discusses the problem of human speech simply as a special case of the master problem of "serial order." Lashley distinguishes two problems, the spatial, and the temporal problem of order. He shows that the master problem is the temporal serial order problem of which speech is but one particular problem in that context. The serial order physiological problems run all the way down to the serial order problem of asking how the control circuits control the legs of the centipedes to the problem of music, for which he cited many examples.

"...the problems raised by the organization of language seem to me to be characteristic of almost all other cerebral activity....Not only speech, but all skilled acts seem to involve the same problems of serial ordering..." (Lashley, 1951, P. 121)

The serial order Phenomenon: the mammalian problem

Lashley throws a great spotlight upon the problem of serial order. One reason I am so impressed with Lashley's paper from the Hixon Symposium is that when he discussed the general problem of serial ordering which covers what Chomsky calls the UG and what I call the vertical components of the nonverbal mind in Chapter 2, the anatomical structures, which made his functions possible, had not yet even been discovered.

The timeline of FIGURE0.01 shows that by 1949, Moruzzi and Magoun had already discovered the first major component of the vertical mind. In 1954, Jasper discovered the second component of the vertical mind, the Brain Stem Reticular Structure. Lashley (1951) while discussing the central topic of serial ordering includes the topic of grammar and syntax of human behavior as special cases of serial ordering, and Lashley's paper was years before Chomsky presented his theory of UG or the organization of language behavior.

Lashley and Beach each addressed a interesting piece of the puzzle of how the evolution of the vertical mind changed the way behavior is organized.

The general problem, which is the topic of this book, is what is an idea? Where do they come from, or in this context: how the problem of serial order is handled or organized by the vertical mind to enable the generic mammals to become the master SOS, which enabled them to fill so many of the niches of the world.

"Temporal integration is not found exclusively in language; the coordination of leg movements in insects, the song of birds, the control of trotting and pacing in a gaited horse, the rat running the maze, the architect designing a house, and the carpenter sawing a board present a problem of sequences of action which cannot be explained in terms of successions of external stimuli." (Lashley, 1951, P. 113)

No one however has addressed the big piece of the puzzle, how the horizontal-vertical, H-V mind creates ideas, which enable the GNMM to make models which destroy the entropy induced by the ancient theories. These mechanisms have served all mammals for over fifty million years. In fact, the mechanisms have served the minds of all the various kinds or species of mammals from bats, whales, moles, golfers, dancers, archers, pole-vaulters, antelope, and apes, etc. It's amazing, but nature it seemed had only to switch fins for legs, for wings, etc. and the same horizontal and vertical circuits did not miss a beat.

"The order is imposed by some other agent. This is true not only of language, but all skilled movements or successions of movement. In the gait of a horse, trotting, pacing, and single footing involve essentially the same pattern of muscular contraction in the individual legs. The gait is imposed by some mechanism in addition to the direct relations of reciprocal innervations among the sensory-motor centers of the legs. The order in which the fingers of the musician fall on the keys or fingerboard is determined by the signature of the composition; this gives a set which is not inherent in the association of the individual movements." (Lashley, 1951, P. 116)

Lashley's concept of a form

Lashley's ability to make inferences based on data, which may seem far afield from some topic at hand, reveals a mind that is a powerful logician.

In the quote above, Lashley (1951) calls our attention to an aspect of rational behavior that brings Kant back into our discussion. Kant (1952) was the first to discuss verbal behavior as a product of two different functions: one-part rational and one-part empirical. The rational element was the form, which he took most literally, while the empirical element was the mud. Kant drew upon a brick maker's model, where the A component or the mold was literally a mold, a wooden box, into which the mud or empirical contribution was packed and left in the sun to dry.

There are two equivalent ways to talk about this model: A can be imposed on B or B can be put into A.

Lashley (1951) conceives of the forms, the A component, in the most abstract or the most general sense. Note in the quote above, Lashley says: the order, A, is imposed by some other agent," Lashley is most helpful here-the form is some dynamic form.

In chapter three on the vertical mind, each of the five circuits of the arousal pathways (D+, D-, F+, F-, and V) contribute some energy component, which is summed into a whole and imposed upon the horizontal components of the mind. This summed energy is precisely what Lashley had in mind when he talked about the order-action taken by some agent of the vertical mind.

In the quote below, Lashley illustrates this last point in a discussion of the language behavior of a polyglot. He notes that each language has a unique form, which can be held in memory and imposed upon different languages.

"The polyglot who has become proficient in a secondary language, who thinks in it and even dreams in it, may still tend to use the grammatical structure of his native tongue. If, as in French, that tongue applies gender to inanimate things, the English pronouns referring to them may, take the gender of the French equivalents, though the French nouns are not thought. The German postponement of the verb or the Magyar use of the past infinitive may be incorporated in the new language. In such cases, the structuring seems to be dissociated both from the content and from the simple associative connections of the words themselves. The ease with which a new structure may be imposed on words is illustrated by the quickness with which children learn pig Latin." (Lashley, 1951, P. 118-119)

Lashley summarizes this last point:

"From such considerations it seems to follow that syntax is not inherent in the words employed or in the idea to be expressed. It is a generalized pattern imposed upon the specific acts as they occur." (Lashley, 1951, P. 119)

The oldest theories of language behavior tried to understand language behavior as overly simple, terms in which one word or phrase was bonded to another of the same kind, words were bonded to words, etc., but the order was somehow embedded in the space where words were associated with words or with one another.

The nativistic nonverbal mind

When we assume that the brain contains a structure that should be called a nonverbal mind, we have already assumed that that nonverbal mental structure has the capacity to elicit the appropriate rational mental functions. The nonverbal mind structure knows how to be a nonverbal mind just as an eye knows how to be an eye or a liver knows how to be a liver etc. Technically, our only problem is our uncertainty about what a nonverbal mind was designed to do. What is a nonverbal mind function? However, our uncertainty about the natural function of a nonverbal mind derives from the fact that for thousands of years we have believed that we had a mind, but we believed there was only one kind of mind, the verbal kind. So our problem stems from the fact that the so-called nonverbal mind is not a verbal mind, which goes against our long time belief about the nature of mind. Therefore, if the nonverbal mind's structure did not come prepared to support verbal behavior, what was it designed to do. It was designed to support nonverbal behavior; for what species? Is it for lions, tigers, or bears, bats or whales? The generic mammalian mind possesses a structure that enables a bat to fly, a whale to swim, a deer to run, or F3 to run the maze.

The role of cortical growth

The philosophical issue behind all these concepts of open and closed, however, is this: what is human nature, and how did the evolution of the massive neuronal cortical structure define human nature? How does the cortex change the nature of mammalian life? The third wisdom is exclusively a function of the redesign of the brain due the emergence of the cortical neurons. Heterostasis asks a series of heretofore inconceivable questions.

Humans have only one mind in their brain, but their one advanced verbal mind evolved perhaps much like modern cars emerged from the first models of cars, with new models evolving but perhaps not every year as is the case with cars. New models of the mind emerged in the human brain over time but with plasticity localizing structures and functions within the brain; reassigning components of the ancillary language inducing organs without concern for being fair to the two hemispheres. In spite of the Hemispheres having different distributions of the components of the mind, and thus different functions when surgically separated, yet the two were designed to function together as one mind.

Emergence Negative: part one

Humans have a unique set of problems created by their having a secondary verbal layer of rationality. The human compound mind is capable of new kinds of conceptual behaviors, the likes of which the nonverbal mammals never had to confront. The problem was that the secondary verbal component of the mind all too often and all too easily got out of touch with the primary nonverbal components of the nonverbal mind. With the evolution of the secondary verbal mind, new kinds of ideas/actions emerged. For example, the capacity of employing verbal metaphors emerged, which has no empirical basis, which generate ideas that have no factual certification. Metaphors are ideas not derived from knowledge but derived by a route unique to the verbal mind; metaphors are ideas parented by words rather than by facts. By contrast, the nonverbal mind is parented by data justified or images certified by experience.

One thing that makes the study of rodent behavior so important is that mammals with only a primary mind never got lost in a secondary verbal-mind created space. Rat’s behavior is always data-driven. With the primary mind, the rat's ideas are concrete domain-specific, species-specific, purpose-specific, etc. In this section, I will begin to let the behavior of the rats be our guide. The nonverbal behavior of the rat is still rational in ways that human behavior long ago outgrew and cast aside. What lessons has the rat's primary mind to tell.

One of the most critical insights humans, intent upon understanding their nature, must acquire is to distinguish between nonverbal and verbal ideas. I have and will continue to supply many examples of this distinction throughout the book. Understanding the difference between nonverbal ideas, the only kind that the nonverbal mammal's mind can entertain, and verbal ideas, the kind unique to humans, is one of the first steps in acquiring a scientific understanding of the advantages and disadvantages of our human nature. The advantage of the nonverbal mammal is paradoxically that by their nature they can only enjoy fact driven ideas! Since only humans have a thirteen station mind, only humans can enjoy (or must endure) those changes in the verbal narrative that are "value-added" thanks to the verbal mind. Metaphors enable humans to go where facts don't lead. My point is that all real ideas and this Plato did not understand begin as nonverbal, concrete, domain-specific, context dependent rational elements of the primary mind.

In these experiments, I will take the reader down into the silent, nonverbal world where the primary mind lives and functions. The nonverbal mind creates the first layer of consciousness. It is invisible to most humans only because the nonverbal mind designed the nonverbal consciousness to be invisible. The nonverbal mind did not design consciousness to be a work hung in the Louvre but rather for inducing action.

Nonverbal ideas cannot safely be divorced from action or from mind. Concrete ideas are the primary kind. With the evolution of the nonverbal mind, all mammalian action was conceived rationally but nonverbally. In such a primary mind, ideas and actions evolved in parallel; rats can take no actions that are not first conceived as ideas, nor are there ideas that can be conceived except within the concrete context of some species-specific action, conceived to achieve some specific goal in some specific domain. The primary rational nonverbal mind conceives of new actions by conceiving new ideas and new ideas by conceiving new actions: they are two sides of the same coin.

The ancients, long before science evolved, had no alternative but to claim they could pull sacred magic out of the sky. Plato (1952), for example, could only claim that the nonverbal ideals created by his nonverbal mind had a magical origin in the highest celestial realms. Moses, on the other hand, at least walked up the mountain and met god part way, but the great lawgiver, all the same, could only claim he got his laws on top of the mountain where god carved them into stone for him. Moses never even admitted to owning a chisel! Supposedly, an important component of the beauty of the Mosaic laws, as well as the beauty of Platonic ideals, was that their authors could claim that they were never touched by human hands; that their origin was independent of human hands! Yet neither Plato nor Moses had enough insight into human nature to claim that the laws they received from a verbal god were never touched by the human nonverbal mind!

This book begins with the emergence of the primary mind, the third wisdom, and the first rational machine. This machine creates species-specific rationality from the bottom up. Rats have no need for the heavens or for the spirit residents of a metaphor-driven world. The rats' non-verbal ideas are concrete, species-specific expressions of rational/mental beings embedded within some concrete niche, to achieve some niche-specific purpose, and in the process rats evolve sets of rational conceptual elements that enable them to conceive and achieve their domain-specific purposes, concrete all the way.

Concepts evolved and still evolve to give meanings to concrete patterns of precepts. Concrete concepts are a complex function of some species-specific mind embodied in some species-specific body, embedded in some concrete niche trying to achieve some domain-specific goal that can only be achieved in that niche.

On the other hand, the emergence of the compound verbal mind endowed mammals with several sets of remarkably adaptive skills. One of the most remarkable was demonstrated by Stratton (1896) who is famous for inventing and experimenting with what today are called Stratton-glasses that inverted the image upon the retina. After Stratton's subjects began to wear his glasses, they became very disturbed by this inverted image as they now saw the world upside down. However, over several days, their visual system had the remarkable capacity to adapt or correct for this inverted image, and they began to see the upside down image, right side up once again. My point is that humans in too many domains over the last two thousand years have been induced to wear Stratton Glasses. For example, Plato with his metaphor of the divided line induces or seduced more human to wear Stratton glasses than any other philosopher in either western or eastern civilization.

My theory of the evolution of the compound mind, NV/V mind, takes the position that all the nonverbal mammals that have ever lived or are now alive, have lived in a world without metaphors. These GNMM species saw the physical world, Spinoza's (1933) pantheistic world, as the sole source of the truth and knowledge, which empowered their SOS1 nonverbal minds to exhibit the most powerful adaptive capacity for dealing with change. This NV mind proved to be "the better way that nature required for survival.

The emergence of the compound mind, the NV/V mind, was at best a mixed blessing/curse. Sadly while all nonverbal mammals were born "Free," so it should have been in the compound-mind, human mammals. Yet Plato's philosophy, and the transcendental religions combined to induce too many humans to wear the glasses, and too soon it became fashionable to profess out of genuine conviction or in too many culture, out of fear, that each human saw themselves as happy sheep, "walking by still waters," or as wretches reveling in amazing grace. Yet the tragedy was that they were no longer allowed to know that they could be the SOS1 mammals nature intended. Metaphors induced too great a difference from what their genetic heritage bestowed.

A theory of the nature of Revelation

Many guests on the Charlie Rose show say that their creative ability is a "gift" from the muses, or they don't know where their talent derives. In religion, these gifts were called revelations. However, my theory of the way the compound mind functions makes another theory possible.

The emergence of the compound NV/V mind suggest that the nonverbal mind creates in the nonverbal consciousness nonverbal images which tell a nonverbal narrative, which the "I" of consciousness believes is the real world occupied by objects of Aristotle's Station One. Yet, the aspect of this theory, which requires a hard sell states that the nonverbal mind is doing: first, all the nonverbal work, and that this nonverbal work is 90% or more of the work involved in speech. Then secondly, the verbal mind again only has one task: to transform the narrative created by the nonverbal mind into memes and package it for export.

I must now call attention to the possibility that, unbeknownst to Plato and Moses, this secret power of the nonverbal mammalian mind allows me to suggest historically the great truth is that the laws and the ideals of both the secular and sacred world were the product of their generic nonverbal mammalian mind. The law's origin was neither the product of some spirit god nor the work of some eternal spirit but had its origin in the silent organ, the nonverbal mind, whose laws the "Law-givers" attributed to some spirit god, rather than attribute them to their silent and invisible nonverbal mind.

My intention, for what it worth, is to do what I can to take back all that has been wrongfully taken from the primary mind and naively attributed to some spirit/verbal god. It is now time to give credit where credit is due. The primary mind is now and always has been the great lawgiver, the great ideagiver, the origin of all mental speciation. Part of my scientific role is to call attention to one route that would explain how these primary nonverbal mind-given laws and ideals probably evolved from the bottom-up not from the top-down.

My point, my goal is to suggest to humanity that it is now time to take off the Stratton glasses, which all the verbal mind advocates are still wearing for all these many years.

Chapter 5 the Technology of the Nonverbal Mammalian Mind

In the fifties, Russia launched Sputnik, the first space satellite and the space race began. The USA Man in Space Program was planned in steps. Before humans could be committed to space flight, the space hardware had to be proven capable of executing its desired function, and furthermore life itself had to prove it was capable of not just living in the conditions provided by the technology, but also prove it could do useful work in space. NASA chose to start proving the technology and life itself by flying smaller and lower forms of life than humans.

Brown and Blodgett

In the mid-1950s, Dr. W. Lynn Brown, the director of the Psychology department of Balcones Research Center, was given a contract by NASA to train monkeys to work in space. Two Rhesus monkeys eventually flew into space, Sam and Miss Sam. The challenge that NASA gave Dr. Brown was to train, actually to super-train, animals so that their failure to work in space would prove that they were somehow incapacitated; they would not stop working unless it was physiologically impossible for them to work.

The mystery

I want to acknowledge my appreciation to Dr and Mrs. Brown who graciously entrusted me with some of their personal copies of these historic events.

FIGURE0.4A

FIGURE0.4A is a picture of a training session, which shows the three main actors in these events: Dr. Brown on the left, Sam in the middle, and Dr. Blodgett on the right.

FIGURE0.4B

FIGURE0.4B is an angelic picture of Sam. This picture leaves no room for anyone to be surprised to discover that for many years he was a star attraction at the Zoo in San Antonio next to the school of Aviation Medicine. His angelic look moreover matched his personality, Dr. Brown noted that Sam never bit him, yet Miss Sam never failed to bite him.

First, I must note that there is something of a mystery about the role of Dr Blodgett in the training of the space monkeys. The mystery did not become apparent until my research assistant called my attention to the fact that she noted while preparing the references for this book that Dr. Blodgett's name was not on either of the two School of Aviation's Medicine publications. Obviously, I never had an occasion to discuss this matter with Dr. Brown, so the exact nature of Dr Blodgett's role must forever remain a mystery.

However, I should say both doctors were on my dissertation committee, and I had taken several graduate classes with both men and that Dr. Brown's choice of Dr. Blodgett as his second, was a wise choice. Dr. Blodgett was a brilliant learning theorist as, of course, was Dr. Brown. Together, the two made an excellent animal training team. The two men had a long collegiate relationship, and Dr. Blodgett accompanied Dr. Brown to Wallops Island to witness and assist him in the recovery of Sam after the splash down.

These two animals were trained, flown, and performed with distinction. (Brown, et.al, 1959) I never talked to Brown or Blodgett, B and B, about these procedures, and I was surprised when I finally read the publications Dr Brown gave me about how little they had to say about those parts of their new procedures, which I found to be the most creative and intriguing

NASA's challenge

However, the NASA specification was an order of magnitude more stringent than those to which any animal had ever before been trained. In order to meet NASA's specifications, they would be required to redesign the classical topic of Transfer of Training, TT.

TT and the problem of change

The challenge was how to train or super train animals on earth to perform in all the unique conditions of space; more specifically, how to train animals in condition A, on earth, to perform in condition B, in Space. In Experimental Psychology, each of the specific differences between A and B is called the TRANSFER OF TRAINING PROBLEM, TT. The TT problem is one of the most common but very real problems humans encounter in all walks of their daily life.

Learning and the problem of change

While the TT problem provides the context in which a particular problem can be conceived, studied, and planned, yet the solution to the Transfer of Training problem is entirely a matter of how the animals are to be trained. This topic of the actual training of animal behavior is conventionally called learning or conditioning. The Topic of TT is about how **training** transfers.

Transfer of Training, TT

Transfer of training is a scientific procedure long used by Psychologists, which occurs in two phases; training takes place in condition A, while performance will take place in condition B. Several examples will be given.

In sports, practice takes place in condition A on the home field when no fans are in the stands, while performance may take place on another team's field with fans in the stand.

The same can be said for choirs, theatrical productions, the Olympic Games, etc.

Our concern now is the fact that B and B were to train Sam and Miss Sam on earth, A, to perform in space, B.

What should make the training of the two space monkeys interesting is that B and B really had the time, money, knowledge, and government support to include in the training all kinds of complex variations on the training procedures that the normal scientist cannot afford.

My purpose, in adding this chapter to my book on the mind, is complicated. Partly it is based on my desire to be a patriot and celebrate the contribution my friends made to advance the space technology of the US, and also belatedly to publish the scientific aspect of their achievement, which was never published for reasons that have no answers.

I call these new procedures extraction. I never talked with Brown or Blodgett about these procedures so I have no knowledge of whom, or how they were conceived or understood.

I have no choice but to describe what I saw and to describe how I understand the scientific significance of their procedure. It would have been preferable that they had spoken for themselves, but I am all that is left, and I feel that it is my responsibility to contribute to the memory of these historic events.

My conceptual point of view calls upon several different areas of psychology.

- First, I will show that Extraction is a variation on Pavlov's classical procedure of extinction (1960).
- Second, Rubin's phenomenon of Figure and Ground also plays a major role in my understanding of the B and B procedures. (Helson, 1964)
- Third, a variation on John S. Mill's (2002) concept of "Varying the Concomitant" also plays a role.

I was awed, at first, by their creative innovations of the classical TT procedures, since I had no way to imagine where their procedures were coming from, or how to understand them.

However, over fifty years have passed, which gave me plenty of time to think about their procedures. I propose that the purpose of their procedure was to EXTRACT the FIGURE component from the GROUND conditions.

Causal Conceptual Conditioning: CCC theory

What are the implications of the experiments described in this book?

The first and most obvious implication is that both a new way to run rats and a new theory of learning are urgently needed. The discovery that mammals have a nonverbal mind, one motivated by Heterostasis, requires that a new theory of conditioning be devised.

The Loop Training Procedures, LTP, showed that the classic way rats were tested in the maze, the Trial Testing Procedures, TTP, were not just wrong, but counterproductive. The TTP by suppressing the retrace behavior was actually suppressing the learning behavior of the rat; they were making it almost impossible for a rat to learn.

The purpose of this section is to define a new general form of conditioning, Causal Conceptual Conditioning, CCC, or 3C theory, or simply Conceptual Conditioning.

The two classical theories of conditioning, both Pavlov's and Skinner's theories, must be revised in the light of the fact that mammals have a nonverbal mind. Neither Pavlov nor Skinner ever considered or conceived of learning as a mental function.

What is required is a unified theory of conditioning, where conditioning is conceived as a function of the nonverbal mind.

- Pavlov (1960) identified the first special kind of conditioning, Classical Conditioning, CC theory.
- Skinner (1974) identified a second kind of conditioning, Operant Conditioning, OC. Skinner distinguished operant conditioning from respondent conditioning, which was his new name for Pavlov's Classical Conditioning. Skinner claimed that Pavlovian CC was autonomic nervous system conditioning, while his own OC was somatic nervous system conditioning.
- Both CC and OC are special instances of CCC theory.

What's different about CCC theory?

CCC theory differs from CC and OC conditioning in two major ways:

- First, the major difference is that both Pavlov (1960) and Skinner (1974) conceive of learning as a function of homeostatic behavior.
- Second, CCC theory proposes that all mammals have a mind, which is motivated not by Homeostasis but Heterostasis. CCC theory asserts that all mental growth is a function of Heterostasis, and that Heterostasis is a function of a technology that enabled mammals to become self-organizing systems.

In this chapter, I take a more systematic approach to Pavlov, and identify and discuss the first four of his test procedures used in my experiments.

Six Essential Types of Scientific Procedures

Pavlov's first four test procedures

Pavlov (1960) established the standards for testing animal behavior, which can be simply outlined as a series of procedures through which each animal must pass in turn. Pavlov identified eight learning procedures, yet only the first four are relevant to this chapter. However, I add two new test procedures to his list, abstraction, and extraction.

1. Deprivation training

- The behaviorists devised several different techniques for creating the ideal motivational state for inducing learning. However, historically the term deprivation has always referred to homeostatic deprivation, which may require weeks of preparation. Surprisingly, Heterostatic deprivation is also a possible experimental concept; experiments studying heterostatic deprivation should prove interesting.

- One of the special advantages of Heterostatic motivation training is that none is required. However, it is still possible. Each rat that has spent its whole life confined to a small wire cage in solitary confinement is more than adequately stimulus deprived without any kind or degree of additional heterostatic deprivation training. Just drop one gently into the Lashley III maze and watch it almost immediately begin to emit spontaneous exploratory play behavior. Almost every rat's rational nonverbal mind is always primed for the detection of anomalies and always excited by anomalies.
- For example, it would be possible to do a heterostatic deprivation experiment by raising rats in home cages with different degrees of enrichment by adding different amounts of play-objects in their cage.

2-Habituation training

Pavlov discovers Habituation

One of the few principles, which almost everyone agrees on, is that a learning session must be preceded by habituation.

Control variables.

Pavlov (1960) was the first to discover this essential step was required when he attempted unsuccessfully to do classical conditioning with his dogs in the colony room. His dogs were too distractible. Therefore, Pavlov designed and built the first isolation chamber where he put each animal until it could settle down or technically "extinguish its freedom reflex." The chamber isolated his dogs from the sounds, sights, and smells of the colony room or in effect, from any change in sound, sight, and smell in the set of "ground" variables. The test chamber produced a wall of constancy around the animal, a set of control variables that were constants or unchanging. Classical conditioning was only possible within this wall of constancy produced by this set of control variables. Let me alert the reader to the statement just above about the control variables, which most experimentalists consider to be throwaway variables, but which are essentially what this chapter is about, so watch for any and every mention of control variables, which will be defined below.

Pavlov called this first step of his training procedure HABITUATION, and it always preceded his conditioning training sessions.

Signal/Noise, S/N

The Habituation procedure was a critical part of preparing a dog for classical conditioning. Pavlov conceived of this stage in terms of what today we call a signal and noise problem. Pavlov discovered that dogs could not be conditioned in the colony room, so he designed a special **Isolation Test chamber**. The basic idea of the concept of signal-noise, starts with the relative size or intensity of his conditioned stimulus, S, relative to the size or intensity of the other stimuli, traditionally called noise, N, in the chamber. If the average intensity of the noise stimuli present in a situation was say 20 and the average intensity of the CS was only 10, then the subject probably would have trouble detecting the CS. But if the CS had intensity greater than 20 say 30, then it would stand out and easily be detected.

In the colony rooms the signal/noise ratio, S/N, which is some large value, is so high or so noisy that the stimulus value of the CS is effectively invisible. Only in the isolation chamber where the noise has been reduced or eliminated can the CS have a chance of becoming visible to the dog. The nonverbal mind must be able to detect the small change induced by some single CS, such as a bell or a light. While Pavlov found that this task was impossible in the colony room; yet, in the quiet of an isolation or test chamber his dog never failed to observe the small change induced by a bell or light.

Step one

In step one; the organism is habituated to the test environment.

Surgery

Each of Pavlov's (1960) dogs was tested after surgery where one salivary gland was externalized. Before a conditioning period could begin, Pavlov put a tube over the dog's externalized salivary gland. Next, each dog was restrained in a stock to prevent them from pulling off the tubing. The tubing allowed Pavlov to count the number of drops of saliva the animal secreted after each CS was presented.

When Pavlov first confined a dog in a stock, it would begin struggling to escape from the stock. In this aroused and agitated state, a dog could not be conditioned because its "**freedom reflex**" was active. Yet at some point, the animal would stop struggling and Pavlov discovered that it was imperative that the conditioning start immediately. FIGURE0.2E shows in which of the four vertical channels, a unit of inhibition grows each time a channel is used. This rule still applies in this stage of training. Pavlov would habituate his dogs presumably until their arousal level drops from C to B on the Y-D law. Pavlov made a point of noting that if the habituation continues then the intensity of inhibition grows so strong that the dog would go into such a deep stage of sleep that it fails to respond to the stimulus.

The use of a special chamber for testing and the habituation of the animal to the test stimuli became a standard for all behaviorists, even though each experimenter had to adapt this phase of the training specifically for his or her experimental conditions.

Cruickshank's isolation chamber

Cruickshank (1961), while working with students with learning disabilities who were unable to learn in the traditional "mainline" classroom, developed a novel variation on this theme. Cruickshank removed his students from the traditional classroom and trained them in a special isolation test environment, just as Pavlov (1960) had done with his dogs.

Ground conditioning or habituation of the ground

Cruickshank's (1961) test environment was a cubicle; a small constrained space that generated a "ground," a small population of control stimuli in which the student was left alone until they could habituate. Once the students were no longer in a stimulus overload situation; (See in FIGURE0.2B) in the red zone on the Y-D Law and finally in a test environment in which they were able to achieve the B-or green zone, did Cruickshank begin to introduce stimuli for the students to learn.

What Pavlov (1960) did, in Cruickshank's (1961) terms, was to discover that dogs are like students with learning disabilities. They cannot learn in the colony room and so Pavlov took his dogs, as it were, out of the mainline classroom and created the first special education classroom for his dogs. However, Pavlov's goal was never as it was for Cruickshank to get his dogs back into the mainline classroom.

Skinner box isolation

Skinner (1974), on the other hand, puts his subjects in a special Skinner box testing apparatus, which is then put in a picnic freezer type isolation chamber with only a small fan in the wall to supply both circulating air, and a small noise, which masks any outside noise before it gets to the animal.

3. Classical conditioning:

Classical Conditioning was the third training session. In this session, Pavlov paired the CS-US, or CS followed by the US. Pavlov believed that learning was due to stimulus substitution. See FIGURE0.2G

4. Extinction

When a CS and a US are paired that produces learning, but then in the next step, when the CS is presented and not reinforced, then extinction occurs; the dog stops salivating to the CS. Extinction is: CS followed by no US.

5- Extraction

See below

6-Abstraction

See below

Abstraction defined

Pavlov's experiments, I finally realized, really only touched the tip of the iceberg of the topic of conditioning. For example, in terms of the figure-ground theory, Pavlov's experiment dwelt only with the figure; he ignored the problem of the ground completely. This chapter is devoted to taking Pavlov's work one more step forward. One central question is where in Pavlov's work the context or the "ground" enters the picture. Where does the ground come from?

If we look at this situation in terms of signal/noise theory, then a concept is defined as a set of features, which are necessary and sufficient for determining whether some object is or is not a member of a class. Historically, Pavlov was creating concepts that were concrete not abstract: in other words, he was creating only a baby concept, first born, concrete, domain-specific concepts, which are far from being abstract concepts.

Concrete vs. Abstract concepts

The original set of features of a concrete concept will be a mixture of two types of features: S and N, where S is the Signal or essential features, and N is noise or nonessential features. Furthermore, the animal does not know which features are signal and which are noise. On the other hand, an abstract concept is one with only S features.

On the practical side, an educator's question is how to create a set of replications of an abstract concept in the mind of each student in his or her class. On the practical side, it is impossible to train a subject to employ abstract concepts in the first step. Concept learning is a long process that must begin by teaching a subject to employ concrete ideas of a triangle, rather than to create an abstract one. If some customer were to order say 25 subjects who can reliably employ abstract concepts in their work, then the trainer must begin by creating 25 concrete-skilled operators, and then transforming the 25 concrete into 25 abstract-skilled operators. So what would an abstract creating skill-training curriculum look like?

How to transform C to A

Causal Conceptual Conditioning, CCC theory, should first contain a theory of how to, in the first step create concrete concepts, and then in the second step, contain a theory of how to convert the Concrete Concepts, C-concepts, to Abstract Concepts; or how to transform C-A.

Why bother, you should ask? There are many advantages that A-concepts have over C-concepts, some of which I will identify in this chapter.

Brown and Blodgett and the transfer of training, TT problem

My first goal is to employ the concept of extraction to explain the TT procedures devised by Brown and Blodgett for training Sam and Miss Sam for space flight. However, there are other situations where extraction is useful, for example, in understanding the abstraction process.

Seven essential sets of scientific variables

In this chapter, all the discussions about how Brown and Blodgett trained Sam for space travel must be, of necessity, conducted within the framework of the unique language of science. There are several technical scientific areas, which will be identified and explained in order for the reader to understand the procedures designed by B and B. The six scientific topics, which provided the context within which the training of Sam and Miss Sam was conducted, are listed below.

Before we discuss these two new procedures, we must first review the three basic variables of science.

This chapter focuses upon the role of the Control Variables, CV. Normally the roles of the CV are most severely limited, or underused, even misused or abused in most scientific experiments. Let me begin to discuss the complex role of the CV in behavioral scientific research by discussing some of the new and different ways we will talk about or name the CV in CCC theory, which will be called Parameters, P. The P variables play such an important role in all learning but go unnoticed.

First, the P variables form the foundation for learning. Metaphorically, learning cannot begin until a **stable** place or region within the rhythms of the cortex, are established, which Ashby (1960) called a "local stability," within the normally incessant action in the canopy of neuronal activity, which is seen in the EEG.

The seven sets of scientific variables, which form the context for the B and B discussion, are listed below:

1. The set of three scientific variables

2A. The particular-general relation in everyday life

2B. The particular-general relation in science

3. The figure-ground relation

4. Triangularity: The concrete-abstract relation

5. Manipulating causality: VARYING the CONCOMITANT

6. Extraction: the role of P- change, making Independent Variables, IV, out of Control variables, CV

This chapter is devoted to a discussion of what I call Parameter Changes, or P-changes, which actually are everywhere. The problem of space forced Drs. Brown and Blodgett to face the P-change problem in a way that was far more serious than science had ever confronted before.

1 - The three scientific variables

All talk about the causal procedures employed by B and B must be discussed within the context of three scientific variables, and the TT Problem is no exception. The three scientific variables are the Independent Variables, IV, the Dependent Variables, DV, and the Control Variables, CV. Together, I will call them the Parameters of an experiment. Usually the problem can be understood in terms of the first two scientific variables, the IV, or the X, variable and the DV Variable, or the Y, variable, so that the results of an experiment can be written or graphed in the form Y = F (X). However in this case, where A and B are so different the third scientific variable, the control or the Parameter variables move to the center stage where they demand and get all the attention.

The IV, the DV, and the CV

- The independent variables, IVs, are those variables the experimenter manipulates to force to change, usually referred to as the X variables.
- The dependent variables, DVs, are those variables the experimenter simply measures, usually referred to as the Y variables.
- The control variables, CVs, are those variables the experimenter must insure do not change. These variables are the neglected members of the set of scientific variables. In fact, to call them neglected is too mild an expression. The general scientific designation for these variables is "the nuisance variables." It is as if the only role of these variables is to cause trouble.
- My new theory of conditioning is called Causal Conceptual Conditioning, or CCC Theory. The role of the mind is to take the uncertainty out of a domain. Uncertainty is induced by anomalies, which are defined as "Effects without a known cause." The responsibility of a rational mind is to find a cause for each of these anomalous effects. The structure of every space is a causal structure. Once an organism has first, identified all the EFFECTs without a known cause, and second, identified the causes of all these Effects, then all the uncertainty inducing events will have been neutralized in that specific domain, and the animal will know the domain.
- The Brown and Blodgett study puts a new meaning into our scientific understanding of the casual structure of the cosmos, or more specifically the causal structure of an earth-defined spatial structure, and a space-based causal structure. Especially, this B and B procedure forces us to evolve a deeper understanding of the role of the nuisance variables.
- For example, there is no convention by which they are called, but in this chapter they will be called the **Parameter variables** and will be abbreviated **P**. In this chapter, their role becomes the prime member of the set because this chapter focuses upon the consequence of changing these P variables from control variables to independent variables. The **P-change** function is defined as what happens when one or more P variables become an X variable.
- The purpose of an experiment is to determine the causal effect of the independent variables, X, upon the dependent variables, Y, **given the conditions of the Control Variables, P.**

The simplest statement of a function can be written Y = F(X) where Y is the dependent variable, and X is an independent variable. However, such a statement is deceiving for the very way the statement is written overlooks the role of the control variables. However, if all a scientist wants to know is how Y varies with X, then often he can ignore the third of the scientific variables, the control variables.

The point of the chapter is to call attention to the critical role of the Parameters or control variables play in controlling behavior. In many respects, most people believe that X, the independent variable or variables, independently cause the changes in the dependent variable Y.

While it may not yet be apparent, the technology of the mind plays the exclusive role in determining the nonverbal actions of the nonverbal mammals. Mammals are self-organizing systems, SOS, which means they must assume responsibility for the consequences of their behavior. The validity of their behavior can be no better than their data; the validity of their self-organized behavior must not be based upon whim, guesswork, or metaphor. The only way that the nonverbal mind can insure the validity of its behavior is to insure that it is closely coupled or tied in with the causal nature of the cosmos. The nonverbal mind directs a sequence of Heterostatic events.

Our question is not where the causes are in the physical technology provided by NASA for Flying Sam and Miss Sam. The question is how to train the monkeys to get their behavior conditioned to the figure-ground stimuli that will cause the monkeys to perform with peak efficiency in space.

The first question is how the manipulation of the figure-ground sets of stimuli can define where the causes, which determine the way the animal's behaviors, are distributed within the animal's nonverbal mind. The Figure-ground phenomenon will be discussed below.

Drawing a graph

In book two, I show over a hundred line graph figures in the result sections of the four experiments, which produced the data for the discovery of the nonverbal mind. These figures visually show that the causal relationship between any numbers of variables can be displayed from main effects to third and fourth order interactions. A discussion of these Anova effects is included in Book 2.

However, the simplest graph, one that plots main effects show the causal relationship between only two variables; one IV and one DV, with the IV plotted on the abscissa, or horizontal (X) axis and the DV plotted on the ordinate, or vertical Y-axis. The control or ground conditions may not be mentioned, but are incidentally specified somewhere in the method section.

On occasion, the X and Y-axes are reversed. In fact, Skinner (1974) perfected this type of graph for use in operant conditioning, and I found this Skinner type graph particularly useful for plotting the FEU (Unit First Entry) data in this book. In Book 2, the Heterostatic behavior will be described, and the new dependent variables will be introduced and defined. FEU and BFEU (Unit Entries between first entries) are two of the new dependent variables, which revolutionized my study of animal behavior. For detailed information on the definitions of these variables and what they mean please see Chapter 1 of Book 2.

The P-change: The last seven words

In dealing with the mind, the verbal, or the nonverbal, the last seven words in the statement of the purpose of an experiment become of paramount importance. "The purpose of an experiment is to determine the effect of the independent variables, X, upon the dependent variables, Y, **given the condition of the control variables.**"

In this section, seven topics in science form the context within which B and B can be understood. In 2A, the second of the seven will be discussed.

2A--Particular-General Statements in everyday life

For over two thousand years, the distinction between two kinds of statements, the particular, and the general has dominated logic. Aristotle (1952) recognized these two statements as the heart of logic.

There are two kinds of logic: Deductive and Inductive, where the action in both cases is all about the relation between particular and general statements. Both kinds of logic have one particular and one general.

The logic in deductive statements goes from the general to the particular, while the logic in induction goes from the particular to the general.

- The classic deductive syllogism says:

All men are mortal.
Socrates is a man.
Therefore, Socrates is mortal.

The syllogism contains three elements:

The major premise
The minor premise
The conclusion

The major statement is a GENERAL statement and the cue is the word "All." The major premise is a general statement because it attributes some feature to all the members of some population.

The minor premise is a PARTICULAR; Socrates is a man.

If all men in the population share some attribute as asserted in the major premise, then any man that is a member of the population will share that attribute, so it can be concluded that Socrates is mortal.

However, for our purpose it is only important to be able to recognize the way the two statements, the particular and general are used in these classical instances of logic.

- The classic inductive statement says:
  If anyone is concerned about the color of crows, then they should go look at crows. When they see crow one, they observe its color:

Crow one is black
Crow two is black
Etc., Etc., Etc.
Therefore, all crows are probably black

After observing a number of crows, randomly sampled from the Ground population, and if all those crows are Black, then probably all the crows in the population are black.

Inductive logic starts with observations about a particular individual, in the case above, a particular crow. After observing any number of individuals, randomly drawn from a population, some inference can be made. If all the crows randomly sampled from the population are black, then probably all crows in the population are black, for random sampling insured that all the crows in the population had an equal chance of being sampled and observed.

2B-Particular-General Statements in science

What is a particular: Math

In this section, I will discuss once again the distinction discussed in section two, the distinction between the particular and the general. However, in the section above, the distinction was very simple and ordinary; the distinction this time is a bit more complex, but is particularly relevant and insightful.

There are two options here: a quick and partial one, or a carefully stated full argument.

First, the quick, and partial:

Let Y = F (X) where Y and X are the "Figure" components of the causal equation and we include nothing about the "Ground" component of the equation.

Second, Concrete # 1:

Let Y=F (X) where Y and X are a statement exclusively about the "Figure" and where the parameter values of the Ground are: A=3, B=7, and C=2.

What this equation says is that Y= F(X) only when the Ground parameters have these particular values: A=3, B=7, and C=2.

This particular function, where Y =F(X) is most vulnerable, for it only works when each of the ground conditions have their original concrete value, 3, 7, and 2. This concrete function has profound implications. If we take Parameter A, for example and start changing its original concrete value from three to other possible values, then External Inhibition, EI, will emerge on every P-change.

So now, we have two very different instances of a particular. In section one above, "Socrates is a man" was our particular. However in section 2, we have a particular that states that Y=F(X) only when the ground parameters have particular values.

Y = F(X) is an illustration of a particular function that is very labile. However, this statement is most useful in understanding behavior for it explains what Pavlov (1960) called External Inhibition and furthermore proves to be the master reflex of the mind. This equation will prove to be most relevant to our discussion of the B and B procedure.

This is true in closed sets such as a deck of cards. All decks of cards consist of four suits (hearts, diamonds, spades, and clubs) with 13 cards in each suit from Ace to King. Because the deck of cards is a closed system, anyone who can count as the cards are being played can know what is still available. Incidentally, this is why most casinos use a six-deck shoot; it is much harder to count cards with multiple decks. Also, a set of dice is a closed system. A dice player can only roll a specific set of numbers when two dice are used. Open systems are much more complicated and cannot be counted upon to give a closed set of results.

2B-What is a general: Math

What was there about the above statement that doomed it to a particular status?

In order to transform the Y = F(X) particular statement to a general statement the component statement about the Ground parameters must be written, not as particulars but as generals. What does that mean?

In the illustration above, we stated that the Parameters: A, B, and C had values of 3, 7, and 2, respectively, which says that the original values of these three parameters were constants. This is but another complicated way of saying that the Y = F (X) statement is a particular, or that the Y= F (X) is only true when the three parameters have these three particular fixed values.

Changing a control variable to an independent variable

Now consider the following statement that is more interesting:

Let Y=F(X) when A=m, B=n, and C=o. In this statement Y=F(X) for every possible value of the A, B, and C, as control variables.

Under these conditions, our knowledge of Y is general or abstract, for the simple reason that m, n and o as variables can take any value and still have no effect on the behavior of the Figure. However, we must go slow here and note that if the control variables; A, B, and C are defined as a constant, then the control variables A, B, and C are indeed control variables. Our definition says that the control variables are variables that the Experimenter, E, controls, i.e. that the E does not allow to change.

So if at some point in the training the E starts treating these former "control" variables differently; E starts to change them, then they are no longer Control variables, but now they are independent variables. The P values are now being changed. These former P variables are now Independent variables.

Why would someone change a CV to IV? What difference does it make?

Let's look at the B and B procedure and see what happened when they shifted control variables, CV's, into IV's.

Heterostasis could well be called the physiology of change. Heterostasis can be understood as a physiological technology with a love/hate relation to change. Heterostasis seeks anomalies only to destroy them.

Two kinds of change: active and passive

So what is change? Two kinds deserve our attention; the active kind happens where the organism seeks change. The passive kind occurs when change just happens, the nonverbal mind was designed to enable an organism to be active and to adapt to change in the External milieu by Heterostasis, which enables a self-organizing system to control how much and when it encounters change.

In the Brown and Blodgett procedure, they decided what kinds of change the monkeys were exposed to in order to prepare them for space. As noted above, the space age TT Problem presented a challenge because the change between the training conditions on earth and the test conditions in space were so extreme. Moreover, no monkey chooses to go into space. This is clearly an example of a passive kind of change.

Similarly, rats are not asked if they want to run the maze, and once they are in the maze, they can sit and do nothing; but most rats chose to explore the maze. This is an example of what I call active change. The difference between active and passive change conditions is seen in the look and feel of the P-change in the two conditions.

Passive change

The change that brought on the great extinction was clearly not sought by rats, any more than the monkeys chose to go into space.

What kinds of changes are there? In this chapter, we will be looking at both kinds of change parameter, P-change.

Active changes

In the earlier chapters, I pursued the role of the active P-change induced by rats exploring within the maze as the nonverbal mind evolved a set of coherent ideas as a model of the maze. I asked where ideas come from, and I proposed that all behavior was an expression of the ideas or the knowledge the nonverbal mind stored within the model of its current domain and its purpose in this domain.

What is a P-change? When the P-change is active, the emergence of an anomalous event forces the rat to retrace, reexamine, and reconceive both its behavior and its evolving model.

The nonverbal mind was designed to deal with physical events that are P-changes. The nonverbal mind evolved, at least in part, because environmental events occurred which changed the face of much of the earth, and thus destroyed the closed simpler forms of life by destroying their niches.

The consequence of a P-change can require that life enter a new niche and create an entirely new life space. Such a change activates the creative function of the mind, and is one of the critical aspects of the NV Mind's chief technology, the creation of an idea. Ideas, at least nonverbal ideas, are not something printed on the page of a dictionary or some book, but ideas, or sets of ideas are elements of the programs or scripts, which create the domain-specific narrative playing in consciousness. Nonverbal Ideas are expressed as nonverbal images, which give meaning to the physical objects or things, which occupy the physical world, alongside the physical nature of the organism's body.

The nonverbal mind creates and shapes each of the exemplars in consciousness by interacting with nature, either actively or passively. However, in the case of B and B, the monkeys were confined and primarily immobilized within a space chair, within a mockup space capsule i.e., the P-changes were passively induced. Passive or active, the P-change procedure B and B designed for training them, however, still altered the ideas or models evolving in their minds.

TT (Transfer of training) and change

In a sense, the significance of TT is that it offers the experimental psychologist a paradigm for studying change where the scientist can pick what kind of change and how much change he wants to induce and study.

3--The Figure-Ground phenomenon

The two features of this overall master series, which particularly deserve attention, are B and B's original use of the old concepts of figure and ground. In order to understand the originality and the significance of these procedures, I will begin by reviewing the concepts of Figure and Ground.

Rubin is famous for his discovery of the figure-ground phenomenon, which the Gestalt psychologists quickly picked up. Helson (1964) in the quote below from Rubin states some of the roles that figure and ground play in behavior.

"The distinction between focal and background stimuli is related to Rubin's distinction between figure and ground in perception (1921). Focal stimuli stand out, are figured, and are distinctive, while other stimuli in the field of view act as background. To be focal, stimuli must differ from level, in either a positive or negative direction. Thus stimuli below as well as above level, and weak as well as strong stimuli, may be focal since it is degree of difference from level that determines their distinctiveness. Stimuli become focal for a variety of reasons; among the causes are the "determinants of attention" of the older psychology, e.g., size, intensity, position, novelty, quality, movement, and internal predisposing factors such as attitude and set, which we have classified under residual stimuli. Mere instructions to "attend to" a stimulus may change it from background to focal status, while instructions to "neglect it" may change it from focal to background status." (Helson, 1964, P. 323)

Almost every general psychology textbook for seventy-five years devoted several pages to illustrating and discussing the phenomenon of figure and ground. What was generally interesting was that they never said why this topic was in a general psychology book in the first place.

FIGUREo.4C

FIGUREo.4C is called Rubin's vase. It was the first instance of the discovery of the figure-ground phenomenon by Rubin in his Dissertation in 1915. This figure is a great illustration of both figure and ground and is a reversible figure-ground relation. While this figure is called Rubin's Vase, the vase image is but half of the possible images it affords. It could have been just as surely called Rubin's faces. As Helson (1964) noted above what you see is partially a function of where you look. If the focus is on the black then the two black faces pop out as figure, which appear in front of a white ground, but if the focus is upon the white center feature of the figure, then the vase image emerges, and the black retreats into the ground, behind the white.

The technology of the mind has this powerful hardware function for processing spatial images in three dimensions, yet it can only exhibit, one at a time, the competing images that are possible within the data set.

FIGUREo.4D

The interposition cue: 1-The tree and the barn figure

Figure and ground generally and most simply may be defined by asking students to draw a picture of a tree in front of a barn, that is with the tree as figure and the barn as ground. So how would you draw a tree and a barn on a flat piece of paper, so that the mind's eye first sees the tree in front of the barn, and then second sees the barn in front of the tree? What we want is first for the tree to appear as figure and the barn to appear as ground, and then second for the barn to appear as figure and tree to be ground.

FIGUREo.4D shows how to draw both of these figures. The first figure-ground relation is drawn on the left, while the second is on the right. Technically the principle at work here is called the interposition cue. The brain or the perceptual mechanisms of the mind know that the object, of which we can see in its entirety, is in front.

The fact is that the technology with which the mind was endowed was programmed to see depth as a part of space perception. Depth is what separates the figure and ground, and produces perceptions in terms of figure and ground; figure objects are seen in front of ground objects. To induce the mind to see the drawing of a tree in front of a barn, simply draw the entire tree, but do not draw that part of the barn behind the tree. The brain in these circumstances draws the whole image of the tree, but only a partial image of the barn. The NV mind employs the rule that figure objects are displayed in front of ground objects.

2-The old woman/young woman figure

FIGUREo.4E

FIGUREo.4E is another of the classic reversible figures, which means that it can be viewed as two images, one an old woman, and another as a young woman. This figure shows a completely new side to the figure-ground phenomenon. As always, figure is in front, and ground is behind. However, in this case there is a bit more than is usual to the figures. In this figure, in order to get the figure to reverse the gaze also has to be refocused. In one case, the old woman's mouth is a strip of black ribbon about the young woman's neck. In the other case, the young woman's nose is a wart on the old woman's nose. However, the point is that there are two images and presumably as each set of cells habituate or become fatigued they take turns entering and leaving consciousness, or as they become fatigued, they lose control of consciousness. This is like what happens in an ambiguous figure-ground situation.

3-Necker cube

The famous NECKER CUBE is seen in FIGURE0.4F. This figure is also a reversible figure as was the old woman/young woman of FIGURE0.4E. However, the two figures have more differences than they have in common. The cube, for example, was drawn with only 12 lines on a white sheet of paper, and yet these twelve lines play powerful tricks with the nonverbal mind. They obviously activate some deep and basic technology of the nonverbal mind for the three-dimensional image they induce the mind to insert into consciousness has a command quality that shows that creating and injecting a three dimensional object into the External milieu is serious business. The mind is active, and its technology is demanding; no amount of conscious effort will make the third dimension go away. Yet it is important to note that the technology that creates and sustains the third dimension of consciousness comes with a cost, which may not be apparent immediately.

Years ago, I made an 18 by 18 inch mosaic of the Necker cube mounted on a piece of plywood with red tiles creating the cube upon a white tile ground and hung the piece on the wall above and behind my TV. The strength of the three-dimensional image created by my mind was even more powerful and insistent than the figure found in the textbooks. In fact, at first I glanced at the mosaic perhaps a few times a day, but over time the mosaic began to flip faster and faster each day, even though the mosaic was not directly in my gaze when I was watching TV. Yet the habituation physiological mechanisms described by Gastaut in FIGURE0.2E were executing their prescribed function. As soon as one of the two reversible images entered consciousness, habituation went to work and when some threshold of inhibition was achieved, the second image replaced the first. Each reversible cycle got shorter and shorter, or the time between two images reversing in consciousness got faster and faster, until it soon became so annoying that I was amused as I was forced to take the mosaic down.

The nonverbal mind has been endowed with some powerful technology, yet every technology comes with a cost, has its quirks, and if we watch we can catch the technology in action.

This figure shows that the nonverbal mind has the technology to take a flat or two-dimensional image drawn on a simple sheet of plain white paper and display it as a three-dimensional image in consciousness. The nonverbal mind has the technology to make us see two-dimensional data as three-dimensional objects in this case as a cube, moreover, as a set of reversible cube images. Each image takes turns being now a figure and then a ground, and the cycles go faster and faster, round and round. Each of two sides takes turn being in the front and being in the back, or between being the figure and then the ground.

FIGURE0.4F

The mind, wherever possible, always gives a three-dimensional or nested meaning to flat figures with the nonverbal mind supplying the third dimension thus making room for distinguishing figure from ground.

The nonverbal mind converts the flat images, which leave the senses, and are dropped, as it were, into what should be experienced as the flat world of the senses. Yet the mind cannot or will not process or experience the world as flat; the nonverbal mind instead adds the third dimension. Out of this third dimension, the mind is then able to separate two images, the figure, and ground spatially.

The central technology the nonverbal mind must provide for the nonverbal mammals is a three-dimensional life space in which each mammal can move in three-dimensions. Does anyone know how many words there are in English that describe the motions or directions of motion options that can only occur in a three-dimensional world: right, left, up, down, get on, under, beside, above, below, on top of the table, Etc.?

Perhaps we should borrow a line here from hard science: physics says, "nature abhors a vacuum." Should we dare offer our own version of this statement; the nonverbal mind abhors a two-dimensional space. For only a three-dimensional space provides consciousness with a spatial model that is isomorphic with physical space, and thus provides a conceptual place with a full set of degrees of freedom for nonverbal narratives to play out. There is no room in a two-dimensional world for three-dimensional objects to move, to act, to serve as a stage for the narratives to be enacted.

4--Triangularity: The Concrete-Abstract, C-A Phenomenon

In this fourth set of figures, another technological phenomenon will be discussed, which plays a central role in understanding the B and B phenomenon.

Let's begin by identifying and illustrating the set of triangles.

FIGURE0.4G shows six kinds of Triangles

There are three ways to classify triangles. There are three kinds based on their sides; three kinds based on their angles and seven when the two parameters are combined.

- There are three types of triangles based on the measurements of their sides:

Equilateral – all sides are of equal length and in this case, all angles are equal
Scalene Triangle – all three sides are different lengths
Isosceles Triangle – two sides measure the same length

- There are three types of triangles based on the measurement of their angles:

Right Triangle – has one angle that measures 90 degrees
Obtuse Triangle – has one angle greater than 90 degrees
Acute Triangle – all three angles are less than 90 degrees

- There are 7 types of triangles based on two parameters: they are:

Equilateral Acute
Right Isosceles
Acute Isosceles
Obtuse Isosceles
Right Scalene
Acute Scalene
Obtuse Scalene

The concept of a concept

Everything acquires an identity only after station seven or the sixth sense of the nonverbal mind confers a class identity by checking the fit or the match between a set of features in the input stream, and a set of features, which the nonverbal mind has created or identified to characterize the concept identified by the mind. When the mind obtains a match, which proves the pattern of features in the input stream is empirically certified, it justifies the exemplar's class image being inserted into the narrative running in consciousness.

Our question in this context goes beyond the story line described in Aristotle's model. In this context, we look at the next step and ask how the nonverbal mind knows where to insert an image or exemplar into the three-dimensional space that surrounds the mind within the brain within the head.

Every anomaly becomes something only by acquiring a class identity by proving to the match-mismatch mechanism, MMM, that it has both conceptual and empirical validity; that its empirical features match the features of some concept. Only exemplars of classes of things have status within the mind; only exemplars of things with class have status within the mental life of the mind. Only after acquiring class status, can a syndrome of features become alive or real and enter consciousness as an exemplar in the place in three- dimensional time and space occupied by the physical object, which is the source of the class.

When anomalous things acquire membership in a class and become a named being, they enter into two life cycles, one physical, and one mental. The latter cycle involves the mind and even as a nonverbally named object, it can play several roles in the behavior of the mammal in whose mind it dwells. For example for Sam and Miss Sam, the light in their space capsule that served as the CS for the shock was always an important event in their life. However, no matter how important it was, B and B showed that the causal significance of the light could be manipulated; especially to have its significance enhanced ten or a hundred fold. When a named mental thing enters the life space of a mammal, it can begin to play a role in mammalian endeavors such as space travel. The question is how much involvement does each object upon entering consciousness acquire, and how can the role it plays be manipulated, which brings us to the door of the B and B procedure.

Every class of being can grow as a member of a domain specific set of hierarchically organized ideational structures. My first task is to show how the vector identity of a class grows as it becomes a member of each successive complex nested place within a hierarchical structure.

But our question is how can we scientifically manipulate the causal structure of an exemplar of a class as it evolves its vector identity?

FIGURE0.4G Triangularity

The nonverbal mind employs Mill's (2002) relationship in an infinite number of ways. For example, any feature of our triangle concept that has an invariant relationship with food becomes a causal feature of Triangularity in our example. FIGURE0.4G shows six of the seven types of Triangles. The rule is that **only a figure, which shows one of the seven types of Triangles, will cover the apple in the food wells of the WGTA (Wisconsin General Testing Appartus).** This rule insures that there is always a concomitant relationship, a causal relation, between the apple and Triangularity.

In the next stages of training, every feature of the triangular figures that are associated with the apple reinforcement will become features of Triangularity.

Each figure consists of two elements: A, those that belong to Triangularity, and B, those features that don't belong. Call the A features the potential "figure" features, and the B set of features the "ground" features. The problem is that every feature that proved to have an invariant relationship becomes a defining feature of Triangularity. In step one, both the A and B features were constants. What is wrong about the concrete method of creating abstract Triangularity is that the animal will recognize only the original concrete figure as a triangle.

So what must be done is that all those B features like the color and the size of the triangle that in the first step of conditioning were constant must now be varied so that their initial invariant relationship with Triangularity is destroyed. The varying the concomitant rule has two sides: first, in step one, any B feature that varied with the A features of Triangularity become a cause.

Now beginning in phase two, those B features, which are not signs of Triangularity, must have their causality destroyed by converting all the B features into IV and varying them and thus making them no longer have a causal relationship with the A features of Triangularity. In phase two, the B features must have their concomitant status destroyed.

The Procedure of Abstraction

The central concept required to understand the B and B phenomenon is the distinction between a concrete and abstract class, for example a triangle, which parallels the distinction between concrete and abstract figure and ground.

Where does the abstract concept of Triangularity come from? Can an abstract concept emerge or evolve in a material nonverbal or verbal mind?

The concept of Triangularity, the prototypical abstract concept, is defined as a geometrical figure with three sides and three angles, which add to 180 degrees. The major part of the difficulty of training the mind to move from the concrete to an abstract concept of Triangularity is that as FIGURE0.4G shows, there are six different types of triangles.

FIGURE0.4G

There are two questions, one easy and one hard. First, how can verbal mammals conceive an idea of an abstract triangle? Second, the hard one: how can a nonverbal mammal conceive an idea of an abstract triangle? For a pragmatist, these questions resolve to an operational question; is there a procedure by which every mammal can be taught such a concept. The purpose of this section is to propose such a procedure, and as long as we stick to the operational definition of the procedures, the two questions are the same.

Transforming an idea of a concrete triangle to the idea of an abstract triangle

In this section, I will propose a procedure for creating in a mammalian mind the abstract concept of Triangularity. For starters, we must get over the spirit concept of an idea, which has dominated our Western imagination for centuries. A named class is a pregnant concept that can **assimilate** presumably an infinite set of particulars, but every new exemplar begins life as a baby, i.e. a concrete object, and as it matures and grows, it can evolve in several directions, but for our purposes, only one of these growth potentials concerns us, the concrete-abstract dimension.

In the first phase, a subject shall be trained on the Object Quality Problem, OQP, upon the Wisconsin General Test Apparatus, WGTA, to discriminate between a concrete triangle and other concrete geometric figures. The details will be described below.

FIGURE0.4H

The WGTA described

The WGTA is a test apparatus designed by Harry Harlow for testing Rhesus monkeys and a diagram of the apparatus can be seen in FIGURE0.4H. (Andrew and Harlow, 1948) The working component of the WGTA is a wooden box with two open sides as the test area. The human experimenter, E sits on one of the open sides and the monkey, S for subject, sits on the other open side. Between the E and S is a movable table top with two holes about a foot apart in its surface where rewards such as a small piece of apple can be hid.

The Object Quality Problem, OQP.

FIGURE0.4I testing materials

FIGURE0.4J identifying pairs

When S is tested on the OQP,

The first thing the E does is to lower the guillotine door on the monkey's side so the S cannot see where the food is being hidden.

Then the E has a lever by his/her side which enables them to draw the sliding table back to their side so that one of the food wells, which is chosen randomly on each trial can have a piece of apple placed in it.

Next, a figure from the triangle set (see the left side of Figure 0.4I) is placed over the apple and a different figure is placed over the other hole.

Next, the tray is pushed forward to the S's side.

Then the door is opened, and S is given a choice, using the non-correction method; if the monkey makes an incorrect choice, the door is dropped immediately so that S gets no second choice. The subject must learn to choose correctly on the first try.

The training must be accomplished in several stages, with the first stage discrimination being relatively simple; generally starting with a concrete triangle and a concrete non-triangle. But with each new stage, the difficulty becomes more abstract and complex.

Stage one

The subject will be taught to recognize the concrete triangle A: a red wooden **right angle triangle** with the hypotenuse two inches long.

Notice that the figure features, the type of triangle, are shown in Bold Type. The irrelevant features are shown in regular type. The idea is that as the S learns the concept of Triangularity, the figure-ground elements of the test object will begin to separate into occupying the forward region of three-dimensional space, as we saw in Rubin's vase-faces figure.

Food, the reinforcing stimulus, will always be found under triangle A, which is randomly placed over one of the two food wells, left, and right.

- In the first phase, the subject learns to discriminate between concrete triangle A and any one of the concrete figures of different geometric types: circles, squares, and rectangles.
- In the second phase, all the concrete irrelevant features of the *figure,* the ground features of the first triangle, which were constant during the first phase, will be systematically varied, one at a time until the organism will eventually be able to recognize any of the seven types of triangles, independently of the ground features. The concrete features of the first triangle: color, size, and type of triangle were the first set of features, which defined the first concrete concept of the triangle but most of the features in this concrete phase were ground features. But in the second phase, each of the initial set of irrelevant control features, the ground features, were made independent variables until a concept of Triangularity was obtained. The goal in developing the concept of Triangularity is to teach the subject to distinguish the figure, A and the ground, B conditions.

  As the training progresses, the subject's mind should ignore the irrelevant ground, B features and only attend to the figure, A features, which are features of Triangularity, so that the figure's features will be seen in front of the ground features.

In step two, all the control variable features will, one at a time, be made into independent variables, which will destroy the causal role in their evolving figure model of Triangularity. For example, the hue or red feature of the triangle will be systematically varied so that the animal is taught that red is not an essential feature of a triangle. Additionally, in the meantime, the S is learning that the features, which are the essential elements of Triangularity, are the invariants. The invariant features of a triangle are those that do not change as different triangles are presented, and gradually over a long series of trials, the abstract concept, the features that are the invariant features of every exemplar of a class of concrete figures will emerge within the animal's nonverbal mind. Each such trial will have been one of a series of two-way discriminations.

Piaget

The features, which defined the initial concrete concept of the first triangle, consisted of two sets, the essential features, and the irrelevant features.

In order to teach a mammal to achieve or recognize the abstract concept of Triangularity, the individual must be trained to ignore the irrelevant features of every concrete triangle and pay attention to signal, to attend to the signal, and ignore the noise. This discrimination can be achieved by the process or procedure I call extraction. The signal is extracted out of the noise by systematically making each of the irrelevant variables of the figure into independent variables; which makes them irrelevant or no causal elements of the figure, thus making the critical causal elements of an abstract concept of Triangularity become independent of the initial set of concrete features of the first triangle.

5--Manipulating causality: VARYING the CONCOMITANT

The need to know causality

The human species has a vital need, as do the nonverbal species, for knowing the causal features of each hierarchically organized domain they occupy, for example the matrix maze. Historically wide varieties of means have been proposed for dealing with Causality:

- Aristotle (1952) recognized four kinds of empirical Causality.
- Plato (1952) was a dualist. He insisted that matter was unreliable, and therefore no causality could be found in matter.
- Statistical methods of many kinds have been identified, but only ANOVA provides valid evidence of causality. The ANOVA statistical tests, F-ratio, are used to certify which effects are significant, and at what level they are significant, before graphs are drawn.
- Before there were computer statistical packages, there were other ways to evaluate the causal nature of statistical functions--one was known as "varying the concomitant," which the British Philosopher John Stuart Mill (2002) made famous.

- In the plot of a graph, often the way the curves look allows an inference of causality to be made. The phrase, varying the Concomitant refers to whether or not "Y varies with X" on a graph this generally means a causal relation exists between Y and X.
- Today figures and statistics serve different purposes. Figures are useful for displaying the nature of the causal relationship that statistics identified as significant; they enable the Experimenter to understand the nature of significant relationships identified by a statistic and then seen in a graph.
- However, there is one more meaning for the phrase "Varying the Concomitant," the one I most use in this chapter.

The types of causal relationships that exist between IVs and DVs, varies with whether Y and X are both in the figure or one is in the figure and the other in the ground.

Varying the Concomitant

CCC theory or Causal Conceptual Conditioning is a theory at the intersection of three topics: Causality, Concepts, and Conditioning, yet for the moment I will argue that the theory starts as a theory of Causality, which is a concept that is probably as large as it is important.

The nonverbal mind is conceived as a conservative adaptive organ, and that the conservative consciousness of the NV Mind has successfully directed the behavior of mammals for over 50 million years because it was designed to enable mammals to act only on exemplars in consciousness, that earned their way into consciousness by being certified by the mind.

In this part of the chapter, however, the focus is upon another important way that causality can be manipulated by conceptual conditioning.

The title of this section is called "Varying the Concomitant." My purpose is to explain the meaning of this phrase. John Stuart Mill (2002), the philosopher made this method for identifying a causal relationship famous. The logic was as follows: If Y varies with X then there is a causal relationship between the two variables. The concomitant relation can have a positive and negative aspect.

"Whatever phenomenon varies in any manner whenever another phenomenon varies in some particular manner, is either a cause or effect of that phenomenon, or is connected with it through some fact of causation." (Mill, 2002, P. 287-288)

One of the major changes required in proposing a nonverbal theory of mind is making a giant leap from Homeostasis to Heterostasis, which I have noted before but without calling attention to one of the major differences between them.

Hull's concept of learning

In Hullian learning theory, the central factor that caused learning was called Reinforcement, Homeostatic need reduction, or need induction. (Hull,

1943) The maximum learning was determined by the intensity of the subject's need. A subject can only be in training as long as it is motivated, or un-satiated. Therefore, Skinner (1974) reduced the quantity of food reinforcement to a pellet smaller than an aspirin size tablet, which was so small that an animal would not soon get satiated by eating these pellets.

Heterostatic motivation can in many respects function much as it does in Homeostasis. Uncertainty, or ignorance in a closed grade, can on occasion be a finite measurable quantity, but generally, it would be folly to try to estimate how much ignorance is in the universe, however, much less is in a closed domain such as the matrix maze. However, I will show in Book 2 that one of my most useful conceptual tools was my discovery of the nature and role of one of the special Heterostatic dependent variables; The BFEU score, which makes a reasonable fair estimate of how much uncertainty a particular anomaly induced. Each anomaly or unexpected effect will introduce a finite degree of uncertainty; however, The BFEU score measures how much time and how many moves each rat required to find a cause that took the uncertainty out of the effect that induced the anomaly?

My bedroom

The method for identifying causality has a long history and many if not most people instinctively know how to use it when they need it for troubleshooting. For example, in my bedroom I have a wall plate with five light switches. How can anyone that understands concomitant variation map the five switches into the five sets of lights? The sets of lights are the outside light, the closet light, the overhead light, the wall plug light, and the hall light.

Troubleshooting requires that the function, the causal association of one switch and one light circuit be discovered at a time. First, let me note that all the lights are off when the switches are down. What we need to do is to use the logic of varying the concomitant and map the five switches into the five sets of lights. Number the five switches from left to right, with the left most switch being number 1, etc. thus matching the lights with the switches. Mill (2002) probably understood this logic to apply to variables that were a continuous function, but we are applying the logic to a set of binary causal circuits, which works just as well.

The matches are shown below:

1-outside light
2-overhead light
3-hall light
4-wall socket light
5-closet light

Each of these causal electrical circuits can be identified by varying the concomitant; turn switch #1 on and off a few times. Each time #1 is down the outside light will go off, and every time the outside light is on, the switch is in the up position. The two-state causal relationship is easily identified and verified by

Mill's logic for all five circuits. I will suggest that B and B may have employed this logic in training Sam and Miss Sam.

6-Two new Paradigms: Extraction and Abstraction

In order to understand the B and B procedure there are two paradigms in addition to the four from Pavlov's (1960) procedures that must be added, which add to the conceptual repertoire of CCC theory. The chief insight of the extraction phenomenon is that now with the model of causal conceptual conditioning, we can understand the role of both the figure and the ground in the causal control of signal-ground phenomenon.

Once we understand extraction in the B and B context, we can then understand the concept of abstraction of Triangularity, the training procedure of Cruickshank (1961), and Pavlov.

The creative function of the mind requires a major refocusing of consciousness, which is induced by every parameter change. The P-Function turns on consciousness.

- In the first step of conditioning, both the figure and the ground begin their mental life as concrete.
- In the second step, the E must begin the process of transforming the concrete figure and ground into abstract. This second step is accomplished by systematically making each of the concrete control variables into independent variables. In the first step of conditioning, the food was associated with all the features of the figure and the ground. This positive stage or the concomitant phase establishes causality, by establishing an invariant relationship between all the features of the figure, which at this first stage includes both the essential A, and nonessential B features of Triangularity. The process of extraction is a process requiring several steps, but the goal of the process is to destroy the causal power of the B features.

A-Kind one: Triangularity: Extraction of the figure

If we continue our discussion in the context of figure and ground, then there are two kinds of extraction: first, extraction of the figure, and second extraction of the ground. In both cases, **the causal features must be extracted from the control variables**, until in both cases the causal features of the figure and the causal features of the ground will have become invariant. The invariant features are those that did not change during change. Once the nonessential features have lost their original causal value because they proved by not being an invariant that they were not real; they thus lost their authority to cause or influence behavior.

How do we sample pairs of figures to train subjects to master the concept of Triangularity?

FIGUREo.4I

FIGURE o.4I shows four populations of stimuli, one in each quadrant that could be used on the WGTA, which is seen in FIGUREo.4H, to train a monkey to conceive the abstract concept of Triangularity.

The four quadrants will be referred to as:

- 1-Top Left

  The TL stimuli are the six types of triangular stimuli. Since all these are all red and roughly the same size, they are intended to represent the population of concrete triangle stimuli. These stimuli will always be reinforced, in the first phase of training.

- 2-Top Right

  The TR stimuli are the population of non-triangles. On each trial of testing on the WGTA, one stimulus from the TL and one from the TR will be paired. The food will be under one of the TOP left population. The point of the training is to teach the animal to discriminate the two populations.

- 3-Bottom Left

  The BL stimuli are all triangles, except these are one-step closer to being abstract, they are no longer red proving that the control feature color has become an independent variable, so that the color of each of these triangles can be varied, which destroys it concomitant relation with color and food.

- 3-Bottom Right

  The BR stimuli are all non-triangles, note that their color varies just as does the color of the BL population of triangles. It is important that the colors never match (red and red) between the BL and the BR test figures.

FIGUREo.4J

This Figure shows how to arrange pairs of test stimuli.

The two upper quadrant populations provide the object quality problem with pairs of figures, which covers the two food wells on the WGTA in the first phase of training. The quadrant on the left, supplies the six types of triangles which will always be reinforced, the positive member of the pairs. The quadrant on the right supplies the six types of non-triangles, which are never reinforced, the negative members of the pairs. The six figures in all four-quadrant sets are numbered in the center to provide a simple way for the E to select the pairs of stimuli to match up and present on each trial. With six figures of the two types, there are 36 ways that two sets of six objects can be sampled taking two at a time. Figureo.4J shows how to identify these 36 combinations. The only thing that is missing is a way to randomize the sequence with which these 36 combinations are presented.

The bottom row provides two sets of figures for testing in the final phase of training. Other sets of figures can be prepared to give each monkey more training samples.

The set of six figures in each of the four quadrants are a very limited set of training figures. First, they are planometric or two-dimensional figures. Some labs have hundreds of three- dimensional objects to be used for training. Three-dimensional objects can be different in size, shape, color, material (wood, metal, etc.).

They are also numbered so that pairs of figures can be drawn randomly by putting six numbers in two hats and can be selected by randomly sampling each hat.

In designing these populations of training figures for use on the WGTA, the most obvious feature to check and avoid are to design each pair of population stimuli to avoid the possibility of accidentally creating artificial sets of cues.

For example, our purpose is to train monkeys to pick a triangle out of the two figures that are displayed on the tabletop of the WGTA. One obvious artifact would be to color all the triangles red, the A set, and to color all the non-triangles, non-red, the B set. Then the monkey would simply have to discriminate Triangularity by simply learning to approach red, the A set are red.

The actual series of steps by which abstraction is accomplished is more instructive in the abstraction of the figure than in the ground. For example, the abstraction of the figure in the case of Triangularity requires first that the causal features never change. However all the ground features, those that were constant on the initial trial and are not essential features of a triangle, which includes color, size, or the material of which a physical three-dimensional triangle can be constructed, must be converted from control variables to Independent variables, which will destroy whatever causal value they had acquired. Over such a series of trials, the technology responsible for the Causal Conceptual Conditioning of the nonverbal mind will learn which features are invariant and which are not. The essential features of any concept are those features, which prove to be invariant or causal features. By manipulating the control variables, which are irrelevant, the causal features will become recognized by the mind's eye, and they will have been extracted from the concrete features of the figure, until only the abstract features of Triangularity, which are the causal features of the figure concept, are left standing. In that state, the figure elements of Triangularity will seem to be in front of the ground.

I also should mention that the study of this concept has a long history in psychology and rightly enough being called concept formation studies. (Andrew and Harlow, 1948)

B- Kind two: Sam and Miss Sam: extraction of the ground

The abstraction of the ground is a procedure, which is seldom if ever, practiced; however, it would provide an experimenter with some useful knowledge.

Space monkeys

The simplest way to state the problem is as follows. Of course, the monkeys are going to be trained on Earth, condition A, to perform in space, condition B. The transfer of training problem is concerned with what factor determined how well the training in condition A transfers to condition B. It has long been known that transfer of training is a function of similarity. In other words, if the conditions in space, B, are the same or similar to the training conditions on earth, A, then the monkey in space should perform as it did on earth.

The reason NASA was concerned and sought a team of professionals to train the primates was that they knew they were going to have transfer problems for the simple reason that the similarity between earth training conditions and the space performance conditions were going to be close to zero.

Extraction: The seven Step B and B Procedures

A completely new aspect of the transfer problem was opened up by space travel, which forced researchers to consider a set of potential problems that heretofore for practical or operational reasons, simply had never been considered by science. Of course, there were major kinds of dissimilarities that would exist between the earth training and the space testing conditions, but what can be done to enhance space performance. B and B ingeniously showed how the animals could be trained on earth to make up for all the dissimilarities. The normal theorists; the normal earth based scientist, had no operational capacity to study or to design a special procedure to manipulate each of the ground conditions to bring out their best; the space-age variable differences. Of, course B and B had the advantage of having the full resources of the U.S Air Force behind them. They were not only able to identify an entirely new set of variables that had to be manipulated to make their training a success in their space-age mission, but they had the resources to implement them.

Step one: Pavlov's Habituation

When one of the space-monkeys was first restrained in its chair and then put into the space capsule, everything was too anomalous and the monkey was too aroused to learn efficiently. So as Pavlov showed that in this state of arousal, the organism must be allowed to stay in this environment until it adapts or habituates to this state, which means its arousal level calms down.

In terms of the Yerkes-Dodson law, upon entering an anomalous domain the vertical arousal system becomes too aroused and moves the subject into the C region of the Y-D law. By simply leaving a subject alone in this domain, it will habituate and its arousal level will drop down into the B region of the Y-D law.

The space capsule served as the equivalent of Pavlov's special isolation chamber in which he habituated his dogs. Only after the monkey quit struggling did B and B begin the conditioning.

Step two: Aversive conditioning

In the B and B test procedures, the monkeys were trained to press a lever to avoid a shock administered through electrodes on their legs, the figure element in these procedures would be the CS-light and the US-shock elements. The ground element would be all the other objects and events inside the space capsule. All of these ground events, which are constant, would qualify as control variables or as parameters of the ground.

After the first round of aversive conditioning, the light and shock elements of the domain had become firmly established as the Figure component of the test conditions, while the chair and all the other objects in the capsule would have become a part of the ground. However, what is critical to understand in terms of the figure-ground metaphor is that the figure components would have become a domain-specific element of the test conditions, which means that the monkeys would respond appropriately to the light only in the presence of the same ground. In other words, the figure at this time would be a particular or a concrete figure.

Only after a monkey had habituated and perfected its model of the set of particular-ground parameters of the space module would the conditioning begin. The set of particular-figure variables is added, which are the set of light-shock causal relations, which can be defined as some operant aversive conditioning reinforcement schedule.

Sam was then conditioned by aversive conditioning to a light-shock contingency. He learned to avoid the shock by pressing a lever

The monkeys were motivated in this study by pain induced by administering shocks through the electrodes placed on their legs. This training procedure is called aversive conditioning. The only work the monkeys were asked to do in space was to press the lever soon after a light came on and before the shock was applied. Monkeys soon became most proficient at performing this task, and thus only rarely get a shock. The intensity of the shock could be controlled remotely by the experimenters so that even after the monkeys were in space the intensity of the shock could be controlled. However, the rule was to keep the shock level administered as low as possible as long as the subject was performing.

SKINNER PUNISHMENT

Skinner's (1974) brilliant study of punishment noted that punishment induces pain. The induction of pain always causes fear, which is always associated with the object that caused the pain. This means that since the monkeys were punished by a shock administered through the electrodes on their legs; they quickly learn to be afraid of being put in their chair and thence into the space chamber. Each monkey is handled by at least two strong men at all times, who put the monkeys into the chair and into the chamber, yet each day the monkeys became more combative about being placed into the chair. This behavior quickly became more and more dangerous, for monkeys are very strong for their size.

B and B solved this problem by having the monkeys anesthetized by the vets before they were brought into the psychology lab space. Additionally, the subjects were put in the chair and put inside the mockup space module while they were still asleep. Therefore, they woke up inside the test chamber without ever knowing how they got there. This procedure allows the subjects to wake up inside the chamber without having any fear associated with being put into the chamber each day. This procedure totally avoids the war that otherwise would be going on between the monkey and its trainers, with the fringe benefit that without any fear being introduced by the daily preparation, the whole training session moved more quickly.

The External Inhibition Phase, EI

One of the interesting innovations of B and B was that they extended the traditional aversive conditioning paradigm in ways that show that they tacitly recognized that the CS light itself was never the sole cause of the CR after the first phase of the training as Pavlov supposed. The CS light, after the first phase of training, was a necessary but not a sufficient condition for causing the animal to press the lever. B and B showed that each change in the environmental test procedures; first, produced **External Inhibition** and secondly required that the monkeys be retrained to press the lever to criterion all over again. This happened repeatedly, external inhibition was elicited every time a P- change was made in any one of the original environmental test conditions or parameters, which had to be followed by a retraining period. The physiology of External Inhibition, EI, is interesting for several reasons.

First, EI is a pulse of energy from one of the four vertical pacemaker pathways seen in FIGURE0.2A, specifically from Pathway 3+, which delivers a diffuse pulse of energy to the whole cortex. A pulse of energy in the brain is analogous to a pulse of energy in many physical domains: a water pulse, a tsunami; a geological pulse, an earthquake; an electrical pulse, a sunspot or the Aurora Borealis.

However, in this case this pulse has the effect seen in FIGURE0.2E in the upper left quadrant labeled N. The pulse has a diffuse desynchronizing effect, D+, upon the whole cortex. This pacemaker induces a high frequency low amplitude sine wave across the whole cortex. Analogies are dangerous, because EI is not like an earthquake of eight on the Richter scale; it does not rip out any physical structure in the brain. An EI event is a pulse that temporarily disrupts the cell assemblies, which house the program that is running in the cortex. A pulse destroys the partitioned structures within which programs run. Thus, EI according to Pavlov (1960) produces an immediate and absolute disruption of all the programs that are running in consciousness.

The exact effect upon the cortex, upon consciousness, and upon behavior depends upon the intensity of the pulse, which varies with the strength of the anomaly, or P. In the case of the training conditions used by B and B or in the case of transforming a concrete triangle into an abstract triangle, the pulse generally should be only strong enough to restore the arousal level of the cortex into the B zone of the Y-D law. Ideally, the strength of the anomaly should cause those logic centers, which record the causal value of each stimulus to note the logical implication of the fact that some causal stimulus lost its causal value by the fact of varying the concomitant.

The P-Change

Unfortunately, in this area of science the names are conventions. In FIGURE0.2E Gastaut labeled this arousal channel from the Brain Stem Reticular Formation responding to a Novel stimulus, N, which is the same as the stimulus I am calling P for parameter. However, by definition a Parameter P-change event must have the logical effect of causing a ground control variable, which had never changed to be processed as a causal concrete stimulus and to be made into an Independent Variable, IV. This effect is called varying the concomitant and has the effect of destroying the causal effect the control feature had been exerting.

Step three: classical conditioning

The classical behaviorist's assumption was that the only learning that had occurred was to the figure CS-US relation. In other words, they ignored the causal role of ground.

This classic assumption was the first assumption that had to be overturned. On the contrary what Brown and Blodgett showed was that the only way to insure that the CS would always cause the CR was **to extract the CS from the original set of particular-ground** conditions. The CS, the figure, must be made independent of the original ground. What the training procedures they invented did, in my words for they provided no words, was to extract the figure from the ground.

Step four concrete figure

The concrete skill stage: At this point in the training somehow Brown and Blodgett knew that Sam and Miss Sam had only a domain-specific knowledge of the light-shock causal relation or what will be called a concrete skill.

Step five- varying the concomitant

Abstraction/extraction training: One by one, B and B varied each of the control, P variables that were an element of the concrete ground training procedure, until the light-shock causal relation became independent of the ground stimuli, until the figure's causal relation had been extracted from the ground.

A-Chair

Say, one of the ground conditions, which were always constant, for example, the fact that the subject is tested sitting upright in a chair is varied, the capsule is turned upside down, but the figure condition is kept the same; now the monkey does not press the lever to the CS even though the figure similarity is 100%. The classical theories of transfer failed to consider the role of the ground condition, the g forces, the position of the body with respect to gravity, the sound level in the test environment, etc. B and B showed that the classical tests of transfer ignored the role of the ground stimuli in transfer of training tests. However, the point of the B and B experiment is not to challenge the conclusion of the classical transfer of training theories that transfer in a function of similarity. Rather, it is to show that the classical treatment of this law was too simplistic and that the law now must be expanded to include a large range of test conditions, which the classical theorist never recognized as a legal test of the law of transfer.

The learning of any skill, like the lever press in aversive conditioning, is not a flat phenomenon but is a nested phenomenon, and the point of this section is to distinguish how the law of transfer based on similarity works in nested test domains. I am not challenging the law, but only showing that the law works in ways that the classical theorist never thought to test.

B and B: concrete to abstract C-A

In the B and B procedure, NASA required Sam to be so well trained that he would be immune to EI. The animals were to be trained so well that it would be an abstract function, not a particular function. What B and B did was to invent a training procedure that took the mind all the way from a Concrete, C, concept, or a domain specific skill to that point where the skill was an Abstract, A, concept. They took all the control variables of the ground, all the P-variables, and converted them to independent variables, and they were no longer constants. Now Sam's mind was immune to change, and thus immune to EI and was incapable of being surprised. The mechanism of the mind had now come full circle.

Historically, evolution created the nonverbal mind at a place in the emerging cosmos when Nature needed a better way for life to obtain the critical knowledge it required to survive. In particular, nature needed a technology designed to deal with change, but one central feature of such a machine was that it could also, at times, when it was rational to do so, be immune to change: or come to recognize that there are different kinds of change. The invariant kind is the good kind and is the kind that doesn't change during change.

The C-A transformation: Concrete-Abstract

The transformation of a C-state mind into an A-state requires a broad series of training sessions, where each T-session involves at step one, the shifting of one ground parameter into a IV and patiently exposing an animal to some or all of the possible variable's states until it habituates to this new set of ground states.

As I will try to show, when B and B executed the C-A transformation for Sam and Miss Sam, the series of training sessions required days, weeks, and possibly months.

What this shift from concrete to abstract means is that a change with profound consequence for behavior has occurred; in particular, the transformation completely altered how the figure is causally controlled, since the figure will have been extracted from the ground!

Change one: At the concrete stage of training, the figure is subject to EI on every occasion of a P-change. In fact, each Context P-Change induces EI, which then requires another series of Habituation sessions.

The most significant transformation is that once the training has reached the abstract stage, the figure will no longer be susceptible to EI since the ground conditions will now be defined by variables not constants, as previously discussed.

C-Chair main abstraction

The first and most important insight of Band B was their insight into the opportunity provided by the nature of the test conditions; the fact of their training an animal for space flight mandated that the animal be confined within a chair. The conditions of the test almost forced B and B to take the steps they took, the monkey was in a chair, the chair was in Space Capsule, SC , and the SC was in the lab, and the whole thing just begged to go round and round.

The chair was confined within a mockup of a space capsule and the whole package, monkey and space hardware, were small and light enough to be lifted and manipulated by one animal handler. If the total package had weighed say a ton then today I probably would not be writing about the training procedures; they were possible at least in part because of the small size of the space capsule. This small size set the precedent for all the extinction steps that followed. The whole of the B and B procedure was made possible by the fact that the monkeys were strapped in their chair inside the space capsule.

The extraction procedure began when the animals within the space capsule were turned upside down the first time. EI occurred so the animals were retrained to work the problem in this second position, again to criterion.

Subsequently, the animals were trained to perform to criterion in all the possible positions of the capsule, on their back, face, sides, etc. and continued training until they could again avoid the shock under all these transformations. The space capsule was rotated in all possible combinations until the figure component of the situation became independent of the ground. Adaptation required that all the parameters, which defined the ground, had to be transformed into independent variables and extinguished. At that point, in the training when the Figure was no longer susceptible to EI, then the animal knew!

State-dependent to state-independent learning

Another interesting way of describing this C-A transformation is to say that in the C-state, the behavior was state-dependent learning and after the figure had become abstract, it could not be said to be state-independent learning.

How do we know when an organism knows? There are several interesting answers to this question, yet in this context, the answer is simple: a mammal knows some function X, when it can no longer be distracted by any anomaly, when it is free of External Inhibition, EI.

D-Sound levels

Then, each animal was trained to perform to criterion on several light levels within the capsule, and then several sound levels. The animals were then trained to perform to criterion in spite of the sound levels produced by a jet fighter during “take off” at one of the local Air Force bases.

E-weightlessness

The animals were taken to Ohio where they were trained to perform to criterion during the weightlessness of space flight. They were then trained to work under several levels of g-forces on a military flight at same military base. (Brown et al., 1959)

This set of super-training procedures produced animals with a performance skill, I dare say, greater than any ever achieved before. However, the space chimps and the human astronauts were trained to this level of performance or greater.

In effect, Extraction required that the subject learn to reach criterion once again after every possible P-change. Each time the position of the space capsule relative to gravity was changed, the local stability of the cortical matrix environment was destroyed within which the figure relation had been established. With the parameters changed, the animal was no longer able to respond for a while. However, after a series of these experiences with the direction of the G-force being varied for all possible conditions of the gravity variable, there eventually came a time when the monkey was able to perform the figure effect under every possible contingency of space.

The B and B procedure was characterized by a large original set of Ground or Parameter conditions. The large set of causal parameters, P conditions, which were extinguished were gravity, sound, light, G-force, etc.

Figure-ground

One of important aspects of the B and B training procedure was that they operationally recognized that the behavior of the organism consisted of two parts, which can be characterized as the figure and ground components of behavior.

Skinner (1974), Pavlov (1960), and all the other animal experimentalists recognized only one kind of extinction, which I will call the extinction of the figure effect. However, what the B and B procedures showed is that other kinds of extinction also can occur. "Extraction" or "extinction of the figure from the ground causes," expanded the conceptual repertoire of the transfer of training psychologists.

Heterostatic 3C (Causal Conceptual Conditioning) theory initially recognized the parameter variables, the ground parameters as the primary causes of behavior and the CS figure variables as the secondary causes of the conditioned response. The beauty of this extraction procedure is that by extinguishing all the primary ground causes of the behavior, the CS, the figure cause can then be transformed into the sole cause of the behavior.

Step Six: Concrete to Abstract C-A-- The girl.

External Inhibition EI then Habitation H

In this stage of training however, all the ground control variables are transformed into Independent variables, and as long as the monkey is awake, each time one of the control variables is converted into an IV then External Inhibition emerges. EI reenters the training sessions and the animal must now undergo habituation once again. At the end of this EI-H stage of training, all of the control variables will have been transformed into independent variables, IV.

The nonverbal mind evolves concepts as coherent sets of hierarchically organized conceptual systems. However, conceptual evolution can be studied two ways, first the evolution of hierarchically organized sets, or secondly, each concept or idea can evolve individually and independently in an experimental setting.

Some of these concepts are abstract, so I will take a moment here and illustrate these concepts with an example that may be more interesting; let's talk about the concept of a girl.

Girls: Domain specific concepts

Concrete concepts are designed to work in only one context. They offer only the lowest level of recognition for they categorize an object with a set of attributes some of which are valid only for the current context. On the other hand, an abstract concept enables an individual to recognize a class under any conditions. For example, as an abstract concept of the class "girls" are potentially recognizable by the nonverbal mind as an invariant, by a syndrome of features that does not change with each particular girl. An abstract concept of a girl would endow the nonverbal mind with a set of features, which are independent of particular girls.

What is a concept? A concept is a set of features that enables the sixth sense to recognize an object as a member of a class. Every real thing is an object that can be experienced by at least one of the senses, but ideally by all the senses combined.

On the first occasion of the nonverbal mind encountering an instance of some class, the likelihood is that many of those first features will not be the essential features of class, and thus the mind could not possibly conceive a valid abstract from the data provided by one instance of a class. Concrete concepts are conceived from patterns of precepts that are domain-specific, not abstract or domain general.

The purpose of the nonverbal mind is to enable life to adapt to change. Concrete concepts work in only one context, and so in order for the mind to adapt to change, concrete concepts are useful but for only one particular context.

B-Girl

Only an abstract concept allows the mind to recognize the same object as an exemplar of the same class under all the possible sets of conditions in which the object would ever be observed. For example, an abstract concept will recognize a girl under changing environmental conditions but also independent of the actions of any one girl. The sixth sense can evolve causal conceptual abstract concepts, which with the proper kind of experiences cannot be fooled by any changes as long as the test object contains the entire set of features essential by an invariant or abstract class.

The girl as an abstraction, an invariant

When the concept of a girl is an invariant, the girl is that which doesn't change in spite of all the changes of every kind that may affect the precept pattern entering into the mind. The primary mind can evolve an abstract set of causal attributes that can always recognize any girl as an exemplar of the set of objects called girl. However, all concepts begin to evolve in the primary mind as concrete domain-specific concepts.

NV Communication girl

The ideal abstract concept must be able, no matter what the pattern of stimulation produced by the movement of the girl, or whatever the point of view from the front, back, or side, be recognized as an invariant pattern that can always be recognized in any domain, which will allow the mind to recognize each girl.

The sixth sense creates consciousness, a field of awareness, which includes all the exemplars of all the ontological real classes of things within the field of action.

The nonverbal mind is the primary language machine. Nonverbal communication is the nonverbal mind interrogating the physical world with all the five senses after which the sixth sense then identifies all the invariants in the field. Each exemplar that passes all the checks of the sixth sense and enters consciousness next becomes a member of those things the mind certifies as real.

The conception cycle:

The creative or speciation function, Plato claimed, could not be executed by any being made of matter. Yet this book's origin was in my insight that the rats' spontaneous play-behavior showed that the matrix maze stirred some deep epistemological need of rats to conceive ideas or execute the speciation function. By recording and analyzing the behavior of the rats exploring the maze, it has been possible to evolve a theory of how the nonverbal mind conceives ideas or executes the speciation function and under what circumstances.

The nonverbal mind conceives ideas in a five- stage conceptualization cycle composed of: Habituation, Parameter change, External Inhibition, Consciousness, and Habituation.

The concept of a universal knowledge-creating cycle

Classical learning theorists believed and still believe that adaptation or learning in both animals and humans was driven by homeostatic deprivation; nevertheless, the experiments in this book show that in rat's learning is based on Heterostasis not homeostasis.

The creation of ideas

Historically learning theorists have tried to explain learning without evoking the concept of mind, and surprisingly, for over one hundred years, they have failed. Yet, they still are not ready to consider the possibility they failed because there is a mind.

However, the rats and mice in the four experiments in this book make it abundantly clear that rats, mice, and probably all mammals have a nonverbal mind, which creates coherent sets of idea-driven objects, which populate consciousness, and that induces them to move appropriately as specified by the scripts in which these objects are embedded.

The nonverbal mind creates concepts and exemplars of ideas. This idea-creating function, both the concept and exemplar creation, are the primary function of the nonverbal mind. The nonverbal mind is a mental speciation mechanism, which conceives ideas or sets of coherent hierarchically organized ideas by rational selection. By contrast, Darwin's evolution mechanism creates new biological species by the biological speciation mechanism of natural selection.

The mind emerged because it proved to have a better way to supply the knowledge mammals needed to survive in times of change or in conditions of change. The virtue of the nonverbal mind was that it enabled mammals to self-organize. Where does a self-organizing cycle begin? How does a nonverbal mind create sets of coherent hierarchically organized ideas? How does a biological machine create sets of ideas?

On the negative side, the primary creative function of the nonverbal mind begins with the discovery of an anomaly, which induces uncertainty. On the positive side, the creative function of the nonverbal mind creates ideas/ knowledge, which destroys doubt, ignorance, or uncertainty.

The creation of a virtual life space

The emergence of the nonverbal mind enabled mammals to create their own mental or conscious life spaces, which enabled them to move into anomalous niches and create a complete life space not designed by their genome. With the emergence of the nonverbal mind, mammals became the precursors to frontiersmen such as Daniel Boone and Davy Crocket, or even Star Trek who could enter regions of space where no man had gone before, and make a model and then populate that space.

What was the primary skill of frontiersmen? One of their roles in taming the wild frontier was to create mental maps or models, which could turn an (unknown x square miles of wilderness area) into a known life space. The primary role of the nonverbal mind has always been to turn the unknown into the known. It is interesting to note two kinds of frontiersmen, the low, and high technology types. Lewis and Clark were the latter kind. Boone and Crocket's technology was closer to F3 than to Lewis and Clark. Boone and Crocket did not create a physical map; they solely used their minds to lead people into the frontiers, while Lewis and Clark used technology like compasses and surveying equipment to create physical maps for others to use to find their way around these new regions.

Step 7-The last big theoretical issue: P-change

The last theoretical issue to get my attention was whether I could in fact make my desired theoretical argument that "all learning begins with a P-change." Two domains must be covered by my intended law: extrinsic and intrinsic rewards.

Ashby

One inspiration for the law: "all learning is triggered by a P-change" was Ashby's (1960) statement that "change can only come from change," and its everyday use is probably not obvious; how many humans go around each day asking themselves where change comes from?

Ashby's law is too broad for me to authenticate in general, but in science, I am confident of the law's usefulness if we revise the law thusly: **"Change in a theory or paradigm can only come from a change in the facts supporting the theory,"** which is what Kuhn (1962) called an anomaly. This will be illustrated in the next section with Galileo's astronomy experiment.

Kuhn: Galileo

If we borrow a page from Kuhn (1962), it is obvious that new paradigms owe their emergence to an anomaly, which is an event that induces surprise, and in science demands that our old invalid paradigm, which allowed the anomaly to emerge, be retired and replaced with a new paradigm. For example, Galileo's discovery of the moons of Jupiter invalidated Aristotle's model of the universe and logically showed that a new paradigm that could assimilate the new knowledge had to be designed to take its place. The new facts about Jupiter were a change of the facts about astronomy that required a change in our model of astronomy. The change in the paradigm had its origin in the change of the facts. The P-change in paradigm A was caused by the P-change in the facts, B; or change B caused change A.

Mill: B and B

In order to change a monkey's behavior, something about the variables that controlled its behavior had to change. Consider an example closer to home. Consider Mill's (2002) "vary the concomitant law." In order to change the causal effect of the concrete ground variables in the B and B study, it was necessary to change the causal control variable, which by definition never changed into an independent variable. Then by varying the concomitant relation between the concrete ground conditions, we caused the causal effect of these IVs to be destroyed. If we had not changed the concomitant relation that existed between the CS and ground control variable then the control variables would have retained their causal influences over the monkey's space behavior.

Christian beliefs

Or, if we go a bit further afield, why do Christian parents not want science taught to their children? They know that change comes from change, but if their children never hear anything but the "old, old story"; if they never hear the facts of science, then the Christian "facts" taught by the parents will continue to have causal control over their children's beliefs or paradigm. The only thing that can change a child's Christian belief paradigm is exposure to the reproducible facts of science. Ashby's (1960) law "Change can only come from a change" is a fact of which many are tacitly aware. If these parents can prevent their children from being exposed to the facts of science then their Christian beliefs will be safe.

The extrinsic reward domains

B and B's training was 100% successful, both animals performed with distinction in space. Miss Sam failed to respond for only a few seconds while the capsule was under an extreme level, 20 g's, which far exceeded what had been programmed. Her behavior proved the success of the B and B training procedure; she performed 100% of the time under conditions where it was physically possible for her to move her arms. She failed to respond only when the g-force made it physically impossible for her to move.

However, the motivation for these two monkeys was extrinsic motivation; both animals were trained to work under aversive conditions with electrodes on their legs. The motivation for these animals was extrinsic, i.e. fear of pain or shock, which is a most stable form of motivation. Intrinsic motivation is much too labile to work under these conditions.

The B and B procedures provided an excellent testing paradigm to enable the extraction of the figure, the light, from the causal control variable of the ground. However, every time one of the ground features was changed, which was a P-change; it was followed by External Inhibition, EI, and then was followed by Habituation.

Meaning, metaphors, and consciousness

FEMALE 3

The first and primary act of communication I call nonverbal communication, which in the nonverbal mammals is a matter of mammals interacting with nature, the material cosmos, or the material god. For example, Female Three in the matrix maze achieved her goal of growing the knowledge-based skill of running the shortest path in the shortest time by probing nature and forcing nature to give her the "nonverbal knowledge/ ideas" she required to create a valid model of the maze. The model of the maze contains the Mammal-Nature, M-N knowledge that allowed her to create a plan, which allowed her to run the maze without error.

When Female three was placed into the anomalous hierarchically organized matrix maze, she was forced to communicate or probe the maze; there was no one to ask for directions. Google hadn't been invented. She had no choice but to interrogate nature. In Mammal to Nature communication, a mammal must interrogate the nature of the spatial domain, in this case the matrix maze, in order to create a set of ideas that model the matrix space.

Her problem, which involves communicating with nonverbal physical nature or Spinoza's pantheistic god, does not consist of rationally selecting from a closed set of English alphabetic characters that reduce uncertainty, which is an example of the H-H communication or a language function.

Female Three's problem is an example of exercising the nonverbal speciation function, which is a creative act that her nonverbal mind must accomplish. It is a task so improbable that some philosopher would classify it as the equivalent of turning water into wine or perhaps something even more exotic.

The nonverbal mind, when exploring the matrix maze, must interact with the maze nonverbally, which means the nonverbal mind must do three things.

- First on the behavioral side, the nonverbal mind must cause the somatic motor system to emit a pattern of rational behavior, which moves the rat skillfully through the maze that exhibits Heterostatic behavior: seeking and destroying anomalies.

- Second on the rational side, the pattern of behavior, which moved the rat through the maze, consisted of a complex mix of behaviors. These behaviors included probing; running up and down; and running back and forth within the rows of the maze, exploring and retracing. All these behaviors or somatic responses cause a vast stream of neural impulses from the five senses to flood into the sixth sense of the nonverbal mind

- Third, the nonverbal mind must now encode, recode, and transform the unique pattern of behavior, which consisted of four unique sets of responses, which the nonverbal mind transformed into four ideas. At the outset of the exploration of the maze, the rat had two subsets of behaviors, ideas, and knowledge: A and B.

- A, an old set of ideas/knowledge and responses that had been constructed and used to solve earlier problems

- B, a new unique set of **ideas**/knowledge and responses that must be designed and emitted in response to objects in virtual space, objects that are the creation of the 12 station nonverbal mind; objects that embody the unique spatial features of the maze. The matrix maze is a complex hierarchically organized spatial structure, yet Female 3 invented the four new **ideas**, which together enabled her to understand, to see the hierarchical structure of the maze in less than thirty minutes.

- Female three's nonverbal mind identified the four features which are the invariants and the essential features of the maze, which make the maze a reproducible spatial unit. Female three first discovered these features as four **anomalies** she encountered on the way through the maze. The task of the speciation function of the nonverbal mind was to discover a rational, lawfully predictable, causal way around these four **anomalous** places within the maze. The nonverbal mind was capable of **turning these four anomalies into four ideas,** from which she constructed a model of the maze, from which she can construct a plan, etc.

- The critical insight that we humans must achieve if we are to understand F3's behavior and ultimately our own is that her nonverbal mind didn't reduce the uncertainty in the maze by creating information, but by creating **knowledge/ideas**. The Bell Labs mathematicians defined i**nformation as that which reduced uncertainty,** which worked for them. (Shannon, 1951) But F3 was not a Bell Lab mathematician, and her nonverbal mind replaced information with ideas. She defined **ideas/knowledge as that which destroys uncertainty.**

In the above section, I proposed a theory of how a nonverbal mind conceives ideas or knowledge. Specifically, I propose that ideas are created by a new brain/mind physiology. Heterostasis consists of a technology that creates ideas by employing a circular or four-step cyclical function.

The universal rule for all mammalian explorers

Ashby's (1960) rule is still working: "Change can only come from change."

Mammals from F3 to frontiersmen have been exploring the unknown, going where no mammal had gone before.

Rats exploring the matrix maze are exploring much like Crocket and Boone explored the frontiers. Both rats and their human counterparts evolved a nonverbal map or model of an anomalous domain, which is a causal conceptual system that as it grows allowed both kinds of mammals, rats, and humans, to interact with an ever-larger knowledge-based percentage of their respective domains.

The rats cannot experience units of the matrix that their conceptual system cannot represent as a part of a system of vectors. The capacity of the primary mind to conceive causal concepts is the limiting factor for experience. Experience is not a passive phenomenon but an active one. An organism cannot experience any complex spatial domain without having first created the concepts or forms that allow the mind to predict and assimilate each object.

The same is true of all domains; a nonverbal mind can only communicate with an object occurring within that portion of space/time, for which the nonverbal mind has a complete and valid conceptual system, which enables a mammal to know and not to be surprised

The primary mind organizes and represents things in space or time. The behaviorist would have us believe that we see something in space and time and then we recognize it. That is backwards. We cannot see something in space and time until our primary mind can represent or can assimilate it by matching its form. What we see in space and time is a product of our primary mind and cannot materialize until the appropriate conceptual symbols have evolved. Perceptions are the primary symbols of our primary mind, which represent things in space and time. This is a nonverbal function. We have no memory of the first few years of life because the nonconventional way our nonverbal mind conceptualizes the space and the objects that occupy those spaces that we employed as a child were not the same we employ at adults. Those that enable us to know as adults are completely different from the nonverbal skills we used as a child. We start life nonverbal and only after a few years of nonverbal knowing do we switch over to the adult form of language. As a child, we obviously had not perfected the requisite skill employed by adults. When I was in the first grade, I thought as a child not as an adult. It takes perhaps 15 years to master all the grades of any domain. When we were young, we had not yet mastered the forms that are the conventions of the adult language community. The things experienced by children are experienced in non-conventional forms. Only after conventional conceptual systems have evolved can experiences be remembered in forms that will be recognizable to our later adult self. When experiences are not conceived in conventional forms, they cannot later be retrieved for the old forms, which created those memories are no longer available for retrieval. Memory mechanisms can only retrieve memories that were stored with the forms that are now in use. Until stable conceptual systems evolve, stable memories are impossible.

Metaphors: two kinds: nonverbal and verbal

Exemplars: two kinds; physical and mental

Unfortunately, there are two kinds of metaphors, the nonverbal, and the verbal. Nothing could be more important than understanding the differences between the two.

Additionally, the problem is that there are also two kinds of exemplars: the physical and the mental. The rose we hold in our hand is a physical exemplar of a Mr. Lincoln rose, a physical rose. Physical species are only species if they reproduce true, and that is true for birds, roses, humans, and for every biological or mental kind of species.

The law holds for mental species as well. What is the image of the rose I see in my hand? First, the rose I hold in my hand is physical, and I can see no exemplar of any physical thing until my nonverbal mind provides a valid mental image. A mental exemplar will make anything become visible. A physical exemplar only seems to become visible when the nonverbal mind drops on top of some physical object a valid image.

As each class of object becomes a known class, they acquire a nonverbal metaphor that gives meaning to the objects of the physical world. The physical rose in our hand is an invariant example, an exemplar of a physical rose. Year after year, the physical Mr. Lincoln rose on the bush is an invariant physical example of the class. But the image of the mental rose we experience in consciousness, B, is an exemplar of the mental class, B, which is a data- driven exemplar. B is also the meaning of C, the word rose.

Verbal metaphors, however, such as god the father, are metaphors that are not data-driven images created by the senses. Verbal metaphors are created by the logical skills of the mind.

Moreover, a verbal God is not something we can see before our eyes. When we think of a verbal god, this is the product of the mind's eye; it does not have its origin in the five senses.

But by way of contrast, every nonverbal image is a data- driven image of some physical object in Station A.

Spirits, in general and spirit gods in particular, do not stimulate any of the five senses. The god of Christianity is not a thing with a physical nature, as a spirit, it has no physical nature that can excite the five senses etc.

The sixth sense creates all the nonverbal images we experience as objects in our life space. The image "exemplars" are created by the nonverbal mind. The sixth sense does not create a nonverbal exemplar without having a syndrome of sensations, which justify the exemplar image. Each nonverbal metaphor is certified, but no verbal metaphor is certified.

The primary mind conceives concepts, which are more than recognition mechanisms. They are the necessary causes of our experiences of things. What we experience in the nonverbal consciousness are exemplars of our conceptual systems, which are our knowing systems. They are the primary metaphors of the mind, the nonverbal metaphors of the mind, but they must not be confused with the verbal metaphors. When I see a car ahead of me on the highway, what I am seeing is not the car but an image or the representation but the exemplar of some model of car. Yet the detail of the car I see only has the detail, no more or less, than the detail of my concept of cars permits. I see no more detail of that car than I hear of music. The music I hear and the car I see only embodies the features that exist in the conceptual system, which my degree of expertise in the two fields reflects. A conceptual system determines the level of expertise with which I can see the car, hear a piece of music, or recognize a bird, etc.

The mind is a not creator of physical nature, but its conceptual capacity represents our conceptual pool from which the exemplars that enter consciousness are drawn. The image objects we experience in consciousness are simply a manifestation of the conceptual structure created by the mind; they are not nativistic concepts as Kant proposed. The primary mind's creative capacity places limits upon the features observed in consciousness. There are profound differences between the physical and our mental world, the one within consciousness. The physical world is what is independent of our awareness of it; it is independent of us. Yet the world in consciousness is a subjective world, and our experiences of the events in any domain are limited by our knowledge of that domain. My ability to enjoy tennis and golf is perhaps even more limited than my ability or inability to hear music.

Five examples of extraction

The same technology that B and B employed to train Sam and Miss Sam also works on a wide variety of other conditions.

1-The abstraction of light

There is one possibility that Brown and Blodgett over looked. While they made the light the CS for pain, which motivated the animal, the figure in this situation was the light-shock contingency. The light used was a concrete light. While they were pragmatically justified in not having taken the time to abstract the light, it is still important to recognize that it is always a possibility to make any stimulus abstract. The abstraction of the light would simply require that the concrete feature of the light generated by the control variables be varied by converting them into independent variables. So with an abstract light, the CS light could be any light.

The theory of causality conditioning suggests operational test procedures that were never before recognized. Heterostasis is the study of the variables that are and always have been the cause of learning, but because these heterostatic variables were confounded with the classical Homeostatic variables, no one was ever able to see the Heterostatic variables. My task is to call attention to these causal but invisible variables that were there all the time.

2-The evolution of the concept of food in the Life space of rats

The Homeostatic model

Hull (1943) advanced a "Goal Gradient" theory to explain maze learning. The behaviorists divided an experiment into several phases: deprivation, habituation, learning, extinction, etc. In first phase, a deprived rat moved from the start box to the goal box randomly on the first trial. In the next phase, in the goal box only after the rat was reinforced would its behavior begin to be organized; the random behavior was gradually transformed into nonrandom behavior. According to Hull's Goal-Gradient theory, learning grows from the goal box back to the start box. This theory predicted that the rat's rate of motion within the maze would get faster the closer they got to the goal box.

According to Hull (1943), learning is not something that is constantly happening, but only happens at those moments when a homeostatically deprived organism eats food or in the maze when the rat is reinforced in the goal box.

Hull was famous as a learning theorist because he was the first to make Cannon's (1932) concept of homeostasis into the primary component of his theory of learning, his Reinforcement theory of learning.

From my point of view, there are at least three things wrong with Hull's theory:

- First, the idea that in learning that a change in behavior can only occur at the exact moment a subject receives reinforcement
- Second, the above idea conceives of the rat as passive with no intrinsic motivation.
- Third, Heterostasis assumes that the rat is active, that learning is always happening, time wise, and place wise.

The TTP concept of food

In running rats, there are two concepts of food that must be distinguished; first the model of food that the rat evolved in its home cage when food was at first freely available and second, the experimental deprivation schedule of food. In running rats using the TTP (Trial Testing Procedures), it may take up to a month to get rats prepared for learning to run the maze.

Phase One: Home-cage motivation

The concept of food evolved by the rat in the home cage is a domain specific, concrete, particular figure within the ground of the control variables of its home cage.

Phase one-Heterostasis is the nonverbal wisdom of the EMm and is "on" as soon as a subject opens its eyes in the morning and consciousness begins. All rats' minds once they are awake and consciousness is turned on are always creating dynamic nonverbal images, which constitute a narrative representation in whatever domain they find themselves. Obviously, the mind of the ninety-day-old rat that was born and raised in the home cage domain has created a model of its home cage. The rat has made a model, which represents every causal aspect of the probability space that is its home domain. This model is the core element of its life space. (Lewin, 1943) Yet let me be clear, where I disagree with Hull is that I assume the nonverbal mind of the rat is constantly making a model of whatever is going on; creating a series of images, which are experienced in/as consciousness whenever the organism is awake. This image-based consciousness is always on and records events independently of any of those events, which Hull classified as essential: homeostatic deprivation and reinforcement.

Phase two: Deprivation training

The first stage of training requires getting the rat ready to run the maze in terms of motivation. There are several ways to deprive the rat and get it ready to run. For example in phase two, put the rat on a deprivation schedule in which food is available in the home cage for only one hour a day from noon to one PM. The rat is a nocturnal creature and food has always heretofore been freely available during its waking night hours. Now food is no longer available during its normal waking night hours. This is an anomalous event and requires accommodation. The rat must change its whole circadian rhythm; it must learn to stay awake during the feeding hour and furthermore it must learn to eat its full day's supply of food in one hour! This is not an easy task for the traditional Purina pellets are hard and it requires a lot of energy to eat a day's supply of pellets in one hour, particularly when this is the middle of the "night" for the rat. After a few weeks of training, rats can adapt to this schedule; they can learn to eat their day's ration in one hour each day.

Phase Three: Goal box motivation

Next, in phase three, the rat is run in the maze and food is now found only in the goal box. This is where I first begin to become aware that something was seriously wrong with the classical learning theory. In the very first study I ran on the Lashley III maze, some of the students testing the animals reported that on the first trials, the rats did not eat the food in the goal box. This fact was to me, an anomaly, as I would now call it. It induced uncertainty in me, which asked at least two different kinds of questions? Why did the rat not eat? How can rats learn when they didn't eat? Now these animals were hungry, I know both how to make rats hungry and how to test for hunger. My standard test is to offer a rat that has completed its deprivation-training period, a small Skinner sized food pellet on a pair of tongs through the wire front of the cage. A rat is ready to run when it eats the food so fast that I cannot see it disappear. Yet when the rat runs through the Lashley III maze and gets to the goal box, it may be several days before that rat will even begin to nibble on the food.

The solution to our enigma is now obvious. However, it is essential to distinguish between two concepts of food: the home cage concept, and the special way that motivation was operationally defined in the TTP motivational maze concept of food.

The Behaviorist assumes food, for the rat, is an unconditioned stimulus; food is innately a transituational reinforcer. If these assumptions were true, then it would not require a month or more of deprivation before rats are run on the TTP.

Consider these facts. A mammal that never experienced food in any other context than its home cage has not had an opportunity to evolve a domain general concept of food: in short, based on the rat's traditional training procedures it is simply not capable of recognizing food in the goal box.

My first theoretical problem based on my first anomaly was this: If rats with only a concrete or domain-specific concept of food they acquired in their home cage do not or cannot even recognize food in the goal box, then why are they learning the maze? It became obvious to me in the early sixties that something was seriously wrong with the textbook reinforcement-learning theories. However, it took me over thirty years to develop an understanding of how Heterostasis learning works.

One of the difficulties humans have doing animal research is that even though they demean the mental powers of the rat, yet at the same time they find it impossible to understand how different our experiences are from that of a rat nor for that matter to understand how similar they are to ours. As adult humans, while our concept of food is domain general and abstract, yet most of us are not conscious of these facts, and so we naively assume that rat's concept of food is, like our own, also general and abstract.

Grade one-domain specific, concrete

Heterostasis, however, begins with the assumption that the NV Mind of animals and humans both evolve their initial concept of a thing, in this case, food, as domain specific, and concrete. By the time humans are adults, they have naturally had their original domain specific, concrete concept of food transformed into a domain general abstract concept without ever being consciously aware of the transformation occurring. Humans are without any awareness of how limited was the nature of our first domain specific concept of food, nor the general powerful nature of our adult domain general abstract concept of food. In fact, most humans are not even aware that they have something called a concept. When Pavlov (1960) called food an unconditioned stimulus, he was talking as an ordinary naive human not as an expert on the topic of concepts. Pavlov was naively attributing to his dog the same type of conceptual sophistication that he had because he was not aware of the fact that even he had this mental creation called an abstract concept.

Scientific Psychological advancement, truly, is a function of the rate of consciousness rising. Psychologists who believe in mindlessness are not aware of either their own mind or the rat's mind. One of the first things the NV Mind does is to create models, conceptual systems, or taxonomies of objects and events in a domain, which allows all the objects and events of the domain to be recognized. The behaviorist, who did not even recognize the mind, neither the human verbal nor the prehuman nonverbal mind, obviously also did not recognize the products of the NV Mind we call the concepts of the mind. The Behaviorists, and even Pavlov, tacitly believed that every stimulus was automatically abstract and domain general. They believed each animal was capable of recognizing any edible object as food at any time and at any place.

Heterostasis is a theory that there is a mind and that the organism's capacity to experience an object must grow from an initial domain specific concrete to a domain general abstract recognition conceptual mechanism. Initially the mammalian concept of food is neither domain general nor abstract. The NV Mind evolves sets of coherent domain specific concepts that enable mammals to function as self-organizing systems. The nonverbal mammals are not born with advanced concepts that are general or abstract; advanced concepts are not nativistic experience--they are the mental tools that account for the "God-function." Sets of coherent concepts are not given but rather, evolved over time as a function of a heterostatic experience.

Nutrition

The domain general abstract things we adult humans experience when we experience the pellets that the rat eats is an experience very different from what the rat experiences when it eats a pellet. Food, for example, to humans primarily involves the idea of nutrition, several food groups, chewing, saliva, swallowing, peristaltic movements of the stomach, weight gain, starvation, experience, the production and distribution, buying and selling, and preparation of food, Etc. The rat that has spent its whole life confined within its home cage with Purina pellets available on the front of its home cage could never have evolved an abstract concept of food. The rat's concept of food is very narrowly focused on the concrete features of the pellet; the immediate effect it can have on its senses, and the feedback generated by its interaction with this object. Yet all experiences, which collectively could enable the NV Mind to evolve an abstract concept of food, are beyond the possible experience of a rat confined for life within a small wire cage.

The primary mind must create just the concrete domain-specific concepts of food objects that will allow the organism to interact with that object in that particular domain, and that is all. The concept of an object will be limited by and specific for the somatic purposes, the specific kind of actions or interaction that the organism has with that object in that domain.

3-Paper: concepts are partial and concrete.

The pragmatists claim that all concepts are partial. This was one of William James's (1950) major points. Consider the thing called paper. Show a student a piece of paper and ask what they can do with it. Most of students, say, "Something to write on." This is what James predicts. An organism's concept of a thing is determined by how it interacts with that thing and differs in every domain; the concept of anything is concrete, domain specific, and partial. Students write on paper. A student's concept of paper is determined by what they do with it in the classroom domain. And yet, students are able in different contexts: to burn, to cut, to color, to glue, to fold, to wrap, and to smoke paper, but when confronted with a list of other common domain specific concepts of paper they are surprised. As we change domains, our concept of paper automatically and unobtrusively changes to whatever domain specific concept is appropriate for that domain. The interesting thing is that humans may very well never develop an abstract concept of paper. They simply have, perhaps twenty-five or more, domain specific concepts that they use one at a time in different domains and never consciously ever think about paper as a single thing with multiple uses.

James (1950) noted that every one of these partial concepts of paper is unfair to paper, but to a pragmatist that is no problem, for the mind creates concepts for the sake of action, and actions are always domain specific. As long as our partial concept gets the job done, who is there to complain if these working concepts are unfair.

4-Abstraction: What is a goal box?

What is a goal box? It is where reinforcement is found in the maze, a very domain-specific, species-specific, etc. concept.

For adult humans, the goal box is a domain general abstract concept. We all know what a goal is. A goal is an abstract domain-general concept: abstract because a goal can be anything an organism seeks and domain general because a goal is a goal in any domain. The goals in sports and business are domain specific in each case, but each of these specific goals can be recognized as members of the goal set.

Now the question is can a rat have a concept of a goal box? Certainly not, rats don't normally have the concept of a goal, not because they are incapable of conceiving this concept, but simply they probably never have occasion for evolving this pragmatic concept from a series of goal experiences. Yet the following procedure should allow a rat to evolve a concept of a goal box.

Stage I. Run the rat to food in the goal box. The Goal box will, however, become in stage one only a pellet box.

Stage II. Run a rat to water.

Stage III. Run the rat randomly to either food or water. Now the goal box will be either food or water box. This is the first step in evolving the abstract concept of a Goal box in rats.

What is the operational test of an abstract concept?

If the rat is deprived of successively different things and the rat learns that the goal will always be found in the somewhere, X, in the maze, then X is an abstract goal box.

All mammals from rats to Plato are rational beings whose behavior is a function of their respective species-specific conceptual systems. However, Plato imagined his secondary or verbal concepts were eternal and perfect Ideals; rats function with primary concepts that are pragmatic, concrete domain-specific types that I call Pragma. Pragma concepts are rough and ready things, whose only virtue is that they get the job done.

Plato was incapable of imagining that ideal concepts, fit only for the use of a philosopher king, could have evolved; could have begun life as pragma. Our thesis is that all concepts evolve; all are created as elements of concrete domain-specific conceptual systems. Plato was an elitist, while the rat is a pragmatist.

5-The three bucket of water model

John Locke's problem

It is interesting to realize that John Locke (1914) one of the three famous British empiricists, Locke, Berkeley and Hume, who began to defend matter against the attack of philosophers like Plato, recognized that matter had problems. Locke was the first to describe one of the best empirical examples of what Plato (1952) characterized as the "appearance" nature of matter in what I call the three-bucket of water model.

"Ideas being thus distinguished and understood, we may be able to give an account how the same water, at the same time, may produce the idea of cold by one hand and of heat by the other: whereas it is impossible that the same water, if those ideas were really in it, should at the same time be both hot and cold. For, if we imagine warmth, as it is in our hands, to be nothing but a certain sort and degree of motion in the minute particles of our nerves or animal spirits, we may understand how it is possible that the same water may, at the same time, produce the sensations of heat in one hand and cold in the other; which yet figure never does, that never producing the idea of a square by one hand which has produced the idea of a globe by another. But if the sensation of heat and cold be nothing but the increase or diminution of the motion of the minute parts of our bodies, caused by the corpuscles of any other body, it is easy to be understood, if a body be applied to the two hands, which has in its minute particles a greater motion than in those of one of the hands, and a less than in those of the other, it will increase the motion of the one hand and lessen it in the other; and so cause the different sensations of heat and cold that depend thereon." (Locke, 1914, P. 177)

Plato (1952) hated matter with a passion no less than a racist hates some races. Plato's whole philosophy pivoted upon his hatred of matter. Why Plato was a nativist and a rationalist and not an empiricist can be explained by three words;

Plato hated matter. On the positive side, Plato's passion was to know what was real. One thing Plato believed he knew: matter was not real because matter could only be known empirically. Guess what; Plato hated empiricism. Matter simply was not real; empiricism simply did not work. Matter empirically proved it only gave "appearances" matter never seemed to give the answers that Plato believed were the right answers.

Plato dealt with this problem many times and many ways. Earlier, I mentioned his famous or infamous slave in the cave metaphor. Generally, Plato explained his contempt for the nature problem with matter empirically, by saying that matter is not real, for all things made of matter die or fade away, even mountains and rocks.

Plato (1952) hated matter so intensely that his criticisms of matter were amateurish. However, in the quote above from John Locke (1914), the first modern empiricist, we finally find what could have been the first serious rational empirical indictment of matter, which I call the "the three buckets of water model."

Locke was an empiricist, not an experimentalist, so in his description of his problem with matter he did not attempt to include a method section that would allow anyone operationally to replicate his experiment.

Adaptation Level Theory, AL Theory.

One of the most impressive technologies with which the nonverbal mind is armed was first described by Harry Helson (1964) who called his theory, Adaptation Level Theory, AL Theory.

While this theory is seldom discussed, yet it plays a ubiquitous role in the behavior of the nonverbal mind. Two examples will be described to give the reader a taste of the role this technology plays. The first technology demonstrates how AL theory explains Locke's problem, the three bucket of water problem.

While Locke (1914) was an empiricist, he was not blindly so, and his acknowledgement of the three bucket of water problem is a testament to his capacity to objectively deal with facts. Yet the relevant physiological and psychological facts did not appear for several hundred years until Helson (1964) developed his AL Theory. With FIGURE0.4K below, I try to show how Helson marshaled the relevant facts and the theory.

FIGURE0.4K

FIGURE0.4K illustrates how Helson's AL Theory, explains Locke's Three Bucket of Water problem. This Figure shows three scales, each with a unique relative zero. Scale 01 on the left in the Figure is a large temperature scale that goes from -50 to 100 degrees, and provides a reference for the other two smaller scales. I will use Scale # 01 to discuss what happens in the other two smaller scales, Scale 0.02 and Scale 0.03 which are shown on the right in the figure.

Scale 0.02 is the scale that explains the temperature behavior of the left hand and scale 0.03 explains the temperature behavior of the right hand.

METHOD SECTION

Imagine that we are doing an experiment on the behavior of the Temperature Receptors of both hands. Below are instructions to follow.

1. Take three buckets of water: 1, 2, and 3.
2. Let the temperature of the water in the three buckets be 40, 70, and 100 degrees Fahrenheit respectively. (Please note: This is a hypothetical experiment and the temperatures chosen were purely for mathematical convenience in calculating the differences in Adaptation level, other temperatures could be used in an actual experiment.)
3. Put your left hand in bucket one, the right hand in bucket 3 and leave them long enough for the temperature receptors in the skin of your hands to adapt to the temperatures of the two buckets of water.
4. Next, take both hands out of their respective buckets, and plunge both of them into bucket #2 at the same time. The question is how will the two hands experience the 70-degree water of bucket two?
5. Results: the left hand, which was adapted to 40 degrees, will report the temperature of the middle bucket "warm" or "hot," while the right hand will report the water in the middle bucket "cold."

Adaptation Level Theory

Each of the skin temperature receptors adapts to the temperature of its ambient environment.

At any moment, the mean temperature of the skin receptors equals relative zero.

The strength of every stimulus is measured relative to the Adaptation Level (AL) of the measuring receptor or is judged by a receptor adapted to some temperature, which is zero or the Adaptation Level (AL). (Helson, 1964)

First, the temperature receptors in the skin passively adapt to the temperature of the ambient environment by assuming the temperature of their environment.

Second, the receptors in the skin each measure the temperature of their environment or the things in their environment by taking the difference, D, between their own AL temperature and the temperature of the objects in the environment. What the brain experiences is the difference or D. What do the skin receptors tell the brain? The knowledge the temperature afferent nerves communicates to the brain is D.

Third, the most important physiological fact is that the temperatures of the receptors themselves always establish a zero point, a relative, or floating zero point. Every nonverbal message by which the temperature receptors inform the brain is always and only a D.

The receptors adapt to the ambient environment and their momentary temperature, whatever it is, is always for the purposes of reporting to the brain, zero. The receptors, in other words, which measure some condition in their environment always measure starting at their own temperature, which is a relative zero, or a dynamic or moving zero, or a floating zero.

The implication of AL theory for the Three bucket of Water

model

In other words, after the two hand's receptors had adapted at 40 degrees and 100 degrees respectively, then the left hand whose temperature receptors are adapted to 40 degrees, which means the temperature receptors in the left hand have established a new floating or relative zero value now at 40 degrees. Scale one on the lower right shows that when the temperature receptors of the left hand adapted to the 40 degree water of the left bucket, the zero AL temperature adapted to zero. When the left hand is now plunged into the 70° water and its receptors are adapted to 40 degrees, which is now the new zero point, then the left hand will tell the mind the appearance of the skin temperature of the water. The temperature which Plato called the "appearance" simply reflected the difference between the 70° water and the 40° temperature of the AL level of the skin receptors. The D the skin receptors will report to the brain is D = 0 + 30, or simply +30. On the other hand, the right hand's receptors will report to the nonverbal mind that the skin temperature is 0° - 30° = -30°

The left hand will experience the water as hot, while the right hand will experience it as cold.

The skin receptors of both hands when adapted to zero, measure in both directions around it. All the temperatures below the adapted level zero will be recorded as minus, while all the temperatures above it are experienced as plus.

The concept of a relative zero means that the temperature value of the "same" temperature is always determined by the direction and distance it is from the temperature receptor, which is doing the recording but the temperature of each measuring receptor itself defines zero.

The facts that Plato (1952) would never have even bothered to observe are, according to Helson (1964), that the skin receptors don't measure the absolute temperature but a relative temperature. Put more simply, they experience a difference score, the difference between the adaptation level of the skin, which is always a relative zero and the temperature of the environmental water at time two.

Figure0.4K

Adaptation Level Theory summary

- Each of the skin temperature receptors adapts to the temperature of its ambient environment.
- The temperature of the skin receptor, no matter what it is; is by definition a relative zero. The temperature to which each receptor is adapted is by definition equal to zero.
- Each receptor is a dynamic relative zero or a floating zero.
- The strength of every stimulus is judged by a receptor adapted to some temperature, which is zero or the Adaptation Level (AL)

Book 2 - Experimental

Chapter 1 the Discovery of Heterostasis

Experiment M84

METHOD

Subjects: The Experimental Group

The subjects of this study were sixteen Sprague Dawley rats obtained from The Charles River Breeding Labs. They were 60 days old at the start of the experiment, with the average male weighing 222 grams and the average female 173 grams. Five rats were run the first two weeks; then another five, the next two weeks. Two males were dropped in the first group; one for failing to move on the first day; another for failing to reach criterion before the second group had to be run. The first three male and the first three female rats to reach criterion were included in this study. The rats were individually housed in wire cages 24.14 cm long, 18.1 cm wide and 17.2 cm high. Water and Purina rat chow was provided ad libitum.

The evolution of the physical structure of the Lashley III maze

The purpose of this experiment can and must be stated and understood from at least two points of view: First, in terms of how the nonverbal mind imposed order, structure, or frames on space; in particular, the terrestrial structure of the matrix maze; secondly, how the mind evolves coherent sets of ideas, scripts, and hierarchically organized sets of cells.

The physical structure of the Matrix maze evolved in several stages:

- One: Lashley's four-row maze
- Two: the four row, sixteen-unit maze
- Three: the six row, twenty-four unit maze
- Four: the six row; sixty-four unit maze

PHASE One: Lashley's three mazes

FIGURE 1.1.1 Lashley Mazes I, II, and III

FIGURE 1.1.1 shows the three mazes Lashley designed: Maze II, Maze III, and I. (Lashley, 1963, P. 31)

Maze I was a modified T-maze with a standard start-box, but with the cross bar of the T modified into two mirror image J-shapes. Food, F, is shown in the left arm of the Maze.

Maze II had an identical start box, but was followed by a long arm with two doors on each side, food was placed in the cell behind the second door on the right.

Maze III was a four-row maze with many subtle features, which will be discussed at various places in the text. Lashley assumed that the mastery of this maze required only a series of alternating turns, RLRLR, between the start and goal box. Lashley assumed that requiring the rat to make a series of reversed turns would be the most challenging problem possible. Moreover, he assumed that each of the required turns would involve only a 90-degree turn. In reality, however, the design of the maze was such that at each choice point, rats had essentially to make a 180-degree U- turn. If, for example, the rat was running East in the first or odd-row then in order to run West in the second or even-row they had to make a 180 degree left or a counterclockwise U-turn. Additionally, Lashley failed to explicitly note the fact that both the start unit and the goal units were not centered on a door; actually each of these two units were set off to one side of the door, and in both cases, of course, the offset was positioned so that it also required an alternation or U-turn response. Lashley (1963) consciously or unconsciously operationally designed the maze so that at every-choice point the rat was required to reverse directions. Both the start unit and the goal unit were set off to the right side of the door, in order to require the appropriate alternation U-turn sequence in the series.

Phase two: the four rows, sixteen-unit maze

FIGURE 1.1.2 16 unit maze

FIGURE 1.1.2 shows the sixteen-unit maze, which was my first modification of the Lashley III maze. This conception was flawed primarily because the unit before the door in each row was conceived as divided into two six-inch subsections, A and B, thus the units of the maze were not a standard size. In reality, this maze was a twenty-two unit maze, with five units per row and two additional units, alpha and omega. The two six-inch subsections are seen in this figure as divided with the dotted lines.

Phase three: the six row, twenty-four unit maze

FIGURE 1.1.3 24 unit maze

FIGURE 1.1.3 shows the twenty-four unit maze, the next maze design. This was a six- row maze. Two additional rows were added with the idea that they would increase the probability that rats would master and exhibit the alternation pattern.

In reality, this was a thirty-two-unit maze: five units per row for six rows plus two additional units, thirty plus two. The design of this maze was still flawed-the door units consisted of two six-inch units.

Phase four: The eight-row, sixty-four unit maze

FIGURE 1.1.4 64 unit matrix maze

FIGURE 1.1.4 shows the final modified form of the 64-unit Matrix maze, one, which transformed the physical form of the maze into a proper hierarchically organized spatial structure with the following features:

1-This maze was an eight-row by eight-column maze.

2-The alpha and omega units were the only units within the first and eighth rows respectively, thus effectively converting the eight row maze into a six row maze;

3-Six rows, two through seven, formed the core elements of the maze,

4-All the sixty-four units of the maze had been given a standard four-by-six inch size, including the start and goal boxes, which were originally 9 and 14 inches long respectively—the alpha unit, in effect, became the prototypical unit;

5-Lashley's (1963) alternation pattern of doors was maintained, but in this spatial pattern, four units of each row were defined as the core units.

6-The doors were four inches wide, with slots on the side to allow a piece of wood to slide up and down in to open and close the door. This plain sliding unit was replaced with a slide with a small hole in the center to transform the unnatural nature of the fully open door to one more natural for a rat, a hole-door. The reader needs to remember that these doors remain open at all times; they are never closed. They need to remain open in order for the rat to be able to retrace.

FIGURE 1.1.5 is a photograph of one of the first of the early mazes at Texas Tech.

FIGURE 1.1.5A – guillotine door photo of maze

FIGURE 1.1.5A is a photograph of a Canyon Land Maze at William Paterson University, and at the far right, one of the guillotine doors with a Hole-door on one end can be seen up in the air. Dr. White is shown photographing a white rat running the maze.

The construction of the maze

The matrix maze was constructed of marine grade ¾-in. plywood painted flat black, and was a rectangular unit with the outside dimensions 123.5 by 125.5 cm with ten rows. Only eight rows were used in this study. Dividers 10.5 cm. wide were used to make an alpha and omega unit in rows one and eight respectively. These units were 27.5 cm. in length and 10.5 cm. wide. The doorways between rows were located 30.5 cm. from the outside walls of the maze, and were 10.16 cm in width. The doorways, in the odd numbered rows, were on the left and, on the right, for the even numbered rows. The doorways were grooved to take guillotine sliding doors that had holes 6.35 cm in diameter in them centered in the doorway and 4 cm from the floor. These hole-doors were in place throughout this study.

The test environment

The maze was housed in an acoustically shielded room 579.12 cm. long by 386.1 cm. wide by 240 cm. high manufactured by Industrial Acoustics, Inc. The lighting was three rows of three lights, each with two seventy-five watt bulbs. Because of the depth of the maze that was required to prevent the rats from jumping out of the maze, two additional one hundred watt spotlights were focused on the rows of the maze, one on each side to take the shadows of the maze.

Procedures

Behaviorists generally ran rats using a Trial Testing Procedure, TTP. However, in this series of experiments the subjects were tested using a Loop Testing Procedure, LTP, which is the opposite of the TTP. The LTP was designed as George Kelly (1970) would say, "by picking up the other end of the stick" for all the elements of the TTP. These procedures will be described below.

The animals in this experiment were run once a day at the same time in the evening. The sequence of events was standardized:

1-Rats were picked up from their home cage the same way each time

2-Placed in a carrying cage

3-Taken out to be weighed

4-Again placed back in the carrying cage, and carried to the testing room.

5-The timer used to measure the appropriate time constraint was set;

6-The computer graphic tablet was initialized and the computer programs started;

7-Each rat was removed from its carrying cage, and put in the alpha unit of the maze.

8-When a rat was injected into the alpha unit the hole-door exiting the alpha unit was blocked. The rat was confined to the alpha unit for 30 seconds before the hole-door was opened.

9-The clock was started when the door was opened.

Only if these testing procedures were maintained as an invariant will the start gradient evolve efficiently.

The Behavioral Telescope

The Experimental rats were tracked through the 64-unit modified Lashley III matrix maze using an Apple graphics tablet driven by an Apple II+ computer. The graphic tablet was covered with a plastic sheet upon which a model of the matrix maze had been drawn. As the rat ran the maze, the experimenter tracked the rat on the graphic tablet by moving the stylus to the element of the matrix that corresponded to the current unit position of the rat in the maze. The tracking data was stored in memory while the rat was being tested and printed to a floppy disk at the completion of each run. The experimenter and the tracking system were on the left side of the maze when standing at the alpha unit. A TV-camera suspended over the maze created a clear image of the white rat against the flat black maze on a TV monitor in front of the experimenter, so that the rat could be either remotely tracked via TV or the experimenter could, by turning his head directly track the behavior of the rat exploring the maze.

The normal point of origin of the Apple tablet is the upper left corner of the tablet so that the columns are read from left to right, and the rows are read from the top down. However, since the rat ran the maze from the bottom up, the graphic tablet was rotated so that its point of origin was at the bottom right corner of the maze. This made it easier to read the data since both the unit row and column numbers increased as the rat moved out through the maze.

Graphic tablet

The Apple computer has two kinds of inputs; first the "forms" supplied by the programmer, and the empirical content, the mud that enters from the Graphic tablet. In this section, my task will be carefully distinguishing these two contents, the forms supplied by the nonverbal mind, and the mud from the tablet!

Historically, the two great philosophical schools of epistemology, empiricism, and Nativism, disagreed most importantly about whether the mind was active or passive, with Locke (1914) proposing on the empirical side that the mind was passive, while Kant (1952), on the nativistic side proposed that the mind was active.

It is ironic, yet most appropriate that I chose the Apple Graphics Tablet as a central component of the recording system to collect and store the data of these experiments. The graphic tablet, in a sense, can be conceived as an embodiment of the both the classic passive notion of John Locke on one hand, and the active notion of Kant, on the other. Locke proposed the mind was a Tabula Rasa, a blank tablet, upon which nature wrote empirical data upon the mind. Locke's empiricism imagined the mind to passively record, or copy, the state of the world moment by moment.

Kant's Nativism (1952) countered with the proposal that the mind was active; that it supplied the forms, the concepts, which gave meaning to the mud, the precepts created by the senses; that experience was a synthesis of the concepts supplied by the mind, and the precepts supplied by the senses.

I am a third generation pragmatist, which means I have taken at least one giant step beyond Kant. I agree with Kant (1952) that the mind supplies the forms, concepts, or molds that give meaning to the neuronal sensory impulses supplied by the mind, but where Kant was a nativist, I am an empiricist. Kant was half empiricist and half nativist, the mud was empirical in origin, but concepts were nativistic in origin.

I, however, am a double empiricist; I propose that both the concepts and precepts have an empirical origin that the nonverbal mind creates sets of conceptual systems.

Where does the Apple graphic tablet fit in?

The irony is that both Locke and Kant can claim the tablet as proof of their side of the epistemological argument. Locke (1914) would suppose the Tablet to be passive, as simply a device for recording the position of the rat within the maze. However, Kant (1952) would suppose that the tablet's technology supplies the forms for recording the particular sector of the matrix the rat is in moment to moment.

The purpose of this section, however, is to show that Kant's nativistic concept of the structure of the mind does not work. The fact is that a major part of the creativity required by this research project has been devoted to devising a set of forms to program the tablet.

One of the chief advanced technical features of the graphic tablet is that it is an empirical not a nativistic technology. The chief virtue of the tablet is that it can be programmed to assume any kind of matrix structure. One of my chief problems was deciding how to program the tablet.

In this section, I review some of the many, too many to identify and discuss, matrix designs I empirically tried and rejected.

The point of the programming of the tablet is really a matter of rodent mind reading. My task is to program the tablet to match the way the rat's mind programmed its sixth sense. The rat's nonverbal mind evolves a hierarchically organized conceptual structure to understand the spatial structure of the maze. My problem is to be a psychologist as Titchener prescribed. Titchener (1912) prescribed that the task of a psychologist was "to objectively study the subjective activity of the mind."

My task has been to discover objectively, the subjective conceptual structure the rat's mind evolved, which enabled rats to understand, to predict, and to control their behavior in the maze.

The technology embedded within the Apple graphic tablet makes it possible, by programming, to partition this tablet into an almost infinite number of rows and columns. My first responsibility as a principal investigator was to select the parameters for programming this matrix, which would be employed to track the rat's maze behavior.

My immediate task is to describe the complex series of steps by which I programmed and operated this tablet.

Each of these several designs described below describes a step in the process by which I was finally able to describe the rat's nonverbal behavior in a way that enabled me to infer that the rat's nonverbal behavior was the product of and emitted by a rational nonverbal mind.

The third wisdom's nonverbal forms

Heterostasis, the third-wisdom, was the first of the rational wisdoms designed by evolution. A rational wisdom is one that enabled mammalian species to be self-organizing, to evolve sets of hierarchically organized mental, Kantian forms, which organize experience and thus behavior: forms, which enable life to control its behavior within the domains of terrestrial space. (Kant, 1952) The physiology of Heterostasis enabled mammalian life, rationally, to evolve sets of hierarchically organized conceptual systems required to partition space in the specific way required to achieve some purpose.

Specifically, in this experiment, the question is how the nonverbal rat's mind parses/organizes the spatial structure of the matrix maze.

This section was designed to describe some of the many conceptual forms that were designed in order to understand how the rat's mind organizes space.

The dependent variables:

The computation of the dependent variables was automated with eleven computer programs. These programs created the data for the experimental and the control groups. In addition, two programs created the simulated rat data. Each program created its own file type.

The Experimental Rat programs

1. **The tracking program** tracking the rat on a graphics tablet generates three variables per move, which were stored in a record for each move: 1-record-number, 2- x (rows), 3-y (columns), and 4- an eight bit military time score. Each record, for example, looked like this: 1 2 6 23678789; which is read: record 1, row 2, column 6, and military time. The data generated by the tracking program is stored as the Raw-file type.

2. **The convert program** this program converted the raw data records into records with the number names shown in FIGURE 1.2.4C, and with the military time converted to latency per unit.

3. **The normal program** this program checked the data and identified all tracking errors. The files were then edited and rerun through this program until all tracking errors had been corrected.

4. **The serial program** this program created 32 variables and updated them for each record.

5. **The ele program** this program selected the 51 records from the serial files on which the rat made first entries into units of the Essential path and stored them into a new file.

6. **The cle program** this program selected the first entry records from the serial file for the criterion loop.

7. **The elf program** this program selected the first entry records from the serial data for the full maze, f, emergent loop, el, growth, elf.

8. **The mle program** this program selected the record necessary for identifying all the four types of growths.

9. **The ple program** this program selected the data for the partial loops.

10. **The nle program** this program creates the node\path data and stores it vertically.

11. **The nsa program** This selects variables from the nle program.

12. **The seqwrite program** this creates node data in rows.

The definitions of the matrix dependent variables for these studies are very technical and it would be impossible to define in words how these scores were computed, or to describe how they could be scored manually. The programs listed above compute the dependent variable data sets, which operationally define the sets of dependent variables, which are the heart of the theory. It would be impossible for the phenomena of Heterostasis to be observed without the use of these programs.

Random Simulated Rats

1. The Survey program simulated the random rat data, and it identified all the growths and their properties, identified, and saved all the parameters necessary to replicate any of the identified growths. This program used the SRS representation system seen in FIGURE 1.2.3.

2. **The replicate program** ran the growths selected from the survey data that were chosen as useful.

Graphics

After being converted to "serial" files, all the experimental and control rat data collected in this study can be processed through a graphics program, which allows the moves of the rat through the maze to be viewed as a simulation on the computer monitor. This program moves a symbol of a rat through a computerized Lashley III matrix maze on a computer screen. Different versions of this graphics program display different variables in real time. These graphics programs allow the real and random rat's behavior to be seen and visually compared. These programs make it easy to see the differences between random and rational behavior.

The science of measurement

Measurement is a form of behavior, which helps make it possible to get to "know" a thing, yet measurement is messy. Looking back, it is now clear to me what I should have done, and what I should have wanted to know; and thus what I should have wanted to measure was the behavior emitted by the nonverbal mind. In the beginning, however, I didn't know what a mind was, or what it did, or what kinds there were, or even that a rat had a mind! Only roughly forty years into the study did I know that rats had a nonverbal mind and understood something of the structure and function of such an organ.

How do you measure the function of the nonverbal mind?

The first step involves the design of a maze to measure the unique kind of behavior emitted by an organ you don't know exists and certainly don't have any idea of its behavior. Anyway, I lucked out in doing all these parts.

My topic for the moment, however, is measurement. The behaviorists, for over a hundred years, have run rats and measured their behavior employing the concept of "trials," which I thought was dumb! Therefore, I conceived a new way to measure behavior, "loops." In the studies in this book, the rats are required to run the matrix maze in both directions: out and back. A criterion loop is defined as Alpha, Omega, and Alpha, meaning the rat must leave the alpha start box, travels the essential path to omega then turns around and returns to alpha.

Then the question was how do you measure loops? How do you define a loop through the maze? Unfortunately, there are many ways to measure loop behavior, but which one is the right one. There is no one right way to measure loops. The rule is that there are as many ways as there are questions.

How to measure is a complicated question, which I will have to address in each method section. Essentially, every unique question requires a unique answer.

For example, the matrix maze is constructed within a box with high walls, which was physically partitioned into eight four-foot rows, each of which eventually I partitioned into eight columns. The maze consisted of 8 rows by 8 columns, or a 64-unit maze. However since rats were to run through the maze in a loop, the maze consisted of 2 * 64 units, or 128 units.

Factoring the Matrix

One of the most powerful advantages of representing the Lashley III maze as a matrix is that it allows the data generated by tracking subjects through the matrix to be analyzed in different ANOVA designs and in as many ways as the number of repeated measures can be factored into sets of within-components. This proves to be a mixed blessing, for while on the one hand this affords a rich source of possibilities to explore, yet on the other hand there is no *a priori* way for an experimenter to know which factoring strategy will reveal the most significant effects. The results are that this data requires in the original screening much time and much skill and experience with the various design possibilities, before the advantage and disadvantages of the different factoring options can be appreciated.

The two halves of the matrix maze.

Three types of ANOVA designs are possible: the full maze design, two partial maze designs, and an essential path EP, and a Blind Path, BP, design.

In this first experiment, it was not possible to make a full maze analysis for too many of the blind units of the matrix were never entered on the first growth, and thus no data was available for those cells. This missing data made the full maze and the blind path analysis technically impossible. The full maze and the blind path analysis will be reported in later studies when subjects were run for as many growths as was required to get at least one entry into every cell of the matrix.

The EP half of the matrix

In the first experiment, it is only possible to analyze the behavior of the rats in the Essential Path, EP, half of the matrix.

However, there are four types of designs available for analyzing this set of data. The EP elements of the matrix maze consisted of twenty-four units out and twenty-four units back for forty-eight units. The exploratory behavior of the rats in the Essential Path, EP, subset of maze units generated 48 repeated measure scores, which can be factored into four combinations of the within variables: units, directions, rows, and columns. Four types of designs were used and were tested in the following sequence:

First, a unit-design of the data was made employing only one within variable; 48 units;

Second, a direction by unit design was made employing two within variables respectively, (2 * 24 = 48 units).

Third, a row by column design was made employing two within variables respectively, (12 * 4 = 48 units).

Fourth, a direction by row by column design was made employing three within variables respectively, (2 * 6 * 4 = 48 units)

FIGURE1.1.6 Factors of 48

FIGURE1.1.6 shows the several ways that N=48 can be factored. Which of these designs should we use? Actually, the answer is all of them; for each asks a different question; each test of significance shows which variable or set of variables was a causal factor in the design. Each of these five ways of factoring 48 allows us as scientists to enrich our understanding of the causal structure of this phenomenon.

The Random Number Generator

The Random number generator was designed as described by Whitney (1984). The generator was fed by a table of 3000 prime numbers and had three parameters, which determined the sequence of random numbers: the seed, the cycle length, and the assignment type. The seed could be any integer (usually 1 to 100). The cycle length determined the length of the series that was read by any seed. The assignment type selected one of the eight combinations by which random numbers could be assigned to the probability structure of the maze. The sequences of random numbers are determinate sequences and all the simulations used in this study were reproducible on the computer on which they were run.

Experimental Procedures: defining Heterostatic variables

The experimental rats were run under a three-element constraint set: a step constraint, a time constraint, and a criterion loop constraint.

The step constraint: a Maximum Steps per Run, MSR, set at 228 steps, MSR228.

The time constraint consisted of two elements, 30 minutes for the first run, Maximum Time per Run or MTR30, and 10 minutes for each run thereafter, MTR10.

The criterion constraint was defined as one complete loop per run, which is called the criterion loop.

Each rat was removed from the maze whenever any of the above constraint criterions was reached and from whatever unit it was in within the maze when the criterion was reached. Each rat was then returned to its carrying cage and directly to its home cage. When any one of the criterion features was achieved, the experiment was over for that rat.

Subjects: Simulated Random Rats- two groups of simulated random rats were used, the M-controls and the A-controls

The M-Controls - five hundred simulated random rats were run through a Simulated Lashley III matrix maze at each of twenty- seven levels of the Maximum Step per Run (MSR) variable (from MSR228 to MSR12000) for 13,500 total simulated rats. The 500 simulated rats run at the MSR228 level are the M-controls.

The A-Controls- the M-controls were so different from the experimental rats in efficiency that an additional control group, the A-control, was run, and each simulated rat was matched on several parameters of the experimental rat's behavior. First, after searching through thousands of simulations, it was possible to identify a random rat criterion loop that was close, if not identical to the behavior of each of the E-group of rats in terms of the moves to criterion. Secondly, runs were removed from the selected random data sets in order to match the number of moves for each E rat when an exact match could not be found. Thirdly, runs were edited, where necessary, so that each run was matched in length with its yoked experimental rat. With these controls in place, it was possible to test for differences between the experimental, E, and the Absolute, A-control rats, which were independent of differences based on differences in growth efficiency.

The simulations were run on twenty 386-25mz AT&T PC computers. The simulation programs were written in Pascal 5.5. The largest Maximum Step per Run or MSR levels, 15k, was given 500 runs at a time. The smallest MSR228, were given 10 runs at a time.

The four hierarchical levels of exploratory behavior

One of the interesting discoveries of this set of experiments was that the rat's exploratory behavior is organized into four hierarchical levels: (a) growths, (b) runs, (c) loops, and (d) node/path, with growths being the outer most level and node/path being the inner most level.

Moreover, each of these four hierarchal levels of behaviors must be described with a unique set of dependent variables but only the growth-specific set of variables will be defined and analyzed in this section.

The three sets of growth-level variables

The growth set of dependent variables consists of three major data sets: First Entry, Between First Entry, and Frequency. Each of the three major sets, in turn has a subset of three variables. Compounding the abbreviation of each component of its subset creates the names for the major sets. The first variable in each subset is a measure of activity in the unit space, U; the second, a measure of activity in time or latency, L; the third a measure of speed per unit, L.

For example when the three-subset element names are added to FE, this generates the sets of subset variables names: Unit First Entry or FEU, First Entry

Latency or FEL, and First Entry Speed or FELL. Collectively this set is called the First Entry Unit, Latency, and Speed or FEULL set.

First entry, FE; Each move of the rat from one unit of the maze to the next is both numbered and timed and the successive differences between these two time scores allows the latency per unit to be computed.

The first entry unit, FEU, and the first entry latency, FEL, describe the rate at which the exploratory behavior of the rat is digesting the anomalous features of the matrix and creating a model of the hierarchical structure of the space. The first entry unit (FEU) and first entry latency (FEL) are recording the number and the time in seconds, respectively, on which the rat entered each unit of the maze for the first time.

It is important to note that convention established that both these first entry or FE scores are defined as measures of growth not run activity; these scores accumulate across runs for each growth. The first entry mean (FEM) score (FEL/FEU) was found not to be a useful score. The only useful measure of first entry speed is the first latency, FL. This is the latency of the rat on its first entry into a unit, and FL is the third element of the first entry unit/latency/speed or FEULL data set.

The BFEU (unit entry between first entry) dependent variable

Of the three dependent variables: First Entry or FE, Between First Entry or BFE, and Frequency or FREQ; the BFE score is the only one that is unique to Heterostasis. The other two scores could be used for measuring homeostatic behavior, but the between first entry (BFE) score is a measure of two concepts: retracing and uncertainty, both are uniquely Heterostatic concepts. Actually, retracing and uncertainty are two sides of the same coin; retracing is a measure of uncertainty.

However, the concept of uncertainty in the context of mammalian behavior is a completely new scientific concept. The information theorists from Bell labs first defined uncertainty as a mathematical concept, "information is a measure of uncertainty." (Shannon, 1951)

However, uncertainty in the context of the mind is a measure of the revolutionary arousal state of the mind. My point in this section is to make clear that BFE (between first entries) is an empirical or experimental concept, which is profoundly different from the information concept of uncertainty. FIGUREs 1.3.2AT and BT are probably the two best FIGUREs to illustrate the Heterostatic meaning of uncertainty. Then FIGURE 1.3.3 is also a great place to go if one is trying to understand the empirical meaning of uncertainty.

The BFEU score is a short for the phrase, the Unit Entries between First Entry. For further information on this dependent variable read the text associated with the three FIGURES referenced above.

BFEULM (Between First Entry Unit/Latency and Mean): The three operational definitions of retracing

Three ways for defining retracing were identified; each has different implications. In this section, I will distinguish these three methods; identify when to use each of them and identify the technology required for computing these scores. In principle, it would be possible to compute each of them by hand, but it would be too time consuming and tedious; these sets of scores were all computed by a computer program.

When to use each kind of BFEU-score (Unit entry between first entries):

Heterostasis behavior includes both approach and avoidance behaviors. Avoidance behavior can be classified into two types.

Anomaly induced retracing, AIR, and redundancy induced retracing, RIR. These two forms represent the opposite end of the certainty-uncertainty continuum. AIR (Anomaly Induced Retracing) is induced by uncertainty, which causes organisms to avoid anomalous units of the matrix because those units induce too much uncertainty. They are on the C-section of the Yerkes-Dodson law. AIR is seen in the maze as a function of the behavior associated with the mastery of the EP (Essential Path). AIR (Anomaly Induced Retracing) is discovered within the EP. AIR induced behavior proves that the subjective model that induced the retracing is invalid.

RIR (Redundancy Induced Retracing), however, is retracing behavior induced by too much certainty or redundancy; rats avoid redundant units because they induce too much certainty. RIR (Redundancy Induced Retracing) behavior on the other hand proves that the rat's subjective model is valid, while the AIR (Anomaly Induced Retracing) proves a model invalid. RIR (Redundancy Induced Retracing) behavior is seen only when rodents are probing the Blind-path. Anomaly induced retracing behavior is a kind of avoidance behavior. The properties of this retracing are also measured by the BFE (Between First Entry). This retracing is but one of two kinds of retracing.

Both kinds of Heterostatic induced retracing can be explained in terms of the Yerkes-Dodson law (see FIGURE 0.2B). The central feature of the curve is B, which is the optimal degree of arousal and can be recognized as the Positive or approach section of the arousal continuum. On both sides of B, the optimal central element of the curve, there are two negative or avoidance portions of the curve. These two sections will be called the avoidance on the right and the avoidance on the left.

Avoidance on the left

Events on the left are events that have too little novelty and are boring, that will be avoided when possible, these events are another example of RIR (Redundancy Induced Retracing), or redundancy induced avoidance. Perhaps this is best explained with the following quote:

"One cannot enjoy doing the same thing at the same level for long. We grow either bored or frustrated; and then the desire to enjoy ourselves again pushes us to stretch our skills, or to discover new opportunities for using them." (Csikszentmihalyi, 1990, P. 75)

Avoidance on the right

Events on the right have too much novelty and induce AIR or anomaly induced retracing.

Both events on the left and right can be seen as only requiring retracing action until the anomalous event has been explored enough to no longer induce novelty or news.

"Interest wanes when action begins to have less effect: effectance motivation subsides when a situation has been explored to the point that it no longer presents new possibilities." (White, 1976, P. 329)

Only events in the center of the YD law will be experienced as positive events.

The simple method for defining retracing

The BFE dependent variable data set measures anomalous retracing and estimates the amount of uncertainty induced by the anomalous value of each unit. The original full name of the data set was UEBFE, unit entries between first entries, which were shortened to BFE. The BFE (Between First Entry) was originally defined as a measure of the number of moves a rat makes between successive first entries between units of the full maze. The purpose of this series of studies is to put the psyche back into psychology. Watson (1972) invented behaviorism to get the psyche out of psychology, to get both mind and consciousness out of psychology. Watson, of course, had no ability or desire to measure either mind or consciousness.

Our goal, however, is not just to put Psyche, in general, back into psychology, but rather to redefine the study of behavior as the objective study of the subjective. (Titchener, 1912)

The BFEULM (Between First Entry Unit, Latency, and Mean) score is the operational heart of the Heterostatic paradigm--it operationally and objectively measures or estimates the quantity of the subjective uncertainty induced in the nonverbal mind by an anomalous event. Having such a Heterostatic-specific dependent variable goes a long way in justifying the use of the concept of the nonverbal primary rational mind - the nonverbal psyche in experimental animal psychology, but such a concept needs to be defined operationally and empirically. The BFEULM measures or estimates the quantity of the primary uncertainty induced by an anomaly.

Heterostasis proposes that the Certainty-Uncertainty Continuum, CUC, motivates the primary rational mind. The CUC plays a role in both the central and peripheral nervous system. Both the Central Nervous System arousal and the Peripheral Nervous System arousal are functions of the CUC. Heterostatic behavior is primarily motivated as a function of primary uncertainty inducing events whose goals can be understood as achieving the optimal degree of arousal or uncertainty with arousal or uncertainty having both a central and peripheral nervous system function.

Heterostasis operationally defines rodent behavior as a measure of the CUC (Certainty-Uncertainty Continuum). The BFEULM (Between First Entry

Unit, Latency, and Mean) scores are operationally defined behavioral measures of the CUC. This data set objectively monitors the momentary dynamic pure heterostatic motivational state of the organism.

Hull's (1943) operational definitions of motivation were designed to measure the tissue-need or Homeostatic motivation not a heterostatic or a state of mental need. Hull's operational definition of motivation was capable of defining motivation only as initial states of need or drive in terms of say, hours of deprivation, but his system had no dynamic variable that gave an objective public running-measure of motivation. The BFEULM (Between First Entry Unit, Latency, and Mean) score is a dynamic operational definition of uncertainty, which reflects the momentary state of the CUC (Certainty-Uncertainty Continuum) of the primary mind of the rodent at all times in the matrix maze. Yet, it is at the same time an objective measure of the subjective state of the primary mind of a nonverbal organism.

How much uncertainty is a rat experiencing at any moment? Obviously, it is not possible to ask a rat to introspect and to describe its subjective state. However, the BFEULM (Between First Entry Unit, Latency, and Mean) makes the need for rat introspection unnecessary, for with the BFEULM the rat's behavior objectively generates a running estimate of the subjective CUC state of the rat's primary mind. The BFEULM is an objective index of the subjective motivating state of the rat's primary mind.

The BFE (Between First Entry) scores in effect, make the subjective CUC (Certainty-Uncertainty Continuum) state of the rat's primary mind a public, objective, and dynamic component of rodent exploratory behavior. The quantity of retracing behavior induced by any unit of the EP (Essential Path) is assumed a public and objective manifest report of the rat's subjective readiness to explore that unit. If the rat has resolved the uncertainty, it will immediately explore the unit, or in other words, no retracing will occur. However, if, after a rat sticks its nose into a unit, it begins to retrace, the quantity of the retracing induced by the anomalous unit, the AIR (Anomaly Induced Retracing) is an objective measure of how much uncertainty was induced by the unit that initiated the retracing. When a unit is an anomalous unit, an illegal-event, one that the model of the domain cannot assimilate this anomalous event will induce a degree of primary uncertainty directly related to the quantity of strangeness possessed by the anomalous unit. Heterostatic behavior is a rational process during which the primary mind evolves and invents a coherent closed domain-specific rule system with the quantity of primary knowledge requisite to reduce the primary uncertainty induced by the anomaly. Rats will cease retracing when the model of the matrix evolving in the primary mind has proven capable of reducing the primary uncertainty induced by the anomaly. When rodents finally enter, for the first time, a once-anomalous unit a new exploratory loop begins and the retrace-loop ends. A retrace-loop begins when a rat refuses to enter a unit for the first time and ends when the rat finally enters that unit for the first time. The quantity of retracing, which the anomalous unit induced is the BFEU (Between First Entry Unit) score. The BFEL (Between First Entry Latency) is the quantity of time consumed by the retrace loop. The BFEM (Between First Entry Mean) is the mean amount of time the rat spent for each move of the redundant loop, BFEL/BFEU.

There are three scores in the *BFEULM (Between First Entry Unit, Latency, and Mean)* data set; BFEU, BFEL, and BFEM, but there are three major ways to define this set of scores.

- First, the BFE for all the elements of the full maze, BFE*fm*
- Second, the BFE for only the units of the Essential Path, BFE*ep*
- Third, the BFE for the blind Path, BPE*bp*

The BFE*fm* scores

The simplest and the most obvious way to define the BFE scores is by first listing all of the first entry scores for all the units of the maze, both the EP and the blind units, in the order of their first entries. To compute the BFE*fm* set of BFE set of scores all that is required is to subtract the successive pairs of FE (First Entry) scores. However, after spending too much time trying to make sense of this measure of retracing, I was forced to admit that it is a total mess and after years of working with this score, I had to admit it was useless and abandon this score.

The BFE*ep* data set

Eventually I recognized this measure of retracing is only effective in the analysis of the Essential Path, EP, growth data. This score is computed using only the EP first entry data. The blind path data makes the Full maze BFE (between first entry) data useless.

The BFE*bp* data set

This is the ideal BFE score for describing the blind path behavior of rodents. However, this score is difficult to obtain and the BFE*bp* is but an estimate of the primary uncertainty induced by the blind anomalous units. This score is also a growth measure that accumulates across runs for each growth. BFEM is the between first entry mean (BFEL/BFEU).

The complex method for defining retracing

The behavior generated by the primary mind can be considered either complex or simple depending upon which criterion one uses and why one criterion works and not another must be empirically determined. The definition of the BFEULM (Between First Entry Unit, Latency, and Mean) scores is a case in point. There are many ways to define retracing, and the simple EP (Essential Path) method described above was the last one that I considered but the only one that works in the analysis of EP data. Finally, one day I happened to decide to use the simpler criterion, and finally Heterostatic uncertainty induced behavior became visible.

The art of defining Heterostatic behavior

Heterostasis is a unique pattern of behavior, which is the sign of the primary mind, but this unique pattern of behavior only becomes visible when the appropriate criteria are used for defining behavior, and why the proper definition works rather than the other possible definitions, can only be empirically determined. Why the simple, but least obvious definition of retracing works while the complex obvious definition does not work, I still cannot say for certain, but I know with certainty that I had no rational way to know *a priori* that the complex definition would not work. Heterostasis is a unique pattern of behavior that only can be seen when the Behavioral Telescope, BT, is carefully focused, but the proper focusing of the BT is for the moment more an art than a science. The probability of seeing the Heterostatic pattern of behavior was simply a matter of learning to focus the BT, which is learning to use the proper set of rules to define the behavior. Heterostasis is a unique rational pattern of behavior that is a function of a set of nonverbal rules created by the rat's primary mind; thus, the pattern of behavior it emits can be described by a set of rules. This unique pattern of behavior can only be seen when the technology, which generates the independent and dependent variables are properly focused by the proper set of rules.

One of the hard lessons that had to be discovered was that the EP and BP behaviors are independent elements; each kind of behavior required the primary mind evolve its own model; each of these two sets of behavior requires the evolution of different sets of rules and their own sets of variables. While, the overall Full Maze behavior of the organism in the matrix maze is the product of both sets of rules, the fact is that the blind behavior is learned by a different set of rules than those that govern the EP behavior. Therefore, the two have to be studied separately, but in this first study, only the EP data will be studied.

FREQULM (Frequency Unit, Latency, and Mean)

The FREQU is a dependent variable that measures the frequency of the unit entries per maze unit, whiles the FREQL is a measure of the total latency of the rat per maze units. These scores are measures of how much redundant activity, in space and time, was required to reduce the uncertainty associated with each unit for a growth. FREQM is the mean of the frequency data (FREQL/FREQU) and is a measure of the average time required for each move in each unit.

Summary

Each of these, three sets of three, or nine growth dependent variables, FEULL, BFEULM, and FREQULM, have important different numerical properties, and each of the nine scores plays a different theoretical role in the theory of Heterostasis.

The **FEULL** (First Entry Unit Latency and Speed) data set measures the growth of the rat's efficiency in exploring the maze. The efficiency of the FEULL score is inversely related to the size of the differences between two successive first entry scores.

The **BFEULM** (Between First Entry Unit, Latency, and Mean) set of variables, BFEU, BFEL, and BFEM measures efficiency, but the BFEULM efficiency is inversely related to the size of the BFEULM scores. However, the

BFEULM set of scores is a measure of retrace-loop activity. The retracing phase of the rat's exploratory-play behavior, as was noted in the introduction, is paradoxically the primary creative phase of adaptation, rather than a measure of random activity. The BFEULM score identifies; (1) which units of the maze triggered the rat's retracing; (2) how much retracing was triggered by each unit; (3) estimates how much primary uncertainty was induced by each unit; (4) and estimates the rate that the rat's perception of the difficulty of the hierarchically organized maze was evolving.

The **FREQULM** data set measures the total amount of activity a rat required in each unit of the Lashley III maze before its representation of the hierarchical order in its cognitive map was valid for all the hierarchical orders of the maze for the first growth. An inspection of the FREQU (Frequency Unit) and FREQL (Frequency Latency) scores showed that rats spent a disproportionate amount of time in the early rows, which Blodgett (1929) noted, yet mistakenly believed to be so unfair that he designed the non-retrace doors to "equalize practice in different parts of the maze." The data suggests that the hierarchical structure of the maze requires several revisions of the rat's cognitive maps, one revision perhaps for each of the hierarchical orders of the maze before a conceptual system, which is valid for all levels of the hierarchy is evolved. Thus, for each of the lower levels of the hierarchal level of the cognitive structure, the rat must explore and re-explore each order of the hierarchy in order to discover a causal set of features to represent each successive order of the hierarchy that is consistent with all the subsequent orders.

The **third** score in each of the three data sets, FL, BFEM, and FREQM are the best estimate of the mean latency per unit available for each of the three sets of data. However, the FL score is not a mean score but is a measure of the latency of the first entry in each unit. The BFEULM is the mean latency for all the retracing activity triggered by each unit, by retrace-loop. The FREQM score is the mean latency within each of the 51 units of the maze. The size of each of these scores also directly reflects the subjective primary uncertainty the rat is experiencing.

Nine nonverbal forms

As James (1950) first noted, all (human) concepts are partial but all pre-human nonverbal concepts are also partial. In this section, I will describe a set of fifteen partial forms that I produced for understanding the maze. Each of these models has a domain-specific purpose.

FIGURE 1.1.4 represents the maze in its final physical form, but now this physical maze will be used to create a functionally larger maze, to double its size, by making the criterion unit of behavior a loop, instead of a trial. In a trial, the rat is only allowed to run through the maze in one direction, but with a loop criterion, the rat is required to run in both directions, from Alpha out to Omega and then back to Alpha.

This functionally doubles the size of the maze. There are now the 48 units out and 48 units back equaling 96 units.

In this section, a set of nine FIGUREs and tables will be presented which were designed to illustrate many of the various ways the maze must be represented for different purposes in analyzing the behavior of rats.

FIGURE 1.2.1 the blank tablet

The Tabula Rasa mind and the Rational Evolution of nonverbal ideas

The Blank Tablet first introduced by the British philosopher John Locke (1914) over 400 years ago has become an enduring part of our concept of human nature. However, while the concept has changed little over the years, this reflects the fact that the whole concept of human nature has stagnated.

It is my intention to break free of the bonds (the rut) that has immobilized the minds of theorists for all these centuries. Most simply the problem has been that while the theory of human nature has partially been data driven, yet no one was able to take the potential science of the human nature beyond the naked eye stage until now. My purpose in this book is to do for mind what Galileo did for astronomy to provide a behavioral telescope so that new facts or anomalies can be provided which will allow a revolutionary scientific theory of human nature to emerge and thrive.

Yet paradoxically, the potential revolution, which should begin broadly with the invention of the behavioral telescope, more specifically must begin with an updated concept of the Blank Tablet. In this section, two concepts of a blank tablet will be described and discussed; first, the historic concept, and a new one, the one used to collect the data for the experiments of this book.

Locke's (1914) blank tablet was literally a blank slate, a purely passive object. On the other hand, the blank tablet of the Behavioral telescope is a dynamic input element of computer technology, a computer graphics tablet. An input device is the critical first element in any empirical system whose survival is a function of the validity of its interaction with its environment.

My task now is to describe the two sides of the historical debate concerning the nature or the origin of knowledge namely to describe the differences between the empiricists and the nativists.

Locke

John Locke (1914), the British empiricist in the seventieth century, conceived of the mind as a blank tablet, a passive object upon which nature could write. Locke trying to understand the mind, yet working several hundred years before the modern digital age, employed the best metaphors of his day. Today, everything is much more complicated. For example, now we must break his tablet metaphor up into three parts; first, there is nature itself, which can still be conceived as active; second, at least for the purposes of this book, I imposed a computer graphics tablet between nature and the mind, and then, thirdly, there is the mind, which I conceive as active.

What is new here is my Apple computer's blank digital tablet, which embodied a most complex digital technology that can be programmed to act like a matrix of almost any complexity. However, my tablet was programmed as an eight by eight matrix.

FIGURE 1.2.1 blank tablet

With this structure, the computer with a graphic tablet can create a set of Kantian concepts, a matrix capable of representing my matrix maze that was used to track the rat through the maze. The chief virtue of the Apple graphic tablet was its capacity to be programmed to represent a hierarchical structure of almost any degree of complexity. The mind itself is not passive, but like our digital cameras today has a structure, which is able to scan and encode the information imposed on the digital tablet by an experimenter tracking a rat through the maze. The rat may explore many units of the maze in each second, and thus as the experimenter tracks the rat, the computer must identify or supply a name for each position of the stylus. At the end of the day, the computer will supply the experimenter with a printout of vast digital raw data streams, which must be saved, which can later then be parsed or factored into any kind of conceivable factorial statistical design.

In this book, I will illustrate some of the many ways such a blank tablet can be conceived and the many ways that order, forms, or names can be imposed upon the behavior of a rat as it moves through such a hierarchical space and how this structure affects the data stream.

The nonverbal mind itself conceived as a blank matrix can probe the maze with a rational pattern of behavior that can allow the Experimenter to determine whether a rat running the maze is actively probing the maze or is creating a model of the maze. Can the rat's mind capture the hierarchical structure of space while the rat is exploring that space? The nonverbal mind is genetically designed to move the body of each mammal to probe the space and take a formal or rational series of behavioral events that in effect take a picture, as it were, of what the rat was thinking.

One of the most remarkable features of the rat's behavior is that it is emitted by the rat's rational nonverbal mind, which can capture the hierarchical structure of the local terrestrial space as a set of ideas. The difference between human thinking and the rat's thinking is that humans can do verbal private thinking, while the rat's nonverbal mind by necessity does nonverbal or public thinking. With the rats thinking public nonverbal thoughts, the Apple tablet allows me to read their minds by recording their active behavior that can be observed externally.

How can the nonverbal mind of a rat probe the maze and capture an image of the hierarchical structure of the matrix maze, and how can a rat-runner objectively recognize in the rat's behavior when and where in the matrix maze each idea was conceived? How can a rat-runner, studying the behavior of rodents in the maze, objectively recognize in the public behavior of rats, the moment, and place of the conception of each idea within the set?

A nonverbal mind rationally evolves nonverbal hierarchically organized sets of ideas, which in turn organizes the public sequence of behavior it emits. My task as an empiricist is to propose a theory that explains how the rat's nonverbal mind creates the nonverbal experience that captures all the nonverbal components of a formal nonverbal rational spatial structure.

The role of the two human minds

In this section, I will take a moment to discuss the role of how each of the two evolutionary stages of the human mind functions to give the reader some sense of the role of the nonverbal component in the functioning of the verbal component, which in the last stage emits verbal words and meaning.

The human body is a function of several kinds of languages: genetic, somatic, and rational and what we may call the machine language of the mind. The nonverbal mind creates, in effect, the ideas and images, which are effectively the machine-driven or the image-driven language of the human mind.

For starters, all that is required is for me to restate Sperry's (1961) statement of what he calls the role of the two minds. He states that the nonverbal mind in the right hemisphere, in the first step, creates a nonverbal element of consciousness, which is then transferred across the corpus callosum into the left hemisphere's verbal mind where it is then transformed into a meme.

The difference between Sperry's and my position is that Sperry recognized humans as having two minds one nonverbal and one verbal. Yet, in effect, Sperry conceives the right hemisphere nonverbal mind as playing only a subordinate, a slave role, while it is executing the recognition function of objects of the physical world.

The rat's mind and the human's mind must be conceived differently for they function in two profoundly different contexts. First, the rat's nonverbal mind, which is the generic mind of all the nonverbal mammals, is the only mind the nonverbal mammals have. At this stage of evolution, we must conceive of both hemispheres functioning as a nonverbal mind.

FIGURE 1.2.1 blank tablet

Two kinds of uncertainty: syntactical and meaning

FIGURE 1.2.1 consists of an 8 by 8 set of 64 unnamed cells. My task in this section is to present various sets of nonverbal names and show how each element of each set of names plays a different functional role.

However, my first point is a nonverbal name plays a very different role than the role verbal names play. Perhaps the basic differences can be identified by suggesting there are two kinds of information and uncertainty. While both kinds of names are designed to provide the mind with information, and thus to reduce uncertainty, yet there are two kinds of uncertainty, and classical theorists of verbal behavior vastly understated the complexity of the nature of human language and the nature of the human mind.

The two classes of uncertainty are verbal and nonverbal, yet there are too many kinds of information of each of these two kinds. I will return this topic many times.

Sperry (1961) distinguished many years ago now, the two human minds, or two mental functions, the verbal and the nonverbal functions, yet language theorists, notably Pinker (1997), continue to ignore Sperry's facts. No doubt, because Pinker has no other choice, his understanding of language has no place for the concept of the nonverbal mind and no place for the concept of nonverbal nouns, nonverbal meaning, nonverbal consciousness, etc.

Both minds conceive nouns, which reduce uncertainty, but verbal uncertainty and nonverbal uncertainty are two different beasts.

Every part of speech, both verbal and nonverbal, plays a role in reducing uncertainty, but nonverbal nouns play a particularly critical role.

Rationality: The value or virtue of class identity

Rationality is not the exclusive function of verbal language. In fact, the primary value of rationality emerged with the creation of nonverbal language behavior. Nonverbal nouns firmly establish the power of nonverbal mind to create ontological forms with the rational power necessary for answering the most basic existential need, "where am I?" Proper Nonverbal Nouns confer place-identity upon each element of a place. Nonverbal nouns are images in consciousness.

One solution to Kant's dilemma

FIGURE 1.2.1 is not an entirely blank tablet; it already consists of a 64-unit formal structure stored within the computer, which is a spatial structure consisting of a set of sixty-four NONVERBAL FORMS or places stored within the program that drives the tracking behavior of the computer. The computer adds these forms to the data as a rat is tracked through the maze, much as Kant (1952) supposed the mind imposes a set of concepts upon the precepts or mud. Every nonverbal place within the set has been given its own unique name by the programmer. Each time the probe of the graphic tablet makes contact with some one of the 64 places on the tablet, the computer supplies the vector name to the data stream stored within one of the computer memory resources. Nonverbal names are required before another computer program can recreate the behavior of the rat and later a nonverbal narrative can evolve. Each of the 64-unit places only becomes functional when it becomes a nonverbal vector type noun, a named place with a multidimensional identity as a complex element of a nonverbal set, with a unique identity. In order for each place to become functional, each spatial vector must have had its identity stored in the tracking program of the computer, which provide that its name be stored on each step of the record when the rat was in that place within the maze. Ultimately, of course, these vector names acquire a nonverbal meaning within consciousness that is a unique meaning, or identity. Each named place assumes a unique role in a complex spatial structure so that it can assume its rightful place within the evolving narrative and become a responsible agent as the cause of a distinct action by having a unique identity. A space without a unique name cannot effectively be attributed actions. Unnamed potential proper nouns cannot become the subject of particular verbs.

While I call FIGURE 1.2.1 "a blank tablet," it is obviously one-step beyond exhibiting that state. While it is true that the names are not written upon each place on the tablet that is but an irrelevant technicality, the name of each place within the matrix is stored in the computer and is stored in the data record as a particular value of X and Y. On the surface, even this FIGURE may seem pointless. However, it brings us to that "Kantian Place" which raises the central empirical issue where nativist Kant (1952) could not go. Where does empirical knowledge begin? The answer is it begins with a blank potentially hierarchically organized machine. How then can a nonverbal mind create a particular set of hierarchically organized forms? In this section, I propose a theory of how a blank generic form creating machine, the sixth sense, creates a particular hierarchical structure that is both a domain and a purpose-specific structure and how forms are imposed upon place.

The nonverbal mind is nativistically endowed with the neuronal technology to conceive sets of particular formal structures that are adaptive in some particular space-time and with the capacity to impose that set of forms upon the raw data of space. How does a matrix maze impose a matrix form upon the nonverbal mind? In the behavioral telescope, one possible answer to that question is demonstrated. How are a set of forms, the concepts, imposed upon the precepts as Kant (1952) asked? The graphic tablet solution to this problem provides one pragmatic answer.

Mind reading

My task as a theorist is, in effect, to read the rat's nonverbal mind to discover how a prototypical self-organizing system functions and to understand how the NV mind creates a recognition scheme, a hierarchically organized causal conceptual system.

What is a noun? For the verbal mind, nouns are the part of speech that first identifies the subject of a sentence. Yet verbal nouns are memes, which the human compound NV/V mind language users employ to identify the actors in their narrative, the characters in their plays, scripts; or their verbal narratives told around the campfire after the hunt.

But for nonverbal mammals, their NV minds were designed to enable individual self-organizing systems to create their own one person hunting script, which means the NV mind must create scripts from scratch, to create original narratives that enable one solitary mammal to achieve some personal domain-specific goal. The point of a nonverbal mammal's narrative is not to publish, not to tell around the campfire, or not even to explain to the wife where it has been but simply to go and do something that essentially, genetically speaking, has on every occasion, effectively never been done before. Their goal is not to solve the task on a one-time basis, but to prepare a script, to create a reproducible script, with a set of recognizable characters, to create a skill that can be used repeatedly and improved each time. The NV mind must create, not a noun for the sharing of a story, but a recognition scheme, a reliable way to define or identify each object in the script.

Place neurons

Recent research has shown that animals have place-neurons in their brain; cells that light up whenever a subject enters a particular place, which the cell represents. The pattern of active "Place" neurons shows where a subject is within a space; the active cells show that the subject recognizes the place it occupies in space. How rats identify these places is not known, but place neuron firing shows when an organism recognizes a place. (Moser and Moser, 1998)

Our particular question is how a rat recognizes where it is within a sixty-four unit matrix maze. How can a nonverbal mind recognize where it is within a matrix? What kind of a scheme; how can a NV mind conceive a causal conceptual system for not only knowing where it is within a matrix, but how can it create a scheme of a structure, which will enable it to conceive a plan for efficiently achieving its goals.

In any case, there are theoretically many ways to create a NV "recognition" function, many ways to identify a place. When working with multidimensional matrix data, each multidimensional space is a vector, which means that any recognition scheme designed to function in such a matrix will allow each place to be understood as a vector. The recognition scheme must consist of a vector with one element of the vector representing each dimension of the formal structure of the space. Each individual vector must enable the rat to recognize where it is in each of the several dimensions of the multidimensional space. For example, the rat must know in each unit of the maze where it is in each of the four dimensions, which level of each variable of the matrix elements that it occupies.

In our matrix, I will show the rat conceives it as an A2, B3, C2, D8 matrix, the question is: how does the NV mind conceive such a Recognition Scheme for this matrix?

Factoring a matrix

In this section, I will systematically illustrate how to identify the various ways a matrix can be factored; in particular how to identify, how to factor the total picture of a 64-or a 128-place matrix. First, I will explain how the set of 64 unit places were assigned. This set of names, which I call the SRS data set is the product of the conventional western style of scanning a page employed by the English language community. SRS data will be explained below.

After years of experimentation, I finally settled upon conceiving both the matrix maze and the graphics tablet as an 8*8 matrix. In other words, I operationally began to conceive the nonverbal mind of rodents upon entering the matrix as a blank 8*8 tablet, a Tabula Rasa. (Locke, 1914) Embedded within the Apple graphic tablet was a software technology capable of being programmed to embody and impose upon the sensory precepts a matrix structure of any degree of complexity as the empirical constraints of the experimenter's experimental needs demand.

The first phase of my research involving the twenty years from '61 to'81 were required to evolve the final operational model of the maze. The steps, which transformed the Lashley original four-row model of the Lashley III Maze into the final 128-unit version, were described in the section above.

In this section, I will describe some of the steps by which the conceptual matrix representations of the maze were evolved.

Locke (1914) began empiricism. He is famous for stating the role of empiricism in knowing by stating, nothing is in the mind that is not first in the senses. Locke supposed that the senses passively processed input, yet these five senses were the only source of data.

Kant (1952), on the one hand, was raised a nativist but eventually partially accepted the empiricist's point of view. Next, he advanced the idea that nothing was in the mind that was not first in the senses except what first was in the mind. Locke (1914) had proposed that knowledge was exclusively an empirical function, while the classical nativists took the opposite position, that the origin of knowledge was entirely a mental function. Everything began with or was dependent upon the senses. Moreover, the senses only supply the mind with copies of the things of the physical world.

On the other hand, the mind according to Kant (1952) supplies the concepts, which give form to the precepts, and so the mind transforms precepts into nonverbal ideas, consciousness, and behavior. If we are to understand human nature as a self-organizing system, we must begin by understanding the nature of the sensory systems, which feeds the mind.

What is a blank slate? The reader must begin by consciously distinguishing between a piece of slate and a computer graphics tablet. Today our ability to understand the working of the mind has advanced immeasurably in the centuries since Locke's time. Still this difference is captured in difference between a piece of slate and an Apple graphic tablet.

Apple Graphic's Tablet

What kind of technology is embedded within or behind the Apple tablet? Three particular technologies must be identified and distinguished: in particular: first, the concept of programming, secondly, the concept of a matrix, and third the concept of a hierarchically organized space. The challenge, however, is putting these three concepts together into an operational data processing system. Specifically we must understand how the Apple graphic tablet as an input device can be programmed to read and store hierarchically organized data from spatial objects such as the matrix maze.

The data is created by a series of moves of the stylus across the surface of the tablet, with each move tracking the behavior of the rat through the maze. Each time the stylus makes contact with some one of the many hierarchical cells, which are defined by the software programmers, the programs that drive the tablet record three data elements, X, Y, and time. FIGURE 1.2.4D shows a printout of one of these tracking records. In other words, the tablet is anything but a blank tablet. This figure will be explained later. In one sense, the graphic tablet is a blank tablet for when the computer is turned on at the beginning of each recording session; the computer's memory is cleared so that as much space as possible will be available for the next testing session.

Yet the tablet is, in another sense, anything but a blank tablet. Why, simply because all the person running or testing the rat must do to "track" the rat running through the forty eight units of the maze is to move the stylus across the surface of the tablet which has been programmed to be partitioned into a hierarchically organized model of the maze. For every contact with the stylus, the computer creates and stores a set of three running variables, X, Y, and time. The computer then, is a Kantian type system, which at the beginning of the test session supplies the forms, which give meaning to each contact of the stylus with the tablet. The actual data is simply an electrical impulse created by the stylus making contact with the matrix of sensors embedded within the surface of the tablet.

Programming the Apple graphic tablet

In this section, I will describe a series of steps by which I programmed the Apple graphic tablet to allow me to track, record, and understand the behavior generated by the nonverbal mind engaged with empirically conceiving a particular active Kantian form. I must impose upon the Apple tablet a form, which can read the behavior of a nonverbal mind engaged with creating a particular hierarchically organized set of forms. (Kant, 1952)

The point of this section is to describe the empirical details of how the primary language machine works - how does a biological organ communicate nonverbally with a terrestrial hierarchically organized structure - how does a nonverbal organ interrogate a hierarchically organized space?

Vector names

The names of the cells in the maze are vector names. In this section, I will illustrate some of the many ways hierarchically organized sets of cells can be given vector names.

What kinds of names are capable of playing nonverbal noun roles? Some readers may never have considered numbers as names even though our society employs a great variety of number names such as street addresses, social security numbers, telephone numbers, military identification numbers etc.

The nonverbal mind evolved to capture the physical features that characterize the domains of each mammal's unique niche. The nonverbal mind is a universal machine for adapting nonverbal species, for nonverbal actions in any domain of space-time that any mammalian species occupies.

The primary spatial role of every nonverbal noun is to establish spatial identity as a vector name. Nonverbal names must parse space, must enable the nonverbal mind to partition space-time into recognizable units, classes, or sets. The kind of nonverbal names that I must now provide for the 64 blank spaces of FIGURE 1.2.1 are names that enable the nonverbal mind to recognize each place as a unique member of a set of a 64-unit space.

What is the difference between the conventional names with which the verbal mind most successfully functions, and the nonverbal names, which satisfy the needs of the nonverbal mind? What is the difference between the set of names 1, 2, 3, 4, and the set of names A1, A2, A3, and A4?

Vectors

My question is how best to label any element of a hierarchically organized space so the nonverbal mind can recognize where it is within such a hierarchically organized space, such as a space within the context of this set of 64 spaces.

Designing the Matrix Maze

FIGURE 1.2.2 the 64 unit-SRS-normal matrix

FIGURE 1.2.2 shows how the 64 cells are scanned by the Apple graphic tablet using the built in standard scanning pattern from left to right and from the top-down. The tablet generates a stream of number names, each of which identifies the sequence of units entered in the rat's behavioral path through the 64-unit maze. Each unit of the matrix has a number name, an identification number, which identifies the rat's current position at each step in the data stream. This scheme is called the Standard Representational System, SRS. This set of numbers is for human consumption only.

FIGURE 1.2.3 SRS 64-unit inverted

The cellular names for the 64 units of the maze are the same in both FIGUREs 1.2.2 and 1.2.3; except in FIGURE 1.2.3, the Apple graphic tablet was rotated, which shifted the point of origin from the top left, to the bottom right corner. The SRS names now run: right to left, and from the bottom up. The tablet was rotated so the rats would run through the maze with the start box, or the alpha unit, on the bottom and the omega unit on the top. This way both the alpha unit and the Experimenter, (E), tracking the subject was at the same end of the maze. In this rotated scheme, the rat ran away from the E on the way out and toward E on the way back.

FIGURE 1.1.4 the 50-unit "Canyon land" Maze

The purpose of this series of FIGUREs is to prepare the reader to understand some of the complexities of the nonverbal mind. Unfortunately, all these FIGUREs are two-dimensional while the maze is a three-dimensional spatial structure, which makes it difficult to call attention to some of the critical features of the maze. For example, Lashley designed maze III with one-foot high walls to discourage rats from climbing out of the maze and to insure that rats will confine their attention and exploring to the floor of the maze. Now empathize with the rat. Imagine how a small creature like a rat experiences this three-dimensional spatial structure with one-foot high walls. Imagine how the rat must feel upon suddenly being dropped within a great canyon - finding itself confined, or trapped within, a series of deep canyons - with walls too steep and sheer to be climbed or scaled; so high, in fact, they hide the outside world. Of course, it will take some time before a rat discovers how many canyons there are and exactly how the series of canyons are arranged.

I do not mean to suggest, however that this Canyon-land maze is a scary space that intimidated the rat. On the contrary, the rats' behavior shows that most rats find this structure to be an irresistible challenge. To capture some of the feelings this maze induces, I call it The Canyon-Land maze. Eventually the rat should recognize it is exploring a series of six parallel canyons with one crawl space (one hole-door) between canyons.

Kantian forms convey purpose and value

The Canyon Land maze's anomalies trigger the equivalent of the autonomic response that Cannon (1932) made famous, the so-called "fight or flight" reflex, but now in this context it has an entirely new meaning. Its purpose in this context is not fight or flight, but to explore, to know, or to model. But the creative functions of the mind are always turned on by anomalies by Pavlov's (1960) Orienting Reflex, "What was that," which triggers a deep rational epistemological curiosity. Anomalies induce uncertainty. The nonverbal mind was designed to immediately begin the Heterostatic cycle of behavior explore retrace explore to create a model that will reduce or destroy the uncertainty induced by the anomalous "bear."

Most rats responded positively to the anomalous events of the Canyon Land maze; they immediately began to explore this space with a self-imposed purpose that dictated that each of the 64 spaces was conceived as either salient or not according to its purpose to discover the shortest path through the maze.

Several features have been added to the standard Lashley III maze. First seven doors have been added, one between every two rows or canyons. Second, the doors in this study were not the standard door used by Lashley (1963), but a new kind of door, the hole-door; a concept proposed by a student, Denis Ilchism. Lashley had originally designed his doors as simply a four-inch wide cut in the wall between the rows. Denis first noted that many rats had trouble detecting and responding to this wide-open space as a door so he proposed that a hole-door should be seen as more natural for a rat; his idea was adopted. In this new figure, the hole-door units have been drawn in, one door between rows. The first hole-door was between unit 6, the alpha unit and unit 14. The second hole-door was between units 11 and 19, etc. Since 14 of the 64 units were not accessible to the rat, the maze was effectively redefined as a 50-unit maze as shown in FIGURE1.1.4. In rows one and eight, only one of the eight units was accessible to the rat, unit 6 (or the alpha unit) in the first row, and unit 62 (or the omega unit) in the eighth row.

How many rows are in this maze?

The question of the size of the maze may seem a trivial question, but even so, this question continued to plague me until I finally recognized that I had to formalize the distinction between the two alternative maze sizes.

Two ways to conceptualize the maze size are:

First, for the carpenters who were contracted to build the maze, it was an eight-row maze.

Second, for behavioral reasons, however, the maze as seen in FIGURE1.2.4A is shown as using only one unit each in row one and row eight, Alpha and Omega respectively. So is this maze an eight-row or a six-row maze?

Third, for the great statistical god, ANOVA, who always has the last word about what is real, in terms of cause-effect real, the maze was a 48 unit or six-row maze. While the complete loop through the maze required 51 steps: 48 units in the EP + alpha-one + omega, + Alpha-two = 51 or (48 + 1 + 1 + 1 =51). However, the boss, ANOVA, will only accept as the value of "N," the total number of observations in an experimental design, a number that can be factored. Fifty-one cannot be factored, but 48 can (6 * 8= 48) or six rows by eight columns equal N or 48.

Look once more at FIGURE1.2.4A, but this time note the two sets of column numbers labeling the rows of the matrix maze.

First: On the left in red, the maze rows are numbered 1 through 8. But then:

Second, on the right, in green, the rows are numbered 1 through 6.

The conclusion was that there are two ways to talk about the maze row size:

First: when we include the alpha and omega rows, the maze has eight rows.

Second, without alpha and Omega, the maze has six rows.

FIGURE1.2.4B 50 unit canyon land maze

FIGURE1.2.4B shows a plain maze, a cleaned up maze with nothing added or subtracted for the sake of explanations.

FIGURE 1.2.4C Vector numbering

FIGURE 1.2.4C show the maze numbered with two dimensional vector names, with rows by columns vector names. The alpha vector is now (1, 6), while the omega unit is (8, 6), or column eight and unit six. This FIGURE shows the system of numbers that was used by the computer to identify the series of units the rat enters as the experimenter tracked the subject through the maze.

This FIGURE shows only one set of the two vector names of the units. The rat, to master the maze, must conceive the maze as a set of four-vector names. The four vectors correspond to the four elements of the matrix maze – direction, replication, row-type, and unit or A2, B3, C2, D8.

FIGURE 1.2.4D Printout of tracking data

FIGURE 1.2.4D shows one page of a computer printout of the tracking data recorded by the computer. Note that this print out shows that the computer printed three numbers for each unit the rat enters X, Y, and time. For example, note that the first row shows the rat was in unit (1, 6) and time. This data is a record of the rat's behavior in the matrix maze where the units are identified by the names shown above in FIGURE 1.2.4C

FIGURE 1.2.5 the Essential Path

FIGURE 1.2.5 identifies the 51-steps of the Essential Path, which satisfy the rat's self-imposed goal of running the shortest round-trip path through the maze. The 51 steps consist of 25 steps out and 26 steps back. The "out-steps" are shown in red; the "back-steps" are in black. The 51-step path takes rats through all the central units of the maze twice, once on the way out, and once on the way back.

FIGURE 1.2.6 SRS-PURPOSE (Structural Representation

System)

FIGURE 1.2.6 uses color to divide the 51-unit maze into three zones, the Essential path subset in green, the Blind Path subset in red, and the inaccessible units in black. The colors were selected to call attention to the implications of the rat's purpose for its understanding of the units of the maze. All rats spontaneously chose to discover the shortest round trip path through the maze, out and back. Given that purpose, the green units are those the rats must enter twice, once on the way out and once on the way back in order to achieve their goal. The red units are Blind path units that rats must learn to avoid if they are to achieve their goal.

While we are speaking of purpose, it is a good time to distinguish the two concepts of purpose, the scientist's objective purpose and the rat's subjective purpose.

The difference between FIGUREs 1.2.5 and 1.2.6 is that the former, drawn in black and white represents the view of the scientists whose only purpose is objectively tracking the behavior of the rat through the maze; while the color image of FIGURE 1.2.6 represents the way the units of the maze are valued given the subjective purpose of a rat.

The purpose of this experiment from the scientific point of view is to understand the subjective view of the maze as conceived by the rat's nonverbal mind. I also should note that while I am studying the rat's behavior, I am not a behaviorist. On the contrary, my position is closest to that of Titchener (1912) who defined psychology as "the objective study of the subjective." The nonverbal mind is not just a passive organ, but also an emergent, a new kind of organ, one not at the mercy of the blind forces of nature. The nonverbal mind emerged as an organ designed to create a new class of rational beings, the nonverbal mammals, and a new kind of force in the cosmos. The nonverbal mind conceives its own purposes, creates its own values, and conceives the means to satisfy its ends. The nonverbal mind's purposes are domain-specific. My human purpose is to understand how this rational organ conceives and executes its subjective functions.

FIGURE 1.2.7 the MPS

The task of the rational nonverbal mind of the rat is to evolve or conceive a domain-specific, purpose-specific set of forms, a causal conceptual system that enables each organism to achieve a specific goal in a specific domain.

When a scientist looks at spatial structure, such as the matrix maze, they ask themselves what features of the structure could be of service to the mission of the organism in question.

A model

The modified Lashley III maze is a six-row spatial structure with each row four feet long. Assuming that that the rat's mind parses each 48-inch row into eight equal units, then each unit would be a six-inch unit and the whole maze would be conceived as a 48-unit space. At any moment, the rat can only experience one of these 48-units.

Lashley (1963) looking at the maze from outside was able to see all the units of the maze at one time, but rats confined within one unit of the maze at a time, can experience only one unit at a time. The question is CAN THE NONVERBAL MIND OF A RAT CONCEIVE a big picture of the whole space as it were, A BIRD'S EYE VIEW OF THE WHOLE.

Lashley looked at the maze and saw its critical features as a series of alternating right-left turns starting at the alpha unit. Lashley conceived of the purpose of the maze only from his own human point of view.

So look at the basic 64-unit matrix maze and ask yourself, what are the obvious features a rat might use to recognize where it is within the matrix? How can a rat recognize each unit of the maze?

One of the obvious sets of features of the maze is its probability structure. The probability structure answers the question: how many ways can a rat move in each unit of the matrix? The number seen in each unit of FIGURE 1.2.7 answers that question. This number identifies the number of moves or unit transitions possible by the rat while in each unit. Within this scheme, all units of the maze were classified into three types.

First, units without a choice, in other words, units in which only one move or unit- transition was possible. FIGURE 1.2.7 shows that both the Alpha and Omega units, as well as all the outside units of each row are such units; they have a probability value of 1.

Second, units with a two-way choice and there are four such units per row. In the second row units: 10, 12, 13, and 15. The corresponding units of each successive row also have a probability value of two.

Third, units with a three-way choice, the two door-units of each row are all three-choice units. There are two door units per row, one pre-door unit, and one post-door unit. In each of these pre-door and post-door units, rats have three possible moves: right, left, or through the door.

The maze conceived as a Probability Space generated the first feature, which allowed us to classify each unit as belonging to one of three sets, and this scheme formed the basis of our simulation of the random motion of a rodent in the matrix maze. The simulation program will be described below.

It is also possible to conceive of each of these three types of units, as having one additional move, a repeat, or non-move, and this possibility is an important option that would indeed make it possible to simulate time. However, I chose not to use this option in this series of studies, for even without such option, the complexity of the simulation programs was almost beyond the power of the computers I had available. See the discussion below.

In FIGURE 1.2.7, the reader should note that the eight column units in each row have the same probability; the probability structure for each of the rows is the same.

This set of features is called the Maze Probability Space MPS, and is one of the most basic set of causal features of the Matrix maze. These features are the backbone of what rats learn in the lowest or innermost level of the hierarchy of the matrix, but there are other features to be recognized.

FIGURE 1.2.8 SRS-FUN 3-WAY

FIGURE 1.2.8 introduces another set of unit-names, the Functional Unit Names, FUN. While computers have no difficulty tracking the behavior of a rat through the maze using the SRS (Structural Representation System) number-names, humans find it difficult to employ effective number-names. The FUN was designed as a more human-friendly set of names. In FIGURE 1.2.8, the FUN consists of two sets the EP units in green and the BP units in red. This continues the color scheme of FIGURE 1.2.6.

The EP or green set of FUN are labeled: E1, E2, E3, and E4; with E1 always the ingress unit of each row and E4 is always the egress unit of each row; E2 is the second unit in the EP in each row, and E3, the third unit of the EP. The four EP units are the center units of each row. However, to my great distress I eventually realized that this alternation of rows presented a problem that was almost conceptually lethal: it made the data almost incomprehensible. Fortunately, after fretting over this difficulty for over a year, I finally discovered a way to surmount this problem. Notice in FIGURE1.2.8 that the sequence of the FUN units alternate in successive rows, they are numbered 1-4 in the direction the rats must move in each row when running the most efficient path right, left, right, etc.

The BP unit names in red are subdivided into two kinds: the Forward, F, units and the Alternation, A, units. The F units are those that are entered when rats fail to exit the row, when they fail to respond to the door. The A units, on the other hand, are entered when a rat makes an incorrect turn upon entering a row. In order to stay within the subset of EP units, upon entering a row a rat must make essentially a one hundred eighty degree turn after going through a door. In each row, the rat must alternate the direction of their turns in order to stay within the EP.

The alternation BP units get their names from the fact that the A units in the odd row are on the left side of the maze, while those in the even rows are on the right side of the maze. In other words, the Alternation units alternate from side to side.

The Alternation Blind Path units are renamed the A1 and A2, while the Forward BP units are F1 and F2. Both A2 and F2 are the deeper of the two BP units of both types.

Perhaps the most important and the most obvious feature of this FIGURE is that simply put there are two kinds of rows: odd and even. There are three odd rows 1, 3, and 5, and three even rows 2, 4, and 6.

It is important to notice that all the odd rows have unit A1 as the first unit on the left, while all the even rows have unit A1 as the first unit on the right.

FIGURE 1.2.90 the design of the FRS-FUN names

Compare FIGURE 1.2.8 and FIGURE 1.2.90; but ignore the obvious difference in colors. Specifically notice that the column names in the two FIGURES are ordered differently. However, the difference between the two FIGUREs is simple. In FIGURE 1.2.8, focus on column six, the column with both the alpha and omega units. Look at the FUN units in the column containing

ALPHA and OMEGA--note that they alternate, EI, E4, E1, E4, E1, and E4. Now look at FIGURE1.2.90 - how are the FUN ordered? In FIGURE1.2.90, all the names in column six from ALPHA and OMEGA are E1's. What I have done is to reverse the sequence in which the unit names were read in the even rows, so that all the unit names in each of the eight columns are now the same. Why would I do this? Let me explain.

FIGURE 1.2.90 introduces the last, and the most difficult design scheme, the Functional Representation System, FRS, yet one of the most critical distinctions. What I had to learn was to distinguish between the SRS and the FRS names, where the SRS is one of the standard human ways of scanning units of a matrix. What made this distinction difficult for me to recognize, and by hindsight, this should have been obvious, I had to become consciously aware that rats do not observe human conventions; they do not scan the rows using the English standard, which creates the SRS; instead, they scan using their own standard, the FRS. Of course, it should have been obvious that rats would not use the English verbal scanning convention - they are neither British nor American!

The English verbal convention was designed to scan a whole object in each row, with only our eyes moving from left to right, and then top down. Rats do not or cannot scan the maze by only moving their eyes; rather they scan the maze from inside. It is difficult for us to think like a rat, or to imagine what the rat experiences. If anyone wants to try to imagine what a rat experiences in a maze, then they should try to imagine that they are smaller than a book, smaller than the letters of the pages on the book, and that they have to move around the letters to distinguish them. So what is the scanning rule used by a rodent? Essentially rats scan the units of the maze not by just moving their eyes but rather by moving their whole body inside each unit of the maze. They are scanning one unit at a time in the direction they run, in this case from the bottom up and then they alternate the running in each row; right, left, right, etc. in the successive rows. Rats confined within the matrix, scan only one unit at a time and in the direction they run, which alternates from row to row. Unfortunately, all this is obvious only after the fact, when I was finally forced to think concretely about such things.

However, I will present the scientific statistical justification for distinguishing between SRS and FRS in the following three FIGURES1.4.15T, 16A, AND 16B.

The orientation of the maze

Compass Orientation

For discussion purposes, let us suppose the maze is aligned with the compass, so that: the Alpha unit is in the south; and the Omega unit is in the north. Then in the first row, after the rat leaves the alpha unit and enters row one, it must run west to east; and then in the second row they run east to west, etc. This alternating row sequence, west, east, west, etc., reflects the way Lashley (1963) designed his third maze; the doors in each row alternate from side to side in successive rows. Lashley designed this feature as part of his third maze because he believed this alternation feature created a spatial structure that would require rats to emit the most difficult response sequence, an alternation sequence: right, left, right, left, right, or East, West, etc.

Lashley was thinking like a behaviorist; that rats learn to respond or to emit an alternating sequence of behavior. Historically three kinds of learning have been recognized stimulus learning, response learning, and stimulus-response learning.

However, in this book I assume that the nonverbal mind of the rat must learn to distinguish both stimuli and responses. First, what initiates a growth cycle is an anomaly, an effect without a cause. The nonverbal mind can only reduce the uncertainty the anomalies induced in the maze, by discovering a cause for every effect. The goal of the nonverbal mind is simply, once an anomaly has been encountered, to destroy the anomaly and destroy its power to surprise. In this case, we can slightly modify the following statement to "the rat infers."

"The human operator infers the need for adaptive change in his internal model of the control situation from errors in prediction. Since predictions are based on the operator's internal model of the situation, errors in prediction are evidence of errors in the model."(Kelley, 1968, P. 227-228)

The errors or surprise are evidence that the rat's model of the maze must be updated to account for the anomaly.

Each anomaly begins a new growth cycle, by initiating a retrace cycle. The rat retraces until it finds a cause that destroys the power of an anomaly to surprise. During retracing, the nonverbal mind must learn to recognize each new anomalous feature of a space. In the matrix maze, the NV mind must learn to see or to recognize the hierarchical structure of the spatial structure, the (A2, B3, C2, D8) formal structure. To accomplish this purpose, the NV mind must conceive a set of four ideas, or a set of four forms, which form a coherent model of the maze.

To be concrete, the NV mind must conceive of the eight units in the first row as consisting essentially of rooms in a house with two bedrooms on each side of the house in the cul-de-sacs and a living room in the ingress position, E1, and a kitchen in the egress position, E4. The first room the rat enters in each row is the living room, E1, and the last room from which they exit the maze is the kitchen, E4.

In the odd rows, 1, 3, and 5, the living room is the first room on the West side of the house, while in the even rows, 2, 4, and 6, the living room is on the east side of the house.

In FIGURE 1.2.90, I have rearranged the sequence of rooms in the even rows so that all the rooms of the same kind are in the same columns they occupy in the odd rows. In each of the rows, all the living rooms are in the same column, etc., that is all the rooms of the same kind are in the same column in each of the six rows.

Interestingly, Lashley's (1963) hypothesis was correct but for reasons Lashley never could have ever imagined. The challenge this alternation sequence poses for the rat's nonverbal mind is not a simple motor challenge, but rather a perceptual/conceptual challenge; how to rationally conceive a set of factors or ideas to conceptualize or rationalize the hierarchical structure of this spatial structure. As we study the behavior of Female 3 in the next section, we will discover that in order for rats to "learn" to run this maze, their rational nonverbal mind must conceive or model the hierarchically organized maze space as a four-factorial design. This design consists of four ideas, namely an A2, B3, C2, D8 design, not a three-factorial design.

Replication

The most difficult of the four ideas in this four-factor design for Female 3 to conceive was B3, the concept of replication, with three levels. In order for rats to achieve their self-imposed goal of running the maze in the shortest path, they must conceive the matrix maze's spatial structure as a coherent set of four nested concepts or ideas. A set of four conceptual elements must be rationally conceived and assembled to construct a coherent hierarchically organized behavioral model, software program, script, or scheme, before a rat's nonverbal mind can extrapolate and rationally emit a set of responses that are a manifest expression of a mental plan, which consists of the requisite series of alternating motor responses.

How to read a rat's mind

In this section, I will show how it is possible for an experimentalist to interrogate, infer, and read the nonverbal mind of a rat by studying the behavior the rat spontaneously emits as it explores the maze during exploratory play. I will show how it is possible to record, score, and then objectively discover how the rat's nonverbal mind conceived or evolved this four-factor model during spontaneous exploratory play and how this model causes the rat's somatic motor system to emit the requisite alternation sequence required to run the shortest path through this maze without an error.

In order to learn to read the rat's mind, I first had to be able to infer from the rat's behavior how the nonverbal mind of the rat solved the puzzle generated by the hierarchical structure of the maze. The set of cues consisted of answers to these questions where within the 96 units of the maze were the anomalous units; how many anomalous units were there; how anomalous was each unit? To answer these questions I had to invent a set of scientific variables that allowed me to operationally ask and answer these questions. I was then able to infer how the rat's nonverbal mind was able to invent a set of concepts that could reduce the uncertainty induced by this set of anomalies.

How can a set of anomalous units objectively be identified? How can a nonverbal mind conceive a set of coherent nonverbal ideas? How can a nonverbal mind conceive an idea such as the formal concept of B3?

Look closely at the difference between FIGURE 1.2.8 and FIGURE 1.2.90. In the former, the row unit names are sorted by the SRS, while in the latter, the row unit names are sorted by the FRS. The FRS sorts or organizes the data in the order the nonverbal mind of the rat encounters the units of the matrix. Once I achieved this insight, I was able to understand how the nonverbal mind works, how Heterostasis works.

FIGURE 1.2.91 PARTITIONING

FGURE1.2.91 was designed to expand our capacity to talk about the row structure of the matrix maze, by adding new ways to label the rows on both the right and left sides of the maze. On the right side of the maze note that there are three columns of labels each showing how to conceptualize the row structure of the maze. From left to right note that the first column on the left numbers the rows from 1-8 from south to north with row one the alpha row, and with row eight the omega row. This scheme will be called the row names with alpha and omega.

The next two columns that name the rows from 1-6 will be called the row names without alpha and omega. The second column labels the six out rows while the third column labels the back rows.

FIGURE 1.2.91 on the left side of the figure illustrates two factors of the Four-Factor design. The first column on the left labels the rows using the row-type factor name, while the second column labels the rows by their replication variable names. The design for this four dimensional maze is (D2, R3, RT2, D8 = 96.) where D2 is Direction, R3 is Replication type, RT2 is row type, and D8 is the label for the eight columns. In the first column on the left, the Row Type variable is simply labeled as TYPE. The next row to the right is labeled Replication for the replication variable.

In the first row, the two row-types are labeled E for Even and O for Odd. Do not be confused. This point can be confusing, but the Even and Odd labels refer to the six- row scheme for labeling rows, i.e. without alpha and Omega. Moreover, the O's and E's only label the six rows "out." The back six rows are not labeled by the O - E scheme.

Notice that there are three pairs of rows or three replications of the row types. There are three east and three west rows.

Lashley conceived of the rat's problem as conceiving of the maze as a series of alternating rows, RLRL. Lashley (1963) knew nothing nor cared how the rat's nonverbal mind thinks. The data of these experiments show that the nonverbal mind conceives the “row” difference in terms of the product of two ideas: replication and row type. In the first grade the rat’s mind masters the first dimension of the four-dimension maze, columns, from left to right; first it must conceive the first variable column, with eight levels, then in the second grade, it must conceive of row type with two levels, odd and even. In the third grade, the mind then must evolve the concept of replication with three levels of each of the two row types. Then in the fourth and last grade, the rat’s mind must conceive the idea that there are two directions to run in the maze, out and back.

The issue can be conceived abstractly in terms of A and B. The rat in the row grade must distinguish between the two row types, with A representing the first row and B the second row. Then the difficult question for F3, for example, becomes what will the third row be A or B? Therefore, the options for the rat are either ABA or ABB. What makes this maze difficult to master as we will see later is that rats assume that the pattern is ABB whereas replication requires that the rat conceive of the difference as ABA.

In Lashley's day, these statistical or conceptual tools had not even been developed; they were not yet available in Psychology. Thus, Lashley was incapable of understanding exactly what kind of nonverbal conceptual system the nonverbal mind must conceive in order to execute this series of turns.

Results: Female Three

Purpose

In this section, the behavior of one of the six subjects will be described; in the next section, the combined behavioral data of six subjects will be described. The purpose of this section is to describe the behavior of one prototypical mammal, a 90-day-old female rat, the third Female, Female 21, or Female 3, whose behavior most clearly exhibits the fundamental behavioral properties of Heterostasis, especially retracing and extrapolation. Additionally, I will begin to distinguish Homeostatic and Heterostatic behaviors.

Run Data

The behavior generated by the nonverbal mind is a revolutionary form of behavior, which science has never before seen. In spite of the fact that untold thousands of experimenters, behaviorists, and untold tens of thousands of rat experiments were published, yet no one ever saw the natural behavior exhibited by the nonverbal mind because they tested the animals using the TTP (Trial Testing Procedure), not the LTP (Loop Testing Procedure) used in this series of experiments.

However, the first step in understanding the nature of the nonverbal mind is to learn to see this unique pattern of behavior. This behavior is most complex and each of the several experiments presented in this book will describe new features of this behavior, but the first step is to learn to recognize the central feature of this behavior, the 51-step criterion loop through the Essential Path (EP) of the maze.

Two rats' data will be examined in this first section, one real rat, Female 3, and one random rat.

With the real rats, the reader will be given an opportunity to see and recognize the basic feature of the real rat's behavior to distinguish:

1 - A growth and a run;

2 - Two kinds of loops: FE (First Entry) and Retrace loops;

3 - Two phases of Heterostatic behavior: modeling and extrapolation;

4 - Four grades or nonverbal ideas; Direction, D; Replication, R; Row Types, RT; and Columns, C; or A2, B3, C2, D8

With the random rat, the reader must learn to recognize the unique features of random behavior particularly those features, which distinguish rational and random behavior.

Female three's behavior was selected as the best exemplar of nonverbal rational behavior, first of all, on the basis of the Run-data seen in FIGURE 1.3.0T, which compares the run data for the six subjects of this experiment, (rats 19-24). This run behavioral-data measures and compares the six rats' skill in reaching criterion. Three scores are shown for each of the six rats: the run length in steps, the run latency in seconds, and the average run speed. Female Three (or F21 in the figure) reached criterion in the fewest number of runs, and thus, required the fewest number of steps to reach criterion — 675 steps.

FIGURE 1.3.0T Run data

Tracking data: two forms

As Female 3 ran the maze, she was tracked through the maze. FIGURE 1.3.0T shows that she required 3 runs and 675 steps to reach criterion. Her 675 steps of data are displayed in five pages in printed form. See FIGURE 1.3.1—FIGURE 1.3.2AT.

FIGURE 1.3.1 screen data: the full set 76-unit data

FIGURE 1.3.1 is an image of the final set of FEU (First Entry Unit) scores generated by the Graph53 program used in these experiments on the computer screen at the end of Female 3's first complete loop. This set of data exhibits some hints of the critical new properties of mammalian adaptation, which are the experimental evidence for Heterostasis, the third wisdom. However, this data is still not in a very useful form, it requires further refining, but it is still worth studying. Notice that Female 3 did not enter all the units this first time through the maze. Actually, the set of units entered consisted of only 76-units of the possible 96 units. The full matrix consists of 24 units out and 24 units back for 48 units. The fact that Female three failed to enter 20 of the 96 units is interesting; Question, did Female 3 skip these 20 units by chance or for rational reasons?

What can you see in these 76-unit scores?

After examining this recreation of Female 3's first growth of 675-steps, and after seeing the 76-unit image of this behavior how would you characterize this behavior? Does it look like the embodiment of nonverbal rationality? Does it look like language behavior?

What kinds of behavior are there? What kind is this kind? How does mind-induced rational behavior differ from other kinds?

Random behavior

The behaviorists believed that the natural behavior of all animals was mindless behavior and was characterized by random behavior; that mindless behavior before it was reinforced was random behavior. Only after random behavior was reinforced would it begin to exhibit nonrandom properties.

Hull (1943) advanced a "Goal Gradient" theory to explain maze learning. A deprived rat moved from the start box to the goal box randomly on the first trial. Only in the goal box after the rat was reinforced would behavior begin to be organized. Learning grew from the goal box back to the start box. This theory predicted that the rat's rate of motion within the maze would get faster the closer they got to the goal box.

I begin by rejecting both the assumptions that mammals have no mind and that their spontaneous behavior is random. My first task is to show that Female 3 does not emit random behavior even though she is never deprived nor reinforced. The behaviorist had no objective understanding of the nature of random behavior, and thus, they were free to characterize the rat's behavior as random.

While Hull's (1943) goal gradient theory predicts that learning moves from the goal box back to the start box, I will show that the nonverbal mind's mastery of the maze moves from the start box to the goal box. Furthermore, I will show that Female 3 has conceived a valid model of the maze before she reaches the fourth row of the maze on her first time through the maze. She masters the maze long before she ever reaches what Hull calls the "goal box', which I call the Omega unit, since this unit contains no food and is not the rat's goal. The self-organizing rat's goal is to run from the alpha unit to the omega unit and back to the alpha unit. In other words, Hull's start box but in the matrix maze is both the alpha unit and the goal box, since the rat's purpose is to perfect the skill of running through the maze, out and back, as fast as possible.

Another way to state the rat's self-defined purpose is to run from alpha to omega one to omega two, with Hull's (1943) goal box omega one and his start box, first alpha and then omega two, first, alpha, second omega one and, third omega two.

Our examination of Female three's behavior evolves in two stages:

Stage 1: The 76 scores final screen data

Stage 2: 52 scores for three dependent variables. The dependent variables are FEU (Unit First Entry), BFEU (Between First Unit Entry), and FREQU (Frequency). Two tables of this 52 step-3 variable data set have been included, one for the real rats and another for the random rats. These two tables describe the special properties of the rational behavior emitted by the nonverbal mind. They define what rational behavior is and is not.

FIGURE 1.3.2AT, the Real rat data

FIGURE 1.3.2BT, the Random rat data

These two tables, in particular, the differences between the tables, provide the basic data essential for understanding the basic nonverbal rational narrative, which identifies the adaptive advantages a rational nonverbal mind confers.

Before I formally introduce and define the data of these two tables, I quickly want to make the most important distinction. My first task in the section is to define, objectively and operationally, random behavior and to show that mind-induced behavior is not random behavior.

	Random		REAL	=	Advantage
FEU	6,858,807	/	675	=	10,161

RUN	29,821 /	3	=	9,940
BFEU	6,485,981 /	431	=	15,048

The table above shows that by having a nonverbal mind; real rats had an adaptive advantage averaging about 10, 000 times over random rats. This table shows that this advantage holds for all three sets of scores.

In row one, the criterion scores for the FEU data shows that the real rat's advantage for this score is 10,161.

In row two, the criterion score for the run data shows that the real rat was advantaged 9,940.

In row three, the criterion for the BFEU data shows that having a mind gave the real rat an advantage of over 15,000. In other words, on the average having a mind provided mammals with a profound adaptive advantage — making this task about ten thousand times easier.

These scores can be found in the two FIGUREs in the following places:

In FIGURE 1.3.2AT and 1.3.2BT, note the last value in the third column: 675 in A and 6,858, 807 in B. The former number you have seen several times and is the number of steps Female 3 required to reach criterion: the latter is the corresponding number for the Random rat.

In summary, having a rational nonverbal mind gave Female 3 an adaptive advantage of about ten thousand times on the average, but so far we have found no hint of the nature of rationality, which confers this advantage.

Definitions: the three new Heterostatic dependent variables

Both FIGUREs 1.3.2AT and BT consist of tables of data consisting of 51 rows by five or more columns. Each of the five columns contains, either one of the two independent, or one of the three dependent variables. Each row vector describes the complex first entry state for one of the fifty-one units of the Essential Path, EP, through the matrix maze. Column 1 provides an index of the 51 rows. Each index shows the ordinal position of each of the 51units of the EP within the Criterion loop. Column 2 shows the names for each of the 51 units of the EP. For example, note that the 26th unit of column one is unit 62, which is the omega unit. The last three columns, three through five, contain the data for one of the three dependent variables: FEU, BFEU, and FREQU respectively.

These fifty-one scores consist of three subsets of scores: first, the 24 units out; second, the 24 units back; and third the alpha-one, omega, and alpha-two units: 24 + 24 + 3 = 51 units. Note that unit 6 appears twice in column B--as unit one and unit fifty-one. The omega unit is named unit 26 in column A and named unit 62 in column B. Each row of the data contains a snap shot of the momentary causal state of behavior associated with each of the 51 units of the Essential Path, EP. Jointly these scores describe the complex roles each unit of the matrix played as the nonverbal mind directs the search for the shortest path. These dependent variables begin to reveal the workings of the nonverbal mind.

One of my most useful achievements was the design of this set of three Heterostatic dependent variables, which uniquely describes the manifest rational features of Heterostatic behavior.

FEU real

The FEU variable, seen in column three of FIGURE 1.3.2AT, is the first of the three dependent variables. The First entry unit, FEU, records the step number upon which a rat first entered each of the fifty-one units of the maze. The FEU scores consist of two subsets; the Essential path units, EP, and the blind path units, BP, but only the EP data is shown in this FIGURE. One of the many difficult things that I had to discover the hard way was that the rat's rational mind creates two different models in order to understand the spatial structure of the maze. In this first experiment, I only discuss the evolution of the first model, the model of the EP behavior. In later experiments, I discuss the growth of the BP model.

The basic critical fact in this regard is that an ANOVA analysis of a set of data cannot be made until essentially all of the cells of a table have at least two scores. If the reader will look once again at FIGURE 1.3.1, they should see two things; first, that the entire EP cells have scores; second that only about half of the BP cells have scores. The BP data can only be statistically analyzed after each rat is allowed to run for several growths, enough to get two scores in each of the BP cells.

The Essential Path, EP, subset consists of 51 units, (48 units out and 48 units back) plus the alpha unit accounting for two units; one on the way out and one on the way back and the omega unit counting for one unit. This 51-step round trip through the EP of the maze defines the criterion loop. A Computer program was written to select and extract the values for the 51-FEU criterion steps from the original 675-step loop. For example, row 29 shows that Female 3 first entered omega, unit number 62, on step 632.

The same computer program also then computes the values for the other two sets of 51 scores, for both the BFEU and FREQU scores.

In order to understand the nature of the FEU behavior, the reader may want to look once again at the numbers on the computer simulation generated during the first growth of Female 3. The FEU data in column three of FIGURE 1.3.2AT are a subset of the numbers on the screen seen in FIGURE 1.3.1 but again only those scores from the EP. For example note that in FIGURE 1.3.1 the number 122, 123, 124 and 555 are all from the EP set of scores—NOTE-The four BP scores: 154, 155, 125, and 126 are not included in FIGURE 1.3.2AT.

The FEU data shows up several times in this section, but each time in a different context suggesting that this type of dependent data is most useful, the reader will learn to recognize this FEU score in all its guises.

- It first emerged on the computer monitor as Female 3's data was displayed by Graph53.exe.
- FIGURE 1.3.1, is an image of the final screen data left on the monitor when all of Female 3's data had been displayed. This image is of a subset of the 96 scores of the Full-maze set of FEU data.
- In FIGURE 1.3.2AT the Essential Path, EP subset of the FEU data is seen as the third column of FIGURE 1.3.2AT. This data is a subset of the 96-element subset of 1.3.1. Next, this EP data is seen in FIGURE 1.3.3 as the prototypical Heterostatic growth curve.

- Finally, this EP data is seen in FIGURE 1.3.6 combined with the Female 3 loop data.

Each growth is a master loop; an alpha-omega-alpha loop, not a trial loop; but a criterion loop. Each subject is removed from the maze upon reaching the criterion of making one round trip through the maze: from alpha to omega, and back to alpha.

BFE real

In FIGURE1.3.2AT, the fourth column shows the BFE (Between First Entry) score values for F3 and in FIGURE 1.3.2BT, the sixth column shows the BFE score values associated with a random rat with each row or unit of a criterion loop. Look closely at the data for Female 3. In the BFE column note that there are only three large scores and all the rest are essentially ones. Look closely and see if you can figure out from the data in the table: one-which maze unit the large scores were associated with and 2-next look at the score in the previous column, the score preceding the large BFE score. How was the BFE score computed? In order to compute the BFE score, one simply needs to take the FEU in Column C and subtract two successive scores. For example, for units 30 and 29 the scores are 555-124, which equals the BFE of 421 shown in Column D for unit 29.

The three units with the large FEU scores are the three anomalies of the maze according to the mind of Female 3.

These three large FEU scores deserve to be recognized as the central characters in the narrative being written by the nonverbal mind.

FIGURES 1.3.2AT and BT

FIGURE 1.3.2BT is a six column by fifty-one-row set of data for a random rat. The extra column is the fourth column, which shows the run number. For Female three who only required three runs to reach her criterion, it hardly seemed worth the effort to include this score, but a random rat's skill or lack thereof justifies the inclusion of this score. The run number score identifies the run number upon which a rat entered each unit for the first time. For example, note that the random rat entered the first eleven units on the first run, but only entered two new units on the second run, none on the third run, then one on the fourth run. This score is just one more way of showing how much harder it was to make each new unit first entry. The random rat finally required six million four hundred and eighty five thousand nine hundred and eighty one runs of 228 steps to reach criterion. In the 1990's when these computer simulations were run, I would take over all the computers in the college computer lab and run these simulations overnight. Some of the runs required several days over the weekend.

FIGURE1.3.3

FIGURE 1.3.3, shows a prototypical heterostatic growth learning curve. The data points plotted in red in FIGURE1.3.3 are the first entry unit, FEU, data for F21 (F3), for the 51 units of the Essential Path of the matrix maze, 24 units out and 24 units back plus the alpha, omega, and alpha units scores. On the ordinate, please note that the Alpha unit is both the first and the last unit. Because of the font size not all the 51 units are printed. Please note that the numbers on the ordinate correspond to the units of the Essential Path, they do not include the Blind Path units. Only half the unit names are plotted, with the unit names that are plotted being those between the tic marks. Omega, for example is one of the units not printed and would fall on the line between the two 54's. There are four EP units per row for each of the six rows in both directions. The data is plotted as the rate of exploration of the rat in the Essential Path, in the style of Skinner. With his cumulative recorder, Skinner (1974) was able to graph the operant data, rates of lever pressing, as the slope of the line drawn by the cumulative recorder. In FIGURE 1.3.3, the FEU data is graphed so that the slope of the line is proportional to the rate of exploration. The steeper the slope, the faster the rat is exploring.

The FEU data is very strange compared to the traditional smooth learning curves that characterize the TTP. The curve of 1.3.3, as noted above is a synthesis of two kinds of loop data, the FE (First Entry) and the RD (Redundancy) with the FE data having a steep slope while the RD data has a zero slope. The two most obvious features of this FIGURE are the RD2 and the FE4 segments shown labeled in FIGURE 1.3.6 as R2 and F4. The former is the long Flat-Place in the center of the record, while the latter is the steep slope of the last segment. The difference between these two phases seem almost like what the gestalt psychologists called an insight curve with the rate of exploration seen in FE loop 3 showing a sudden acceleration, following the long "flat place" in the center of the curve caused by RD loop 2.

According to the theory of Heterostasis, when a rat is evolving a mental map, it creates primary knowledge, which is stored in a nonverbal map as opposed to genetic knowledge, which is stored in the genome. Primary nonverbal rational knowledge is a measure of the cognitive order created within in the mental map in order to destroy the primary uncertainty induced by some anomaly. Female three had to spend 431 of her 675 moves, or sixty four percent of her total maze activity, to accommodate this lower level of the hierarchy before she was able to enter unit 29. Once F3 had created a representation in her mental map that was valid for this particular order of the matrix maze, the puzzle of the maze was essentially solved as evidenced by the fact that she suddenly stopped retracing. The third phase of the FE data was almost a vertical line.

This data of FIGURE 1.3.3 shows that a real rat's exploratory behavior is not the product of some internal random number generator as the classical trial and error learning theory assumed. The primary mind of the rat actively emits somatic responses, which are directed by the knowledge stored in spontaneously evolving mental models of the matrix maze during exploratory-play. The purpose of this set of exploratory-play mechanisms can be stated, either cognitively or behaviorally. The cognitive purpose of the Heterostatic mechanism is to create a valid cognitive map of the maze, one that can generate efficient exploratory behavior. The behavioral purpose of this mechanism is to perfect the criterion loop. The degree of perfection of the criterion loop is an expression of the validity of the mental model generating the loop. It is assumed that a valid mental model perfected during exploratory-play could have a critical survival value later. Both the brain mechanism, which evolves play behavior, and the evolution of play behavior by the brain, would have evolved by natural selection.

The active rat is purposive, having the capacity to organize its behavior into a series of short and long-term goals. The goals evolve or unfold, from the short to the long term, beginning in the alpha unit at the inner most level of hierarchy, and working their way through the outer most levels of the hierarchies.

These three outer levels (units, row types, and replication) are integrated instinctively and quickly into organized exploratory-play behavior in the Lashley III maze. This complex set of behavioral features show that the rat's map or model of the maze evolves in stages, beginning at the alpha unit. The active rat has the capacity spontaneously to constrain its behavior to subsets of the domain of the Lashley III matrix maze, which are hierarchically appropriate to both its long and short-term goals and sub goals. It has the capacity to retain, to promote, or to demote itself to the grade of the domain hierarchy that it needs to explore in order to evolve in its mental model of the maze the particular hierarchical level that requires development, at any moment. The hierarchical representation of the perceptual and motor units of behavior evolves in parallel. As was noted above, the degree of perfection of behavior at any moment is an expression of the validity of the model at that moment. An invalid model is obviously not going to generate appropriate behavior. Exploratory-play behavior will spontaneously oscillate between exploring and retracing. Valid models will terminate retracing and initiate exploring. Invalid models will terminate exploring and initiate retracing.

The purpose of this section was to show how the outer three hierarchically organized behavioral levels are translated into experimentally defined Lashley III maze variables (A2, B3, C2, D8) and how these sets of variables collectively reveal and describe the hierarchy of purposes that evolve as the active rat explores and models the Lashley III Matrix maze. This data also reveals what units of the matrix maze were anomalous and induced uncertainty. During retracing, rats create a mental model that reduces the uncertainty induced by these initially anomalous units.

FIGURE 1.3.3 Heterostatic growth curve

FIGURE 1.3.3 is a plot of the 51 elements of FEU scores, which were emitted within the Essential Path units of the maze, out and back. In other words, this FIGURE shows the progression of Female 3 through the Essential Path units of the maze. This static curve freezes the dynamic pattern of FEU behavior and reveals that this data consists of two periods of behaviors. First, there are several periods of "flat-place" behavior, and then several periods of "steep" behavior. During the flat-place periods, the rat's behavior shows no improvement. On the other hand, during the steep-periods of behavior, the rat's behavior manifests great improvements. What do these two periods of behavior tell us about the nature of rational behavior? How can nonverbal rational behavior be explained as a product of these two periods?

FIGURE 1.3.3A-Two classic "learning" curves

The purpose of this book is to describe objectively the structure and function of the nonverbal mind. Several metaphors will be employed to guide the reader's imagination to achieve various insights. First, in order to understand the nonverbal mind, it helps to be able to "read the rat's nonverbal mind." In this section, I invite the reader to compare three sets of data, or three kinds of learning. First, consider two classic types of learning curves Hull (1943) and Kohler's (1925) curves. Hull's curve is an S-shaped or negatively accelerated curve. Mathematicians have been debating whether a learning curve is an S-shaped curve or a negatively accelerated curve. The Gestalt psychologist, Kohler (1925), created what is known as an "all or nothing," or "one-shot" learning curve. These debates are irrelevant since the testing procedures of TTP (Trial Testing Procedures) are an artifact.

FIGURE 1.3.3A shows these two kinds of curves: Curve A is Hull's curve; curve B is the Gestalt curve. Which of these two classic curves does the curve of FIGURE 1.3.3 most resemble Hull's or Kohler's?

Kohler's curve also consisted of two components one flat and one vertical. The flat segments of the curve were understood as indicating either that no learning is occurring, or at least that no learning was being manifest, while the vertical component indicates first, that learning does not occur incrementally as proposed by Hull (1943), but rather that learning occurred all at once, as they say, "in one shot." Kohler (1925) called this kind of behavior "Insight Learning" or "one-shot learning" stressing the all or nothing nature.

In the first flat-place phase of the Gestalt curve, no learning is manifest, but at the end of this period, suddenly an insight occurs. In insight learning, the improvement occurs suddenly and completely. Kohler (1925) claimed that Sultan initially made no progress during the training sessions, but then suddenly had an "insight" and thereafter performed the task without error. His point was that learning occurred all at once, and not incrementally. Sultan's problem was to get a stalk of bananas that was hung from the ceiling. First, he tried jumping and various other means with no success. Finally, all of a sudden, he had the idea to use several boxes that were scattered around the room. Sultan proceeded to stack the boxes and climb on top of them to get the bananas. After this insight, he was always able to get the bananas.

Flat-Place Behavior: What is Flat-Place Behavior?

Of course, none of the classical learning theorists claimed that learning was a rational function. A primary goal of this section is to define rational behavior to, at least, operationally define what nonverbal rational behavior looks like. Which of the two classic types of learning curves best represent the behavior of Female 3 which is seen in FIGURE 1.3.3A? The similarity between the two periods of Female 3's behavior and Sultan's behavior are, at least to me, most obvious. The data looks alike; however, Sultan's problem was simple and only required one insight, while Female 3's problem was a compound problem. Each level of the hierarchy required an independent insight. So in FIGURE 1.3.6, each flat place represents one of the insights she achieved. There were three insights; each flat place was followed by a steep slope shown in the compound curve.

FIGURE 1.3.3A shows that the behavior of the rats consisted of two kinds of segments a fact-finding period when the rat's nonverbal mind is evolving a model of the maze and an extrapolation segment in which the animal no longer has to think but only to act with conviction. During the extrapolation period, the primary mind emits a trajectory of behavior with the full confidence that it knows what it is doing; in this period of behavior, the mind is acting without fear, without hesitancy, without uncertainty, and with self-confidence.

In other words, Female 3's behavior looks like a complex version of Kohler's insight curve. One difference is that Female 3 required a series of insights to solve the problems presented by the four-factor hierarchically organized maze, yet each insight consisted of two components one flat and one vertical section. When rats retrace, their behavior obviously does not manifest any forward motion; or mental growth but the longer they retrace, the longer the flat place. Since Female three made three major retrace loops, her behavior reveals three flat sections in the record. Each of these retrace-periods ended with an insight, which is manifest as a sudden spurt in performance; but for Female three, each insight ended after some limited degree of progress; ended abruptly when the next anomaly was encountered.

However, at the end of the third and longest retracing period, after the 431 steps retrace, Female 3 then proceeded forward swiftly and surely without error for the rest of the way through the maze. This manifestation of the final insight can be called an extrapolation.

What is an extrapolation?

Rational behavior consists of three components:

- In the first period a model, program, map, or script is written, evolved, or constructed.
- In the second period, once a domain-specific purpose has been identified, a plan is generated; a plan is an extrapolation from the model.
- In the third period, Female 3 emits the plan, which consisted of a trajectory of almost error-free responses or behaviors.

In mammals, nonverbal rational behavior is a manifestation of such a plan. Rational behavior is a unique kind of mammalian nonverbal behavior that expresses or manifests the knowledge contained in a model. Nonverbal rational behavior is a product of a rational nonverbal mind.

The mammalian nonverbal rational mind is the product of the second kind of evolution. Biological evolution produced what Darwin (1859) called the "The Origin of the Species." Darwin's name for his book, however, overstated the case of this first kind of evolution. Darwin was correct, biological evolution is responsible for the original set of genetically designed species of life. However, Darwin failed to recognize there are two kinds of species, biological and mental.

The Heterostatic behavior of the nonverbal mind, however, is responsible for a new set of rationally designed mental species, ideas, or ideals. Biological evolution is the product, as Darwin (1859) noted of "natural selection," while mental evolution is the product of "rational selection."

The purpose of this first experiment is to show systematically how the nonverbal mind creates a coherent hierarchically organized set of forms, ideas, or causal concepts by rational selection, how a nonverbal mind creates or evolves new species of mental beings, new species of ideas. In a philosophical context, a being simply plays on the most fundamental word in the English language "to be." When something is being, it simply is. Plato (1952) uses the word becoming, if something is becoming it does not exist. A being, on the other hand, is something that is or exists. A spiritual being can only be taken seriously on faith, but a material being can be known by science.

This extrapolation phenomenon gives new meaning to the concept of rationality. Plato long ago assured humankind that they were special because they alone were rational, that humans alone have the power of verbal rationality, but the behavior of Female 3 now shows that humans have no monopoly on rationality; moreover, we now know that verbal rationality is not the only kind or even the primary kind of rationality. The nonverbal rational behavior of the nonverbal mind is the only organ that can give birth rationally to nonverbal rational forms. The nonverbal mind is the womb of the soul!

Nonverbal extrapolation behavior is the *sine qua non* of nonverbal rational behavior. Extrapolation is the product of a plan. Rational nonverbal behavior is the product of the execution of a plan. The advantage of rationality is that it empowers mammals to emit trajectories or scripts of behavior that are extrapolations of some rule, which can be executed without error.

Some may find it paradoxical or ironic in a sense when I say that, "as a human being I feel jealous of the nonverbal mammals." On the one hand, they have a disadvantage, they have to learn on their own, but at least they never have anyone to fill their heads with metaphors like parents, churches, or schools. The advantage of the rat is it gets to learn firsthand, while the disadvantage for humans is that we have to learn second hand. Human's minds are filled with what we are told by our schools, by our teachers, or by our parents, preachers, etc. which are often myths. They are filled with inaccuracies and falsehoods. Alternatively, more simply, the difference is the nonverbal mammals live in a nonverbal world while humans are raised in a verbal world. The verbal world is filled with often well-intentioned myths and inaccuracies that fill our mind; the rat's nonverbal world, on the other hand, is incapable of lies or deceit.

The GPS model

When we humans ask our computers to print a copy of a proposed route for a trip, where we specify the start and endpoint, the computer printout is a plan. What we do on our trips is to execute the plans our computer program generated. A plan is an extrapolation of the model. The amazing rational feat is that our computers have the ability to print out a plan, but they only have this power because some programmer initially gave them a model of the domain we intend to cover on our vacation. However, no one gave Female three a model or map, yet her behavior in the maze clearly revealed that her nonverbal rational mind conceived a map, conceived the extrapolation, and then executed the plan.

Plato (1952) was correct. Rational behavior is a three-stage function. The first stage is the creation of the model; Plato called this, the god function. The second step, the preparation of a plan, Plato called the human function. The execution or third stage is, of course, also a human function.

Plato (1952) imagined that humans were limited in their rational ability, as today we know that computers are limited. Human programmers execute the god-function for computers. Computer behavior is limited to the execution of the programs.

Our car's navigation systems get their geographic or spatial knowledge, their map, on a DVD. The computer program in our navigational technology computes a plan. All we humans have to do on our vacation is to execute such plans. Female three, a ninety-day-old rat, is equally proficient at executing all three of these functions!

The difference between Sultan's behavior and Female 3's behavior is that the maze presented the rat's mind with a complex problem; a series of problems, the solution of each required an insight. Each of the four factors in the hierarchically organized maze required a separate two-phase solution, a retrace period followed by an insight period. After each insight, Female 3 proceeded forward through the maze until she encountered yet another anomaly and once again, she was unable to move forward until she had retraced and found the causal conditions that took the uncertainty out of the last anomaly encountered, etc.

The BFEU data

The BFEU (Between First Entry Unit) dependent variable is defined as the difference between two consecutive FEU scores. For example, in FIGURE 1.3.2BT the BFEU for the random rat in row 12 is 253, which is the difference between the two FEU scores in rows 11 and 12 (356 – 103 = 253).

The BFEU score is a measure of how many steps a rat had to take in order to move between each pair of consecutive maze units. In FIGURE 1.3.2AT (the information for Female 3) the maze unit names are shown in column B. The maze unit of row 10 is 30, while the maze unit of row 11 is 29.

Female Three's BFEU score for unit 29 is 431, which says that after entering unit 30 an additional 431 steps were required before entering unit 29. The 431-step difference between unit 30 and unit 29 were spent retracing.

Random rat retracing

The rule for random behavior is that the difficulty of a run is an exponential function of the length of the run. The Random rat BFEU data of FIGURE 1.3.2BT demonstrates this rule. For the longest run of 51 units the random rat required 6,485,981 steps, Six million four hundred eighty five thousand nine hundred eighty one steps. For a run length of 12 units, this random rat required 253 steps, but generally, all the subsequent BFEU scores kept getting larger and larger.

Two BFEU graphs

This section focuses upon the second of the three dependent variables, BFEU, of FIGURE 1.3.2AT. In the last section, the focus was upon the FEU score in the fourth column of this FIGURE; the BFEU scores are in the fifth column. Two FIGUREs display this BFEU data: FIGURE 1.3.4A and 1.3.4B, each in a different way.

FIGURE 1.3.4A BFEU

FIGURE 1.3.4A places the large BFEU scores within the units of the maze where they occurred. It is critical that each of these scores be understood as a function of the specific unit element of the hierarchically organized structure of which they are a part. This FIGURE shows where the three largest BFEU unit scores occurred within the maze, and thus set the stage for a discussion about what features of these maze units provoked the retracing; what features could have caused the uncertainty triggered by these large BFEU scores in a nonverbal mind. The placement of these scores provides a foundation for our attempts to read Female 3's mind. Two questions are paramount; first, what features of these units surprised her, and secondly, what kinds of ideas did the nonverbal mind have to conceive in order to destroy the capacity of these units to induce surprise? Why did these scores occur in the particular units of the maze where they did? What were the features of these units of the maze that the primary mind found so anomalous?

FIGURE 1.3.4A identifies the three particular BFEU scores that Female 3 found anomalous; so difficult to digest; these scores are 101, 431, and 53.

The first anomaly occurred in unit 19, which is unit E1 in row two; E1 is the unit behind the door; the unit in which each row is entered. The UEBFE score for this unit is 101, which means Female 3 had to take 101 steps after she first had a chance to enter this unit before she was actually willing or able to do so.

Question: why did unit 19 initiate retracing, and why did it initiate just this amount of retracing?

FIGURE 1.3.4B BFEU CURVE

FIGURE 1.3.4B plots the BFEU values of FIGURE 1.3.2AT column four. This FIGURE shows that almost all the BFEU scores has a value of one, which says essentially that the transition between most of the units of the maze was made without any retracing required; retracing was only initiated by three units. This FIGURE highlights these three points and shows that all three retracing scores occurred in the first half of the maze, actually the first quarter of the maze. Furthermore, the three points 101, 431, and 53 reveal that retracing behavior is a curvilinear function. Retracing increased up to a maximum of 431, then quickly decreased to 53 and then to one.

The comparison of FIGUREs 1.3.5T and 1.3.6

In the first lesson on mind reading, the FEU behavior of FIGURE 1.3.3 was compared to two kinds of classic learning curves in FIGURE 1.3.3A.

In this second lesson on mind reading, the loop behavior of FIGURE 1.3.5T will be compared to the combined FEU + Loop behavior of FIGURE 1.3.6. This comparison should provide a deeper understanding of how the rational nonverbal mind works.

Female three solves the nonverbal problem induced by the anomalies of the maze by engaging in exploratory-play behavior employing two kinds of loops - retrace and exploratory. The paradox is that an anomaly induces in the nonverbal mind an almost overwhelming state of uncertainty, which usually is a positive; it is perceived as an opportunity to grow. Rather than inducing depression, uncertainty induces a state of excessive excitement or puts the animal in the C-state on the Yerkes-Dodson Law. The usual rational response to an anomaly is for the mind to enter a state, which begins a growth cycle; what may seem paradoxical is that mammals in this state begin to retrace. The classical behaviorists despised retracing; they conceived it a waste of both the animal's and the experimenter's time. In a military term, this behavior could be conceived as retreating, but in this context, the question is what opportunity does a state of uncertainty induce?

The basic fact is that an anomaly is an effect without a cause. The only rational way to eliminate this state is to discover a set of causes for this unpredicted effect. The nonverbal rational mind is nativistically optimistic; while the effect was unpredicted, there is no reason to believe it is unpredictable. So where can the animal go to find a cause? Retrace! Retrace! The only rational option of the rat is to retrace. During retrace loops, the nonverbal rational mind searches and re-searches the region of space prior to the anomalous unit; the rational mind must find a cause or set of causes that will take the uncertainty out of the anomaly by supplying a cause for every effect. The nonverbal mind nativistically understands the cosmic nature of hierarchical spatial structures. In a causal universe, every effect by its very name must have a cause! The prime directive of the nonverbal mind is "discover a cause for every effect." Every effect initially emerges in consciousness as an unexpected effect, or an anomaly. The mind understands the hierarchical structure of the maze when it has found a cause for every effect, when no anomalous effects exist within the maze.

Just as the eye was designed in the image of light, and the ear in the image of sound, just so the nonverbal mind was designed by natural selection in the image of god or the cosmos or nature. The rational mind emerged to create a new, better, or faster way to create survival-knowledge, to create or to replace the role of natural selection in the design of the genome. What we need is a name for the place in the mind where this knowledge is stored; we need a name for the home of our mental gene pool.

Therefore, in the first phase of a growth cycle the nonverbal mind induces the somatic structures of the body to retrace so that a cause can be found to take the uncertainty out of the current anomaly. Once the anomaly has been destroyed, and a revised model has been created, the nonverbal mind issues an all clear signal, so that exploring can begin again.

Two questions: why retrace - because that is where the causes are to be found. Why explore - because that is where anomalies are to be found.

FIGURE 1.3.5T LOOP DATA

FIGURE 1.3.5T shows that Female three made seven loops; four First Entry (FE) and three retrace or redundant, RD, loops, and shows how many steps Female 3 made in each of these seven loops.

In three of the FE loops, the numbers of first entries made were small: 4, 5, and 3 respectively, but on the fourth first entry loop the number of FE scores made was 36! The total for all four loops is 48. This data shows that Female 3 required four insights to master this maze. The consequences of each insight were small for the first three, but the last insight really paid off big.

The steps made during each of the retrace loops were respectively:

RD1 - 101
RD2 - 431
RD3 - 53

The numbers have been encountered before but in FIGURE 1.3.4A these numbers were called BFEU scores, or the large anomalies scores. Therefore, the large BFEU scores, and the length of a retrace loop scores are different names for the same thing. Just the second redundant loop total steps (431) were far larger than the other two scores.

Run, Loop, and Growth Interrelations

FIGURE 1.3.6 the FEU-loop data combined

FIGURE 1.3.6 combines the data of FIGURE 1.3.3 and FIGURE 1.3.5T. The line graph is from FIGURE 1.3.3. The data shown above and below the line is the loop data of FIGURE 1.3.5T, with the numbers above the line the three RD loop totals and the data below or to the right of the line the FE totals.

The central feature of the FEU curve is the great central Flat Place or Plateau generated by RD2, which lasted for 431 steps. In the FIGURE, note that the end points of this RD loop were units 30 on the left and unit 29 on the right. In other words the great third row mystery is: why did female three find the distance between unit 30 and 29 so difficult as to require 431 steps of retracing: what feature about this place in the maze required so much mental effort, so much computing to understand. Why did unit 29 induce so much uncertainty why did it require 431 steps to assimilate or accommodate?

The loop data of FIGURE 1.3.6 is included to show how the FEU growth data is related to the loop data. This FEU growth data is the product of the next two innermost aspects of the hierarchically organized behavioral data, runs, and loops. With an inspection of this FIGURE, the reader will observe that this FEU data obviously or naturally is divided into seven segments.

Recall there are four EP units per row for each of the six rows, in both directions for a product of 48, that one complete loop through the maze required

48 First Entries. Recall that the steeper the slope of the curve, the faster the rat is exploring and conversely a flat place in the FIGURE reflects the fact that a rat is not moving forward but retracing. Therefore, the large Flat Place in the FIGURE reflects the behavior of the Retrace or redundant loop two, while the steep progression in the last third of the FIGURE reflects the behavior of FE4.

FIGURE 1.3.6 makes the point that the emergent growth curve of FIGURE 1.3.3 is the product of two different types of loops: FE-loops, and Retrace-loops. This growth curve is the embodiment of nonverbal rational behavior. The FEU growth curve is the prototypical nonverbal rational Heterostatic growth curve emitted by the primary rational mind. This curve is independent of rat behavior; it is the prototypical Heterostatic FEU growth curve of mammals. This curve should be observed in every mammal given an analogous problem. Moreover, this curve should also be independent of any rational wisdom. This curve reveals the essence of the rational behavior of any mammal mastering an anomalous domain; it should also be observed by humans mastering any wisdom; such as learning a piece of music, etc.

Any mammal exhibiting this behavior can be called a “Heterostat.” The unique pattern of behavior is the prototypical Heterostatic behavior. When any mammal acts this particular way, they are acting as a Heterostat.

This concrete pattern is but one concrete instance of an infinite set of abstract Heterostatic patterns, just as a red right triangle on paper is a concrete instance of an abstract triangle for which there are an infinite set of particulars?

This same pattern should be seen in humans in music, sports, science, and dance or in any domain specific behavioral domain where mammalian behavior occurs. But likewise, it should be observed in any species of nonverbal mammal's behavior, rats to bats to whales, etc.

The growth cycle

The growth or insight cycle consists of two phases with a retrace loop followed by a FE loop. FIGURE 1.3.6 shows that F3’s FEU curve consisted of seven loop elements. The seven loops of F3's behavior, which are shown in FIGURE 1.3.5T are listed below:

Cycle one
FE1. First-entry loop #1 with 4 FE's

Cycle two.
RD1. Redundant-loop #1 with 101 steps
F2. First-entry loop #2 with 5 FE's

Cycle three.
RD2. Redundant loop #2 with 431 moves with run-two and run-three ending at move 234 and 468 respectively
FE3. First-Entry loop #3 with 3 FE's

Cycle four.
RD3. Redundant loop #3 with 53 moves

FE4. FE loop #4 with **36 FE's**

FIGURE 1.3.6 shows how Female 3's loop cycles alternated between RD and FE loops, with the number of FEs per FE loop taking a dramatic increase over time: 36 FEs occurred on the last FE loop, while only four FEs occurred on the first FE loop: (4 + 5 + 3 + 36 = 48).

An examination of the RD loops shows that the number of moves per retrace or redundant loops for Female 3 is an inverted u curve: 101, 431, and 53.

This data shows how the three levels of the hierarchal structure of the exploratory data are interrelated, the loops are within the runs, and the runs are within the growth. Unfortunately, it is not possible to show the node\path data within the loop data in this FIGURE.

The two most dramatic features of all four-growth cycles are RD2 and FE4. The former is the long flat-place in the center of the record, while the latter is the steep slope of the last segment.

The difference between these two phases seems to capture what the gestalt psychologists called an insight curve with the rate of exploration seen in FE loop 3 showing a sudden acceleration, following the long "flat place" in the center of the curve caused by the RD loop.

On the other hand, the flat places in this exploratory data looks very much like Bryan and Harter's data for their original human learning curves, which was based on skill learning curves of telegraph operators. Bryan and Harter (1897) called the Flat places in their data "Plateaus." A plateau or a flat place in our data reflects a period of behavior during which the rate of progress is zero. The length of a Plateau in Female 3's behavior reflects the duration of this period of retracing.

Coincidently, while I am giving credit to those who anticipated some aspect of this heterostatic behavior of the nonverbal mind, I should also mention the concept of "punctuated equilibrium" made famous by Stephen Jay Gould (2002), the Harvard evolutionary theorist who rightly enough did observe in his own behavior many phases of punctuated equilibrium. His phrase "punctuated equilibrium" is another interesting variation upon, "an all or nothing behavior," or "one-shot" learning. Gould's idea that follows could also be used for F3's learning.

"Only years later-and perhaps serving as a spur to my later interest in punctuated equilibrium-did I conceptualize the possibility that plateaus of stagnation and bursts of achievement might express a standard pattern for human learning, and that my previous frustration (at the long plateaus), and my exhilaration (at the quick and rather mysterious bursts), might only have reflected a false expectation that I had carried so long inside my head-the idea that every day, in every way, I should be getting just a little bit better and better." (Gould, 2002, P.957)

The rational growth curve of FIGURE 1.3.3 then can be considered the product of a series of Gestalt insight segments. Kohler's (1925) chimp "Sultan's" insight curve had only one flat place followed by an insight. Female three's problem was too complex to be solved with one insight; since the maze is a four dimensional hierarchical structure, theoretically it should require an insight to conceptualize each grade or dimension. Remarkably, Female 3's growth curve consisted of only three flat places, thus her rational mind abductively solved the puzzle with only three major insights. The conceptualization of the direction dimension she seemed to solve effortlessly.

Two metaphors; the grade school, and the Digestive model

There are many useful metaphors for understanding the Heterostatic behavior of mammals. Another one to be considered in this section is the most important grade school metaphor.

Pestalozzi (1859) designed grade schools by dividing the matter into subject domains like reading, writing, and arithmetic, etc., each of which he divided into grades, which were divided into lessons.

The plan was to organize the material in each domain into a kind of linear sequence, from simple to complex, through which each student could pass. Exactly how and why each grade's material was conceived is not clear to me, but in the case of the maze, conceived as an A2, B3, C2, D8 conceptual structure, the domain-specific material of the maze can be parsed into four grades.

The first grade: the evolution of Factor Four D8

The nonverbal mind begins its mental digesting behavior in the alpha unit, and the first grade mastery of the maze took place in the first eight units of the first row before unit 19 was ever encountered by the rat.

The first grade question, then, is what rule was controlling Female three's behavior in unit 19, why was she not able to turn left? Why did she have to retrace for 101 steps before she was able to conceive and emit a left response?

Hypothesis one: Unit 19 was anomalous because it was the first instance of the West-type, or left-turning row-type. On entering the first row, the correct response was a right turn in an East-type row and thus in the second row rats tended to make the same response that had been successful in the first row.

Pavlov

One of Pavlov's (1960) basic concepts was External Inhibition. The principle of External Inhibition states that a novel stimulus has an absolute power to immediately and absolutely interrupt or disrupt behavior in the face of surprise. Pavlov's concept was the scientific version of Naomi Klein's (2007) "Shock Doctrine." Surprise is *prima face* evidence that the nonverbal mind was not prepared for the event that induced the surprise. External Inhibition requires rational beings to at least temporarily stop and examine a situation in the face of novelty before resuming exploration. The concept of retracing is sort of a follow up on Pavlov's idea of external inhibition.

Once an anomaly or a novel stimulus has produced External Inhibition and stopped a line of behavior, what is an animal to do next? Retracing is an overt manifestation of the mind's recognition that it is not prepared to process or assimilate some novel or surprising event. The surprise induced by an anomaly is an overt manifestation of the arousal response, which activates the abductive logical functions of the nonverbal mind. (Conkin, 1976) The nonverbal mind must retrace until a set of conditions can be identified that destroy the uncertainty induced by the anomaly. In order to destroy the power of an anomaly to surprise, the nonverbal mind must discover a set of causal conditions that predict the anomaly and destroy the surprising power of this event.

Grade One:

In the first grade, or in row one, for example, the mind was required to evolve the lowest level factor, D8; to evolve a causal set of concepts that can reduce the uncertainty induced by the eight units of the first row, or the eight cells of the first row.

The discriminative powers of the mind in the first row were exclusively focused upon distinguishing the eight units, of discriminating say eight rooms of a house. However, notice in the first row the idea that this is the first house, house type A, never enters consciousness. In the first row, the mind had not yet been required to focus upon identifying this feature, which will become the subject of the second grade, which will be to distinguish the two row types, house types A and B.

Every anomaly, Lindsley (1960), and then Gastaut (1957) showed desynchronizes the cortex and arouses the organism by activating the SNS (Sympathetic Nervous System). This activation turns on the abductive, problem-solving function of the mind. An aroused mind must discover a set of causal features that can reduce and/or prevent the arousal response induced by the anomaly.

Each relative anomaly ultimately must induce a new grade, the mastery or successful modeling of each dimension of the hierarchy requires a dimension-specific grade, each dimension contains a unique relative anomaly. What is the rule? Where do the features, which define each new anomaly, which require the mind to conceive a new idea, a new dimension of the causal structure of the maze, lay in wait for the NV mind?

Grade Two

The BFEU (Between First Entry Unit) score of 101, just so happens to be the unit, which first required F3's mind to begin to evolve the idea of C2, the row-type dimension. Unit 19 is the unit where her mind first encountered a set of facts that forced her to recognize that her model of the maze was invalid. She was forced to consider there was another kind of row, and thus her mind was forced by the anomaly to conceive a more complex model of the maze, to conceive a row dimension for the hierarchical structure of the maze. Since there were eight rooms of columns in the first, or east row, then there must be eight columns in the West type row; in other words, there must be two kinds of houses A and B with eight units, the A8 dimension.

Grade Three: Grade three-Factor two; the replication problem

In grade three, the mind must conceive a solution for the anomaly that lay in wait in the third row, an anomaly, which would require the evolution of an additional factor, a third factor; B3; the Replication factor.

The anomaly of the third row required F 3's nonverbal mind to evolve, to conceive a new grade-specific idea, new base-place factors, to remove the uncertainty induced by this specific anomaly. (For a review of place-base, factors please see Book 1) However, the rule is that every relative anomaly encountered at every new anomalous choice point requires a new idea, a new dimension to be added to the form. An anomalous choice point is defined as a place where the organism does not know how to respond; which way to turn, or cannot even recognize how many ways there are from which to choose. Ideas are domain-specific concrete conceptual structures, which take the uncertainty out of some domain-specific concrete space induced by the local anomaly.

The largest of the three BFEU scores was 431 in unit 29. Unit 29 is in the third row. This BFEU score of 431 is almost two thirds of the 675 steps Female three required to complete her first complete loop through the maze. What is there about the third row that made it so hard, almost impossible to assimilate? What is the third row mystery-feature? One of the chief challenges in understanding the nature of mental rational evolution of the nonverbal mind is solving this third row mystery.

Rats found the third-row problem by far the most difficult of the four cause-effect puzzles to solve — solving the grade three problems required almost two thirds of the total steps needed by Female 3 for the mastery of the maze.

The nature of the problem induced by the anomaly of the third row will be discussed at length, in two sections after we quickly consider the fourth grade problem.

Grade Four: the direction problem

In the fourth grade, the mind must find a solution to the direction problem. D8 is the innermost factor, the first grade factor; A2 is the outer-most idea, the fourth grade factor; the idea at the top of the pyramid A2, B3, C2, D8. A2, the fourth grade factor has two levels out and back. The direction effect will be discussed in detail in a later experiment in this book.

The strange fact about how the nonverbal mind works is the fact that rats find this fourth grade problem, understanding the direction distinction, A2, almost no problem at all. All rats found conceiving a solution to A2 very easily. What is so strange to my human mind is that this problem is for me by far the most difficult. One of the facts that show that rats find this no problem is that no large BFEU score that can be attributed to this factor. FIGURE 1.3.4A shows that Female three encountered only three anomalies within the maze, 101, 431, and 53; there is no fourth anomaly that can be attributed to a direction anomaly.

In this section I have proposed that the behavior of the nonverbal mind can be understood as being the product of F3 having gone to school, specifically four grades of schools.

The unit 29-30 problem

FIGURE 1.3.4A shows that unit 29 triggered the longest period of retracing, 431 steps and was the most anomalous unit of the maze. The task of the nonverbal mind is to destroy the uncertainty induced by each anomaly by identifying a set of causal conditions, which make this effect predictable.

FIGURE 1.3.6 shows that redundancy loop RD2 is one of two remarkable loops in this FIGURE, RD2 loop, and First Entry, FE3. What was remarkable is that Female 3 invested 431 of her 675 moves on RD2 - she spent sixty-four percent of her total allotted exploratory-play moves on this one retracing loop. Her primary rational mind required 431 moves to get prepared to enter this unit.

However, once she had resolved the uncertainty induced by this unit she suddenly stopped retracing. Once she had discovered a cause of this anomaly she had, in effect, conceived a valid model of the hierarchical structure of this maze. At that point, there was essentially no significant source of uncertainty left in the matrix, and so she began what was to prove to be the last FE loop, FE3. This last loop is seen in the almost vertical line in FIGURE 1.3.6.

Replication

I assume that the difficulty F3 encountered in the third row with Unit 29 was because she found the concept of Replication particularity difficult to conceive. One of the chief challenges in proposing a theory of mind is to explain the nature and origin of ideas. F3's problem with unit 29 affords me a way to explain the nature of an idea, by comparing two graphs: Figure1.2.4C, and Figure1.3.7.

Figure1.2.4C represents a two- dimensional matrix, row by column, an 8 x 8 matrix, with 64 units or cells, The Alpha and Omega rows have been included in the figure.

While Figure1.3.7 represents a three-dimensional matrix: Replication 3, Row type 2, Columns 8. (3 * 2 * 8) = 48 Notice that FIGURE 1.3.7 does not include the Alpha and Omega rows. Compare these two figures. What is different about the two? What does Figure1.2.4C say about ideas? Notice first that the column effect is the same in both figures. The idea of Replication has an effect only on the way rows are conceptualized. In 4C, the eight rows are represented simply as eight rows. Note in column two that the rows are numbered or named as simple rows, numbered (1 to 8). However, in Figure 1.3.7 the row variable is not observed. In the first figure, there are two variables: rows and columns. In the second figure, there are three variables with column the same. The row variable was replaced with two variables: row type 2 and Replication 3.

What is the idea that distinguishes these two figures? In 4C, all the rows are the same there are eight of them. So the nonverbal mind of the rat only has to evolve an invariant concept of rows; a concept that doesn't change as the rat runs through the rows.

Now, what does Figure1.3.7 say about the row ideas? The row concept has been dropped and has been replaced with two ideas: row type and Replication. Row type we discussed in grade two. What is new here is Replication. What is the difference between the idea of row and replication? Figure 7 has been color coded to assist your mind in making this critical discrimination. Figure 7 divides the Six rows of the matrix up into three replications: Green, red, and purple. Each replication consists of two row types: east and west. But the unit 29 that was so anomalous for F3 is named 215 in this scheme.

Consider cell 212, which in FIGURE1.2.4C was a member of row four, in this new scheme it becomes the second instance of the first type row. The response behavior that the nonverbal mind should emit is RL, RL, and RL. The challenge for the rat's mind is to see this row as an exemplar of the concept of row type. Where "row type" is an invariant, but now the mind must be much more discerning and see the third row as the second instance of the first type row. When row is the concept then the differences between the east and west are irrelevant. However, in the second design the row type becomes most important, and the rat must make a point of paying attention to which of the row types it is in, and whether it is it is in the first, second, or third replications of one of the two row types.

FIGURE 1.3.7T

This unit 30-29 problem is interesting for it can give us a broader insight into the nature of an anomaly.

Kuhn's (1962) two most famous examples of anomalies were Galileo's observation of the moons of Jupiter and Darwin's observation of the Galapagos Island finches. In both of these cases, the anomaly was an event that in effect jumped out and said, "Boo." In both cases, a set of objects appeared where they did not belong within the current paradigm.

Generally, I define an anomaly as a subjective illegal event; an event that induces surprise that the mind did not expect. An anomaly in other words, tells us something about the validity of the current state of the working model being used by the nonverbal mind; it is invalid for if it had been valid then the current anomaly wouldn't exist.

However, with regard to the unit 30 - 29 anomaly the events are different in an interesting way. First, let us examine the facts.

FIGURE 1.3.7T shows the technical nature of Female Three's unit 29 - 30 problem — Female three's problem was she entered unit 30 ten times as an E1 unit, but on the first nine times she kept making the same error, she turned left into the blind units. Only on the tenth occasion was she finally able to make the correct right turn into unit 29. For nine consecutive times, F3 made the incorrect turn, a left turn into the Alternation Error Blind path units, cells unit 31 and 32. So where was the anomaly? Female Three's model was obviously invalid at that place within the maze, for she kept making the same error and was unable to correct her behavior.

So what was wrong with her model? Female three, it seemed, assumed that the third row was another instance of the second row, a left-turn row. Repeatedly upon entering unit 30, her nonverbal mind kept emitting left turns and was not able to change its mind. In psychology, this behavior is called preservation. Only after nine "wrong turns" was she finally capable of conceiving that she needed to do something different, she needed to make a right turn.

The anomalies of this maze proved not to be absolute but relative anomalies. Relative anomalies are not too exciting, but on the contrary are interesting events. Relative anomalies are strangely exciting unexpected events not paralyzing strange events. Relative anomalies are capable of being known and understood and are not experienced as domain-specific terrors.

This problem shows us just how hard it was for this 90-day old female rat to learn what my mommy used to say, "Take turns!" The record shows her nonverbal mind required 431 steps to acquire the concept of taking turns, as Lashley (1963) first intended, this maze requires rats to learn to "take turns" R, L, R, L ,R. The record shows she made nine wrong turns: A1, A2, A3, A4, A5, A6, A7, A8, and A9 before she was able to emit her first B1! FIGURE 1.3.7T lists the ten times Female 3 entered unit 30.

If we look at the big picture for Female 3, she turned right after leaving the alpha unit in the first row and so in the second row when she went through the door she wanted to turn right again, and she did that for a few times but always ended up in the blind path units. Therefore, she had to learn to turn to left in the second row, which she accomplished with some difficulty, but in fact, she learned to turn left too well because she kept turning left in the third row. She was unable to conceive that she needed to turn right again. She had to learn to turn right again, which took her nine attempts. Her first hypothesis was that she turned the same way every time she went through a door (in her mind she repeated the last turn direction over and over) until she learned that it was necessary to alternate turns, RLRLR, etc. every time she went through a door into a new row.

Are loops real; are they special? The origin of the loop concept

The new procedures I invented for the study of Heterostatic behavior called the Loop Testing Procedure; LTP was developed as an alternative to the classical Homeostatic test procedure, the Trial Test Procedures, TTP. The LTP was essentially the only logical alternative to the TTP, which only allowed subjects to run through the maze in one direction, and each one-way trip was called a trial. At the end of a trial, the Experimenter had to pick the subject up and return it to the start box if it was to make another trial.

Everything about the TTP seemed to me to be arbitrary, if not somewhere between absurd to silly. The unnatural aspect of this procedure should be obvious: how many rats in the real world, in the morning, run from home to work but then in the evening have a human to carry them home!

One of my first decisions was to allow or actually require rats to run both to and from work -this round trip is called a loop. A new loop starts every time a subject leaves the alpha unit, and ends when it returns. Therefore, in the LTP loops replaced the trials of the TTP.

Is unit-6 special?

However, to the point, the question is this: is the loop procedure just as arbitrary as was the trial procedure? Operationally, how can we test this question?

The practical question asks is the alpha unit special to a rat? Did I as an experimenter arbitrarily define a loop as part of my test procedure, or does the alpha unit means something special to the rat? Do rats have a concept of a loop?

FIGURE 1.3.8T Alpha Unit Latencies

FIGURE 1.3.8T is the record of the latency of each of Female 3's 23 alpha unit entries. Technically there are only 22 alpha latencies scores since, when any rat reached criterion they were removed from the maze; the testing was over. For this last alpha unit entry, there could be no latency since she was never allowed to begin another loop.

The interesting point is that these average latency scores for these 22 alpha units are many times larger than the average latency scores for all the other units, the non-six units.

The average for the 22 alpha unit latencies is 16.28 seconds, while the average latency for all the other 651 units is 1.43. In other words, the average latency of the alpha unit scores is 11.38 times larger than the average latency of all the other scores. The larger size of the alpha unit scores suggests that the alpha units were special for Female three. If I may be anthropomorphic for a moment, this data suggests the hypothesis that after getting home after a long trip, Female Three felt like kicking off her shoes and taking a break before going out for another trip. These facts support the idea that a loop is a real entity for a rat and not just a figment of my scientific imagination. This supports the idea that a loop is a conceptual tool employed by her rational nonverbal mind to create the nonverbal narrative that organized her behavior.

The several evolutionary stages of the conceptual models that evolved as she explored the hierarchically organized spatial structure of this maze were hinged, attached to, had the alpha unit as the zero point. The basic units of her exploratory behaviors were organized into loops, each of which began and ended in the alpha unit.

FIGURE 1.3.9A FREQU data

FIGURE 1.3.9A shows the FREQU score for each cell of the SRS matrix. Each FREQU score is a count of how many times Female 3 entered each unit of the matrix maze on the first growth. This FIGURE shows that the FREQU scores in the last three rows were ones or twos. One of the remarkable facts that constitute a tribute to the power of the rat's nonverbal mind is that the majority of the units Female 3 entered on more than one occasion were in the first three rows of the maze! This fact contributes a critical piece of knowledge about the nature of nonverbal rational behavior. This feature of nonverbal rational behavior is a consequence of the most impressive feature of rationality; the capacities of the nonverbal mind to first make and then extrapolate from a model. The nature of rational behavior is essentially a function of its ability to extrapolate. If you do not understand extrapolation then you cannot understand rationality!

FIGURE 1.3.9B FREQU in graph

FIGURE 1.3.9.B plots the FREQU data as a line graph. This FIGURE shows in another way that Female Three spent almost all her allotted moves exploring the first three rows of the maze, which shows she was able to reduce or eliminate all the uncertainty induced by the maze within only the first three rows. The creative power of the nonverbal mind is its power to conceive a coherent set of ideas and this power exists in a 90-day-old rat, which was one of the first mammals! This fact makes it clear that the fundamental feature of nonverbal rationality is the profound adaptive advantage of the nonverbal mind's capacity to extrapolate. The whole point of creating a model is that a model then allows the mind to extrapolate.

In the case of Female 3, she found empirical answers to all her questions induced by the three great anomalies of the matrix maze within the first three rows! Female Three was able to reduce all the uncertainty induced by these three anomalies of the maze within the first three rows! Her primary nonverbal mind swiftly and effortlessly accomplishes what Plato insisted that no human rational verbal mind could do. Female three's nonverbal mind with only a relatively small cortex accomplished what Plato insisted only the mind of god could do. Perhaps Julian Jaynes (1976) was correct, perhaps we should call the right hemisphere the "god-side." The following quote exemplifies that perfectly:

"For in many respects the right hemisphere acts in god-like ways." (Hampden-Turner, 1981, P. 92)

Female Three's behavior shows that within the first three rows, she was able to conceive the four basic ideas, which composed the A2, B3, C2, D8 model of the maze. The rational power of the generic mind to evolve or conceive a coherent set of ideas, a model, emerged in her rodent head only within the time it took her to explore the first three rows. Her extrapolation upon entering unit 29 proved the muscle power, the intelligence her rodent mind already had evolved.

Female Three's behavior shows that even this rather simple mammal had a remarkably powerful ability to self-organize; her rational nonverbal mind enabled her to 1-conceive a goal 2-conceive the means to achieve that goal 3-to conceive maps, models, or scripts necessary to support extrapolation by exploring only a small portion of the whole maze!

The power of the nonverbal mind is the power to create domain-specific coherent sets of ideas, which support extrapolation. This is the essence of the generic mammalian adaptive advantage of having been the first form of life to evolve a mind, but this first mind was a nonverbal mind; a kind of mind certainly Plato (1952) never could have imagined. However, let us not be too harsh on Plato; today over twenty-four centuries after Plato, the best minds at our best universities still know no more about such a nonverbal mind than did Plato! They still have no idea where ideas come from. When will we persuade humans to distinguish between word-based communication and image-based communications?

The latest theories of the human verbal mind designate words as memes, which are only possible in the human mammals. Female three was alone in the matrix maze when she conceived the four ideas required to conceive a valid model of the maze, one that supported extrapolation. She never submitted this set of four nonverbal ideas for peer review, had them published, nor waited upon some committee to grant these four essential ideas the status of a set of memes.

FIGURE 1.3.9A: The FREQU data

The FREQU variable is the third element of the set of Heterostatic Dependent variables shown in FIGURE 1.3.2A.The FREQU variable is seen in the fifth column of this matrix of fifty-one units of the EP by five variables. This dependent variable answers several interesting questions, which reveal the interesting features of Heterostatic nonverbal behavior.

Why does a rat retrace? It must begin retracing anytime an anomaly is discovered with the purpose being to discover a set of causes that can take the uncertainty out of the anomaly.

How far back must a rat retrace. Every new anomaly requires that retracing be absolute, back to the beginning. Each unit is a vector, but most importantly a dynamic vector, which has its name constantly in flux, until the dimension of the whole has become established as a constant. Each time a new dimension is added to a hierarchy each unit vector must be updated. With each new anomaly, as each new dimension is added to the whole, just so each vector's dimensional properties must be updated.

So why is the FREQU distribution shaped as it is? The answer simply restates what I have been saying various ways in this section is that the paradox is the nonverbal mind does all its thinking, executes the creative rational evolutionary function while in the retrace mode. Thinking can be defined as a nonverbal function that destroys empirically induced uncertainty. Thinking for the nonverbal mammals is a function that creates a set of coherent causal-conceptual ideas that constitute a valid survival-certified model of a territorial space.

Results: Group Data

Section 2-the group data; and Discussion

In the first section of the Results section, Female three's data was described. In this second section of the Results section, the group data will be presented.

Two kinds of hierarchies are distinguished and discussed in this series of experiments. First, the hierarchies of the matrix maze itself, for which the nonverbal mind evolves a set of coherent hierarchically organized ideas or models, one level or one grade at a time as discussed above, and secondly, the hierarchies of the rat's behavior.

A Review of the hierarchies of the maze

The matrix maze can be thought of as consisting of a set of grades, which the rat explores at its own pace; promoting, retaining, and demoting itself using criterions it defines for itself. A major part of the theoretical task of the experimenter is identifying objective criteria for the identification of how many levels the rat recognizes; objectively identifying the subjective criterion by which rats distinguish these levels; and objectively discovering how the rat organizes its behavior to master these several hierarchically organized levels of the matrix.

The hierarchies of behavior

The behavior of the rats run in the Lashley III maze is hierarchically organized into four levels: (A) growths, (B) runs, (C) loops, and (D) node/paths. Growths are the outer loop and node/paths the inner loop. While the classical learning theorists conceived of learning as one-dimensional and thus capable of being described with a single growth curve, hierarchically organized learning behavior is multidimensional and requires at least one or more growth-curves at each of the four levels of the hierarchy of behavior.

The three sections of the group data

In the **first** section: The E-Group Effects

In the **second** section, the M and the E-M data will be presented.

In the **third** section, the E-A loop and node data will be presented.

The E-group data

In this **first** results section, the set of variables that define the first three levels of the hierarchy, growths, runs, and loops of the rat's behavior will be described for the E-group. The purposes of this first section are first, to establish a gestalt of how the first three levels of the behavioral hierarchy fit together. Second, the multivariate analysis strategy for the analysis of the data will be described; third, several methods for factoring the data will be illustrated.

Section one: *The E-Group Effects***.**

In this first study of Heterostatic behavior, only one growth-cycle of the rat's behavior was observed and recorded. Each cycle is called a Growth. The Growth cycle is the outer-loop or the highest of the four orders of the hierarchically organized behavior. The results for the hierarchical behavioral orders will be described in the reverse order that is from the inner to the outer: loops, runs, and growths. The interrelation of the behavior of these three orders can be seen in FIGURE 1.3.6, which was selected as the prototypical Heterostatic growth curve. This FIGURE shows a growth, the FE curve, then above and below the lines you have the knowledge about the type of loops and below the line very faintly you see the marks indicating where each of the three growths start and end.

Loops

Loops are a complex element of the three levels of the hierarchal organization of Heterostatic behavior. In this section, the eight types of loops will be defined, and their role in Heterostasis will be introduced. How these three elements are related to one another will be shown below.

Alpha and Omega Loops

The first dimension for classifying loops focuses upon the origin of the loops - Alpha and Omega loops. Every loop that both begins and ends in the alpha-unit is a forward or alpha loop. Loops that begin and end in the Omega unit are backward or Omega loops. However, due to the extremely small MSR228 used in this study, neither the real nor the simulated rats make enough omega loops to justify talking about them. However, in simulated random behavior, the number of omega loops equals the number of alpha loops when the MSR is 1500 or greater.

First Entry and Redundant Loops

The second distinction between loops is the First Entry, LFE, and the redundant loops, LRD. An LFE is one in which the rat enters or explores any unit of the maze for the first time. A FE can be measured in, (a) length in units entered, LFEU, (b) length in time, LFEL, and (c) in the number of first entries per loop. RD loops are loops in which the rat does not explore any new unit but spends its time re-exploring the units it had previously explored on earlier loops. Only length-in-steps and time are measures for this loop type.

There are three kinds of LFE loops: (A) the criterion loop, the last LFE in a growth, (b) complete LFE, and (c) incomplete FE loops. Complete LFE begin and end in alpha or omega. Incomplete LFE loops are produced when a rat, real or simulated, is removed directly from whatever unit of the maze the rat is in when any of the three constraint criterions is reached.

The Criterion Loop

The third distinction for classifying loops is the criterion loop. Each rat was run in the maze until it was able to achieve some criterion of exploratory excellence. The criterion for this study was set by the limits of the technology in 1984 when this study was done. Our Apple II+ computer would only accept 228 steps of tracking data before it involuntarily "took a break" so that its "garbage collection routine" could do its thing. This was such a severe criterion of excellence, that it took another 6 years of programming (and the development of the 386 series computers) before the quantity of data could be processed that would make it possible to run simulations which could actually simulate the behavior created by this criterion of excellence.

The Emergent Loop

The fourth type of loop is the Emergent loop and is the loop-type that best reveals, at a single glance, the rate of the exploration of a real or random rat. However, it is an artificial loop created by extracting all the first entry scores from each run and combining them to create one Emergent loop for each growth.

The criterion loop, the emergent loop, and the FE loop are not necessarily, independent loops. The dependent variables for all three loops are the same:

First-entry unit, FEU, and the first entry latency, FEL. The correlations between the redundant loops, the emergent loops, and the criterion loops are directly related to the amount of exploring that occurs on the last first-entry loop. If, for example, a rat did all its exploring on one FE loop, then the FE loop, the criterion loop, and the emergent loop would all be the same. For Random rats, the three types are usually independent. For Real rats, they are usually dependent.

The Retrace Loop

The fifth type of loop is the retrace-loop, of which there are two kinds: first for the Essential Path, EP, secondly for the Full Maze, FM, which includes both the data for the EP and the Blind Path, BP. Only the first kinds are described in this study. The second type of loop requires many growths to complete, and since this is a one-growth study, no blind path loops were observed. Retrace loops are different from all other loops in this respect; their origin is never an alpha unit. Retrace-loops play a special theoretical role in our understanding of rat behavior. A retrace-loop can begin and end on any unit other than an alpha unit, but they must always begin and end on the same unit. A retrace-loop begins when a rat breaks out of the exploratory mode on a FE loop and shifts into the redundancy or retracing mode. The retrace-loop is a measure of how long it takes a rat to get ready to start exploring again, or it is a measure of how long it takes the rat to get rid of enough uncertainty so that it is able to begin exploring again.

There are, technically speaking, three kinds of EP retrace loops, although we only score one of them. A Type-One retrace-loop occurs when retracing takes place within a single retrace-loop. A Type-Two retrace-loop occurs when retracing begins on one FE loop and ends on the next FE loop without one intervening redundant-loop. Both types one and two retrace loops, technically qualify as retrace loops, but they are weak types and, so far as I can determine, without any special theoretical significance. A type-three retrace-loop is the strong type and begins on one FE loop, ends on the next FE loop, and has one or more intervening redundant loops between the two FE loops.

Retrace loops are measured by three special dependent variables, unit entries between first entries, BFEU; unit latency between first entry, BFEL; and mean BFEU, BFEULM. The BFEU and BFEL variables for the essential path are computed by subtracting the first entry and first latency score from successive units of the essential path. If, for example, a rat entered unit 11 on step 10 at the elapse time of 25 s. but did not enter unit 19 until step 110 at the elapse of 125 s then the BFEU would be (110 - 10 = 100) and the BFEL would be (125 - 10 = 115). These scores show that the rat took 100 steps and 115 s. before it entered unit 19 after it had its first opportunity to do so from unit 11. The BFEU for most of the maze units will generally be one and the BFEL will generally be only a few seconds. However, these scores become very large at times, in both real and random rats, but especially so in random rats. In random rats, the distribution of these large scores should be random. The important question, of course, is this: How are these large scores distributed in real rats? These loops will be seen in FIGURE 1.3.0T.

Runs

FIGURE 1.3.0T shows the run data for each of the six subjects, (19-24) of the E-group of rats. Three run scores are shown for each of the six rats: first, run length in steps, second, run latency in seconds, and third, average run speed.

FIGURE 1.3.0T Run data

Individual data is shown because the theory of Heterostasis stresses the importance of individual differences; it is the strength of Heterostasis to study how individual play ability differs. Two of the three female rats showed consistently high rates of play performance, Females 19 and 20, as measured by the number of steps per run. Both females used the maximum number of steps allowed each day, 228, except on the last day when they reached criterion with fewer than 228 steps. Female 19, however, showed great variability, as did all 3 males, 22, 23, and 24. The behavioral variability might be described in terms of good and bad days of play. The bad days when the rats did not fully utilize their opportunity to play could be measured three different ways: First, in terms of the number of steps taken per day; on day 7 rat F19 took only 106 out of a possible 228 steps. Rat M22 on days one and three took only 96 and 36 steps respectively. Rat M23 took only 105 steps on day 4. Rat M24 took only 113 on day two. Please note that the 114 steps for rat F19 on day 10, and the 194 for rat M24 on day four are not signs of bad days but good days. These are the days they reached criterion and the lower the score the greater the efficiency. All the above bad days were also marked by maximum latency scores; either 1800 s (30 min. times 60 s.) or 900 s. (15 min. times 60 s.) The five bad days were also marked by low speed scores, 8.4, 17.9, 24.3, 8.8, and 7.9 s. per step. The good day scores when the rats took all 228 steps allowed never had an average speed score larger than 3.5 s. per step. The fact that rats have a bad day, behaviorally, does not necessarily imply, however, that the rats did not receive as much novelty as they could digest, nor that they did not spend their time in a cognitively efficient way. Bad days are interesting phenomena and whether they inform us concerning, the lack of motivation, the lack of intelligence, the presence of fear or shyness, or whether they imply that the rat preferred to sit awhile and organize their mental maps of the maze are all serious alternatives that will require much experimentation to evaluate.

The Multivariate Analysis Design Strategy

The purpose of this section is to show how Heterostasis can be seen in the **group data** for a growth. The movement of both the real and random rats within the Lashley III Matrix maze created a series of loops that grew in size (number of units per loop), until they finally encompassed the whole maze. This progression was captured in the emergent loop data. Presumably, the difference between a real rat and a random rat's emergent loop is that their growth behavior was based upon entirely different principles. While the random rat's growth was an exponential random function, the real rat's behavior is a special rational function generated by the primary mind. The differences between the three data sets in the real rat's group behavior, FEULL, BFEULM, and FREQULM, and the random rat's group behavior make it possible to describe the primary nonverbal mind that creates this rational form of behavior.

However, the decision to represent the rodent's behavior as a matrix and to describe this behavior in multivariate data sets made it necessary to create new ways to think about animal behavior, one having a double or triple aspect. First, the activity of an animal from one unit of the maze to another is called a move or a step. The number of moves a rat generates within a matrix is an arbitrary number, which is determined, first, by how dense the experimenter chooses to make the elements of the matrix. Initially the Lashley III matrix maze's four foot long rows were represented as four one-foot units, but in this study, the same space is represented as eight six-inch units, so that making one loop through the maze now generates twice the number of moves than earlier rats made.

The number of moves a rat makes is also dependent upon the reaction time of the experimenter tracking the rat through the maze. While it might be useful to double again the density of the matrix, the activity level of the rat in the 64-unit matrix, at times exceeds the human's reaction time even with the matrix at its present size, so it is not practical to increase further the matrix density as long as humans track the rat's behavior.

The basic assumption of classic science that the dependent variable is simply "something we measure" obviously does not reflect the fact that the dependent variable is something the experimenter creates to represent the behavior of the rodent's activity in the maze. The quantitative value of this dependent variable is, in this experiment, a compromise between: (1) the matrix maze, element density; (2) the rate of activity of the rat; (3) and the reaction time of the experimenter.

The Double-aspect model of the representation of rational behavior

Earlier I introduced the theoretical position that the rat represents each unit of the maze two different ways. In this section, I will call attention to the fact that the behavior of the rat is also represented two different ways by the experimenter. The Double-Aspect Model of behavior formally recognizes the dual aspect of our multivariate representations of the rat's motion; each move of the rat is represented twice: once as a move in space and secondly as a move in time. If a theoretical rodent moves through the matrix maze with its present arbitrary density at some ideal rate, say one move in time for one move in space, then:

1. FEU = FEL
2. BFEU = BFEL
3. FREQU = FREQL

Recognizing these possible tautologies, the three data sets of variables for measuring the emergent loop growth activity are statistically analyzed in a multivariate design. The arbitrary, one-second-in-time for one-move- in- space, assumption is not actually very arbitrary. The 1 to 1 rate assumption turns out to be, under many conditions, an empirical fact. When the 1 to 1 assumption is true, obviously analyzing the nine scores in multivariate designs saves time and publication space.

When, however, the rat moves with something other than a 1 to 1 rate, either greater or smaller, the multivariate analysis is still just as useful, or perhaps even more so, for then and only then, will the independent variable data types be significant and only then is it justified to graph and discuss the different dependent variables individually. Whether the ratio between the two variables is the same or significantly different is one of the most important facts for Heterostasis. The third score, the ratio of the moves to the times scores, is a measure of the subjective confidence of the rat in the validity of its mental model in representing some element of the maze. When the rat subjectively perceives its model to be valid for some element of the maze, the ratio will tend to be 1 to 1 or less. However, when the organism's model is invalid and the rat is uncertain about how to respond in some unit of the maze, the time/move ratio will tend to be high. The time/move ratio is a major measure of Heterostasis. It is an objective measure of the subjective certainty-uncertainty dimension, which a rat has in the validity of its cognitive map. This dynamic variable can vary from one unit of the maze to the next.

The series of moves within the matrix is at some frequency. The slower the frequency of the rat's moves within the matrix, the higher will be the time/move ratio. In other words, the greater the amplitude difference will be between the moves and time scores for all the pairs of the three sets of scores. Certainty, however, tends to get the pairs synchronized. Uncertainty tends to get the scores desynchronized; the greater the uncertainty the more time in seconds will be required for each move in space. As a rat's mental maps become more powerful or valid, the faster the rat moves and the time/ move ratio becomes smaller.

The FEULL analysis

The E-group alone has all nine dependent variables. Each of the three data sets: FEULL, FREQULM, and BFEULM have three scores. These three data structures, each with three scores describe the essential path behavior of the rat in the Matrix maze. A group analysis of the FEULL data will not be presented in this section, since the multivariate analysis of this data does not add anything to the individual difference data.

The FREQULM analysis

The FREQULM (Frequency Unit, Latency, and Mean) data reveals where the rat spent its allotted moves and times within the maze. Each rat received a daily allotment of 228 moves and 600 s for each run, and it could spend this allotment in the maze any way it chose. There are several theories about how a rational rat and even a random rat should spend its allotted time. Blodgett, (1929) for example, recognized that real rats don't spend their time and moves equally in each row of the maze and designed non-retrace doors to force rats to spend their time equally and fairly in each row.

The statistical section of our paper: meet ANOVA and Pascal.

We have now reached that place in our experiments in which what we say is largely determined by the results of our statistical analysis where ANOVA and Pascal reign as King and Queen.

ANOVA, for anyone who has not studied statistics, is the most powerful type of statistics, and generally is the preferred type of statistics especially for those experimental scientists, whose data can satisfy the strict assumptions of ANOVA. One of the powerful advantages of ANOVA is that it can identify causality. However, anyone who never had a year or two of statistics may find that the language of ANOVA is a strange new language. So much so, that I will take a time out and present some of the basic features of ANOVA, all the while hoping that this too brief introduction to ANOVA, will reduce enough of the uncertainty that everyone will be able to continue to read.

ANOVA has several strange aspects, yet I will focus upon one critical aspect, Pascal. As I noted in passing above ANOVA and Pascal are bonded. Pascal (1665) is most famous for his Pascal Triangle. All the data that is critical for identifying the factors in a statistical design of any size is plainly in full view within this table if you know where to look and what to look for.

The mathematicians find all kinds of almost magical numbers in this triangle. However for my nonverbal mind, this table was created to inform students of ANOVA what Sum of Squares can be observed in every possible design size.

It assumes that every whole is a hierarchically organized structure and that every hierarchically organized whole can be factored or petitioned into parts. The Pascal triangle tells us everything that can be known about hierarchical structures:

- how many parts in the whole
- how many kinds of parts there are in any whole
- How many parts of each kind are in any design
- How many or these parts are significant statistically
- It draws a figure for each component part so that the reader can understand the meaning of each part in the whole

Analysis of variance is a complex domain of study, yet for our purposes in this book, we will conceive of ANOVA as having two main functions:

FIRST, the statistical analysis function

This function evaluates every one of the component parts and objectively determines how many are statistically significant. In the text, I provide the reader with the specific statistical information required to judge the significance of each effect. This function involves the integration of too many experimental facts, and it is not necessary to understand how to evaluate the degrees of freedom, the F-ratio, etc in order to understand the significance of the figures.

However, understanding these functions, I will leave to the statistical experts. The reader can still understand the graph without understanding the degrees of freedom shown at the top of each graph.

SECONDLY, the graphing function

For our purposes, the most important component of the analysis of variance program is that it will print out a graph which enables the experimenter and the reader to see and to understand exactly what each independent variable did or contributed in each statistical analysis component of the Whole.

FIGURE 1.4.02A The Modified Pascal Triangle.

In FIGURE 1.4.02A, I have taken the liberty to rearrange the critical data from the original triangle and to create this new version so it presents the critical knowledge about the features of ANOVA in a more user friendly way. Specifically the new table shows: the types and number of types of sums of squares, SSs, and the total of all types of SSs for any design up to one as large as having 18 independent variables or being an 18 variable factorial design.

Across the bottom row in FIGURE 1.4.02A, the columns are labeled. In the language of ANOVA, the two important words are the names of the columns. Under the second column on the left is the words "Main effect" All the columns to the right are called "interaction orders." These orders vary from the 1st, 2nd, 3rd, on the left, etc., to the 17th order on the right!

The first column on the left, lists how many Factors or how many independent variables are included in the design in each row. These numbers vary from one on the top to 18 on the bottom. On the far right, the total numbers of elements per row are shown.

For example, in the Fifth row how many total Sum of Squares are computed? What are the elements of the whole, what is the size of the whole? In Row 5, in the first column it says there are five factors. In the second column labeled "main effects," the FIGURE shows that there will be five main effects, one for each factor in the design. Five factors we are then told will produce 5 main effects and 10 first order interactions, with 10, 5, and then 1 fourth order interaction. So how many SSs are generated by an ANOVA design with five factors? Answer 31.

5 + 10 + 10 + 5 + 1=31

In FIGURE1.4.04, I start showing the data from my experiment that was found to be significant. The Pascal figures are presented to assist those who would read this book who are not statistical experts by explaining the main differences in each of the many figures below; namely, how many Independent variables caused the significant effects described in each figure. The main effect, A is an effect that has only one IV as a cause, while in the interactions, the number of letter symbols identifies the number of causal factors that were acting in some one of my significant experiments. The type of higher order effect or kind of interaction, a first, or a second, etc. is shown. As always, the first thing to notice is that the differences between all the SS parts within a design are simply the number of independent variables, which cause the effect seen in a figure. How many independent variables are in a main effect sum of squares? Answer only one! How many independent variables are found in a First Order Interaction? Answer TWO, 2.

The names of the columns shown under the columns tell us how many interactions of each order are found in each design.

FIGURE1.4.02B Enumeration Table

FIGURE1.4.02B is an Enumeration Table, which shows a set of names for all the component parts of a design of every order up to and including an eight, 8, factor design (or an H design).

On the other hand, this Table can be understood to supplement the Modified Pascal Triangle of FIGURE1.4.02A. For example, in FIGURE 1.4.02A, the column on the far right shows how many parts are in the whole defined by each row. Again, look at row five. The last column for row five shows that there are 31 sums of square parts in a five factor whole.

The enumeration chart also shows how many sums of squares for each type can be observed for a design or in each row. For example, column one says that in a one-factor design, only one Sums of Square will be observed. Obviously in this case, it will be called the A effect, which is also a main effect.

In the second column, this chart shows that with a design consisting of two IV's, three effects must be examined, A, B, AB. where A and B are "main effect" names and AB is the name for a two way or first order interaction effect.

The third column shows how many sums of Squares must be computed for this type of design.
The third column shows the new SS's added by the third IV. This figure shows there are: one main effect, C, and two first order interactions AC and BC, where AC is the effect caused by the interaction of the two independent variables and BC is the name for the effect caused by the interaction of the two independent variables: B and C.

Can you translate what column four says? How many two-way interactions does this design create?

Each of the eight, 8, sets of alphabet names supplied by this table, however, are cumulative. For example, for a five factorial design, which requires 31 names, it is necessary to employ all the names for set five plus all the lower order set. Set five supplies 16 names plus the lower order sets supply 15 names (1+ 2+ 4+ 8 = 15) but 15 + 16 = 31.

The Pascal Triangle rows

FIGURE1.4.03A Pascal's Triangle Rows

FIGURE1.4.03A is another modified version of the Pascal Triangle. This time the FIGURE was modified by having the rows colored.

Definitions:

The reader can also refer to Chapter 5 in Book 1 for more information on these variables and their interactions in an experiment.

- The Independent variables are the variables the Experimenter manipulates or causes to vary.
- The dependent variables are those variables the experimenter measures.
- The control variables are those variables that the experimenter holds constant.
- In an experiment, the experimenter manipulates a set of independent variables in order to determine their effect upon a set of dependent variables given the conditions of the control variables

The first thing a reader should do when reading the various FIGURES in a section like this one is to identify how many independent and dependent variables are discussed in each figure.

The biggest error beginners make is to do "main effect thinking." This error reflects the assumption that each effect has only one cause or independent variable. Experience with ANOVA then enables each student to discover that most effects have several causes. An interaction means that two or more IVs caused the effect seen in some DV.

Again, the purpose of an experiment is to determine the effect of the IV upon some DV given some set of control variables.

What are the IV and the DV that are active in each significant effect you are reading? Which one is the IV and which one is the DV?

In the text below the figures or the effects are collectively organized by dependent variable sets. For example, there are three kinds of dependent variables in these experiments: FEU, BFEU, and FREQULM.

Yet each of these DVs has three kinds of data components so $3 * 3 = 9$

The rule is that every interaction is a whole with many parts depending upon how many IVs are identified as the causes, which are shown in the names of the effect, main effect, etc.

The higher the order of an interaction the further out on the right the interaction is to be found and the more complicated the effect; that is to say, more parts will comprise the whole and more IVs shared the cause of the effect seen in the DV.

As you read the figures in this first chapter, I have made it easy for you to identify what level of interaction is occurring. While reading, please note the following:

- **Main Effects are shown in bold typeface and have one independent variable**
- <u>First order interactions are underlined and have two independent variables</u>

- *Second order interactions are in italics and have three independent variables*

This should make it easier to identify how many variables are causing the effects seen in the graphs. In later chapters, be sure to look to see how many variables are causing the interactions while looking at the graphs. Please note in the later chapters, there will be some figures with third order interactions that will include four independent variables. Good luck!

The DATA

FIGURE 1.4.04 Unit analysis

FIGURE 1.4.04 shows the Multivariate FREQULM (Frequency Unit, Latency, and Mean) data structure, which is the product of the type 1 analysis, the unit analysis. It shows the overall FREQULM data, structure effect, *the second order interaction, data type by direction by unit*, for the forty-eight units of the matrix maze. F (96,720) = 10.269, P<.000000. How many variables does a second order interaction have? A second order interaction has three IVs; in this case, they are data type, direction, and unit. What are the dependent variables? They are the forty-eight units of the matrix maze. In this effect, we will see what causal effect the three IVs had upon the forty-eight units.

This is a complex data structure in which all three elements of the FREQULM data set cycle continuously and in phase. It is a whole, and the purpose of this section is to break down this complex set of waves into their component parts using the other three types of factoring seen in FIGURE 1.1.6, numbers two through four. However, it is important to begin the process by noting the special relationship between the FREQU and the FREQL how these two differ from the FREQM score. As can be seen in FIGURE 1.4.04, the FREQL score has, in effect, the same frequency as the FREQU score, but has greater amplitude. The FREQM does not show the same improvement in the first half of the maze units the other two scores show but does have the same frequency as the other two scores. The significance of all these points will be discussed in the subsequent component analysis.

FIGURE 1.4.05 main effect for rows

FIGURE 1.4.05 is the product of the type 3 (from FIGURE 1.1.6) method of factoring the repeated data scores, the rows by columns analysis, and is the **main effect for rows** (12) for the FREQULM data set. F (11,132) = 28.388, P<.000000. How many variables are in a main effect? Right, only one variable is in a main effect. This is a smooth curve and is the mean of the three sets of scores of FIGURE 1.4.04. This main effect shows that the E-group of rats spent most of their allotted moves and times in the first few rows of the maze just as Blodgett (1929) first noted and complained about. This curve has the same shape, as a traditional learning curve typical of those found in the trial paradigm, except, it is important to note that the rats in this study were not deprived nor reinforced as these rats were run on the LTP instead of the TTP. The data is plotted as if the maze were a twelve-row matrix, six rows out and six rows back. In spite of the fact that this is a traditional looking learning curve, nevertheless, it represents spontaneous exploratory-play learning and is a manifest-learning curve rather than a Latent Learning curve.

FIGURE 1.4.06 data types by rows

FIGURE 1.4.06 shows the first order interaction, data types by rows, for the three-FREQULM scores. F (22,132) = 8.804, P<.000000. What are the variables in this interaction? They are data types and rows for this figure. The FIGURE shows that the row effects for each of the three scores are significantly different in spite of the smooth row effect seen in FIGURE 1.4.05. Generally, the FREQL score was larger than the FREQU score but had the same shape. FIGURE 1.4.06, however, shows that the gradual improvement effect seen in FIGURE 1.4.05 is not true for the FREQM element of the row effect as it is for the other two. The FREQM score did not begin as a large score, which gradually decreased in size. On the contrary, it began small and reached a peak in row six. The FREQU and FREQL do have much the same shape. The FREQM scores reflect the FREQL/FREQU ratio and show that the ratio was close to one to one from the beginning. The ratio went up on occasion, especially at row 6, for example. In this effect, the FREQU and FREQL scores show that while the E-group of rats spent most of their moves and time in the matrix maze exploring the earlier units of the maze, the mean time per move was relatively independent of the rat's position in the maze. In other words, the effect of uncertainty was primarily to induce more short latency exploratory moves, rather than fewer long latency moves.

FIGURE 1.4.07 main effect for columns

FIGURE 1.4.07 shows the **main effect for columns** for the FREQULM data set. F (3, 36) = 23.4566, P<.000000. This data shows that columns 1 and 4, units E1 and E4, received about three times the activity of columns 2 and 3. This bucket shaped curve is not exactly symmetrical, columns 1 gets a little more activity than column 4. This column effect may look plain and unimportant; however, it is the key to understanding the wave-like effects seen in FIGURE 1.4.04. Each cycle of the waves in FIGURE 1.4.04 is the column main effect for one of the six successive rows. If the reader will examine FIGURE 1.4.04 and mark off each set of four units, they will be able to see how this column effect creates a cycle for each row.

The Lashley III matrix maze generates two types of bucket-shaped curves. This first type is the narrow-bucket type, which involves only the four units or column elements of the essential path. The wide-bucket curves will be seen in the analysis of the eight units for the full maze data in later studies. These bucket shaped column curves are powerful effects that explain generally the major proportion of the variance in any design and make it possible, for the first time, for the two-dimensional properties of animal behavior to be recognized. Traditional learning data focused upon the row and ignored the column dimension of adaptation.

FIGURE 1.4.08 data types by columns

FIGURE 1.4.08 shows the first order interaction, data types by columns, for the FREQULM data set. F (6, 36) = 13.188, P<.000000. The FIGURE shows that the depth of the bucket effect differed for each of the three scores. The FREQM variable data seen in FIGURE 1.4.08 has no overlap with the other two variables and its mean values are relatively small. The relation between the FREQU and FREQL, however, overlap in some important ways with FREQL having the greatest value for units 1 and 4, and the smallest values for units 2 and 3. Both FIGUREs 1.4.06 and 1.4.07 show very clearly, where rats experienced the primary uncertainty in the column units of the matrix maze in the relationship between the FREQU and the FREQL scores, the ratio between the FREQL/FREQU scores varies with columns. For the E1 and E4 columns, the ratio is about 3 to 1. For the E2 and E3 column units, however, the ratio is less than one. Primary uncertainty is directly related to this ratio. In other words, the E2 and E3 units of the maze induced very little uncertainty and thus rats spend little time exploring these units. The E1 and E4 units, however, which were the three way choice units, consumed the majority of the rat's exploratory time when it was constructing a valid model of the maze during exploratory-play behavior.

FIGURE 1.4.09 directions by rows by columns

FIGURE 1.4.09 is the product of the type 3 factoring from FIGURE 1.1.6 and is the *second order interaction, directions by rows by columns*, for the FREQULM data set. F (15,225) = 9.91596, P<.000000. The data of this FIGURE is also a learning curve but certainly not a traditional learning curve as was seen in FIGURE 1.4.05, a row learning curve. This FIGURE is a column learning curve and another good example of one of the advantages of the matrix maze. The traditional analyses of maze performance were limited in the sense that learning traditionally was considered a row main effect. Of course, there is no good reason why learning cannot be a column effect. This FIGURE plots the column main effect for each of six rows in two directions. The FIGURE shows that generally for each successive row, the column improvement gets progressively smaller, but that the majority of the improvement occurred in the first half of the maze.

FIGURE 1.4.10 data types by units

FIGURE 1.4.10 is from the type 2 shown in FIGURE 1.1.6, directions by units' analysis, and shows the first order interaction, data types by units, for the FREQULM data set. F (46,345) = 10.4499, P<. 000000. The difference between this FIGURE and FIGURE 1.4.04 is that in this design there are half the units, 24 instead of 48. FIGURE 1.4.10 is a simpler version of FIGURE 1.4.04 and is the average of the two direction effects, out and back, and shows more clearly than did the data of FIGURE 1.4.04 how the three sets of scores are almost perfectly in-phase for the 4 columns for each of the six rows (4 * 6) when direction is ignored. This FIGURE is included in order to emphasize, once again, the importance of the columns in determining the rat's behavior. Note how the

FREQM cycles parallel the FREQU and the FREQL score cycles, and how the FREQL score has larger amplitude but the same frequency as the FREQU score.

The BFEUL Data

The FREQULM and BFEULM data sets each play an important, but different role in the development of our understanding of the rat's exploratory-play behavior. The FREQULM data set reveals where the rat spent its allotted behavioral resources retracing and exploring the maze, but doesn't explain why these resources were spent as they were. The BFEULM data set reveals where the retracing was triggered within the maze, and how much retracing was triggered by each unit.

FIGURE 1.4.11 data type by unit

FIGURE 1.4.11 shows the third overall Essential Path data structures, the first order interaction, data type by unit, for 48 units for the BFEULM data set. F (94,705) = 1.13, P<. 2090. This effect is not statistically significant, which is itself interesting, for it shows that, in addition to the large variance, these three scores are in phase and even their amplitudes are very similar, particularly the BFEU and BFEL scores. The differences between this and the FREQULM data structure of FIGURE 1.4.04, however, are striking, and the source of these differences will be seen in the parts of which this whole is composed. This FIGURE reveals several important facts about the data that are not seen elsewhere: First, the BFEU scores have either a value of 1.0 or some large value. In other words, the BFEU data shows extreme variability as was previously noted. Second, the BFEM speed scores are often under one second per unit, as low as .4 s per unit. This latter point is one of the basic facts about the rat's behavior that should be remembered, the average latency of the rat per unit gets down to .4 s per unit when the rat is free of uncertainty.

FIGURE 1.4.12 Main effect for 12 rows

FIGURE 1.4.12 shows the **twelve-row BFEULM main effect**. F (11,165) = 4.0, P<.0000. This BFEULM row effect is curvilinear and explains a major portion of the variance seen in FIGURE 1.4.11. This FIGURE should also be compared with the traditional main effect for the rows of FIGURE 1.4.05. The BFE row scores gradually increased to a max at row 4, and then decreased until after row 7, the mean scores were 10 or less. The BFE scores are a measure of retracing and this data shows that the rat gradually increased their retracing over the first few rows and then after row 7, they stopped retracing. The numerical values of these scores reveal exactly where and how much retracing was triggered by each row. Rows 3, 4, and 5 each triggered almost an average of 100 steps of retracing per unit. This curvilinear curve, as we shall see later, is one of the most critical and interesting phenomena supporting the concept of Heterostasis. The hierarchal matrix maze presents a puzzle to the rat, which most rats chose to solve, but the complexities of each level of the maze were not the same. The first levels of the hierarchy were quickly solved and thus seemed to be easy, but each successively higher order level of the hierarchy becomes more difficult for the rat to solve as evidenced by the ever-increasing amount of retracing required.

FIGURE 1.4.13 main effect for columns

FIGURE 1.4.13 shows the **main effect for columns** for the BFEULM data set. F (3, 45) =16.75, P<.0000. This is for the functional reading of the data, the FRS shown in FIGURE 1.2.90. Compare this with the bucket effect of FIGURE 1.4.07. This effect shows that exclusively column units E1 and E2 triggered the column retracing. Column unit E3 and E4 never triggered any retracing! BFEU column data does not produce bucket-shaped curves. Only the FREQULM data set produce this bucket-shaped curve. This data reveals which units of the EP of the matrix maze induced primary uncertainty.

FIGURE1.4.14 data types by rows

FIGURE 1.4.14 shows the <u>first order interaction, data type by rows</u>, for 12 rows for the BFEULM scores. F (22,165) =1.90, P<. 0124. This data shows that these three scores differed significantly over rows even though they were not significant over units as was seen in FIGURE 1.4.11 which FIGURE 1.4.14 modified. The BFEU and BFEL scores stay in-phase; however, with the BFEL score always larger than the BFEU score, yet the ratio does not remain the same. When these scores separated, their ratio shows where the primary uncertainty occurred in the rows. The difference between the two scores shows where the primary uncertainty was induced during the exploratory-play behavior in the first part of the maze. These curvilinear curves show that the game of generic science evolving a valid model or paradigm of the hierarchal maze was over before the rats began exploring the second half of the matrix.

FIGURE 1.4.15 T FF-FS DISTINCTION

The Structural-Functional Distinction

Earlier a distinction was made between four types of ways to read the matrix data: F-F (Functional-Functional), F-S (Functional-Structural), S-F (Structural-Functional), and S-S (Structural-Structural). All the matrix data for this paper was read using the F-F cycling rule. However, in this section, some illustrations will be given from the consequence of reading of the matrix data in the S-F format. This distinction is most critical for understanding how rats make a model of the matrix maze.

These two different ways of reading the data, however, only alter the column main effects and their interactions. FIGURE 1.4.15T is a comparison of two of the ANOVA tables generated for the six column main effects and their interactions for the BFEULM data set using the two different cycling rules for reading the matrix data, F-S, and F-F. The within data, in both cases,was analyzed as a direction by row by column design (2 * 6 * 4). The two designs, as was noted above, gave the same results for all the effects that do not involve the column variable, or in other words, the differences between the two types of analysis are seen entirely within the column effects. Table 1.4.15T shows the results of the functional-structural, F-S, analysis on the right. Only two significant effects are seen: rows by columns and directions by rows by columns, and these have F's = 2.12, and 2.03 respectively. However, the Functional-Functional, F-F, reading of the data on the left in Table 1.4.15T produced four significant effects: columns, data types by columns, directions by columns, and data types by direction by columns, and all have F ratios of 6.9 or greater. Moreover, the two effects that were significant in the structural analysis were not significant in the Functional analysis. In other words, the Structural, F-S, analysis totally misses the four major effects that are significant according to Functional analysis and makes two meaningless effects significant! This difference between the functional and structural analysis of the data is proof of one of the most profound characteristics of the rat's cognitive map, namely that the rat's primary mind created a cognitive map of the maze which recognizes the units of the maze functionally not structurally. The spontaneous learning of rats during exploratory-play consists of the evolution of a cognitive map or conceptual system, which allows the primary mind to recognize each of the 100 units of the matrix maze but the conceptual system recognized each unit place functionally.

FIGURE1.4.15 Main effect for Columns

FIGURE 1.4.15 shows the structural **main effect for the columns** for the BFEULM data set, which looks like a modified narrow bucket curve that characterized the FREQULM data set. $F(3, 45) = .98$, $P < .4127$. This effect is not statistically significant. However, the same column effect in the functional analysis (see FIGURE 1.4.13) was significant. A comparison of FIGUREs 1.4.13 and 1.4.15 shows that the two analyses give a very different picture of the column effect. FIGURE 1.4.13 shows that all the retracing was triggered by columns 1 and 2, or E1 and E2. However, FIGURE 1.4.15 in the structural analysis shows column 4 was the major trigger. Therefore, the Structural analysis is very misleading. The functional analysis is the only analysis that shows that the triggers for retracing were always columns 1 and 2 or that these two column units E1 and E2 were always the only two anomalous maze units in all the successive rows of the maze after the first row. This is a most important fact and is missed in the structural analysis. This structural data's bucket shaped BFE (Between First Entry) curve is an artifact of the structural type of analysis. The functional BFEU column curve of FIGURE 1.4.13 is not bucket shaped and is the proper description of the data.

FIGURE1.4.16A rows by columns for FS

FIGUREs 1.4.16A and B show the rows by columns first order interaction for the structural and functional analysis respectively. Table 1.4.15T shows that the structural row by column effect of FIGURE 1.4.16A is not significant while the functional column effect of FIGURE 1.4.16B is significant. The structural effect, however, if it is closely inspected, furnishes one of the best clues that identified the need for the functional analysis. In FIGURE 1.4.16A, the important clues are the fact that the large scores on the left are rows 2, 4, and 6, while the large scores on the right are rows 1, 3, and 5. Yet, the column effect for each of these six rows is the same as the row main effect seen in FIGURE 1.4.13. This effect clearly shows that the odd and even rows generate the same column effect, but each in a different direction. Based on a close analysis of this FIGURE, it became obvious that if the odd rows: 1, 3, and 5 were reversed, the interaction effect would be simpler. FIGURE 1.4.16B shows just how much simpler the functional column effect is over the structural effect, in the functional effect the column effect for each of the rows are now all aligned. The F-F analysis, in effect, simplifies and sharpens the column effects.

FIGURE1.4.16B column main effect FF

The S-F interaction effect of FIGURE 1.4.16A which has an F of 2.125 becomes a **column main effect** in the F-F analysis with an F of 16.74 seen in FIGURE 1.4.13.That is to say that for loss of an interaction with F = 2, I gained a column main effect with an F = 16. However, more importantly, now the data began to make sense.

FIGURE 1.4.16C data types by columns

FIGURE 1.4.16C shows the first order interaction, data types by columns. This effect is included in order to clear up any confusion that may have been caused by FIGURE 1.4.16A if anyone compared it with FIGURE 1.4.14 which showed that column 4 never was an anomalous stimulus, in other words it never triggered any retracing. FIGURE 1.4.16A shows that the BFEULM activity increased at column 4. However, FIGURE 1.4.16C should clear up this mystery. The BFEU score at column 4 was always one, confirming what was shown in FIGURE 1.4.14. The increase in activity in column 4 seen in FIGURE 1.4.16A was due to the increased activity in the BFEL and BFEM scores as can be seen in FIGURE 1.4.16C in the BFEL and BFEM scores.

Section two: The M-group Effect

In the previous section, the E-group data was described for the first three levels of the behavioral hierarchy and for the nine scores of the growth data. In this section, the M-control group data will be described. The size of the numbers and the variability, however, are so large that it is best to show this data in table form. FIGURE 1.4.17T shows the basic data structure for one growth with six variables for subject, M9. Each variable is a column.

1. Column 1 is the index for the 51 units of the FIGURE 1.4.17T ranging 1 to 51 for the units of the essential path for the first growth.

2. Column 2 shows the unit names of the 51 units as shown in from unit 6 to unit 62, and back again to unit 6. In other words, the effect is Alpha, Omega, and Alpha for one growth.

3. Column 3 is the First entry unit, FEU, score. The numerical value of this variable is for growth data not run data and shows the step number on which each unit of the matrix maze was entered for the first time for the first growth. This table, for example, shows that the rat's first entry into unit 6 or alpha 2, the fifty- first unit of the EP of the matrix maze, was on step 6,858,807.

4. Column 4 shows the run number for each score. The MSR228 required that the random rats be removed from the maze after every 228 steps. The data

in Table FIGURE 1.4.17T shows that M9 made eleven first entries, units 6-29 on run one (Note the first eleven 1's in column 4). Unit 29, identified in column 2, was first entered on step 103, both for run and for growth 1. The rest of the remaining steps, out of the 228 allowed, were spent retracing. M9 on run 2 made a first entry into unit 28, on growth step number 356, as shown in column 3.

5. Column 5 is the between first entry unit score, BFEU, which is a measure of how much retracing a subject makes between two successive first entries. The table shows in column 5 that M9 made 253 steps after entering unit 29 before entering unit 28 (356-103). This data shows the variability of the BFEU data, which was previously noted. The data is either one or some large number.

6. Column 6 is the Frequency unit score, FREQU, which is a measure of how many times the rat entered each of these 51 units of the essential path. This data shows the number of times M9 entered each of the units of the EP. This is the way random exploratory data should look. The behaviorists, beginning with Thorndike (1911), assumed that real rats explored randomly, but they failed to recognize the difference between this exponential function and the curvilinear function seen in the E-group data.

The numbers in column six do not add up to the sum in column three because 6,858,807 is the total number of steps required by M9 to reach criterion and includes both the EP and the Blind unit data. The sum of column six is the number of those steps taken in the essential path. The difference is the number of moves spent in the blind units of the Lashley III maze.

Several important phenomena are illustrated, in this table.

FIGURE 1.4.17T RANDOM RAT DATA

First, the table shows that while random rats did make a complete loop through the Lashley III matrix maze under the severe constraint set of 228 moves, it was a difficult task and it did not get any easier with practice as was seen in the E-group, for example in the data of F3 in FIGURE 1.3. 3. The M9 simulated rat of the M-group required many more moves to make 1 growth, 6,858,807, and runs, 29,821 than did any of the other M-group rats. The greater number of moves and runs required by the M-group was, however, a quantitative difference, proving only that the M-group of random rats was less efficient than the E-group of rats. The BFEU scores in Table 1.3.3, however, shows there was both an enormous qualitative as well as a quantitative difference between the simulated rat and the real rat data of FIGURE 1.4.17T. The qualitative difference was that real rat BFEU data was curvilinear while the BFEU random rat data of column five was an exponential function. The curvilinear data of real rats proves they were able to benefit from experience in the Lashley III matrix maze after exploring a few rows they stop retracing. An examination of column 5 in Table 1.4.17T shows that M9 never stopped retracing. On the contrary, the BFEU data shows that the quantity of retracing grows at an exponential rate.

By examining the relationship between the values of columns 4-6 in Table 1.4.17T the reader should be able to develop a better understanding of the relationship between the FEU, BFEU, and FREQU sets of scores and how they relate to the run data of column 4 and the unit names of column 2. The data of

Table 1.4.17T shows the big picture for the M-group data for one random subject, M9, and from this data, it is clear that the random rat behavior in the Lashley III matrix maze and the E-group data are based on very different quantitative and qualitative principles. The M-group data is an exponential function while that E-group data is a curvilinear function.

Additional Measures of Efficiency Differences

In FIGURE1.4.18T, the seven columns show the effect of the MSR (Maximum Step per Run) variable upon five additional sets of dependent variables, all measures of growth efficiency. The first row level seen is the data for the experimental group.

Column A is the index of the 27-MSR levels tested.

Column B shows the MSR values tested.

Column C shows the number of runs required to reach criterion at each MSR level.

Column D shows the number of loops required to reach criterion at each of the MSR levels.

Column E shows the number of steps required to reach criterion at each level of the MSR.

Column F shows the computed efficiency score of the rat's movements at each of the levels where efficiency was defined as the average number of steps required to move through each unit of the EP of the maze.

Column G shows the probability of reaching criterion at each level.

FIGURE 1.4.18T gives the mean values for 500 simulated M-controls rats at each of the 27 levels of the MSR for each of the five measures of efficiency. The twenty-seven levels of the MSR range from 230 to 12000 (see column 2). The five variables of FIGURE 1.4.18T are the mean number of (a) runs, (b) loops, (c) steps to criterion, (d) efficiency, and (e) probability. There is an inverse exponential relationship between the MSR and the first four variables. There is a direct exponential relationship between the MSR and the fifth variable.

FIGURE 1.4.18T MSR

FIGURE 1.4.18T also allows a comparison of the real rat behavior seen in row 0 with MSR228 with the random M-Control rats with an MSR230 for each of the five dependent variables. This comparison is not entirely fair since the experimental rats were run at the MSR228 and the control rats were run at MSR230, so a word of explanation is in order. The experimental rats run at MSR228 were run at a much more severe MSR than the MSR 230. If the reader will study the MSR function seen in FIGURE 1.4.18T, they will see that the exponential function is growing at such a rate that the MSR230 was for all practical purposes the upper limit that could be achieved on the computer available at the time this series of simulations were run. However, the test at MSR228 did not really need to be run. The point had already clearly been made that random rats exhibit an exponential function and the additional two points on the MSR would have only required us to tie up our computer for additional weeks while these simulations were run. If anyone is interested, they can take the data and compute the expected values for MSR228 but knowing the exact values of the expected value at this data point could add nothing to the understanding of the difference between the experimental group and the control group. The control groups simulation was too easy, as noted elsewhere, for the control rats were not allowed to repeat a turn in a unit. This could have been used to compute latency for the control rats, but again this would have consumed too much computer time and would have added nothing to our understanding of the difficulty of random behavior. The experimental groups of rats found the MSR228 test very easy to master. In only 1000 steps, the experimental rat learned to make one criterion loop, while for a random rat a criterion at MSR228 was so improbable that finally, out of consideration for our computer, I settled for the test at MSR230. The effect of the MSR upon the simulated rat's random loops is an exponential function on each of the variables of FIGURE 1.4.18T.

FIGURE 1.4.18T shows that, at the 7 largest levels of the MSR from 6,000 to 12,000, the task of making a criterion loop within the constraint of the MSR 6000 was so easy that the 3500 simulated rats tested (7 * 500) all reached criterion on a single run (see column C), which averaged about 25 loops each, (column D). The probability of a successful criterion loop for these large values of the MSR is .04 or a criterion loop can be expected about 4 times in a hundred loops.

Column D in FIGURE 1.4.18T shows that as the MSR gets smaller the probability of a successful criterion loop gets progressively harder or that the effect was an exponential function. At the smallest tested MSR value, 230, the probability of a criterion loop is .0000028 or 2.8 times in one million loops. The probability of a criterion loop at MSR230 occurs only once in 40,146 runs or once in 352,362.5 loops.

Column E, the loop variable in FIGURE 1.4.18T translates this exponential function into the average number of steps required to reach criterion. At the seven large MSRs only about 2,400 steps are required to reach criterion, while at MSR230 an average of over nine million steps are required.

Column F shows that the efficiency of the random rats at the large MSRs is about 50 to 1. About 50 steps are required on the average for the random rat to move through each one of the 48 units of the essential path of the Lashley III maze. The random rat's exploratory activity is characterized by a lot of noise or jitter. However 50 to 1 is as good as a random rat ever gets. The table shows that the efficiency gets worse as an exponential function of MSR. At MSR230, the efficiency score shows that a random rat required over one hundred and ninety one thousand (191,000) steps on the average to move through each of the units of the maze.

Compare in FIGURE 1.4.18T, the two groups, the E-group and M-control rats, rows 0 and 1 respectively. The mean for the random rats ranges from seven to nine thousand times larger than the mean for the real rats. The mean for each of the five variables are 5.17 vs. 40,146 runs; 40.5 vs. 352,362.5 loops; 1029.8 vs. 9,233,558.8 steps; 21.45 vs. 192,363.8 steps per first entry; .0247 vs. .0000028 probably. While technically the fourth variable is called the efficiency variable, all these variables are just different ways of measuring efficiency.

The E-M FEU Analysis

FIGURE1.4.19 Groups by units

The E-M comparisons only involve the unit data for each of the three types of data, since latency data was not simulated. FIGURE 1.4.19 shows the FEU, group by units, first order interaction. $F (24,336) = 6.528$, $P<.000000$. The data is plotted with the normal x and y-axis reversed, in order to plot the emergent loop data as the rate of exploring the units of the maze, after the operant conditioning tradition of plotting the rate of lever pressing. (Skinner, 1974) FIGURE 1.4.19 is a group mean that includes the F3 data of FIGURE 1.3. 3. The ANOVA interaction on the group data shows that the differences observed in individual rats data seen in F3 and M1 were significant differences. However, even though this is the mean for five subjects, the "plateaus" and the extrapolation effects still are very visible in the E-group data showing that the E-control rat's retrace-loops, which produce the plateaus, were triggered by the same maze units so that in the group means the plateau data was not canceled out.

FIGURE 1.4.20A and B main effect for 50 units

FIGURE 1.4.20A and B compare the **main effect for the FREQULM data** for the fifty units (alpha + 48 units + omega) of the maze for the E-group and the M-group subjects. The data for both the E-group and the M-group shown in FIGURE 1.4.20A were graphed with a linear-linear plot. The convention of plotting the two curves on the same y-axis was not observed in this case because if the E-group data were plotted on same y-axis of the M-controls, the E-group data would simply be a straight line parallel to the ordinate. The interesting similarity to be discussed below between the shapes of the two curves in spite of the magnitude difference could not have been seen. The M-control data shows no plateaus or extrapolation effect at the finish. The M-control group of rat's retracing loops occurred randomly so that in the group data the individual differences in the data canceled the plateau effects for the simulated rats.

The FIGUREs seen in 1.4.20A and B for the E and M groups are paradoxically very similar in shape, except one is ten thousand times larger than the other is! It is interesting that both groups of rats generated the same shape curve when they run the maze, yet are so different in efficiency. The FREQU data shows where rats, real or simulated, spent their moves in the maze. Both clearly do not distribute their moves evenly throughout the maze but spend most of their moves in the first part of the maze.

FIGURE 1.4.21 M groups by rows

FIGURE 1.4.21 is the first order interaction, groups by rows, for the BFEU variable. $F (11,132) = 4.137$, $P<.00003$. The curve for the E-group is curvilinear but exponential for the M-controls. The curvilinear E-group effect shows that real rats found the first few rows increasingly difficult but then stopped retracing by the sixth row! This curve shows that by the time the E-group of rats had explored only the first half of the maze, they had already evolved a valid model of the maze, and thus they were able to assimilate the rest of the maze without making many further errors. The M-controls, on the other hand, never stopped retracing, but instead the retracing continued to increase exponentially. The differences demonstrate that there is a powerful advantage to having exploratory behavior generated by a primary rational mind rather than by a computer program driven by prime numbers.

The E-A Effects

In the above section, the behavior of the E-group of real rats and the M-group of random rats in the Lashley III matrix maze were compared, and it was shown that the differences between these two groups at each of the three outer-loop levels of the hierarchy were either quantitatively or qualitatively different or both. In this section, however, the question to be investigated is purely a qualitative one. The E-Group and A-group of yoked rats, by definition, ran the maze with the same efficiency; the question is whether the two groups can still be distinguished. The primary purpose of this discussion is to comprehensively study of the nature of randomness. In this context, the question is do both groups of yoked real and random simulated rats exhibit the same behavior or can their behavior be distinguished. The many forms of the quantitative efficiency differences observed above between the E and the M-group are certainly the most powerful and obvious aspect of the difference between random rats and real rats with the former being the least efficient, while the latter's behavior showed great efficiency. In this section, however, by definition, any differences between the two groups cannot be a quantitative difference, but any qualitative differences observed hopefully will partially explain the cause of the quantitative efficiency differences observed. In addition, the A-group of random rats was created in order to focus upon the two-inner most levels of rat behavior: first the loop-level and secondly the node-path level of the behavioral data. The fact that the A-group of rats consist of a set of rats each yoked to one of the E-group rats has the effect of quantitatively equating the two groups on both of these inner most levels of the behavioral data. However, the yoking does not wash out the qualitative differences at either level of the behavioral data, if there is any. Yet, as I shall show both the loop and the node-path data tell the same story. So the point of this section is to reveal the way the primary mind of the rat begins to organize hierarchical behavior first at the two innermost levels and then outwardly through the other respective several higher order levels. The implication of this data is that the Heterostatic mechanism of the primary mind that generates hierarchically organized behavior has important nativistic skills, which allow the intrinsically motivated rats to make discriminations between the column units of the matrix maze that random rats can't make, and these discriminations occur at both the loop and the node-path levels of behavior. The E-group of rat's behavior reveals the rat's spontaneous ability to make these discriminations at both these inner levels and thus, reveals the secret of the E-group's success. The matrix maze makes it possible for the first time to show that the serial order behavioral problem is a two-dimensional rather that the classic one-dimensional problem conceived by the behaviorists. It is the mastering of this second dimension, the column dimension of the matrix maze, rather than the mastery of the row dimension, which is the primary challenge in the Lashley III maze. The classical behaviorists were studying the rat's learning behavior only in the easiest dimension for the rat to master.

In this section, as noted above, the search for the differences between real and random rats will be focused upon the qualitative differences between the two inner-loop levels of the hierarchy, the loop, and the node-path levels.

Loops

The E-group of real rat's loop-behavior can be described and predicted by a rule, which will be described below, while the A-groups loop behavior, which was random, cannot be described by this rule. For example, consider loops with a length of six steps. There are four permutations of loops of that length. The E-group made 34 loops of that length, all of which exhibited the same permutation: 6, 14, 15, 16, 15, 14, 6. All the loops of length 12 were also of the same permutation. All the loops of length 16 were one of two permutations, either 6, and 12, or 12 and 6. There are many permutations for loop lengths of 12 and 16 and yet none of these permutations ever occurred in the E Group.

FIGURE 1.4.22 E groups by loop lengths

FIGURE 1.4.22 shows the first order interaction, groups by loop lengths, for the E, and A groups. $F(8, 80) = 10.02$, $P < .0000$. This FIGURE plots the frequency of all the loops in the range from 2-18 steps where loop length is defined as the number of steps per loop. When a rat leaves the Alpha unit and enters unit 14 that is step #1. The loop lengths, by this method of calculation, are always an even number of steps with the last step always the reentry into Alpha. This FIGURE shows that the A-control group of rats had loops lengths that were a smooth continuous curve with the loop length of two steps having a frequency of seven but dropped until the loop length with 18 steps had a frequency of only one. The E-group of real rat data is, however, not a smooth continuous curve. The curve shows three peaks, which are the three loops lengths with the highest frequencies 6, 12, and 16.

The random A-control group of rats exhibited loop behavior, which was the product of the random number generator, which moved the rats through the maze, and so the loop length of the random rats should show a random distribution of loop length, which was observed. The E-group of rats, however, did not produce a random distribution, for their behavior was not generated by a random number generator but by a primary mind. This mental mechanism generated a tri-modal distribution that reveals that real rats have a remarkable capacity, which spontaneously emerges to organize their behavior on their very first loop through the first row of the matrix maze so that rats respond to the Matrix maze's first row structure with three very specific loops with lengths of 6, 12, and 16. These three lengths represent three specific non-random trajectories, which are shown below:

Loop length steps:	0	1	2	3	4	5	6	7	8	9	10	11	12	13	14	15	16
Trajectory # 1:	6	14	15	16	15	14	6										

Trajectory # 2: 6 14 13 12 11 10 9 10 11 12 13 14 6
Trajectory # 3 A: 6 14 15 16 15 14 13 12 11 10 9 10 11 12 13 14 6
Trajectory # 3 B: 6 14 13 12 11 10 9 10 11 12 13 14 15 16 15 14 6

Trajectory # 1 was a loop with six steps consisting of the units of the matrix listed above. While Trajectory # 2 was a loop with 12 steps through the units listed above. The reader should use FIGURE 1.2.4B to trace each of these four types of loops.

There are two permutations of type 3. See FIGURE 1.2.4B for the unit names. The different permutations are determined by which of the two sets of blinds the rats select to explore first, either the A and then the F set, or the reverse.

The fact that the E-groups of rats were able to selectively make loops with only three of the several possible lengths shows that the E-group of rats bring to this task a nonverbal intelligence which gives them the adaptive capacity to respond spontaneously and appropriately to this level of the hierarchy on the first loop without learning. The E-group of rats exhibit what might be called a spontaneous or unconditioned capacity to discriminate and appropriately chunk these first row column units of the matrix hierarchically organized maze, at least in the first row. Either the E-group of rats explored the left blind in the first row, which required six steps, or the right side of the first row in the maze, which required twelve steps, or they explored the whole of first row of the maze, which required 18 steps. The exploring of the complete first row could be accomplished either of two ways, first the left side of the row and then right side of the row or the reverse.

FIGURE 1.4.23T loop behavior for two groups

FIGURE 1.4.23T compares the loop behavior of the E-group and A-group of rats broken down into six types of loops for six types of measures of loop activity. The experimental rats required an average of 40.8 total loops to reach criterion.

For the experimental rats, the total 40.8 loop average breaks down into 35 redundant and 5.5 FE loops. The FE loops break down into 1 criterion loop, 3.8 complete, and sixty-seven incomplete FE loops.

The random A-group of rat's 35.5 total loops breaks down into 30.5 redundant and 4.97 FE (First Entry) loops. The experimental and random rat's percent of random and FEs are identical, 86% and 13% respectively. So while the random A-group rats make significantly fewer total loops, the breakdown of redundant and FE loops expressed in percent of the total is identical. This is only one of many similarities, but these are to be expected since the A-group is yoked controls.

The loop length results were, however, not the same. The experimental rat's average loop-length in steps for the total was 24.3 steps vs. 28 steps for the controls.

Latency scores

As noted above, the MSR228 was such a severe criterion of excellence that it took 6 years to assemble and write the hardware and software necessary to achieve a simulation of a random rat at this MSR. On a 386 PC at 25 MHz, the average simulation required five hours. While in principle, it would have been possible to simulate a random latency by allowing a simulated rat to remain in each unit it enters a random number of times, then to multiply the total number of reentries by some time constant. However, it would not have been technically possible for simulated rats to achieve criterion with our 386 computers if we added this requirement to the program. Therefore, in this study we have simulated moves only. However, by not allowing the random M-control group of rats to reenter a unit of the maze, the random rat's task is much simpler than the task of the experimental group of rats. In other words, the random data is very conservative, in addition to the fact that the M-control group of rats was only run to an MSR230 instead of the more difficult MSR228 as discussed ***above***.

The Node-Path level of the Hierarchy

The Node-Path is the inner-most-level of the hierarchy of the behavioral data and has two elements, nodes, and paths. A node is operationally defined as units in which the rat turns around or reverses its direction of motion, in other words, in the tracking record, the same unit number will always precede and succeed a node; paths are the distance between nodes.

There are four dependent variables for this level of the data: node frequency, NF; path length in unit's steps, PLU; and path length in time per unit, PLT; and unit speed, US (PLT\PLU). For each of these scores a mean is computed: node frequency mean, NFM; mean path length in units, PLUM; mean path length in time, PLTM; and unit speed mean, USM.

The nodal data was analyzed two ways. First, since the random rats do not have time scores, it is only possible to make E-A group comparisons with the three scores, NF, PLU, and PLUM. Second, since only the E-group has time scores, the E-group data will be analyzed separately comparing the time score mean, PLTM with the other two mean scores, PLUM and PLM.

The First Node Behavior

The node-path data is so voluminous and complex that only a few examples will be reported at this time. First, allow me to review. The hierarchically organized behavior of the rat within the matrix maze is organized into growths, each of which can be described as a set of runs; each of which can be described as a set of loops; and finally each of which can be described as a set of nodes\paths.

The shortest loop, the one with the shortest node structure is a (6, 14, 6) alpha, unit 14, alpha loop, and was the most frequent A-group loop and has a one-node structure. Two other single node loops are the (6, 9, 6) alpha, blind path unit 9, alpha loops and the (6, 16, 6) alpha, blind path unit 16, alpha loops, which were the two most frequent E-group loops. Loops with a complex node structure have many nodes. For example, the criterion loop for E-group rat F10 had the following node structure:

Node order	0 1 2 3 4 5 6 7 8 9 10 11
Nodes	(6, 9, 34, 48, 62, 56, 49, 41, 24, 17, 21, 6)

This particular criterion loop had eleven nodes, with unit 62, the omega unit, the fourth order node. Unit 9 is the first order node for this loop and unit 34 the second order node. The node orders are of special interest for they are another important part of the data, which reveal the exploratory intelligence of an organism; they show the difference a primary mind makes.

The E-group of rats had criterion loops with the following frequency of nodes, with the first number being the number of nodes (10-3, 11-1, and 12-1), the A-group (10-2, 11-1, 12-1, and 19-1). One random rat had a criterion loop with 19 nodes, which says it turned around often. Since these are yoked controls, this similarity is not surprising.

Several ways that loops can be classified were identified above. Additionally, loops can be classified according to their node order. The total number of loops made by the E and A groups were 243 and 213 respectively. These loops were all classified based on their node order. First, all loops were described as a set of nodes, secondly, they were sorted on the first order node; thirdly, the first order nodes were then sorted on the second order nodes, etc. The data is interesting but too complex to describe at this time, except for the first order nodes.

FIGURE 1.4.24 shows the total number of first order nodes made in each of the eight units of the first three rows of the Matrix Maze. The solid red line shows the node order data for the A-group and the solid blue line shows the node order data for the E-group. The data shows that real rats are particular about where they turn around within a row, or they are particular about the kind of units, which function as nodes, only eleven of the possible twenty-two units were ever used as first-order-nodes! In the first row, the most used nodes were the deep blind units, A2, or unit 9 with a count of 105 nodes, and F2 or unit 16 with a count of 89 nodes. The real rats explore a row using the mental rule; explore a path to its end. There are some exceptions to the rule in the first row. In the second row, however, the rule was followed without exception. The third row had only one node, unit 27 which is an E4 unit for the third row which is remarkable for this unit is twelve units down the EP of the Matrix maze and shows how far at least one rat was able to run without turning around in the maze. Considering the fact that these rats were allowed to explore for only one growth, a run of 12 units is a significant achievement.

Rational behavior could be caused by any logical fact, but with these rats, we are witnessing one of the simplest possible causes for rational behavior. In the figures to follow note how many of the very different looking effects are still the product of this one simple rule.

Next, consider this group of rats whose behavior was generated by random numbers. So it should not be surprising to note that the nodal data for three rows for the A-group of rats does not show the same distribution of nodal scores as seen in the E-group. The A-group of nodal scores show exactly what should be expected by subjects taking a random walk in the matrix maze, a normal distribution of scores about unit 14, the E1 unit of the first row. The normal distribution curve in a asymmetrical matrix has, of course, its own look with the matrix forcing the distribution of the normal data to appear skewed since the left side of the curve only has two possible units, while the right side has (8 * 6 - 3) = 45 possible units.

The A-group data shows that, in fact, node unit 14 was the mode node unit with a score of 75. On the left side of the distribution, the total nodal activity was (34 + 31 = 65). On the right side, the total activity was (36 + 13 + 5 + ... = 73). The distribution was close to normal (65, 75, and 73).

FIGURE 1.4.24 First Order Nodes

The E group data, however, is not a normal curve, with the nodal mode at unit 14, but is multi nodal. The E group of rats have a natural ability for selecting the natural nodes of the maze- the deep blind units 16, 9, 17, 24 and 27(A2, F2, A2, F2 and A2 respectively) which account for 91% of the first order nodes for the E group, but only 20% of the first order nodes for the A group. These four units are natural nodes or nodes-by-necessity, since; when they are entered, the rat must reverse its direction of motion. All nodes other than A2 and F2 are arbitrary nodes. The E-group nodal behavior shows that intrinsically motivated rats spontaneously recognized or discriminated the logical nodal structure of the maze. Alternatively, to put the point another way, real rats do not arbitrarily turn around in just any unit of the maze, and thus create an artificial node: real rats, when they first enter a row, explore the row to its natural end, the deep blind unit that is the natural node of the row. This spontaneous behavior of rats, to select the natural nodes of the maze, in the early part of their exploring the maze, shows that when rats enter a blind unit it is not an error as the Behaviorists have always characterized this behavior. Rather, it is a logical necessity: a temporary strategy for empirically discovering the limits of the path of the rows. The early blind entries for real rats are clearly not a part of Trial and Error random behavior. The E-group of rats run initially with the rule, explore to the end of the row. What does random behavior look like?

A group node behavior

FIGURE1.4.24 Node Behavior

The A-group of rat's behavior demonstrates what random behavior looks like at the node\path level of a hierarchal behavioral organization. FIGURE 1.4.24 makes it possible to distinguish the random node pattern of the A-group of random behavior and the non-random pattern of the E-group. The distinguishing feature is that the deep blind units, which are seldom parts of a genuine random node behavior, are the preferred nodes for real rats. Deep blind entries or nodes scores, therefore, are not the sign of errors or mindless random behavior as the behaviorists have always insisted but are, on the contrary, one of the unique signs of the rational behavior generated by the primary mind, which I call Heterostatic behavior. When rats are exploring a row for the first time real rats explore with a rule, which says, explore a path to its end or natural node. This rule-generated behavior generates the unique loop length data seen in FIGURE 1.4.22 and a corresponding unique node behavior seen in FIGURE 1.4.24. The entries into the blind units of the maze are not something that random rats do and so the behaviorist in characterizing these entries into the blind as errors is an error on the behaviorist's part not on the part of the animals. This behavior is not a sign of random behavior but the sign of rule-behavior.

The power of natural nodes

The power or the role of nodes can be show in two sets of figures:

First: Compare FIGURE 1.4.22 with FIGURE 1.4.24, for the two figures complement each other. Each is a different side of the same coin. FIGURE 1.4.22 shows the E-group of rats' preferred loop lengths, while FIGURE 1.4.24 shows their preference for nodes. The loops of LL16 are the same loop with a node in unit 9, etc.

Second, compare FIGURES 1.4.25, 1.4.26, and 1.4.27, these three figures, which will be discussed shortly, all show in their own way the role, the power, or the natural nodes on the Behavior of rats.

Simulation of the data

The difference between the way real and random rats move at these two innermost hierarchy levels, the loop and the node-path levels, is most obvious when the exploratory data is viewed "live" on a computer monitor. As a simulated rat moves through a simulated Lashley III maze, the random behavior that characterizes the random rat's moves seems to jitter, Random simulated motion, in other words, looks like Brownian motion. The real rat's node\path moves, on the other hand, show no such jitter, but are smooth and continuous in appearance. This difference makes it possible for the human eye to learn quickly to distinguish a real and random rat's moves when observing the data on a computer monitor. The continuous and smooth appearance of the real rat's behavior when it is exploring a row is an expression of the rule, which governs its row exploratory behavior.

Section three: The Total Node-Path Behavior

Earlier in this book, the two outer- loop levels of the behavioral hierarchy, growth, and runs, were used to describe the EP (Essential Path) behavior of rats. The two innermost levels of the hierarchy of behavior, however, allow us to begin to describe the BP (Blind Path) behavior of rats. The EP and the BP behavior are learned independently. That is they each require the rat to evolve a separate model to remove the uncertainty induced by their respective anomalies. The BP behavior of rats cannot be fully described in this study since the rats in this study only make one growth, and many growths are required before all the cells of the BP are entered for the first time and before the data can be analyzed by ANOVA.

However, the two innermost levels of the behavioral hierarchy in this first experiment give us a first glimpse at BP behavior.

The behavior that rats exhibit as they adapt to the two inner-most levels of the maze hierarchy, the loop and the node\path level, are also full of anomalies as was the EP data and each anomaly requires us to rethink how rats learn the maze. These BP anomalies challenge the imagination.

In order to understand BP behavior, two stages of pure Heterostatic adaptation must be distinguished. The first stage can be characterized with several different terms: the pure accommodation phase, the preformal stage, or the sympathetic dominant stage. The second pure Heterostatic stage can be characterized as the pure assimilation phase, the formal nonverbal stage, or the parasympathetic stage.

In this first experiment, it will only be possible to describe the behavior of the first of the Heterostatic stages but the loop and node path behavior can only be understood in the context of the distinction of these two stages.

The accommodation stage

The first stage of pure Heterostatic adaptation is called the accommodation stage because this stage of adaptation is the nonverbal equivalent of Kuhn's (1962) revolutionary stage of science. During this stage, rats evolve a conceptual system that is capable of destroying the anomalies of the domain. This stage is both an ontological and an epistemological stage. It is during this stage that the nonverbal primary mind must invent a set of nonverbal ontological universals, or classes of events, that can assimilate the anomalies of the domain. The primary nonverbal mind of rats evolves a set of classes that can assimilate the 96 units of the matrix maze domain empirically. The point is that the only way a rat can evolve a set of valid concepts to recognize the classes of the blind path units is by entering the blind path units. Initially, the entry into these BP units is not an error, but on the contrary, a logical empirical-ontological necessity. It is impossible to evolve a valid concept for these BP units without entering these units several times. The first phase of pure Heterostatic adaptation is the accommodation phase where a conceptual system is evolved that destroys the anomalies of the BP domain. During the first phase, when the rat is exploring the anomalous BP unit elements of the matrix domain for the first time, the unexplored unit will be positive in value.

In the second phase, however, the BP units, once they have been discovered to be redundant, will become negative. However, this topic is the focus of later chapters. In this first experiment, it is important to note that only the properties of the positive stage will be described, but the reader must understand that this positive phase is the first of two phases, with the second stage a negative stage.

As noted above, during this first accommodation phase, the BP units of the maze are positive until the rat evolves a valid conceptual system for recognizing these units. During this first phase, the exploration of the BP unit is not experienced by rats as an error or as a negative experience to be avoided; that only occurs in the second stage.

The primary mind of rats brings to the matrix maze such an impressive set of conceptual tools that all rats spontaneously adapt to the first row reflexively, as the data in this section will show. The primary mind of the rat enters the matrix domain with an unconditioned rule that guides the rat's behavior in ways that will be described in this section for the first time.

In the previous section, the node- path data for the first node of each loop was described. In this section, the total node path data for all the loops will be described. Some of the questions are how many nodes did rats make per growth and loop? Which units did rats select as a node? And on what basis do the E and A-group of rats make their selections? The average E-group rat made 184 nodes per growth, while the average A-group made almost three times as many. The row exploratory rule of real rats minimizes the number of nodes.

FIGURE1.4.25 Node Frequency

FIGURE 1.4.25 shows the 48 unit data for the E and A groups of rats for the mean NF (Node Frequency), PLU (Path Length in Steps), and PLUM (Mean Path Length in Units) scores. $F(96, 1440) = 5.87$, $P < .0000$. The difference between the E-A groups is most obvious and dramatic. Yet the Experimental group's behavior is the product of the real rat's absolute reliance upon one simple rule use only the natural node units, as nodes, or in this maze only turn around in the deep Blind units, F2, and A2.

The data of this Figure shows that the shape of a real rat's behavior can be determined by a very simple inner loop rational rule that historically has gone unnoticed

FIGURE 1.4.26 group by node frequency

FIGURE 1.4.26 shows the <u>first order interaction, group by node frequency</u>, NF, for the first 48 units of the maze. $F(47, 1410) = 11.72$, $P < .0000$. The data for the 48 back units was not included because too many of the scores were zeros. The most obvious difference is the eight-unit rhythm of the E-group contrasting with the uneventful almost straight-line function of the A-group data, at least for the first 24 units. This eight-unit rhythm is the product of the real rat's row exploratory rule. This rule was described above for the data for the first node behavior, but now this same rule is seen in the node data for the whole maze. The real rat's behavior can once again be seen in data of FIGURE 1.4.25 to be very different from the random behavior the behaviorist always claimed. The real rat's behavior is rational behavior that consistently is seen to be an expression of a rule. The behavior of rats that are spontaneously exploring this maze is rational behavior determined by their primary rational mind.

The data of FIGURE 1.4.25 must be distinguished from FIGURE 1.4.24. The former is the first node data per loop. The latter is the total node frequency for growth one for all loops. The E-group data of FIGURE 1.4.25 shows that the effect seen in FIGURE 1.4.24 for the E-groups for the first order loops in only the first few rows held for all the rows of the maze. The E-group of rats in every row of the maze selects as nodes the deep blind units. The A-group data in FIGURE 1.4.24 shows that the first-node was unit 14. However, in the total data, Unit 14 was not exceptional.

FIGURE 1.4.27 groups by columns effect

FIGURE 1.4.27 shows the first order interaction, groups by columns effect, for the average for three scores, NF, PLU, and PLT for the E-group and the A-group. F (7,210) = 29.97; p<. 0000. The column effect for the A-group is essentially flat, which means that the random rats have no power to discriminate the column unit's fitness for being a node. The FIGURE shows that every one of the eight column units had, what every random event is supposed to have, an equal chance of being a node. The E-group data, however, is bimodal, with two sets of scores, those over twenty, A2 and F2, and those under five, A1, E1, E2, E3, E4, and F1. The E-group of rats is particular about where they turn around within the columns of a row. They use the rule explore to the end, and thus only the natural node units, F2 and A2 are used as nodal units. This data was analyzed in the functional not the structural format.

The E-group data is obviously the product of the cyclical behavior of FIGURE 1.4.25, which looks like sine waves with the latency score having greater amplitude than the unit score.

FIGURE 1.4.28 Row main effect for 12 rows for node data

FIGURE 1.4.28 shows the **row main effect** for twelve rows for the node data. F (11,330) = 35.01; P<0.0000.This curve is a node learning curve; it shows that the number of nodes per row decreased over rows at least for the first six rows. As rats learn to explore more efficiently, the length of their loops increase and the number of nodes per row decreases.

The structure of behavior

The organization of this section is designed to show that each of the four levels of the animal's hierarchically organized heterostatic behavior makes a unique contribution to our understanding of how the nonverbal mind spontaneously makes models of a hierarchically organized spatial domain during exploratory-play. The four levels of the hierarchically organized behavior are growth, runs, loops, and the node\path from the outer to the inner loop respectively.

Theoretical

The "Out" behavior of the rat

On each run, the rat is given an allotment of behavior both in times and moves that it can choose to spend any way it desires in the matrix maze. In this first experiment, there were three ways for the rat to get "out" of the maze. First, it can complete a criterion loop, Alpha, Omega, and Alpha. Secondly, the rat can choose to use all of its allotted moves within the allotted time and be removed from the maze when all its moves were used; and third, the rat can get out by using up all its time.

Growth

A growth is defined as a complete loop through the maze in both directions. The spontaneously defined Heterostatic goal of the rat during exploratory-play in the matrix maze is to perfect its ability to emit growth behavior. Not all rats choose to accept the challenge of the maze, and why some do and some do not is not known. However, for rats that accept the challenge of the matrix maze, their exploratory-play behavior shows that their goal is to perfect their ability to make a growth with the shortest path. Rats spontaneously enjoy perfecting their behavioral ability to run through the matrix maze with ever improving competence, even though there is no objective Homeostatic incentive, no reinforcement, to reward their performance.

Work vs. play behavior.

One of the most interesting features of the rat's behavior is its similarity to human behavior. For example, I should note that heterostatic behavior only works when an experimenter is willing to totally put the subject in the driver's seat namely to allow the potentially self-organizing organism to self-organize.

What I am talking about most simply is the difference between work and play behavior.

Play behavior refers to those kinds of purposes that are intrinsically motivating for an animal. One of the most interesting and most important phenomena I discovered about rational behavior was about devious behavior, but what was so remarkable is how complex devious behavior can be when the organism's first choice of playing is punished.

Work behavior was not studied in this set of experiments. When rats are allowed to choose their own purpose or purposes as they were in these experiments, the behavior the rats emit will be play behavior.

One simple rule: Mammals will pay to play, but must be paid to work. Play behavior is intrinsically motivating, while work behavior requires extrinsic motivation/reward.

The shortest path learning as Homeostatic Theories

The classical behaviorist learning theorists tested rats using the TTP (Trial Testing Procedure) and observed that rats as they learned to run through the maze would improve their behavior in several ways. First, eliminate entries into the blinds units; second, stop retracing, third, the overall effect of all the rat's improvements had the effect of allowing the rat to run through the maze with the "shortest path." Classical theories of learning all attributed this trend in the rat's behavior to the effect of reinforcement. The rat, so the story went, ran the "shortest path" because every digression from the shortest path, in effect, delayed the rat by some increment of time from the consumption of food and cost the rat some additional increment of work. The experimental studies of the rat's behavior for a hundred years using these Trial Testing Procedures, TTP concluded that the purpose of the rat in the maze under these test conditions was to find food with a minimum amount of time and effort. This homeostatic assumption "explained" the shortest path behavior of the rat.

The shortest path learning as Heterostatic behavior

In this study, the rats were run in the matrix maze and were neither deprived nor reinforced and were run on the LTP (Loop Testing Procedure), instead of the TTP. The TTP allowed the rat to run in only one direction through the maze. However with the LTP, the rat is allowed to explore freely in both directions, there are no non-retrace doors to limit the rat's ability to retrace. The non-retrace doors were eliminated in the design of the LTP because it was considered most important not to limit the rat's ability to retrace in the maze, which the behaviorists called "random" exploratory behavior. The purpose of this experiment was to study the "random" retracing behavior of the rat, and it would have been counterproductive to use the non-retrace doors, which were designed to limit the very behavior, which the experiment was designed to study. The changes introduced in the design of the LTP were made to optimize the ability of the rat to emit "random" behavior, so that a full evaluation of the rat's ability to select the "shortest path" could be evaluated. The ability of the rat to select the shortest path is, in effect, a large measure of the rat's ability to stop emitting this "random" behavior. The presence of the non-retrace door arbitrarily limited the amount of "random" retracing the rat is allowed to emit as it explores repeatedly every unit of the matrix, both in the essential path and the blind units. The practice of allowing the rat to run in both directions also, of course, doubles the length of the shortest path the rat is required to master, but the increase in the length of the shortest part that defines a loop does not entirely measure the increased difficulty added by the looping requirement. The loop requires the rat to run in two directions in the maze, out and back. There could be a real difference between doubling the length in one direction and doubling the length by requiring the rat to run in two directions, but I have not yet made such a comparison.

The changes made in the LTP were selected to maximize the difficulty of the rat to learn, to run the shortest path, and yet the rats are able to discover a set of rules to describe the hierarchically organized matrix maze, which make it possible for them to run the shortest path before they have even explored half the maze. The behavior of the rats learning to run the shortest path in the matrix maze using the LTP clearly cannot be explained within the traditional context of Homeostatic motivation and reinforcement. The shortest path behavior seen in the LTP in the exploratory\play behavior has a completely different set of behavioral properties than the shortest-path behavior traditionally seen the TTP in the work context: The shortest path in the LTP evolves as a start gradient, while the shortest path in the TTP evolved as a goal gradient. The different set of properties of the two behaviors shows that the behaviorist's explanation of the shortest path behavior in the TTP work context was an artifact produced by the confounding of the work and play variables in the classical test environment. In this experiment, the confounding between the work and play variables was eliminated, and the rat's behavior reveals that the shortest path behavior is emitted even when the homeostatic conditions are not acting in the experiment. The rats of this experiment show that rats spontaneously emit the growth behavior under these test conditions simply because it is fun.

The growth behavior of the rat in the matrix maze that evolves when rats are tested using the LTP show the rat's spontaneous ability to enjoy learning to run the shortest path even when there is no physiological reinforcement for the perfection of that skill. The perfection of the skill itself is an intrinsic goal of the organism. The loop testing procedures allow the rat to demonstrate its skill to master the challenge presented by the anomalous hierarchically organized matrix maze.

The Heterostatic or psychological goal of the rat can be distinguished, for the first time, from the Homeostatic or Reinforcement goal the behaviorist claimed dominated the rat's adaptive behavior.

The inner levels of the hierarchy

The central issue of this experiment is whether or not real rats allot their time and moves in a way that is different from the way virtual random rats spend their allotted times and moves.

There are two different issues and each issue is tested by a different group of virtual rats. The M-group of virtual rats was designed to test the meaning of "random" behavior in the two outer loops of the behavioral data, the growth, and the run data. The A-group was designed as a control for testing the meaning of "Random" behavior in the inner two loops or behavior.

The node\path level

At the inner most level of the behavioral hierarchy, the node-path data shows that the rat's heterostatic mechanism has the capacity to adapt so quickly to the first row that the rat, in effect, adapts in real time to the units of the first row. The node-path behavior shows that the virtual random rats, but not the real rats, respond by making random distributions of path lengths with a node length of two. Real rat's exploratory-play behavior, in the first row especially, but in all rows, in general, follows the rule "explore any path to its end." In short, the nodal unit in each path of the row is the natural node. In other words, real rats do not arbitrarily or randomly turn around in a row of the maze and thus create an artificial node, as do virtual random rats. Real rats, once they have committed themselves to explore a row in some direction, will explore that path to its end, which is the natural node of the path. Even at this lowest or innermost level of the hierarchy, the rat behavior reveals how the rat's mental mechanism, the Heterostatic mechanism, acts to generate behavior that is not random.

The loop level

At the next level of the hierarchy, the level of the loop, the rat's behavior reveals that loop lengths occur only in certain lengths with lengths of six, twelve, and eighteen being the most frequent length. These loop lengths are those that are produced by the rule stated above, "Explore any path to its end." This loop supports the node\path data and shows that real rats do not randomly explore a row in the maze.

The loop behavior of the rat changes over time. The initial loops are either short exploratory loops or long retracing loops. However, there comes a time when the reverse occurs, the rat stops retracing and stops making short exploratory loops. The rat suddenly makes one long exploratory loop, which completes the exploration of the whole matrix maze. The run data shows that the rat's Heterostatic mechanism is a conservative mechanism that does not explore beyond the range of convenience of its model. The rat does not explore at a rate that takes the rat into units of the maze where it will be lost. The rat's rate of exploration is no faster than the validity of its representation of the matrix will allow.

The growth shows in a many ways that the exploratory-play behavior of the rat is not random.

Confounding

The result of this experiment is a complex set of data all of which is completely anomalous within the context of classical learning theory. The learning data generated by the TTP has for over one hundred years been confounded, yet, the behaviorists, of course, were not aware of this confounding. The LTP used in this experiment for the first time has eliminated the confounding by creating a pure exploratory-play test environment in which the work variable was eliminated from the test procedure. However, while the work variable and the confounding were eliminated, nevertheless, the animals tested in the LTP learned every bit as swiftly as the learning that occurred in the TTP. Thus, a completely new understanding of how learning occurs must be created. The anomalous data generated by the LTP requires a revolutionary new theory to explain the data. Heterostasis is a mechanism, which can take this complex and strange data and make it simple and familiar.

Learning is a trick, but Growth is not a trick

Thorndike (1911) conceived of the rat as a creature that had to be tricked into learning first, by depriving it of food for several days, secondly, by putting food into the goal box. The mindless rat has no intrinsic interest in learning the maze but because of its intense need for food, it learns the quickest way through the maze. Thorndike's concept of trial and error learning became the central myth of the behaviorist.

The alternative

Heterostasis conceives of the organism as an active not a passive being. Every mammal is an active organism with its own set of autonomous purposes. In the design of the cortex, the gene or nature redesigned the whole brain, keeping and building on the old mechanisms, but integrating them into a new system, which had a new set of priorities. The active organism, the heterostatic organism has a completely new set of needs. The Homeostatic and the Heterostatic mechanism are two different mechanisms and each has a different set of needs.

This first experiment of the series shows that the non-deprived rat has a set of Heterostatic needs that are independent of the homeostatic set of needs. The Heterostatic mechanism does not require a tissue need to motivate its action nor a physiological reinforcement to close the adaptive loop. The tissue variables that the behaviorists imagined as the *sine qua non* of learning are not even relevant to the discussion. The Heterostatic organism is motivated by anomalies, news, and redundancy. The data generated by the LTP require a completely new model of the rat. The shift from mindless to a mindfulness theory of the rat requires a paradigm shift.

The results of this experiment have shown that, in general, Behaviorism's theory of mindlessness, in particular, behaviorism's concept of random behavior, Trial and error learning, Homeostatic deprivation, and reinforcement are all myths.

Trial and error learning the source of the myth of mindlessness

Descartes (1946) set the stage for the Classical Learning Theory, but Thorndike (1911) wrote the play. Descartes on moral ground had refused to conceive of the mind as a passive mechanism, but Thorndike broke with Descartes and created the first formal theory of the mechanical organism. However, our quarrel with Thorndike is not the fact that he conceived of the learning as generated by a machine. Heterostasis is also a mechanism. Our quarrel with Thorndike is with the too simple nature of the mechanism he conceived. Thorndike's machine produced learning by trial and error, Homeostatic motivation, Reinforcement, Random Behavior, etc. Trial and Error learning was the mother of all the passive myths that were the foundation of Behaviorism. Hull (1943) rewrote the Trial and Error story and refined the myth but retained the classic passive conception of the organism. Watson (1972), of course, made a religion of this entire passive model of the organism and claimed the environment was not just the primary but also the only actor on the stage: the organism was not really an actor; nature not nurture was cause of all that happened on the stage.

The purpose of this book is to challenge and replace all these myths. The purpose of this book is to challenge the whole family of myths by focusing upon one particular member of the family of myths, the concept of Random behavior.

The primary purpose of the design of this experiment was to make it finally possible to examine, formally and experimentally, the myth of random behavior. The result of this experiment makes it unmistakably clear that Thorndike's (1911) assumption of "random" behavior, just does not have any basis in fact, and with its demise logically the whole theme collapses.

However, the data of this study shows that real rat's behavior does have some interesting parallels with random behavior, but they are coincidental and superficial. A virtual random rat can run the Lashley III matrix maze in both directions, out and back, as required by the loop paradigm, as can a real rat, but every one of the many criterions of efficiencies computed showed that the real rat was more efficient than a random rat by several orders of magnitude. The FREQULM row main effect, for example, had almost the identical shape for both the E, and M groups, but the M group mean was four orders of magnitude larger than the E group mean. The BFEULM main effect for rows furnishes the most critical differences between the E, and M groups. The BFEULM E-group of rat's row effect is curvilinear indicating that they stop retracing before they reach the omega unit for the first time, while the M-control rats curve is an exponential function indicating that they never stop retracing.

Most, but not all rats, that are not homeostatically deprived nor reinforced, emit high rates of exploratory-play behavior in the Lashley III maze. As a product of this Exploratory-play behavior, these rats were able to create efficiently, a model that takes the uncertainty out of the maze, and they generally are able to complete this model so efficiently that they essentially complete the model several rows before the rats reach the omega unit for the first time. The data shows that the retracing behavior that the classical behaviorist called random behavior, and found so irritating, is a complex sophisticated intrinsically motivated form of behavior. One that the Behaviorists, with their trial paradigm could never have imagined, given their primary assumption of the passive rat emitting random behavior prior to the action of reinforcement upon its behavior. However, it is during this despised retracing behavior that the rats are actively constructing both the most difficult central and peripheral form of learning or adaptation. This retracing behavior is not random, but this so-called Random behavior is actually the form of behavior during which the rat makes the most difficult form of adaptation, accommodation. The rat is not randomly retracing but systematically retracing, which must be thought of as a set of demotions, retentions, and promotions until it is able to evolve a model that takes out the necessary degree of uncertainty required at any particular level of the hierarchy until the rat is able to resume exploring. The retracing rat confines its behavior to just those sections of the maze, which contain the novelty that must be modeled in order to create a model that will take the uncertainty out of the subsequent higher level of the hierarchically organized matrix maze.

"It is useful to recall the findings about novelty; the singular effectiveness of novelty in engaging interest and for a time supporting the persistent behavior." (White, 1976, P. 329)

The rat is going to retrace only for as long as is necessary to remove the uncertainty induced by the anomalous event.

The pre-reinforcement behavior of the rat is not random but a new nonrandom form of behavior that the behaviorists, who prided themselves on their experimental rigor, failed to recognize and measure for almost one hundred years. This nonrandom form of behavior has finally been distinguished from the behaviorist's "random" notion of behavior.

Theory-The certainty-uncertainty dimension

Heterostasis is a theory based on the assumption that the cognitive and behavioral activity of an organism shares a common dimension, the certainty-uncertainty, **CUC**, dimension. In other words, Heterostasis assumes that both the rate of the central operations of the cognitive apparatus to reach a decision, and the rate of the peripheral operations of the autonomic and somatic nervous systems to act are both an expression of the subjective CUC the organism experiences in a domain. The cognitive-behavioral activity of an organism reflects the status of this joint cognitive-behavior control system at any moment in any domain. Any event that tips the balance in either direction either slows or increases the rate of activity in both the cognitive and behavioral dimensions of the system. What I have tried to show in this book is how to identify and objectively measure the several ways that this dynamic shifting of the balance of the certainty-uncertainty dimension in the organism's Heterostatic control system. I was able to develop a theory that explained this Heterostatic dynamic control system's activity with a new set of conceptual terms that made the understanding of the mind independent of the concept of Homeostasis that has dominated Behaviorist's theorizing for 50 years. The active organism's assumption assumes that the organism has a primary mind. It is this primary nonverbal mind that controls the CUC balance in several ways: primarily, by the organism promoting, retaining, or demoting itself to explore new or re-explore old levels of a hierarchically organized system or just stick with the present bit of data until it is fully digested.

Growth data-The dual aspect, dynamic ratio feature of Heterostasis

The CUC dimension can be measured in several ways. First, the dual aspect theory of measurements within the matrix domain shows that the time/move ratio is a dynamic measure that reflects the momentary subjective status of the organism's cognitive-behavioral control system, with large ratios objectively reflecting the subjective state of uncertainty. As the organism evolves and grows a model that is valid in controlling the activity within both the central and the behavioral system, CB system, within some specific domain of action then the time/move ratio decreases. The BFEUL data set scores generate a particularly powerful BFEUL/BFEU ratio that reflects the status of the CB control system. While the BFEUL scores inform us, both of the elements of the domain that are the triggers of uncertainty, and the quantity of the uncertainty elicited by these triggers, the ratio itself reflects the degree of uncertainty the organism is experiencing in accommodating to the level of the hierarchy that is being actively modeled at any moment.

The FREQULM data set shows how much cumulative exploratory activity was actually required in each element of the maze before a model was created. A model that was not only valid in one level of the hierarchy of the maze, but how much exploratory activity was required before a model was evolved that was valid in all the levels of the hierarchy of the maze. The FREQULM score measures the amount of the behavior that the behaviorist had called "Random" was required before the rat stopped retracing. In other words, the FREQULM scores measure how much exploratory behavior is required to create a model that is perceived as valid by the rat, so that effectively all the anomalous events within the maze have been converted into "effects with causes." The rat retraces until it achieves a subjectively valid model of the maze; one that takes the uncertainty out of the originally anomalous domain and converts the once anomalous domain into a domain that seems simple and familiar. Retracing is not mindless random behavior. The FREQULM score is a retracing measure of how much exploratory behavior is required for the mind to destroy the anomalies lurking within the hierarchies of the domain.

Heterostasis says that the cognitive behavioral, CB, activity of an organism is dynamic and reflects the momentary UC (Uncertainty-Certainty) status of the organism. The CB control systems of Heterostasis are genetically designed to create a model of a domain that is valid in the logical/behavioral sense of being able to generate a solution to the serial order problem of behavior that can accomplish, first of all, the organism's cognitive/behavioral purposes, namely to move effectively through a domain. By animal logic, however, I am not referring to something that verbal logicians invented or something that only they use. The logic of the CB system is concerned with the nonverbal logical or causal validity of the representation, which generates the behavior in a domain.

How does the behaving nonverbal organism logically "know" or evaluate the nonverbal validity of its CB action system? How does a rat know or determine how much UC is appropriate for its model at any moment? There are several ways, but one way is based on the behavioral efficiency, which is a measure of the validity or tautological status of the model online. As a model of a domain evolves, its CB validity increases as a system, and this should be reflected in the efficiency of the behavior generated by the model in its domain. For example, the real rats of this study differed from the M-control group of rats in several ways: first, in their efficiency in exploring the maze. The M-group efficiency scores show that they required 191,000 steps on the average to move through each unit of the maze, while the real rats require only 50 moves on the average to move through each unit. This efficiency difference can be measured in dozens of ways. Some of the most obvious significant effects, which reflect this difference in efficiency, are shown in FIGURE 1.4.21. The random rat's efficiency represents the worst-case efficiency. In addition, the real rat's efficiency in the first row of the maze is, as was observed in this study, objectively no better than a random rat, and in fact, even a little worse. However, while the random rat's performance score grows exponentially showing zero efficiency, the real rat's efficiency score as measured by the BFEUL data set suddenly improved. The E-group of rats stopped retracing before they reached the omega unit for the first time.

Theory-The start gradient

Second, the behavior is revealed in their rate of exploration of the units of the essential path. The real rat's exploration rate was a positively accelerated curve, while the M-group's exploration rate was negatively accelerated curve. This difference shows that the real rat's model of the maze developed during play-learning and started to evolve in the alpha unit and became more efficient the further the rat explored the maze. This phenomenon of play-learning evolves in exactly the opposite direction than work learning. Hull's (1943) Goal Gradient Hypothesis describes work-learning as evolving backward from the goal box to the start box. Play-learning, then begins thankfully, where logically all things should begin at the beginning. Play-learning begins in the Alpha Unit and evolves as the rat explores outward to the Omega Unit and continues as the rat continues to explore its way back to the Alpha unit to complete a loop.

Loop data-The plateau phenomena

Third, the plateau phenomena, as measured by the Retrace Loops, shows that the plateaus of the M-control rats continue to become exponentially larger, right up to the last unit of the 99 unit maze. (There are 96 units in the body of the maze, 2 * 48 =96 = alpha + omega + alpha = 99 units). The random rat never stops retracing. Furthermore, the plateau never gets any shorter. On the contrary, the plateaus just keep getting longer and longer. The real rat's plateau grows at about the same rate as the M-Control's plateaus for the first four rows, but then the plateaus extinguish. The real rats, in other words, extinguish retracing. This phenomenon, the extinction of retracing, explains the rate difference in the way real and random rats explore, the second difference above. The real rat's exploration is positively accelerated as the retracing behavior extinguishes. The significance of this difference is pivotal in understanding the nature of play learning. The profound question we have to ask is this: How can we explain the fact that real rats have the ability to stop retracing? This phenomena is most certainly not an aspect of random behavior, random rat's plateaus grow exponentially. While there are many differences between the behavior of real and random rats, this one phenomenon, the fact that a real non-deprived and non-reinforced rat can stop retracing, is probably the central fact that mindlessness-learning theoreticians must explain. This phenomenon is, in my opinion, the primary anomaly which classical work learning theory can't explain, and which logically should force the behaviorist to enter the revolutionary phase of science and create a new paradigm of how rats learn.

The node\loop behavior

The random behavior of rats in a Lashley III Maze exhibits one universal sign, a normal distribution of nodes.

Real rat's path\node behavior, however, shows that this inner most level of the rat's hierarchically organized behavior is organized by mental rules, which spontaneously appear when the rat is exploring the very first row of the matrix maze. The first rule is "explore a row to the end." Rats do not arbitrarily turn around just anywhere within a row. The behaviorist considered a blind entry to be an "error," however, an entry into a blind when the rat is first exploring a row is empirically necessary. How else is the rat to know that a blind is a blind unless it explores it? However, more importantly, entries into blinds are not necessarily pointless. Behaviorists were not known for their capacity to empathize with a rat. From the rat's point of view, the rat's first task is to classify the units of the matrix into functional classes, and one of its prime values is to find, recognize, and return to units that are "safe places." The rat is a prey organism and for such a creature, a large measure of paranoia is rational. The blind units of the maze should be called secondary safe places. The primary safe places are the two end units, the alpha and omega units, which are closed on all four sides and have only a hole-door, which has to be defended. The blind units have three sides. The question to keep in mind when evaluating any unit's safety quotient is this: from how many directions is an organism vulnerable within any unit? The answer is the fewer the better.

One of the rat's first goals is to discover and classify all units in terms of safety. This innermost level of the rat's hierarchy of behavior allows the experimenter to observe the rat's nodes or to identify units of the maze where the rat turns around. The rat's natural and spontaneous ability to recognize and to take advantage of the natural nodes of the maze, the deep blind units, F2 and A2 is not a sign of mindlessness but on the contrary is a sign of mindfulness. Only a real rat can emit path lengths within a matrix maze that reflect the natural structure of the matrix, which were in this matrix, path lengths of 6, 12, or 18.

Anomalous vs. Novel Events

In order to understand Heterostatic behavior, it is necessary to distinguish two types of events, novel and anomalous, and two types of mapping behaviors, assimilation, and accommodation. (Piaget, 1952) The units of the maze, which triggered retracing, were E1 and E2. Why do these particular units trigger retracing? Depending upon the degree of knowledge that has been incorporated into the model of the maze at any moment, the E1, and E2 units may be experienced as either novel or anomalous events. An anomalous event is one, which the model in its momentary state of evolution cannot assimilate. Thus, the rat must begin or continue to retrace to evolve further its cognitive map of the maze until its map can assimilate the E1 and\or E2 event, which triggered the retracing. Retracing is the time during which the rat accommodates or evolves its cognitive map. Paradoxically it is during retracing, which the behaviorists called random behavior, and which Blodgett (1929) invented non-retrace doors to eliminate, when the serious adaptive activity of play-learning takes place. Retracing behavior is not random or purposeless behavior that deserves to be eliminated by nonretrace doors.

A novel event is one, which is new but legal or recognizable: one that can be assimilated by a cognitive map in its present state of growth. A novel event is an interesting event, one the rat will approach and explore. An anomalous event is an illegal and aversive novel event; one the rat will avoid. The degree of the anomaly of an event can be estimated by the quantity of the retracing required to evolve the model of the maze until the assimilation of the event in question is possible. The real rat's retracing in this experiment increased through the fourth row. The difficulty of the Lashley III maze pattern only gradually becomes obvious to the rat. The highest order of the hierarchically organized Lashley III maze required the greatest amount of retracing. More exactly, the inverted "u" BFEU curve shows that the difficulty of the several hierarchical orders was directly related to the order of the hierarchy. The exploratory-play behavior of the rat in the Lashley III matrix maze that generates the maps of the successive higher hierarchical orders cannot begin to evolve at each higher order until the lower orders of the hierarchy had evolved and assimilated their respective features.

The cognitive mapping of the rat is accomplished with the evolution of the somatic and autonomic behavior of the rat in parallel but interacting. The somatic component is not random behavior but is and can only be a manifest expression of the knowledge embodied within the cognitive map at any moment. The primary mind which generates the somatic behavior, when directed by a model that does not incorporate some component of the hierarchical features of the maze, will in the presence of those features of the maze which the model does not represent, experience these illegal events as anomalous events which will induce emotional or autonomic behaviors, both arousal and surprise. Logically, we can infer, therefore, that not until the fourth row could all the features of the hierarchical structure of the Lashley III maze have been encountered by the rat, since it was only after the fourth row that the rats began to extinguish retracing.

The hierarchical knowledge that must be included within a valid (one capable of eliminating retracing) cognitive map of the Lashley III maze is:

ORDER #1 The Alpha Unit is the prototypical unit and is the first unit of the maze. This unit is the point of origin for the rat's entire activity in the maze and is probably experienced by the rat as subjectively the safest unit of the maze.

ORDER #2 a row is composed of eight distinctive units, which I have represented in FIGURE 1.2.90 and designated as the FUN.

ORDER #3 the 1st order of the units (in the first row) is: FUN1; A2, A1, E1, E2, E3, E4, F1, F2

ORDER #4 the 2nd order of the units (in the second row) is: the same (functional) order. FUN2.

ORDER #5 the sequence of the two row orders is an alternation pattern. Order five is the recognition by the rat of the Order effects in the hierarchy, the alternating rows.

ORDER #6 the Omega unit breaks the pattern of rows and is end of the maze.

ORDER #7 the back sequence of the alternation pattern of rows is reversed.

The rat could not incorporate the fifth order features into its cognitive map until it had modeled the third row of the Lashley III maze, where for the first time it could encounter the second instance of the first order row. The assimilation of the third row sequence by the ORDER #4 hierarchy rule would generate the hypothesis of an alternation pattern, which would become the ORDER #5 rule when row four was assimilated by the ORDER #4 rule.

By the fourth row of the Lashley III maze, a real rat with a nativistic and sophisticated mapping intelligence could empirically have sampled enough of the maze to create a relatively valid cognitive map of this maze, one that would embody the first five hierarchical orders of our theoretical model of the maze.

Once the rat's model of the matrix maze is valid, the rat will experience no more anomalies, and the rat from that point on in the maze can adapt to the maze by assimilation. Assimilation is seen in the behavior of the organism when it stops retracing, in other words, when the rat's model generates no more anomalies, the rat needs no longer to either "stick" or demote itself, so that, in effect, the rat's behavior will advance, advance, advance. This behavior is the nonverbal equivalent of Extrapolation, which is seen in FIGURE 1.3.3. A cognitive map incorporating these five orders could then predict and prepare a rat to assimilate all but the Omega unit feature and the reverse order rule. These two rules are quickly accommodated by the rat.

Physiology

Both novel and anomalous events are, of course, novel events in the strictest technical sense. The latter are, however, qualitatively unique in their role in the mental life. They elicit the most powerful of all the cognitive reflexes Pavlov's (1960) orienting reflex, or External Inhibition. *Anomalous events are illegal novel events, and turn on the sympathetic nervous system so powerfully that they have the capacity to interrupt, absolutely and instantaneously, all learned parasympathetic reflexes. A nonverbal logical rat has every right to avoid novel events in this extreme form and to retrace until their model of the maze has evolved to the point where the features of the maze that at first are anomalous have become simply novel. Novel events are milder than anomalous events. Novel events, as I define them, are legal events that are not too novel and whose challenge the rat feels competent to accept. Play-learning is adaptation to both novel and anomalous events. However, all novelty elicits a sympathetic nervous system response. Heterostasis is the adaptation of the organism to its external milieu by mapping or modeling each domain of the milieu. The cognitive mapping function of Heterostasis is composed of at least three components: a cognitive component, a somatic function, and both of the former are energized and activated by the autonomic nervous system. Anomalies primarily trigger the sympathetic division of the autonomic nervous system. Homeostatic drive states are not required to motivate exploratory-play behavior, which supports Heterostatic mapping activity. The sympathetic nervous system is the most powerful division of the autonomic nervous system and there is no lack of motivation when it is dominant. The problem of motivation has been a problem that the behaviorist could not resolve because historically motivation has been conceived as a Homeostatic drive state. The possibility that this most powerful division of the Central nervous system could function as a source of cognitive drive states has been overlooked. Sympathetic activity was classified as an emotion, which was conceived as illogical rather than as the foundation for logic, and therefore emotions were never considered as a component of learning but instead as incompatible with learning.*

The Classical Homeostatic drive states are mixed autonomic states. They are physiological states in which both divisions are active. Heterostasis is a state in which the sympathetic nervous system is one component.

Motivation and emotion

All the variety of behaviors the rats exhibited in this study was spontaneous behavior. The rats were not homeostatically deprived, nor homeostatically reinforced. The rat's brain is a structure, which generates a function called the mind, which has genetically given cognitive purposes, and these intensively defined purposes are sufficient to keep most, but not all, rats busy and entertained for the average time of this study, about one hour in this maze. This study makes it clear that it is not necessary now, nor probably never was, to deprive a rat to induce it to learn in a complex hierarchically organized domain like the Lashley III matrix maze. Tissue needs neither excess nor deficiencies are required for learning to occur. The rats in this study were motivated to learn by removing them from their home cages, for which they must have a very complete model and suddenly injecting them into the Lashley III maze, which was an enormous anomalous domain. Play-learning exploratory behavior is initiated by uncertainty, and the quantity of uncertainty induced by the matrix maze was so excessive that some rats, either could not, or would not move out of the alpha unit for a considerable amount of time. The cognitive activity of an organism is designed to solve uncertainty problems. A Heterostatically cognitive or psychological need is different from a physiological Homeostatic need. Physiological needs are tissue needs and Homeostasis is a physiological mechanism designed to reduce physiological needs. Heterostasis is a mechanism designed to reduce cognitive or psychological needs, which probably can't be called by a better name than their old-fashioned name, ignorance. When an organism has a problem for which it does not have an answer (knowledge) the organism can be said to be experiencing uncertainty, but that is just a scientific euphemism for what is ordinarily called "ignorance." The awareness of ignorance is an emotion. Uncertainty, the degree of sympathetic activation, is a direct function of the intensity of the organism's state of ignorance. All this follows the Yerkes-Dodson Law.

RESCORLA

The Rescorla-Wagner (1972) rule defined reinforcement as a contingency rather than contiguity. The idea is that learning will proceed until all the surprise has been removed from some reinforcing events. The problem with this concept of learning is that it recognizes only the uncertainty induced by a particular kind of an event, which they call reinforcement, and thus this concept of learning is a particular rather than a general rule of uncertainty reduction. Their statement of the rule in terms of Reinforcement does not recognize that a rat will learn until it has taken the surprise out of any unexpected event, novel or anomalous. The problem with the classical work paradigm is that work and play are confounded. The Rescorla-Wagner rule is a special case of this confounding. Rats in a contingency situation learn the contingency, singular or plural, if it is one they are capable of understanding, and this is true of any kind of event not just reinforcing events.

Once Homeostasis and Heterostasis has been unconfounded, it is possible to see how, in the classical work paradigm, the homeostatic concept of reinforcement and the Heterostatic concept of uncertainty were confounded. Heterostatic reinforcement, if we were to continue to use the term, would have to be paradoxically defined as an anomalous event. The purpose of the heterostatic behavior is to create a model that takes the surprise out of all the anomalous events of a domain, not just the one event in the domain called the "reinforcer." Any event, whether it has a homeostatic value or not, will promote learning so long as it is anomalous.

Work and Play

The phenomenon of play-learning enriches but also complicates the already difficult life of a reinforcement theorist. The definitions of reinforcement, theoretical and operational, devised to explain work-learning just do not work in the domain of play. Two examples will be considered: Skinner's (1974) operational definition, and Rescorla's (1972) theoretical definition.

Pavlov

Pavlov (1960) was the only classical learning theorist that honestly faced the issue of novelty. His treatment produced three concepts: First, External inhibition, Second, "Extinction is inevitable," and third, Reinforcement delays extinction.

Kamin

Pavlov (1960) and Kamin (1969) are very close. Pavlov has considerably more insight into the phenomena than Kamin, but they both wind up making the same mistake. The problem is the idea that reinforcement increases the probability of a response is not the simple idea it seems to be but a compound idea with at least two ideas confounded.

First, there is Guthrie's notion that an event is reinforcement because it turns off the motivating stimuli and so preserves the last critical association. (Hilgard and Bower, 1975) Information is an event that does two things. First, it reduces uncertainty, for that is where it gets its name. Secondly, it thereby decreases arousal, for uncertainty is the Heterostatic source of arousal or emotion. In the Rescorla-Wagner (1972) theory of reinforcement, the max is determined by the amount of uncertainty a reinforcing event induces. It is the decrease in this uncertainty which learning measures. The max is the highest level of associative strength that can occur between the stimulus and the response according to Rescorla and Wagner (1972). However, knowledge, which evolves within the model, is defined as "that which decreases uncertainty" which is the reinforcement. The Homeostatic-reinforcing event, which induced uncertainty, which Rescorla (1972) called the reinforcement, is not the reinforcer. In the work-play paradigm, everything is confounded. In the pure play paradigm, it is possible to isolate these events and thus eliminate the confounding. Knowledge/reinforcement does two things. First, it decreases uncertainty, which uses up some of the max, but in doing so, it decreases the need for learning, the need for knowledge. The consequence of this is that it both increases and decreases the probability of behavior. It increases the probability in the sense that in so far as the organism is a rational creature, the next time it needs to act in this particular domain, it will have the knowledge which was invented to decrease the uncertainty that was just reduced in its cognitive map to use as a resource. The new knowledge added to the map will increase the probability that the organism will make this same response when the same conditions reoccur.

Secondly, knowledge decreases or satiates the need to play. The purpose of play is to invent knowledge, which will reduce the uncertainty induced by the novel and anomalous events of a domain. Knowledge does two things; first it decreases the need to play, but, secondly, it increases the probability that if the organism plays, it will play in a particular way.

The Exploratory-play behavior of rodents furnishes a new and fresh perspective from which to observe some old theoretical problems. The nature of reinforcement, for example, has seen a lot of movement through the work of Rescorla, Kamin (1969), etc., and the thrust of their work has been upon the role of surprise or "surpriseness" in Kamin's terms. These theoreticians, at times, have seemed to be getting close to understanding the role of uncertainty in learning, however, the confounding of work and play that occurs in the TTP has stood in their way.

Novel\anomalous events induce uncertainty and activate the Heterostatic mechanism designed to create models, which contain the knowledge that will take the uncertainty out of the events of the domain that induced the uncertainty in the first place. Novelty requires accommodation for the model that generated the novelty. The online model, the model in charge when an anomalous event occurs, obviously is not valid or else no surprise would have occurred. Anomalies\novel events are *prima face* evidence invalidating a model. However and here is the problem, there are two kinds of novelty; News or simple novelty, which although it is novel, is not so novel as to be fear inducing. Simple novelty is interesting and fun and induces exploratory behavior. Anomalous events are, by nature, too novel, thus inducing fear\avoidance behavior. The problem then is that the two kinds of novel events that surprise may have two different consequences. They may increase or decrease behavior directed at the source of the novelty. The classical consensus seems to be that Skinner's (1974) definition that work reinforcement was "an event that increased the probability of a response" was correct. In other words, reinforcement acts in only one direction. However, in a play situation, novel and anomalous events, which induce surprise and induce uncertainty, have two consequences, one that decreases and another that increases the probability of a response. This dual consequence of the novel\anomalous event in the play-test environment complicates the problem of those that would define uncertainty-inducing events as reinforcing. Either, first reinforcement will have to be defined to allow a two directional reinforcement effect, or the Surpriseness mechanism will have to be abandoned.

The myth of Reinforcement

Reinforcement, from the Heterostatic point of view, if the term is to be retained cannot be defined in terms of uncertainty induction but rather uncertainty reduction, so that the duality of a news event is not a problem for Heterostatic theory. Events, which induce uncertainty, are the heterostatic equivalent of arousal inducing events like deprivation in homeostasis. Thus, just as the quantity of need reducing reinforcing events that could occur on any day would be directly related to the quantity of deprivation that had occurred, so also the quantity of uncertainty reduction that can occur is directly related to the quantity of uncertainty induction that has occurred. Satiation determines the max for any reinforcement equation whether Homeostatic or Heterostatic in nature.

Need, drive, surprise, or uncertainty reducing events cannot reduce more than has been induced. However, the surprise theory of reinforcement misses one-half of the pictures. The extreme surprise of an anomalous event is not reinforcing, but it does determine how much, the max, of information\reinforcement will be required to satiate the cognitive uncertainty induced in the organism by the anomalous event. The heterostatic solution is that Knowledge, or Reinforcement, as a central cognitive event is the other half of the uncertainty induction picture. Heterostasis assumes that the certainty-uncertainty dimension is the central dimension of brain functioning. The Heterostatic brain, made possible by the cortical mantel of the mammalian brain, is genetically designed for one primary task, to induce and reduce the uncertainty of the anomalous\novel events of the empirical world. The brain is the source of all knowledge. Knowledge is a secondary not a primary quality. How many times does this lesson have to be learned? Behaviorism and cognitive psychology have both fallen into the same trap that Johannes Muller got us out of so many years ago. (Boring, 1929)

So what is knowledge? Knowledge is to behavior as the gene is to the body. Socrates long ago said that "bad" men were not bad but ignorant. In other words, all organisms prefer to construct their behavior out of knowledge rather than ignorance. Knowledge, as a thing, is the stuff of which cognitive maps are made. Knowledge is the content of the map. Knowledge is power, and all organisms seek the power that knowledge confers

The models, cognitive maps, engrams, programs, schemata, scripts, or whatever one chooses to call the structure in the primary mind contain the knowledge from which behavior is constructed. Knowledge is the stuff out of which the mental events and behavioral acts of an organism are constructed. A mental model, which contains valid knowledge, will create valid ideas and valid actions that are means to the ends intended. Success is the ultimate reinforcing event for it is the pragmatic validator of the knowledge contained in a model.

Heterostasis, in other words, ends up with two definitions of reinforcement. If it is to retain a word referring to some external pragmatic success, ignorance is surprise then knowledge is the absence of surprise and is probably the best candidate for the new meaning of the word.

The informing mind

The mind is active. The behaviorist, the cognitive psychologist, and the associationists of all types, have in their effort to represent the mind as passive, skipped the first step in the understanding of behavior. They have begun their model of behavior with the assumption that the forms of the mind are object public events in the world rather than subjective events within the mind. Heterostasis is a theory that knowledge is subjective. Knowledge is that which both informs behavior and experiences and reduces the uncertainty induced by the novel and anomalous events of a domain. The forms of the mind, which organize the mind, are not events that nurture supplies the brain; they are not a product of the environment. The forms that inform the mind evolve within the mind. The act of the primary mind is to invent the primary forms that can take the uncertainty out of the uncertainty-inducing events of a domain. The mental events that reduce uncertainty are the knowledge, which is a form in the mind. Knowledge when meager or invalid will prove life threating to an organism; valid knowledge is the only thing that can safely increase the probability of a response. As George Kelly (1970) said, organisms seek to construe the replication of events, and only valid memory-information stored in an organized fashion called a map\model\engram\etc. can insure the organism can construe the replication of events repeatedly.

"A person anticipates events by construing their replications." (Kelly, 1970, P.11)

However, exactly how does a rat create knowledge? The answer is that the primary mind of the rat is a generic scientist, and like all scientists, it must excel in the exercise of the generic scientific method. The essence of the generic scientific method is the manipulation of objects and events by behavior to create sets of experiences from which causal models of the events of a domain can be induced. The generic scientist must create sets of experiences by acting in two ways; one act is the experimental act and the other the control act. The generic scientist must act at least two different ways, and if one act always produces the desired effect and the other doesn't then the scientist infers causality for any behavior that is reliably the means to some desired end.

The different consequences of the two acts allow inferences of causality, which allow the organism to create knowledge, and it is this knowledge creating process which is reinforcing and which motivates exploratory-play behavior in particular and all intrinsically motivated behavior in general.

The name

Heterostasis gets its name from the assumption that any organism with a primary mind and capable of executing generic scientific behavior can dynamically shift its goals or set point from moment to moment, and up and down throughout the set of goals that are required to create a hierarchical organized cognitive-behavior structure isomorphic to the hierarchically organized domain. Therefore, retracing is not some dumb wasted motion that some "superior" being needed to prevent the rat from doing by creating non-retrace doors. Retracing behavior is the essence of scientific behavior. It is during retracing that the rat as a generic scientist acts once, and twice; now this way and then another way to determine which acts work and which do not. Retracing behavior is the behavior that allows the rat to construe the replication of behavior. (Kelly, 1970) Retracing behavior is the animal equivalent of what Piaget (1952) calls the circular behavior of the sensorimotor stage of development. Play learning of arithmetic shows that children will respectively "count out" their numbers so many times that it seems they will never quit. Why has no teacher thought of a way of creating the equivalent of non-retrace doors for preventing this circular behavior in human young?

Naked-eye Behaviorism

Behaviorism had both positive and negative aspects; the positive were important and serious; but the negative were something of a fraud. Behaviorism was and is a convention, a social group that uses group pressure to enforce its concept of standards on psychology and psychologists. On its negative side, behaviorism, particularly Watson (1972), was against consciousness, mind, and all sorts of terms, which he understood as "mental" terms. Watson's negativism was justified by his philosophical positivism; a pragmatic sort of argument in which he rejected all these "mental" terms primarily because he did not know how they could be objectively measured and defined. I have no quarrel with anyone's honest admission that some topic is beyond their scientific competence but here is the where the fraud begins. Neither Watson, in particular, nor the behaviorists in general, honestly studied behavior, which was the sole basis of their claim to be behaviorists. They did not show that these "mental' terms could not, in fact, be objectively defined in experimental Behavioristic data. The fact is that the behaviorists were as opposed to the study of behavior, in general, as they were to the study of mind. The particular bits of behavior they approved for study were carefully selected little bits of behavior, a lever press here, or a salivary response there, a blind entry here, or latency there, but the behaviorist never seriously adopted a program of systematically measuring and describing all forms of behavior. Behavior is a complex phenomenon, and the behaviorist ignored 99.99 % of all this complexity. In the maze, for example, on the first trial as the rat moves from the start box to the goal box, the behaviorists take one or two little measures, but ninety nine point nine percent of the animal's behavior they ignore, and they justify their indifference to all this behavior by saying, like Aesop's fox, "This behavior is Random behavior." The behaviorists did not know this behavior was Random any more than the fox knew the grapes were sour. The behaviorist simply never bothered to measure this behavior, they were content to demean it as "random" behavior and thus justify their indifference to all this "Random" behavior, which should have been their primary goal to study, if they had been serious behaviorists. What is important about their hypocrisy is the important link between their irrational refusal to systematically measure and analyze "all" the behavior in a situation and their denial of mind. For the fact is, as I shall show in this series of studies, the objective experimental behavioral data needed to objectively define all those "mental' terms that Watson (1972) so despised has all the while been contained in and just waiting to be scientifically explored by some Behaviorist in this "Random" data they ignored. The data of this series of experiments will show that these "mental" words are not concepts a behaviorist must avoid. Positivism does not justify the Watsonian naked eye Behavioristic position that "mental" terms cannot be objectively defined in psychology. Mind is an empirically justified concept if the behaviorist will simply get serious about the measurement of all behavior. Behavior is a complex phenomenon and in that complexity exist all the experimental behavioral data needed and then some to justify objectively all the "mental" terms that Watson threw out with his too simple conception of what the behaviorist's responsibility to the study of behavior should be. Heterostasis is the Wisdom of the mind and of a mind that is every bit as objectively based on the data of behavior, as what Watson (1978), Hull (1943), Pavlov (1960), or Skinner (1974) ever introduced.

The naked eye behaviorism begun by Watson (1972) and that has survived all these years failed to find the experimental behavioral data to allow the objective definition of these "mental' terms for several reasons. But the basic fact is that behaviorism did not find a scientific way to introduce these terms into psychology because they did not try, and they did not try because they knew, *a priori,* on non-scientific grounds that the objective definitions of the terms were impossible.

The purpose of this set of experiments is to press the behavioral telescope into the service of a new kind of Behaviorism, one that is willing to go beyond the flat behaviorism of the naked eye behaviorists. A behaviorism that is willing to see, with its eyes and its mind, the multidimensional complexity, and all the implications of that complexity that lies behind and beyond the surface features of animal behavior that can be seen with the naked eye. The behavioral telescope used in this study reveals in the rat's behavior, a world of anomalies that could prove to have a consequence upon the study of behavior, every bit as revolutionary as the moons of Jupiter as seen through Galileo's telescope had upon the science of astronomy.

The purpose of this series of experimental rat studies using the behavioral telescope is to develop a multidimensional behaviorism, one that looks at the behavior of the rat in the maze in depth. In these studies, I have expanded the Behavioristic concept perhaps a hundred fold or more, but I have not even begun to exhaust the possibilities that remain for obtaining all the anomalies that are possibly contained within the behavior of the rat as it runs the maze. The rat's behavior is generated by the rat's mind, a complex machine whose potential has been tragically unrecognized by the naked eye behaviorists. The basic assumption of Heterostasis is that the complexity of the rat's behavior reflects the complexity of the rat's mind: who can say what anomalies remain to be discovered by the behavioral telescope. The data reported in these studies is, as I noted above, many times more complex than ever previously described in a rat study, and yet the complexity reported does not even begin to exhaust the complexity of the mind-generated rodent data.

Watsonian (1978) Behaviorism was primarily a philosophical or top down behaviorism. I have no problem with an empirical Bottom Up form of Behaviorism that "follows the data." Watson's problem was that his objection to mind was not empirical but philosophical or ontological. Watson's conception of mind was ontologically dualistic. Watson was not a monist. Watson did not or could not understand mind as a monistic possibility, and hence he never even needed to examine data to determine whether the data justified the empirical use of the "mental" terms such as consciousness and mind. The paradox of behaviorism was that in the Random behavior they refused to study was the evidence for the "mind' they denied. The problem with the naked eye, ontologically dualistic behaviorists, is that it failed to achieve the potential that is inherent in behaviorism. The purpose of the new behaviorism is to develop a new kind of open behaviorism, one dedicated to following the data.

The problem of Watsonian behaviorism was the *a priori* nature of their denigrating remarks about what was possible by behaviorism. What is needed is a behaviorism, which is willing to let the data decide what can and cannot be achieved by the experimental study of behavior.

An objective estimation of the degree of power amplification by the behavioral telescope is difficult to compute, but estimation can be made by either a comparison of the degrees of freedom available in the experimental designs employed in the TTP and the LTP experiments or in a comparison of the number of data points. By either measure, the power of the behavioral telescope, in the state of development present in this series of experiments, is probably more powerful than Galileo's (1989) telescope but considerably less powerful than the Hubble telescope.

Chapter 2 The Discovery of Devious behavior

Experiment M87

Experiment IIA, a false start – six runs:

In the first study, such a severe constraint set was used for testing the animals that only one growth cycle of Heterostatic behavior was studied. The purpose of this experiment was to observe several growth cycles using a generous constraint set or one with the fewest possible constraints. The question is this: how robust is exploratory-play heterostatic behavior? This experiment was designed with Pavlov's (1960) Chapter 14 as a guide. In this chapter, Pavlov developed the thesis that "extinction is inevitable." By this, Pavlov meant that even with continuous reinforcement, redundancy would eventually produce extinction. Since I was running non-deprived and non-reinforced rats, I obviously did not expect exploratory-play behavior to be robust; in fact, I expected exploratory-play behavior would habituate or extinguish very quickly.

In the first experiment, the animals were tested using a severe three-element constraint set: (a) 228 maximum steps per run, MSR228, (b) 600 s per run, and (c) one complete loop per run. The severe MSR228 constraint was forced by the involuntary initiation of the Apple II "Garbage Collection" routine during which all tracking data was lost. This severe constraint was, nevertheless, an excellent choice because, while it created an almost impossible task for the random control group of rats, the experimental group found the task to be relatively easy.

For this second experiment, the Apple Graphics Tablet tracking program was revised to eliminate the "garbage collection" routine that had forced the severe MSR228 constraint. In this study, a generous four-element constraint set was selected: (1) 2000 moves; (2) 7200 s (two hours); (3) one complete loop per run; and (4) x growths. Since the purpose of this experiment was to discover how many growths the rats would spontaneously emit before extinction occurred the number of growths, x, to be run was left open. In the first study, the average rat achieved the criterion of one complete loop on the average of 600 S., one thousand moves and in five runs. The new and generous constraint sets of parameters were designed to take the pressure off the rats to perform. While the primary or positive purpose of this experiment, as stated above, was to measure the staying power of the exploratory-play behavior under these leisurely test conditions, another way to view the purpose of this study was to view it as a study of factors that caused the extinction of exploratory-play behavior. There were several obvious questions: what happens in an exploratory-play paradigm when the novelty runs out, when rats get bored? Will extinction occur only after a rat has explored the whole Lashley III matrix maze and created a valid cognitive map or before? What does an exploratory-play extinction curve look like?

Method

Maze
The maze was described in Experiment I

Apparatus
The apparatus was described in Experiment I

Subjects
The subjects were four Sprague-Dawley rats, three male and one female, obtained from the Charles River Lab. They were sixty-days old at the beginning of the study.

Procedures
The rats were housed in individual wire cages, had food, and water ad libitum throughout the study. They were maintained on a light-dark cycle of: twelve hours on, and twelve hours off, beginning at nine o'clock. The male rats averaged 249 grams and the female rats, 171 grams at the beginning of the study. Each rat was weighed each day before being tested. Each rat was carried to the test room individually in a carrying cage and placed in the alpha unit with the hole-door covered. After thirty seconds, during which the timer and the computer were started, the door cover was removed. As this was the first experiment to look at the phenomena of multiple growths during exploratory-play behavior, the decision whether to run the rats another day or not was to be made one day at a time.

Results
The behavior of the rats in this experiment was full of surprise: true anomalies; of the kind, that Kuhn (1962) said induce the revolutionary stage of science and is historically rare; yet three such anomalies emerged in this experiment.

The first anomaly: The Discovery of the phenomenon of the Punishment of exploratory-play behavior
The first anomalous discovery of this experiment was the phenomenon of punishment in the context of exploratory-play behavior. My naiveté, unfortunately, made me initially misunderstand the nature of this phenomenon--the possibility that punishment could occur in an exploratory-play environment never occurred to me and, when it did, I was expecting extinction and I confused the two. In other words, the behavior that was the sign of punishment was initially misunderstood as an instance of extinction.

FIGURE2.01 Six Runs for F4
FIGURE 2.01 shows the rats' behavior that I initially misunderstood— six runs for Female 4, F4, are described by two different run scores: FEU (Unit First Entry), and FEL (First Entry Latency). For the first four runs, her behavior reveals an obvious

improvement: First, the number of steps dropped from on run four about 400 steps on run one to 100 steps; second, her latency went from 1709 s on run one to 107 on run four. However, beginning on run five, and continuing through run six, her performance underwent a dramatic change, which I perceived as evidence that her interest in the maze was extinguishing. Obviously, I reasoned, she had lost interest in exploring, since her move and latency scores regressed to their original level on run one, or worse. All four rats exhibited this phenomenon and I assumed they had become bored--I had expected extinction, and this shift in the pattern of the rat's behavior, I assumed confirmed my expectation. Unfortunately, I prematurely terminated the experiment.

Discussion

In hindsight, it eventually became obvious that the data never supported the boredom-extinction hypothesis, and the misunderstanding that occurred was but another example of a self-fulfilling prophecy. Boredom, or extinction, should have produced a very different pattern of behavior than the one observed. First, the rat should have stopped moving; the number of moves should have continued to decrease. Secondly, the latency should have increased. However, in the pattern observed both the number of steps taken and the latency increased! Because I had never had any occasion for distinguishing these two patterns of behavior, they were initially confused. However, the facts were clear, a dramatic change occurred between runs number four and five in the rats' behavior. The rat's behavior obviously proved her attitude toward running changed, but the question is how it changed. What kind of an attitude change would create the behavior pattern in which the size of both dependent variables decreased?

A careful reexamination of the data finally made it clear that the new pattern of behavior that emerged was an unexpected form of behavior, an anomalous phenomenon.

A new hypothesis

This new strange, unexpected, pattern of behavior forced me to evolve a new hypothesis about the nature of her new attitude about running the maze. It finally became clear that perhaps the third element of the constraint set, which had required the rats to be removed from the maze after one complete loop through the maze, might be responsible for this anomaly. Until this new pattern of behavior was observed, no thought had been given to the possibility that this third element of the constraint set; the removal of the rat from the maze upon the completion of a run could be experienced by the rat as punishment. This anomalous pattern of behavior, therefore, made me aware of a possible new dimension to animal behavior that had to be experimentally examined.

A new experiment was designed to test the hypothesis that this anomalous behavior was due to the effect of punishment upon exploratory-play behavior

This book was originally intended to be a one-experiment study, but that was before I discovered how little I understood about this new domain of behavior. As the result of this first group of rat's strange behavior, I designed a new study, which is the second experiment.

Experiment IIB, Twenty-five growths

In the first part of this experiment, IIA, the rats were stopped prematurely, just when the results were getting interesting, after run six or seven. The purpose of Experiment IIB was to run each rat for however many runs were required so that each rat could make at least 25 growths which, hopefully would be long enough to test the punishment hypothesis; namely, that rats were avoiding the Alpha unit. The rats in the LTP are removed from the matrix upon entering the alpha unit after making a complete loop through the maze. By requiring each rat to make twenty-five growths, they could have been punished at least twenty-five times, which should be more than enough time to evaluate, objectively, the consequence of punishment in this context.

Since, by definition, with this three-element constraint set, a rat can only make one growth per run but does not have to make a growth per run, some rats may require more than 25 runs to make 25 growths.

Method

Subjects

Three new rats, two females and one male, were the subjects of this second phase. However, in the Results section where the data from the first growth is being considered, the data from the four rats from the first part of the experiment will be combined with the data for the three from the second part when it is logical to do so.

Procedures

Each of these rats was tested for at least 25 runs, or for as many runs as required to achieve 25 growths.

Results

The behavior of the rat in the matrix maze paradoxically consists of two almost independent parts, the behavior in the essential path, EP, and the behavior in the blind Path, BP. The latter is by far the most strange and difficult to understand. What is paradoxical is that the behaviorists used the BP behavior as an index of the mastery of the EP behavior. However, I will show that the two behaviors: the BP and the EP are essentially two independent behaviors.

The Essential Path, EP, Behavior

In this section, I will only describe the EP behavior. The analysis of the data in this study will again be limited for two reasons. First, technically it will only become possible to describe the BP behavior in Experiment III for only with the design of the third experiment did I finally understand how to fill all the cells of the BP so that there would be sufficient data for a statistical analysis. Secondly, only in the design of the third experiment had I had finally learned enough about Heterostasis to design an appropriate control group that made it possible to compare and discuss the EP and the BP behavior for both the experimental and control groups.

The EP behavior in this experiment will be described in three subsections:

First, the run-data will be described.
Second, the growth data for the first growth will be described.
Third, the data for the twenty-five growths will be described.

The run data

There are three anomalies discovered in this experiment: the phenomena of the punishment in exploratory-play behavior, the long induction phase phenomena, and the phenomena of devious behavior. There was one additional phenomenon, which was not originally analyzed when I first processed the data of this experiment, but much later, I went back and ran a belated partial test on the phenomena of blind path behavior, which will be reported below under the topic of the fourth anomaly.

- **The first anomaly: punishment**

The first anomaly described above was the discovery of the phenomena of punishment that resulted in the design of this second experiment.

- **The second anomaly: long induction phase**

The second anomaly was the behavior that characterizes the long induction phase of the rats in this second study. By the induction phase, I am referring to the time it takes a rat to get involved and accept the exploratory-play challenge the matrix maze affords. The long induction phenomenon was observed in this experiment, IIB for the first time as an unexpected result. This behavior was the consequence of my providing these rats what we considered a leisurely test conditions. Because of these new test conditions, the rats in this experiment required several hours to achieve criterion on the first growth.

FIGURE 2.02: Log plot of F11 data

FIGURE 2.02 shows a log plot of the data for Female 11, F11, for 30 runs for three scores: the number of steps per run, FEU; the time per run, FEL; and the mean time per run, FEM. This female, F11, probably deserved her own chapter as did F3, for she was a remarkable individual in several ways, not the least was that she was probably the most unlucky rat.

The long induction phenomenon is seen in the first five sets of run scores in FIGURE 2.02. The data for FEL shows that for the first five runs, the time required to run the maze was almost eight thousand seconds, which is over 133 minutes or over two hours; these rats were tested by Dave Mosley and I am grateful for, and amazed by the patience he showed in testing these animals.

Female 11 was the slowest in this study to complete a first growth and had the longest induction phase of any rat in this study. Actually, she holds the record for having the longest induction phase of any rat ever run on the matrix maze. FIGURE 2.02 plots runs not growths. The first five runs were all a part of her first growth data. That fact, by itself, is not remarkable. Other rats required as many as ten runs to complete the first growth. What is remarkable and anomalous is seen in the first five move and latency scores. The total number of moves for the first growth or the first five runs was one thousand nine-hundred and sixty-seven (562 + 435 + 546 + 218 + 206 = 1,967), but the total latencies in seconds for the first five runs were twenty-nine thousand three-hundred and thirty-eight (7201 + 7181 + 7226 + 7280 + 450 = 29,338s). Seven thousand seconds is approximately two hours. In other words, F11 spent two hours a day in the maze for the first four days before finally making a normal latency of 450 seconds on the fifth run and finally completing her first growth. That is a long induction phase. Since in the first experiment the average rat ran on MSR228 had an induction phase of only 30 minutes or 1,800 seconds, the eight-hour latency observed in this study for the first growth was something of a surprise. However, what was surprising and anomalous about these phenomena was not just the fact that rats explored for 7,200 seconds a run, for I had allowed for that in defining the constraint set. The anomaly is this: Why did F11 take so long to get involved in creating the shortest-path-criterion-loop, i.e., making the first growth? F11 could have spent this amount of time in the maze and still made one or more growths per run. The anomaly is the length of the induction phase. Why did this rat take so long to get involved with making a model of this maze that would destroy its strangeness?

The leisurely constraint set of the second study shows how agonizingly slow rats can be in making their first growth and how critical the selection of the proper constraint set can be, particularly the time constraint. This phenomenon raises an important question; what variables determine the rate of play induction. The long latency induction phenomena seen in FIGURE 2.02's run data can also be seen in several different hierarchical levels of behavior, and in several different data sets at these different levels, so more data will be presented later.

The major difference between the first experiment from Chapter One of Book Two and this second set of experiments is the difference between the two constraint sets. The first experiment had a constraint set with a short time constraint, an Out time of only ten minutes, and thus no rat was allowed to be slow. In other words, all rats, fast or slow, were removed from the maze after ten minutes, but in this study, slow rats were allowed to be as slow as they wanted, for up to two hours a day for as many days as they wanted. Therefore, this study made it possible to observe, for the first time, just how slow some rats can be in the induction phase when given an opportunity to be slow. In addition, even this eight-hour induction data cannot be assumed as the limit. There is no way to estimate based on the data in hand just how slow a rat might be if the out-elements of the constraint set were extended beyond two hours.

- **The third anomaly: devious behavior**

The third anomaly was the phenomena of devious behavior seen in both the group data and the individual subjects' data. However, only the latter data shows the full complexity of this devious behavior

There are two kinds of retracing, first, Essential path retracing which is triggered by anomalies. The first kind is, "Anomaly Induced Retracing," AIR. The second kind is Redundancy Induced Retracing, RIR. RIR is stranger than most anomalies, which is characterized by rodents, rats, or mice, refusing to enter a unit of the blind path the first time. It was as if rodents knew there was nothing in the blind path units, which was worth exploring. Only much, much later, after they had entered all the EP units many times did they seem to reason that an entry into the Blind Path BP units might now be a source of stimulation that might be more interesting than the EP units might.

The Graph53 program stored on the computer screen the step number each unit was first entered; In Experiment IIB there were two sets of BP units that set the record for making the largest FEU late entries.

FIGURE 2.02A-F11 FEU data

Figure 2.02A shows the FEU data for Rat F11-R1132 of Exp. IIB. Two sets of BP units illustrate RIR, Redundancy Induced Retracing with both sets of unit's members of the alternation blind unit's types of row four, with scores of 3668 and 3669, and the alternation blind units of row one on the way back, with scores of 6439 and 6440. This BP unit avoidance shows most eloquently and powerfully that rats make a model of the Blind Path units, which enable a rat to extrapolate or predict what units await them even though they have never entered these units. For contrast, follow the rat's back-behavior for the first four steps; 1882, 1883, 1884, and1885, which show that she turned right into the alternation units.

Rat extrapolation in the Blind path evolves as rapidly as EP extrapolation. They make a model of both of the BP type of units, the Forward going units and the Alternation units, however, both of these two last sets of units entered were Alternation type avoidance entries.

Devious behavior as seen in individual subject data

The third and in many ways probably the most interesting anomaly observed in this study forced me to recognize yet another phenomenon of mental adaptation, which I call devious behavior. This is a special form of behavior induced by the punishment of exploratory-play behavior. The most interesting aspect of the behavior of FIGURE 2.02, of F11-R1132, is the fact that it shows how complex the exploratory behavior of rats becomes when they are punished. Devious behavior is a complex effect of punishment observed in the context of exploratory-play behavior. The complex pattern of behavior in FIGURE 2.02 forces us to consider several new sets of theoretical questions, only some of which will be considered at this time: but the question which will require most of our time is: what is the purpose of the rat during spontaneous exploratory-play? How many purposes can a rat manage at a time? If rats, in fact, can exhibit different behaviors each with a different purpose, then how can the rats' behavioral priority structure be objectively described?

FIGURE 2.02 shows that the Path goal and secondly, the exploratory-play goal interact.

Goal one-The shortest Path goal

The first and the natural goal of a rat spontaneously running a maze motivated by exploratory-play behavior is to discover and run the shortest path, which in this matrix maze is through the EP. The shortest path is operationally defined, in this matrix maze, as I will explain below, as running a type-eight loop through the maze. This goal effectively represents the behavioral embodiment of the goal of the primary rational mind to be efficient, which is one of the essential components of intrinsic motivation.

"Intrinsically motivated behaviors, which are performed out of interest and satisfy the innate psychological needs for competence and autonomy, are the prototype of self determined behavior." (Ryan and Deci, 2009, P. 65)

"Intrinsic motivation is defined as the doing of an activity for its inherent satisfactions rather than for some separable consequence. When intrinsically motivated a person is moved to act for the fun or challenge entailed rather than because of external prods, pressures or rewards." (Ryan and Deci, 2009, P. 56)

The same can be said to be true of the rat, the rat is intrinsically motivated to achieve its goal of creating a valid map of the matrix maze. The goal of the primary mind of rats during exploratory-play behavior is to design a map or model of the matrix maze that promotes or maximizes its behavioral efficiency. If a rat or any mammal were content simply to wander through a maze, there would be no need for a map. The only way to explain the design of a map is to assume that the primary mind is spontaneously rational and has as its primary goal or as its first priority the mastery of the maze domain: that is to evolve a strategy to run the maze with the shortest path.

Goal two-the love to learn

Aristotle (1952) said in the introduction of his Metaphysics, "All men by nature love to learn." This statement obviously shows that Aristotle was not a behaviorist, for he knew, as the behaviorist did not know that humans do not have to be tricked into learning by manipulating Homeostatic needs. Aristotle's statement was probably the first recorded recognition of Heterostasis. However, Aristotle did not say, all rats by nature love to learn or that all mammals by nature love to learn. Heterostasis, however, is a theory that says, "All mammals by nature love to learn."

Therefore, there are two sides and maybe even more kinds of reasons for a rational mind to explore a maze. The main one is to seek the knowledge that will allow it to run the shortest path behavior. The first goal, as noted above, is the rational goal of the nonverbal mind to evolve a valid model of a domain for the sake of promoting efficiency. Yet behind this first goal is a second goal, which for the moment can be supposed to have one or two possible components: the first of which may be described as the rat's need for a change of scenery. The rats in this experiment, I must remind the reader, spend their whole life confined to a small wire cage, and so it is possible that they derive pleasure simply from the change of scenery.

This hypothesis could also explain, in part, the long induction time noted previously. Simply, the rats were just enjoying being out of their home cages; enjoying a romp in the park; playing in the maze. My basic assumption in asking these questions is that I will be able to answer these questions and that various kind of motivation of behaviors can be identified in the pattern of behavior exhibited in the maze. For example, Heterostasis is both a unique physiological mechanism and technology, and a unique complex pattern of behavior generated by the mechanism. Heterostasis, the mechanism, is the wisdom of the mind; heterostasis, the behavior, is the unique sign of the primary mind and the spontaneous sign of a rational mind at play. The heterostatic mind emits heterostatic behavior. Just as the heart was designed to pump blood: the lungs were designed to breathe, and so, the somatic muscles were designed to move the body, etc.: just so, the primary mind was designed to emit heterostatic behavior. Just as there is a need for wakefulness, there is also a need for play.

However, let me carry this argument one more step: my basic assumption in the book is that there are many kinds of physiological technological behaviors; and that each kind can be identified by the pattern of behavior it creates. However, while Heterostatic behavior is a generic name for the unique kind of behavior emitted by the primary mind, yet there are all kinds of variations upon this generic form of behavior.

The whole purpose behind these two experiments is that one of the first tasks a rat runner must master is to discover an objective way to identify all the various kinds of behavior. The first task of an experimental study of behavior is to create a taxonomy of behavior. I came into the first experiment with an inadequate set of taxonomic tools and I misclassified a kind of behavior because I had never seen this pattern of behavior before and had had no reason for evolving a rule for distinguishing the two confounded behaviors.

The two patterns of behavior, which I confused, were extinction and devious behavior. The purpose of this second experiment was to allow me, finally and objectively, to identify the distinguishing properties of devious behavior.

A rat spontaneously exploring the matrix maze is doing two things. First, it is rationally evolving a model of the hierarchical structure of the matrix maze in order to optimize the efficiency of its domain specific exploratory-play behavior, but the rat evolves this model for more reasons than one. The rats' second goal, in general, is the intrinsic motivation to have fun, but fun does not have to be understood as doing nothing but playing: but rather, must be understood as an essential element of rational behavior. Playing means operationally evolving the skill, the means, to maximize both the time and moves allowed in the maze, and the effectiveness achieved by those moves. The first point that must be formally stated is that rationality and fun are normally compatible goals and are normally two aspects of the same behavior. However, in this experiment, by dumb luck I designed a task for a rat in the matrix maze, in which I unintentionally forced these normally compatible goals to become mutually exclusive events during exploratory-play behavior.

FIGURE 2.02 introduction

The data of FIGURE 2.02 allows me to begin to tell the story of devious behavior. This first FIGURE is not a great example of this phenomenon. The data of FIGURE 4.07 allows me to describe the critical features of this behavior even more effectively. However, let me start by calling attention to FIGURE 2.02. The data pattern in this FIGURE is complex, but the basic elements of the devious behavior pattern can be seen, and this will get us started and will set the stage for FIGURE 4.07. FIGURE 4.07 has the advantage of comparing both a control and experimental mouse's behavior side by side. In this experiment, a control rat was not included since I had no idea that devious behavior even existed.

In the fourth experiment, I finally design a valid experiment. The control mice run the maze with a two-element constraint set; a unit and time constraint, but the experimental animals run with a three element constraint set; unit, time and, a performance criterion.

The simplest way to describe the set of test conditions which cause devious behavior is to consider these conditions as describing the conditions for ending the test session and removing the animal from the maze, let's call these conditions the "out-conditions." The control animals have a two-element set of out-conditions. The experimental animals have a three-element set of out-conditions.

The two-element set exclusively put the focus of the primary mind upon the primary goal, the mastery of the goal of running the shortest path. In addition, I will show that animals with this single-minded goal emit a complex unique pattern of behavior.

However, the animals with a three-out set of test conditions discover that, in effect, they live in two worlds. For example, for the first five days, this Rat, Female 11, in this experiment discovered that this matrix world was a vacation paradise in which she could take her time and spend her time and energy anyway she chose. However, after spending literally hours a day for several days in the maze, suddenly the rule changed. Suddenly after just a few minutes, she was removed from the maze. The shift in the rules of the game she would have found very disturbing. The shift from a leisurely world to a harsh severe world was experienced as an anomaly and this triggered a profound change in her mind. Her mind conceived, during the first few days, her self-imposed task to be a pleasure: to evolve a paradigm to describe this maze world, but her first conception of the rules of the game, did not include the idea that unexpectedly and precipitously she might be removed from the maze. This removal invalidated the old paradigm and the rat's mind had to go back to the drawing board and begin to invent a new set of rules which would fulfill her need in this new world with the rule which suddenly went into effect.

However, the chief feature of this new set of rules was the dramatic difference between the times she was able to spend in the maze. The first set of rules provided her with an abundance of opportunity to spend her allotted opportunity anyway she wanted, but the new set of rules cheated her out of this playtime, which she had come to enjoy.

The behavior of the Experimental rats faced with this three-element set of constraints is dramatically different from the control subjects facing a two-element set of constraints. The data of FIGURE 2.02 reveal how confused the experimental rat's behavior becomes. It is critical to remember that the rat's two rational goals, which are compatible under a two-element set of constraints now have become mutually exclusive so that on any day rats choosing A must forsake B and vice versa. In this new world, if a rat focuses upon maximizing her playtime, then she cannot continue to maximize her exploratory play efficiency.

The reader must learn to distinguish several different behavior patterns, especially the devious behavior pattern that emerges in the face of a three-element constraint set. The rat's behavior reveals the fact that rats experienced the third element of the constraint set as a punishment. In this data, it is possible to watch the two different patterns of behaviors evolving on different days. These two patterns represent patterns of behavior generated by two mutually exclusive goals. The behavior of the rat in FIGURE 2.02 reveals how the rational mind of the rat deals with this conflict between two of its natural goals. Our task is to understand the complexity in this spontaneous behavior that is the product of two mutually exclusive goals.

Serendipity Strikes Again

The phenomena of devious behavior forced me to recognize, for the first time, that rats run the maze with two or more goals. Normally these two goals are compatible- the first goal is to run the shortest path, while, the second goal is to maximize the moves and time it can spend during exploratory-play, which means they can both be satisfied simultaneously. However, when the two goals are mutually exclusive, the behavior of the rat must become very complex in order to satisfy both goals. FIGURE 2.02 shows that when punishment is associated with one of these alternatives, the rat is forced to alternate between the two goals, since the presence of punishment for one of the goals makes it is impossible to generate one pattern of behavior that can satisfy both of these mutually exclusive goals. Therefore, serendipity strikes again. The introduction of punishment into the exploratory-play behavior not only allowed devious behavior to be observed, but also made it possible to isolate the two normal goals of the primary mind. The effect of punishment when applied to only one of the goals puts the primary mind of the rat into a conflict situation. The conflict means that the rat cannot satisfy both mental-needs by growing in a way to run the shortest path. The conflict can only be resolved by the rat defining a new alternative shortest path through the maze; namely to run through the essential path using some other of the seven possible alternative path definitions, other than the Type 8 loop (the criterion loop), a type that is not punished. The loop types will be discussed in FIGURE 2.04AT.

The complex behavior of F11 seen in FIGURE 2.02 can only be understood as the behavior of a rat forced to alternate between two mutually exclusively goals, the shortest path goal, and the devious behavior goal.

"The conscious organism alone is able to conceive of and choose among goals; the mechanism has no goals of its own." (Kelley, 1968, P. viii)

The rat's behavior in FIGURE 2.02 begins as a normal growth pattern with the primary goal being the shortest path goal; the behavior generated with this goal persisted as the dominant goal for the ten runs before it began to "deteriorate." In FIGURE 2.02, the deterioration began on run five. At some point during this time, the rat finally began to recognize the conflict between its two normally compatible goals. Beginning on run 11, F11's behavior shifted to the devious behavior pattern and "regressed" for the next seven runs, through run 17. However, note that on run 18, F11's behavior dramatically shifted back to the normal dominant shortest-path goal; she made a growth in 79 steps and 57 seconds! This data shows that F11 on the 18th run, which was the fourth growth, had both steps and latency scores with values under 100! This normally dominate mode of spontaneous behavior can be seen several times in the data, in FIGURE 2.02. Compare runs 7, 10, 18, and 23. These days are four examples of the normal dominate shortest path goal- all of these days were "under-100 score days." This pattern of alternation between the shortest-path goal and the devious-behavior goal shows that the rats' spontaneous exploratory-play behavior can become very complex, when required when punishment induces conflict between these two normally compatible goals. A theory must be developed to explain this pattern of alternation between these two patterns of behavior.

Each of these now incompatible goals requires its own program, which takes turns at directing and organizing behavior, and each of these goal-specific programs generates its own distinctive growth curve. In the devious or second phase, on trial eight, F11's first "regression" day, she did not "regress" anywhere close to her first day score, as did F4. Only on run 17, did F11 finally "regress" to the number of moves close to equivalent to her first run scores; however, her latency score on day 17 was only half-way back to her first-day latency score. Notice the series of data for runs 10 to 17 in FIGURE 2.02; this anomalous series of behavior is clearly not extinction behavior reflecting boredom. These gradual changes in behavior, which were initially perceived as regressive changes in the rat's normal behavior, are another kind of behavior, devious behavior, and devious behavior has its own growth-learning curve pattern.

The devious behavior seen in FIGURE 2.02 shows the rat to be evolving two kinds of growth curves, one that satisfies the rats' need to optimize its playtime in the maze or to avoid being removed from the maze. This need is satisfied by staying as far away from the alpha unit as possible and for as long as possible. However, how can the rat's attention systematically alternate between mutually exclusive forms of growth behavior? The first seven days of the rats' behavior reveal behavior that had a normal exploratory-play learning curve, one characteristic of the shortest-path goal. However, days eight through eighteen were devoted to the second kind of growth learning behavior pattern; one generated by the devious path goal, and this series of data points shows the growth of this first series of long-avoidance runs. On day eighteen, however, note that the rat suddenly reverts, once again, to the shortest path goal that is the rat's primary goal. The data shows that on day eighteen, the rat ran the maze very fast, setting her best record for any run so far; one with the fewest steps and with the shortest latency. The subsequent data shows the rat alternated several more times between these two goals, the shortest-path and devious behavior goals. Notice how the rat cycles between these fast and slow behavior pattern days in the data. Three days in which she ran the shortest path were followed with run 28 and then one run that was a little slower. A technical reminder is in order here. The data of FIGURE 2.02 is individual data not group data. In the next section, the same data seen in FIGURE 2.02 will be described as group data. The point of FIGURE 2.03 will be to show some of these phenomena survive in the group data, but some do not. However, the group data has the advantage that it allows additional aspects of the rats' spontaneous priority structure to be described. The group data has both advantages and disadvantages. However, the most important point is that the ideographic or individual subject data as seen in FIGURE 2.02 is essential to understand if the full complexities of the Exploratory- play behavior are to be recognized.

Thinking like a rat, a hypothesis of how rats think about the matrix

The primary mind must invent a rational system for understanding the structure of the matrix maze, yet on the way out, it has no way to anticipate how many rows it will encounter in the first half of the matrix. The rat's mind had no way, or no reason, to assume that there would only be three levels to this cyclical pattern. So what if the rat's primary mind was expecting a fourth level of alternating rows, and a fifth, etc.

1. **The Direction effect problem FIGURES2.02A, B, C, and D**

FIGURE 2.02 direction effect

Some rats have a problem in the first row on the way back, which was the sixth row on the way from Alpha to Omega. Additionally, on the way out, the rats in the sixth row moved west through the four E units, 1-4, and entered omega. Now, however, on the way back, when they enter the now first row from omega, it was as if they were entering the first row from alpha, and the rats were set or prepared therefore to turn right from unit 6 and move east through four more E units, etc. But of course, they are not leaving alpha but omega, and this while this first row is also an east row on the way back, yet there is a problem: upon leaving alpha, a right turn in the first row, headed them east: however, now a right turn heads them west and moves them into an alternation, A1 unit. What unit did a right turn in row 1 on the way out move them into immediately; what Blind Path units were they headed for?

The question then is this: what effect could "omega," or should omega, have had upon the rat's assumptions about the structure of the maze. We humans understand the concept of Alpha/Omega as rather conventional elements of the primary path both alpha and omega signals to us that they are the natural nodes of the maze. A natural node is a feature of the maze, which requires the rat to turn around. In the current matrix maze design, rats seem to recognize the maze at first has a natural node.

The facts on the ground are these; with a six-row maze, the strange fact is that the path in the two directions; out and back are not symmetrical. In the first half of the matrix, the trip through the EP requires the rat to alternate between first an eastbound row and then a westbound row for three alternating cycles. I am prepared to assume that the rat's mind was also prepared to assume that this cycle would continue; the rat's mind had no reason to believe that after the omega unit this alternating cycle of rows would end and would have to be reset to some other pattern.

If I were reading the rat's mind correctly then upon leaving the omega unit the rat would expect to move in a pattern that would continue the rhythm of the rows discovered in the first six rows of the maze. In other words, it should be prepared to turn right upon exiting omega. However, the rat's mind now encounters a hard-fact; an anomaly--the row pattern has reversed, and this adds a new higher-order or a new level to the hierarchical structure of the matrix. The primary mind will adapt to this too in time, but do not despair for the rat; what makes life interesting for mammals is not just sex, food, or any other material thing; the primary mind evolved to digest the cosmos's infinite resource of structures of the hierarchical order of the matrix.

However, on the way back the situation is very different. The first row out of omega is an eastbound lane, but omega is located on the west end of this row, so that the rat's primary mind must now conceive a higher-order rule. The rule that the mind must adapt to is the fact that there are six rows in the out cycle and six in the back cycle. However, in the two cycles, the odd and even rows trade places, the rows that had been east bound and entered with a right response, now become westbound rows, which are entered with a left response. What the reader must understand is, as I said, a very technical point, but the primary mind of the rat is designed to deal with the real world at this technical level.

I am predicting that the difficulty the rats have in the second half of the maze is due to this reversal of the east and west bound rows. Therefore, I propose alternate solutions to test this hypothesis.

What structure and function enabled the NV mind to master change

My purpose in this book is to understand the structure and function of the nonverbal mind, which evolved to understand the changes that happen in space.

To make this point most simply, remember that Lashley conceived his four-row maze as requiring rats to make four alternation turns, which he conceived was the most difficult sequence they could be required to make. My extension of his maze could be conceived as requiring a rat to make a sequence of seven rows of alternation turns. Lashley, of course was a member of the classical school of rat testing behavior. In other words, he would have employed a Trial Testing Procedure, TTP, while a Loop Testing Procedure, LTP, was employed in these experiments. So Lashley (1963) was not concerned with the problems, which are the topic of this section. Let's call the sequence on the way out the Alpha sequence, which can be seen below. On the other hand, call the sequence on the way back the Omega sequence. Note however, that the two sequences are opposites, and this difference accounts for the difficulty rats have on the way back. Furthermore, the punished rats have a double problem in the Omega sequence for they are also concerned with avoiding the alpha unit.

However, consider how the rats become aware of the direction effect, perhaps the easiest way to call attention to one of the main consequences is the reversal of the sequences between the alpha and the omega types.

	EVEN- SIX- ROWS						
	1,	2,	3,	4,	5,	6,	7,
E-Alpha	R,	L,	R,	L,	R,	L	
E-Omega	L,	R,	L,	R,	L,	R	
	ODD- SEVEN- ROWS						
O-Alpha	R,	L,	R,	L,	R,	L,	R
O-Omega	R,	L,	R,	L,	R,	L,	R

First, look at FIGURE 2.02A, particularly look for units 55, and 56, which are one of the two pairs of Blind units in the 6th row. Now in FIGURE2.02B note that these two units, as do all the units, have two sets of functional names: F1, and F2, and A1, and A2. Next, look at the data for FIGURE2.02A for the units 55 and 56, which show a copy of the final page of the print out of Graph53, which shows the First entry data for all the 48 cells of the matrix. The important data of course is in red and shows that the last first entries on the way out were in unit 55 and 56, with 6439 and 6440 steps respectively. This data shows in these units on the way out, the rat was the least likely to make a Forward Error; but they were not immune to making a forward error.

How are ideas known or recognized? Are ideas memes?

Perhaps the most difficult aspect of Primary language behavior is discovering or distinguishing the difference between an idea and a word. Simply: words are memes, while ideas are not! Words are conventions while ideas are not. Words are standardized, at least, in the civilized world by authorities, generally represented as book or dictionaries, while ideas are not. Words are public events, while ideas or images originally were designed to work privately and personally. Words are conventions established by common or uncommon agreement between two humans or between linguistic communities

F3 confined within the matrix maze was able to create all the knowledge she required to meet her self-defined goals with the maze by employing only primary language behavior. She was alone with respect to the community of rats. Her only communication was directed at the physical or spatial structure of the matrix maze. The control of a rat's nonverbal behavior in the maze is entirely a rational nonverbal behavioral function.

In order to understand rat's behavior in the maze, we humans must learn to read the rat's nonverbal mind. What propositions were directing or programming the rat's behavior before and after each encounter with an anomaly?

Two questions, First: What nonverbal proposition makes the two units of the Forward pressing Blind Path elements of the 8-unit columns easy to avoid?

Second: What nonverbal proposition makes the two units of the forward pressing Blind Path elements of the 8-unit column hard to avoid?

Meme statement: "In the first row go east for 4 units until you encounter a door. Do not proceed past the door! Go through the door, turn west, and proceed through four units.

The nonverbal communication system is the primary one, but it is a pre-meme form of communication--the nonverbal mammals' communication with nature across the image -driven nonverbal narrative of the nonverbal consciousness. What do these statements mean?

Answer to question one.

The Matrix maze has two structural features, which made these two forward BP units easy to avoid. The door is the most powerful cue working to support the nonverbal avoidance behavior of the Forward going BP units. The Causal Conceptual Conditioning, CCC, explanation for the strength of the forward going blind units proposes that the important rule is "turn at a door" or "never ignore a door." The two Forward Blind Path units are the easiest to avoid, "Watch for the door!" These two blind units can only be entered if a rat fails to exit the row through the door.

Punishment enters the picture

But then a series of anomalies happen, which inject a new twist in the causal spatial structure of the maze; put most simply, punishment has now entered the picture. Punishment alters the rat's perception of the matrix structure in several ways. Let's just focus upon the door, how punishment might have altered the mind-generated-rules about the door. These rules again are the creation of the primary mind. The simplest possible rule for getting a rat through the maze is "Enter every door." This rat followed the rule for almost fifty days; it entered Omega 49 times and thus avoided the forward entry BP in the sixth row for 49 days. But then, on day 50, something happened and the rat failed to enter the omega door and instead entered the two Forward Entry BP units, which had been avoided for 49 days.

If the reader will take a second to look once again at FIGURE2.02, what changed was punishment, which specifically altered each rat's concept of a door. Presumably, in the first stage, this rat had learned that the Forward blinds were redundant and that they could be avoided most simply by entering the Omega unit.

In the second stage, once the rat has been punished one or more times that fear gets associated first with the alpha unit and then devious behavior begins as soon as a significant quantity of fear gets associated with Alpha. However, this statement is crude. Fear could get associated with any part of the alpha or omega unit. But the point of this section is that any amount of fear associated with the alpha or omega unit could cause the organism to fail to enter omega and thus finally enter the Forward Entry BP units in the sixth row.

- **The big-purpose - the Heterostatic theory of Blind Behavior**

The blind-unit data generated by the spontaneous exploratory-play behavior of rats in the matrix maze requires its own set of laws. In other words, in order to explain the blind behavior of rats, a major extension of the classical theories of how rats learn mazes had to be made. The blind-unit behavior of rats in the matrix maze requires that it be recognized that rats not only evolve two independent sets of conceptual models, but two different models explaining how BP and EP fear induced behavior alters the overall behavior pattern.

In order to understand their behavior a new theory of motivation must be invented. The new motivational element that the BP data justifies is **redundancy**. The three motivational elements of Heterostatic theory are anomalies, news, and redundancy. The first two elements were discussed in the earlier experiments.

The purpose of this section is to develop a conceptual system that will allow the blind-unit motivational feature of redundancy to be understood.

Understanding the blind-path behavior

Rats are organisms with their own purposes; they are self-organizing systems, SOS. Historically self-organizing behavior has generally been defined as a form of human behavior.

What is wrong with this human-only theory?

First, SOS first emerged as a property of the generic nonverbal mammalian mind.

Second, SOS behavior is the original form of nonverbal rational behavior.

Third, the mammalian organisms capable of exhibiting SOS behavior are self-organizing because they have the rational capacity to design and evolve their own conceptual systems, to assimilate domain-specific anomalies.

Fourth, mammalian SOS are capable of evolving more than one model or conceptual system at a time: which I call the alpha model and the omega model of the matrix maze.

The primary mind, in fact, can complete no purpose that cannot be achieved by acting upon objects that are exemplars of conceptual structure of its own creation. Rats are not mindless organisms that randomly move through either the EP or the BP. A rat's rational mind constructs two different sets of rules that allow or permit a rat to run the matrix maze. In this section, I will focus upon the rule system, which rats evolve to control their behavior in the blind units.

One of the most interesting properties of this new kind of structure is that it allows rats to enter the BP units only when there is good reason for doing so. Rats do not move randomly. For example, after having entered a blind unit often enough for the primary mind to have evolved a valid model of the classes of events of which the BP units are members they stop entering future exemplars of the redundant blind classes.

Since the matrix maze is hierarchically organized, each BP unit and all EP units are vectors with as many levels in their names as there are parameters of the hierarchical design. Every unit is a member of one and only one level of every parameter, which defines the hierarchical structure of the maze. Since the matrix maze is hierarchically organized, each of the 48 blind units is a member of each of the spatially organized classes, and each element of the 100-unit matrix maze is a unique element of a multidimensional vector, which is recognizable only when defined as a product of all these classes, as in a factorial design.

2. The Dual Names of each maze unit.

The matrix maze has many features, which escape our human awareness because we observe the maze from outside looking down on the maze, while the rats perspective is from inside the maze, one unit at a time, which makes all sorts of features obvious to the rat, which escape our attention. One of these features I call the Dual Names feature of each maze unit.

FIGURE 2.02B Dual Names

FIGURE 2.02B makes this feature of the maze units most obvious. It shows that every one of the maze units has two Functional names, one name on the way out and another on the way back. In most cases, this dual representation feature is a problem/opportunity for the rats in this experiment. In those cases where the dual name made a difference, sometimes the difference is an opportunity and sometimes a liability.

The direction effect: Dual names: the fourth dimension of the matrix space.

Another complication of the matrix maze is that each unit has two sets of names, or dual names, 1 set, 48 for the way out, and 48 or another set for the way back.

The matrix maze is a four dimensional hierarchically structured space. Each of the four dimensions of the maze is an idea. The topic of this moment, the direction dimension, allows me to discuss a most important larger question: what is an idea? If a dimension is an idea, then obviously we must understand the nature of an idea. What is the idea of direction? What is the meaning of the direction dimension? An idea is simply a name calling attention to some feature of an object that requires discrimination be made. The nonverbal mind evolved to detect and deal with change. Every change requires discrimination and a name.

William James (1950) noted that concepts or conceptual systems are partial; or today we would say that concepts are pluralistic: that many alternative, partial, conceptual systems can be designed for any task. I could choose between two possible frames of references to impose upon the matrix maze; the compass or the clock coordinates system. However, in the tradition of Ptolemy, I chose to employ the compass coordinate system.

One of the differences between a solely nonverbal mind species and a nonverbal/verbal mind species is that the former must personally/empirically grow their own understanding structures, their own conceptual systems, while the latter can borrow a metaphor.

The compass model

To apply a compass metaphor to the matrix maze, we first agree to talk about the number of rows in the maze as six. Assume a rat running from alpha to omega is running south to north; the alpha unit is located on the south-west corner of the first row. Additionally, the odd rows 1, 3, and 5, are eastbound lanes, while the even rows 2, 4, and 6 are westbound lanes as shown in FIGURE1.2.4A using the row count in green or those rows without alpha and omega.

The out or alpha-rule

FIGURE REPLICATION

FIGURE2.02D

(To follow this discussion, refer to FIGURE 2.02D) We will be using the row numbers without Alpha and Omega shown on the far right of this figure.

The rule for the rat's behavior can be broken into a two-phase cycle: the eastbound rule and the westbound rule.

- The Eastbound rule says the odd or eastbound rows 1, 3, and 5 are entered on southwest corner into unit E1 and the rat turns right to proceed east for four units.

- The Westbound rule says the even or west bound rows 2, 4, and 6 are entered from the southeast corner and the rat enters E1, and turns left to proceed West for four units and exits and enters the odd rows and the cycle begins again.

The back or Omega-rule

Our task, as experimenters, is to understand the rat's behavior or to "think like a rat." So will the rat still try to apply the Alpha rule? When will it wake up? Our question is will the Alpha rule, which worked on the way "out" also work in the second half of the matrix on the way back. When and how will the rat's consciousness start to get involved?

The first step on the way back requires the rat to turn around 360 degrees and run south for the second lap of the EP run. Generally, we observe that rats conceive the omega unit, which was the last or nodal unit in the 'out path' as the alpha-two unit for the way back. In other words, in the rat's mind an Alpha unit starts the loop in both directions out and back, but alpha one is the first alpha and Omega is the second alpha.

The "out" rule states that in the six-row conception of the maze, the odds rows are eastbound lanes, which are entered on the southwest corner and with a right turn.

This rule obviously no longer works on the way back! Notice that the row which was the sixth row on the way out becomes the first row on the way back. See FIGURE 2.02D.

A new set of laws must be evolved to describe the structure of the maze in the second half of the matrix. On the way back, the odd and even rows retain their east-west bound status; the odd rows are still eastbound lanes and the even rows are still westbound lanes. However, what has changed is the way the rat must turn to enter each row. The rats must now turn left to enter the eastbound lanes, and right to enter the westbound lanes. Additionally, now the rows that were entered on the southwest are entered on the northwest corner. For example for the first row, on the way out, rats had to enter on the south-west corner and turn right to go east. But on the way back, rats in the first or odd row must turn left to go east and enter the row on the northwest corner.

FIGURE2.02C One -unit omega

However, if a rat on the way back leaves alpha two and entered row one with a right turn they will enter unit 55 (See FIGURE2.02C) and will be headed west on this east bound lane. That difference is anomalous, surprises the rat, and justifies the writing of a new nonverbal law. The rat's erroneous behavior shows that the first law was factually invalidated. This difference challenges the primary mind; it arouses the mind to the fact that it must evolve a new higher order variation upon its first set of rules in order to predict or to take the uncertainty out of this latest anomaly in the second half of the maze. The rat will not be able to run the second lap of the EP efficiently until it has evolved a new rule structure, the Omega model, which the rat can employ to plan its behavior efficiently without error.

The difference is that on the way back, the rat enters the now-row 1, which formerly was row six on the way out, on the northwest corner of this now eastbound row; now that's new! However, this anomalous behavior forces the nonverbal mind to add another level of complexity to the hierarchical structure of the maze and creates a new grade in which the new facts justify the expansion of the model of the maze from a three to a four dimensional model.

In a balanced-row matrix maze; one with an even-number of rows, where the count of rows exclude the alpha and omega unit, the rule is that the odd numbered rows are east bound, while the even numbered are west bound, and this is true both when the rat is north bound and south bound. It is also true that all the rows are entered on their south side and exited on their north side. However, the concrete difference is that all six rows are entered on the north side and exited on the south side when the rat is traveling south, but the directions of the turns are reversed. In the next grade level of the hierarchy, this higher-order difference must be written into a new law.

There are two kinds of alternating patterns, an even and odd row pattern. The paradox is that an even-number-of-row matrix pattern is not symmetrical in the two directions out and back. However, in a seven-row maze, or any maze with an odd number of rows, the sequence becomes symmetrical in both directions. However, this is very messy and complicated.

Experimental test number one: the seven/nine row maze

In order to understand this technical point let us consider how to design or redesign a maze in which the same rules would work in both directions.

The problem of the rat is that I am requiring it to make a complete loop, once in each direction, in an eight-row maze. However, two test conditions are possible!

First, the symmetrical or even number of rows in the test maze.

Under such a set of symmetrical conditions, the primary mind must create two sets of domain-specific laws, one for the way out and another for the way back.

However, the first way to redesign the problem is not elegant or clean, but it might work and deserved to be tested.

So to understand this point consider the design of a new maze, one ideally which does not require the rat to conceive two sets of laws. Since the problem the rat faces is one that exists because of the six-row nature of the matrix maze, lets design a nine/seven-row maze. See FIGURE 2.02E.

FIGURE2.02E 9 row maze

Ideally, what we want is a new matrix in which rats can run in both directions out and back, which would allow the rats to employ the same model or set of rules about the hierarchical structure of the matrix maze in both directions.

FIGURE2.02E shows a 7-9 row maze, which is one way to create such a structure in which this purpose could be achieved. First, why do I call this maze the 7/9 row maze? On the right side of the maze, the rows are numbered 1-9 but, of course, Alpha and Omega occupy row one and 9 respectively, leaving only seven rows for the Essential Path, EP.

The FIGURE2.02E the seven/nine row maze

While this is far more complex behavior than the behaviorists would have attributed to any non-human animal or any human for that matter, yet now that I am proposing that rats have a primary nonverbal mind; obviously, there must be some function, some mental function, to justify such a claim. The facts will show that during their exploratory-play behavior, rats spontaneously evolve a compound model of the matrix maze. In the matrix maze, a compound model includes the evolution of two almost independent models, one for the Essential Path, EP and another for the Blind Path BP. The nonverbal mind of rats are so efficient at inventing and perfecting models for these two halves of the Matrix Maze that they soon shift their mode of operation from the model-making to the model-using modes. Rat's rationality is tempered with Pragmatism. While rats enjoy the model-creating phase of rational behavior, yet this creative phase is but a means to an end, their goal is to discover and run through the maze with the shortest path and soon they have created a model that is "good enough" and they begin extrapolating from the model. In the text above, we described how EP extrapolating behavior of Female 3 looks, but in this section, I must now describe how BP extrapolation looks.

The most interesting but originally puzzling aspect of BP extrapolation behavior is manifest by rats and mice refusing to enter redundant blind units that are exemplars of some type of maze unit that they have already successfully modeled. The two kinds of extrapolation: EP and BP, however look very different. The EP extrapolation generally generates long paths, over 50 or 100 steps, while BP extrapolation usually involves only some one or more of the BP blinds, either the Alternation, A, or the Forward Blind, F, units. Those blind units that are the redundant exemplars of the easily modeled BP units are, of course, the first to be avoided. The F units are easiest for the rat to model and thus they are the blind units whose redundancy rats first learn to avoid.

Devious behavior: group data

The group data seen in FIGURE 2.03 made possible a formal statistical test of the hypothesis that devious behavior results from punishment. FIGURE 2.03 shows the first order interaction, data type by runs, for four scores for 26 runs. $F (75,200) = 3.03$, $P < .00000$. The four run scores are first, the average number of steps per run; second, time per run, third, the average number of loops per run, and fourth, the average time or latency per move per run. FIGURE 2.03, first, confirms the concept of devious behavior by revealing for the first time several new features of this pattern of behavior. However, the F11 data of FIGURE 2.02 confirms in the group data the same effect noted in the individual data that the average rat improved its performance for six days before its behavior shifted to the avoidance pattern of devious behavior. However, since FIGURE 2.03 is group data rather than individual subject data, it is not surprising that the complex variability seen in the individual data for F11 was washed out in the group data. The full complexity of the alternation between the two mutually exclusive kinds of growth-curves can only be seen in the individual data.

FIGURE 2.03, however, for the first time reveals an important relationship between the loop and the run data; all four scores decreased for the first 6 or seven days! However, as the number of steps and time scores increased during the first avoidance period, beginning at run six or seven, the loop scores did not show a proportionate increase. On the contrary, the number of loops per run continued to decrease from an average of twenty, on run one, to an average of one or two per run for 25 runs. Each time a rat, because of its fear of punishment, refuses to enter alpha, it also fails to terminate its present loop, or fails to start a new loop. In the normal non-avoidance situations, when rats are making type one loops, a new loop starts as the last loop ends in the alpha unit. When the rats avoid alpha they both fail to end the old loop and to begin a new loop, and this, of course, not only decreases the number of loops per run, but it also increases the size of the loops in effect.

As rats continued to decrease the number of loops, they make per run, while increasing the number of moves and times per run, the length of each loop must increase, and this increase is the manifest sign of avoidance behavior. FIGURE 2.03 shows this increased steps and latency beginning after run 6, which is the sign of avoidance behavior. The drop in the number of loops per run shows that the rats were avoiding the alpha unit since each loop ends with an entrance into alpha and a new one begins when it leaves alpha. This change in the number of loops in the punished rats' behavior is a sign of the evolving devious behavior.

FIGURE2.03 Exploratory play behavior

The group data of FIGURE 2.03, in other words, shows that the exploratory-play behavior of rats in the Lashley III matrix maze was, much to our surprise, a robust form of behavior and that our expectation that it would soon extinguish was ill founded. The exploratory-play behavior of rats in this maze has amazing staying power; so much so, that there must be something about the pattern of this matrix maze, which sustains the rats' exploratory-play behavior once it gets going.

The complexity of the spontaneous exploratory-play behavior of rats demonstrates that rats are not mindless creatures whose behavior is completely determined by the environment. On the contrary, this avoidance behavior forcefully demonstrates that rats are self-organizing systems, and they have a powerful ability to modify their own behavior in complex ways in order to achieve their purposes in the face of obstacles. This punishment induced devious behavior may well prove to be the prototype of human devious behavior. The rational behavior of the rat is not a platonic type of relation but one in which the self is intimately involved.

FIGURE 2.03 shows an important part of the story of how devious behavior evolves. As the rats were punished for entering, or reentering, the alpha unit after having completed a type one or criterion loop, an alpha, omega, alpha loop, their behavior changed. The rats rather than extinguishing their behavior because they were bored, on the contrary, became even more interested in the maze as shown by the fact that their total behavior per day increased per run as measured by the move and time scores. Yet, at the same time as their moves and times per run were increasing, the number of loops per day was decreasing. The rats were obviously evolving a new more complex way to run a maze. As their total activity increased, the number of loops went down and obviously, the mean length of the loop went up proportionally. This data shows that these rats evolved a complex new strategy for running the maze that avoided the particular event that was punished. They stopped entering the alpha unit after completing criterion loops and so the number of loops per run went down even though the total activity (or number of steps in a loop) per day increased.

Devious behavior: Multiple Growths

Devious behavior, however, has several additional complexities, which are the consequence of the fact that exploratory-play behavior is hierarchically organized with four levels; growth, runs, loops, and node-path. For example, the growth is the complex outer-loop phenomenon of which there are several types.

Figure 2.04AT 8 growth types

Figure 2.04AT identifies the eight possible types of multiple growths. The normal type; Type 1 is seen at the top of the table. This normal type is an alpha1, omega, and alpha2 loop. When rats are not punished and explored to satisfy their normal goal to discover the shortest-path through the maze, they normally conceive the shortest path as a fifty-one step loop. This includes, of course, the forty-eight units of the essential path and an entry into alpha twice: once at the beginning and once at the end of the loop, plus one entry into omega. However, when rats are punished for entering or reentering alpha to complete their criterion loop, they have this remarkable rational power to redefine the nature of the criterion loop, the loop, which also, at least in part, satisfies their goal of the shortest-path. Another interesting feature of devious behavior is seen when rats start running one of the other seven types of nested loops; loops which include all the forty-eight units of the essential path, but fall short of executing the fifty-one step loop which is punished. Type eight, for example, which can be seen in FIGURE 2.04AT at the bottom of the table, is; a non-alpha1, non-omega, and non- alpha2 loop with 48 steps. Now please note this important fact. The rat in order to accept this type of loop as the shortest path would literally have had to redefine its operational concept of the shortest path. It would now have to accept a 50-step nested loop instead of a 51-step loop as satisfying the shortest path criterion. Figure 2.04AT shows that there are three of the nested loops that have a length of 49 steps; loops types 4, 6, and 7. Figure 2.04AT shows that there are also three paths with 50 steps: loops type 2, 3, and 5.

The type 1 loop is the normal growth, where the rat enters the alpha and omega units in the pattern, alpha, omega, and alpha. Types 2-8 are alternate possible devious growth patterns that a rat, punished only for type one, could resort to in order to avoid the punishment associated only with type 1. Please be aware that types 2-8 are nested loops within a growth. The rat always starts its initial running of the maze in the Alpha Unit. If rats are mentally capable of resorting to the use of any or all these seven multiple loops to avoid punishment then they would have to have the remarkable capacity to subjectively redefine both their concept of the essential path and their concept of the shortest path these seven alternative paths are all options from which the rat could choose. For redefining the shortest path, all the possible alternatives have in common the essential-path, EP, which includes all the 48 units of the core of the matrix maze, 24 units out and 24 units back. The differences that make the devious path alternatives unique are in the choice of which of the alpha and omega units are to be added to the essential path. The presence of the seven multiple- loop types allows the preservation of the assumption that the rats' primary cognitive goal is to run the shortest path. However, the concept of the shortest path when redefined in terms of one of these nested loops means that the rat has to be able to redefine its concept of the shortest path. All of these seven devious-growths alternative patterns are acceptable alternatives for the type 8 growths because they are all safe growths in the sense that they are free of punishment, and offer the rats, if they have the wisdom to take advantage of them, substitutes for the normal growth type that is punished. These alternate growths allow the devious rat to modify its behavior in the presence of punishment two ways. First, to maximize its exploratory-play activity or to maximize the time the rat is allowed to spend in the maze; secondly, to perfect its ability to run the shortest path as long as the rat can accept the alternate growths as a substitute for the type 1 growth.

FIGURE2.04BT Frequency of growth types by rat

Figure 2.04BT shows the frequency of the multiple devious growth types for seven rats, which are listed in the four columns from the first experiment and three from this experiment. While nothing is significant for this small sample data set, yet the data is suggestive. The female's dominant multiple growth type was type 3, with both female 10 and 11 making 21 and 22 loops of this type respectively, but they made no loops of the devious type 3. The males however all had one or more devious type 3 scores. The devious type-3 growth pattern, if it were chosen rationally rather than accidentally, would represent the most insightful devious growth behavior and would raise a most important theoretical question. The significant aspect of the type-6 growth is that the choice of this devious growth once again allows a rat to enter the alpha unit without being removed from the maze! (See type 3 in Table FIGURE 2.04AT.) In a type-three loop, the rat only avoids the omega unit, yet this allows the rat to enter alpha two with impunity. However, in order to assume that rats are rationally capable of taking advantage of this option requires that as scientists we attribute to rats a new cognitive ability, namely a particularly powerful ability to detect **causality**. Scientists are well aware of the fact that the power of the experimental method is its ability to detect causality, and I am now suggesting that at the heart of the power of the primary mind is its ability to detect causality. If we assume the rat to be a generic scientist capable of evolving nonverbal paradigms or models of a domain by representing the causal structure of a domain, then causality detection replaces reinforcement as the *sine qua non* of learning.

What would make the discovery that some rats are able rationally to select a type-3 loop interesting is what logical capacity this would require we assume of the primary mind. The critical question is how much of a time delay would the detection of causality take?

The only comparable situation is the famous question about the role of time delays in reinforcements. Just as the efficiency of reinforcement is an inverse function of the delay of reinforcement, the question could then be raised about how delays of causality affect learning. At first glance, it might seem that causality should be affected by delays just as is reinforcement. However, if this is so, then how could a rat ever learn to avoid the omega unit in the Lashley III matrix maze, for the causal consequence of avoiding omega could not be experienced by the rat immediately after avoiding omega under the loop test paradigm? On the contrary, the consequence of the act of avoiding omega cannot be experienced until after the rat reenters alpha and discovers it was not removed from the maze! The quantity of the delay of the causality consequence, the time it takes the rat to traverse the 24 units of the essential path of the maze from omega back to alpha, will almost certainly be as much as 12 seconds, at a minimum, even when the rats' average speed is at a fast .5 seconds per unit. These numbers would require us to assume that causality in the rat may not be as sensitive to delay as reinforcement has proven to be. Then the work of Garcia, Ervin, and

Koelling (1966) has shown that even rats, under special circumstances, can also take long delays of reinforcement. The rat as a generic scientist must then be capable of comparing the differences between the set of two consequences, the experimental AOA-P (Alpha Omega Alpha), and the control A-A-NP conditions. Causality, in other words, is not a property of any single contingency but is the property of a set. The possibility of the animal mind dealing with causality of this degree of complexity raises many important questions and makes the animal mind very unlike what Thorndike (1911) assumed for the classical mechanical rat, which supposedly learned by trial and error. Much more research will be required to resolve these possibilities.

FIGURE2.04C Growths and loops

FIGURE 2.04C shows a critical bit of data, which demonstrates the existence of the phenomenon of punishment by describing the relationship that emerges as the rat evolves its capacity to maximize rationally the amount of devious behavior it can emit. FIGURE 2.04C shows the evolving relationship that emerges as the rat increases its capacity for emitting devious behavior between the number of **loops and the number of growths.** FIGURE 2.04C shows the first order interaction runs by scores, for two scores, the mean number of loops per run and the mean number of alternate devious growths per run. $F(25,100) = 11.96$, $P < .00000$. The loop scores seen in this FIGURE were also seen in FIGURE 2.03. An ordinary loop, a type one loop, begins and ends with alpha, it begins when a rat leaves alpha and ends when it returns to alpha. However, in addition, a normal type-one criterion loop requires in addition to the two alpha entries, an omega entry. The third element of our constraint set required that rats be removed from the maze for retracing into alpha after making a type one-criterion loop. If, however, rats find exploratory-play behavior "fun," then the consequence of the third contingency the removal of the rat from the maze will be experienced by such rats as a punishment. Since, punishment always produces fear and this fear is always associated with the object, which causes the punishment, it can be inferred that the fear in this situation apparently gets associated with the proper stimulus; that is the rat will begin to fear retracing into the alpha unit, which is then punished. FIGURE 2.04C shows that the average rat, by the fifth run, has decreased the number of loops per run from the initial twenty to only somewhere between one to three loops per run but has increased the number of growths per run to an average of two or three a run. This is what would be expected if, by associating the fear created by punishing the rat upon entering the alpha unit, the rat evolves devious behavior and completely redefines its operational conception of its goal in exploring the maze. The run data of FIGURE 2.04C shows the mean number of growths per run, which includes both the normal and devious growth types the average rat made per run for 25 runs. On run 1, the average number of growths was .4, but by run 26, the average rat was making a mean of 3.5 growths per run. The two curves of FIGURE 2.04C together show that as the punishment contingent upon entering the alpha unit took effect, the rats stopped making the type 1 loop, which was punished. Instead, the rats started emitting one of the seven other kinds of possible multiple loops that were not punished. These new loops are examples of devious behavior. Devious behavior is behavior that has been altered because of punishment and has been modified in order to sustain some optimal level of exploratory play behavior as defined, or redefined, subjectively by the self-organizing rat itself. In other words, the data shows that over runs the rat was evolving the capacity to emit more and more nested growths within each of the ever fewer loops it was making.

FIGUREs 2.02, 2.03, 2.04, 2.04AT, 2.04BT and 2.04C, jointly reveal several of the rat's strategies and capacities for generating devious behavior. The rats' strategy for extending their playtime required that rats develop growths as nested loops. As the number of loops was decreasing, FIGURE 2.03, the number of growths per run was increasing, FIGURE 2.04, and the average loop size was growing, except on days 7, 10, 18, and 23; the "under 100 score day." Female 10, for example, on run 23 made six nested growths in one long loop of 627 steps and 3097 seconds! This shows this rat was a creature that did not passively accept its fate. Can rats have an internal locus of control? This rat had its own purposes and in the presence of obstacles was very capable, at least in this experimental test, of modifying its "means" to obtain its "ends."

Devious behavior: the run and latency data

Figure 2.05AT runs to criterion for 6 rats

Figure 2.05AT shows three run scores: moves or steps, the number of runs to criterion, and the time to criterion for six rats for the first growth. The mean number of steps and latency to criterion was 966, and 10,033 respectively. The mean number of runs to criterion was two. This table shows, for example, that F11 required the most moves and times to reach criterion for the first growth of any subject in this study, 1973 moves and 29,339 s. While this data might suggest that F11 was a maze-slow rat, all the other data shows that such an inference would be superficial. Especially, we must remember how hard she worked and very effectively I might add, to perfect her deviant behavior, fifty-five growths, as seen in FIGURE 2.04BT, which is the record; the next best rat was female 10 with fifty-four growths. Another interesting aspect of her behavior is the fact that she had a counterpart in the first experiment, F3. Both female rats took almost twice as long to master the first growth, as did any other rat in their respective study both took 1973, and 2057 moves respectively. Yet F3 required 10 runs at 10 minutes each (only 100 minutes or 5400 s) Vs. 29,339 s for F11. The next part of this story will be told in the next section, The First Growth.

The behavior of the rat in the Lashley III maze is hierarchically organized. In the above section, the data for the run levels was presented. In this section, the data for the first growth in the essential path will be presented. Our purpose, in this section, is to show the various ways that the rat's first growth data in this second experiment is both alike and different from the rats of the first experiment, which were run on different contingency sets. In this section, the data for the first growth before the rats experienced punishment for the first time will be examined. This first growth pattern is the base line or control data for the subsequent punishment growths. There are several strange differences in the animal's behavior produced by the extreme differences between the two constraint sets, particularly the results of the generous nature of the newer constraint set. The latency effect difference seen above in the run data is also seen at this level of analysis, first in terms of its effect on the BFEULM plateaus at row 5, and secondly, upon the sitting behavior that occurred in the alpha unit.

The FEU effect

Exploratory-play behavior is motivated by the generic rational emotion, uncertainty that is the *sine qua non* of exploratory-play behavior. The leisurely constraint set designed for this study had an unexpected effect upon the induction phase as noted above. Uncertainty is seen in each of the different data sets in different ways, but in the FEU data, uncertainty is made manifest in the **plateau** phenomena, which is one of the most obvious elements of the emergent or criterion loop. The FEULL (First entry unit, latency, and speed) data set, in particular, reveals the most basic set of facts about the retracing behavior generated by any particular constraint set. Three elements make up the plateau pattern: the number, size, and distribution of the plateaus.

FIGURE 2.05B Emergent Loop Data

FIGURE 2.05B shows the FEULL emergent loop data set for the first growth for this study with the three scores: FEU, FEL, and FL. $F(94{,}705) = 3.78$; $p < .0000$. This FIGURE is a standard plot of the data. The FEU and FEL data are roughly parallel data sets, but with FEL, about two orders of magnitude larger than FEU. The data of FIGURE 2.05B is a log plot of the data and four steep periods of vertical change can be seen in both sets of data, but the four are more clearly seen in the FEU data set. These four vertical periods in the record are the four plateaus plotted vertically rather than horizontally. The FL data shows a rhythm that is not seen in the other two scores. FIGURE 2.05B is the group's mean log data. Figure 2.05B has the same curve as Figure 1.3.3 for Female 3; it is just that the axes have been switched for Figure 2.05B.

FIGURE2.05B FEULL data set

In FIGURE 2.05B, the FEULL data set is plotted in the standard format with the conventional X and Y values. However, this FIGURE sadly fails to reveal the significant features of Heterostatic behavior. However, I eventually realized that Skinner's (1974) style of graphing his data revealed the critical features most effectively; his style reversed the X and Y coordinates.

FIGURE2.06A Mean Male FEU score

FIGURE 2.06A and FIGURE 2.06B are linear plots of the FEU data seen in FIGURE 2.05B for the male and female rats respectively, however this FEU data is graphed in the style of operant conditioning so that the FEU data is plotted as the rate of progress through the matrix maze. In this type of plot, the plateaus look like plateaus.

FIGURE2.06B Mean Female FEU score

The linear plot also makes visually obvious what was only logically recognizable in FIGURE 2.05B, that the enormously different lengths of the four plateaus, especially the long fourth one. Actually FIGURE 2.06A and FIGURE 2.06B show the four plateaus seen in FIGURE 2.05B were an artifact produced by ignoring the sex effect. The female plateau effect of FIGURE 2.06B is most dramatic. Actually if the reader will look closely at the FIGURE they will notice that the female rats had not one long plateau, but two plateaus, back-to-back, with the two plateaus broken at unit 17. Plateaus are the result of retracing which are measures of the quantity of uncertainty induced by anomalous units of the maze. FIGURE 2.06B shows that the female retracing was induced by units 16 and immediately again by unit 17.

If the reader refers back to Figure 1.3.3 from the section on Female 3 and compares it to Figure 2.06B they will notice a remarkable thing. The shape of the curve for Female 3 and the group data for the female rats in this experiment are remarkably similar.

Two phenomena must be distinguished in FIGUREs 2.06A and B: first, the late appearance of the plateaus in this study. This is another of the interesting and surprising results of the data of this study. Experiment one showed that a plateau developed for each of the first few rows and that the plateau length was a curvilinear function with the size increasing from rows one through four and then suddenly stopping. In this study, the curvilinear effect was so extreme that it appears as if only one plateau occurred, the plateaus at row 5. This shows that the difference in the constraint sets of the two studies made a difference but not why? Second, three anomalous plateau patterns can be distinguished. Plateaus are triggered by units of the maze that are so novel they initiate retracing, thus these anomalous triggering units of the maze become nodes. These anomalous units cause the rat to begin to retrace those units of the maze of which the rat had already evolved a valid model.

However, the chief causal feature of space is its hierarchical structure, and the mind evolved primarily in response to the features of nature. The whole concept of Heterostasis is that hierarchical problems are so complex they must be solved grade by grade. Which means, that a Heterostat at grade X evolves a solution which is valid for the data at grade X, but upon entering the next grade, the Heterostat will encounter the next anomaly which will perhaps invalidate the X solution, and so the organism must now evolve a new solution for grade X+1. Heterostasis literally means stopping at a different point. Homeostasis means stopping at the same point. The physiological technology of Homeostatic organisms evolved within a world where every problem has a single solution, each regulator was given a nativistic set point. However, a new physiology was required for a world consisting of hierarchical structures. Each grade requires a different solution, a different set point.

Kinds of hierarchical structures

There are two kinds of hierarchical structures: open structures with an infinite set of grades and closed structures with a finite set of grades. The task of a mentor is to create a finite safe structure, which their charges can master during exploratory play—these were the insights that Pestalozzi (1859), Montessori (1965), and Herbart (1901) first achieved.

The scientific significance of rational behavior is that empirical truth is contained within a hierarchical structure, with so many possible levels that a final truth can never be achieved; there are always new grades to master. In our educational system we graduate individuals, not when they know all there is to be known, but rather when they know enough to take charge of their further mental evolution.

What does this abstract argument mean in this context? The matrix maze is a kind of special world for rats that human schools or mentors must create; the matrix maze is an artificially finite hierarchical structure, a six-row maze. What these FIGURES show is that the rats in the hierarchically organized school, in the matrix maze I constructed, were able empirically to discover the spatial truth about this maze.

The behavior of the rats in my school can be seen to consist of two components: A and B. In A, the discovery phase, the rats explored the levels of the maze until they evolved a valid model of this finite hierarchically organized structure. In phase B, once their model was valid they had only to employ the information stored within this mental genome to evolve a plan for achieving their goal in the maze and then execute this plan.

In phase A, the primary mind of these rats creates a finite set of laws. In phase B, they extrapolated! The Flat places in each grade are a measure of how much uncertainty had to be reduced in each grade. The slope of the line through the matrix maze reflects the length of the extrapolation.

Nodes

There are two kinds of nodes, physical and logical. One of the paradoxical features of heterostasis is that an anomaly is always a logical node. An anomaly is not a solely unexpected event; but is one that, in effect, is a wall that a rat will not or cannot go through. Rats will not enter an area where they do not know what they are doing, since they are rational creatures. Then the second kind of node occurs in a row, for example, when a rat is starting at the alpha unit and running right, when the animal hits the wall in the cul-de-sac, that is a physical node, the animal must turn around. Therefore, the anomalies are nodes because the animal immediately turns around and starts retracing at the discovery of that node. What we are discussing in this section are anomalous nodes, nodes induced by anomalies. All the units labeled one in Figure 1.2.7 are physical nodes.

The functional anomalous nodes occur in three types. They are always: E1, E2, or both. In addition, the units selected as nodes are one of the few differences between the sexes seen in this maze. There are rats that only have E1 anomalous nodes, and in this study, these are the male rats. Female rats normally made an E1 or an E2 unit as their first anomalous node. But in addition, females also make the E2 unit a second anomalous node. So females create three patterns, first, E1s only, secondly, E1, and then E2, third, E2s only. E2 was the anomaly that caused Female 3 from experiment one to make 441 steps of retracing. This shows that the female rats on this leisurely constraint set did not discover an anomaly until the third row! Why should this constraint set delay the rate that rats detect the first anomaly in the maze? The difference once again, could be another example of the effect of massed and distributed practice, with the massed practice in this study delaying learning**.**

FIGURE2.06A Male rat FEU scores

FIGURE 2.06A shows that the male rats did not make these back-to-back plateaus; they made only one very large plateau.

The interesting theoretical significance of these large late-appearing plateaus is this: why did this leisurely constraint set seem to delay the rats from discovering the complexity of the hierarchically organized maze? Is there something about the constraint set that both delays the induction of the exploratory-play behavior and delays the rats from discovering the complexity of the maze? Both phenomena seem to have something in common, something is delaying the rats' ability to discover and thus get involved with making a model of the hierarchical complexity of the maze.

The FREQULM analysis: The First growth

FIGURE 2.07A and FIGURE 2.07B show two views of the unit main effect for the FREQULM data set for forty-nine units of the maze with FIGURE 2.07A a linear plot and FIGURE 2.07B a log plot. Usually, in all the other FIGUREs, only the forty-eight units of the essential path data are analyzed and presented in the FIGUREs (the twenty-four out and the twenty- four back units) but with this FIGURE an exception was made in order to show the special role and value the Alpha unit has relative to the other forty-eight units. $F(48,864) = 5.23$; $p < .0000$.

FIGURE2.07A Induction phase linear plot

The linear plot of FIGURE 2.07A seems unremarkable, since only the first unit score the alpha unit data seems to be different from all the other scores. However, this appearance is deceiving, yet it makes an important point. For the first growth, as was noted above, the rats under the constraint set of this study had a long induction phase. The FIGURE shows that all the extra time spent in the **induction phase** was spent in the alpha unit! FIGURE 2.07A shows how the alpha unit data was remarkably different from all the other forty-eight units' data.

FIGURE2.07B Induction phase log plot

Now look at FIGURE 2.07B, which is a plot of the same data, only as a log plot. The data of FIGURE 2.07B, shows that there was more going on than was visible in the linear plot of FIGURE 2.07A. The next FIGURE will shed some light on what was happening here.

FIGURE2.08 data types by unit

FIGURE 2.08 shows the first order interaction, data type by units, for the forty-nine units' FREQULM data set. F (94,846) = 3.87, p< .000000. This FIGURE shows that, indeed, there was much more going on that FIGURE 2.07B revealed. This data is the first growth data for the forty-nine units of the essential path. This FIGURE reveals the complex rhythmical structure of the three elements of this data set, which were completely hidden in FIGURE 2.07A, but only partially revealed in FIGURE 2. 07B. The data in FIGURE 2.08 shows that both FREQU and FREQL cycle with almost the same frequency but differ primarily in amplitude with the FREQL having the greater amplitude fluctuations. The FREQM data has obviously a simpler structure. This effect is a whole composed of the many significant parts, which will be shown in FIGUREs 2.09 to FIGURE 2.12.

FIGURE 2.09 main effect by rows

FIGURE 2.09 shows the main effect by rows, for 12 rows, for the first growth for the FREQULM data. F (11,165) = 9.16; p< .0000. This curve shows that the rats, on their first time through the maze, after spending most of their times and moves in the first few rows of the maze, were able to move through the last rows of the maze with great efficiency. When one considers that these rats were neither deprived nor reinforced, this curve has to be seen as impressive proof of the considerable spontaneous ability of the rat to make models of the matrix maze during exploratory-play in the absence of deprivation and reinforcement.

FIGURE 2.10 data types by rows

FIGURE 2.10 is the first order interaction, data type by rows, for the FREQULM data set. F (6, 45) =4.457, p< .0012. This data expands the detail of FIGURE 2.08 and shows that both the FREQU and FREQL data take the same shape seen in FIGURE 2.09, but that the FREQL scores were considerably larger than the FREQU scores. Again, both scores show that during exploratory behavior rats spend most of their allotted unit and time activity in the first few rows of the maze. The FREQM score, however, does not show the same improvement over rows. There is some small improvement in the time per unit score but with the range varying from over 2.5 for row 1 to just under .9 s per unit by row 11. The curve for the first two scores is of a type that was seen in classic learning curves, which shows that rats during exploratory behavior will learn the maze even though they are not deprived or reinforced, but it does not answer the obvious question. Why do rats spend such a disproportionate amount of time exploring in the early rows? However, the simple and most obvious answer is this: they spent their time in the first few rows because that was where the uncertainty was. Once the rats had explored these rows and created a valid model for these rows, all the subsequent rows are relatively simple thanks to the extrapolation powers of the model developed in these early rows. The basic assumption of heterostasis is that the size of the FREQU and FREQL scores in any unit is a direct function of the amount of uncertainty the rat experiences in that unit.

FIGURE 2.11 main effect for columns

FIGURE 2.11 shows the main effect for columns. $F(3, 45) = 6.19$; $P < .0013$. This is the standard column effect for the essential path data, the bucket effect. This FIGURE shows that rats allot most of their time and moves in the matrix to the E1 and E4 column units. E2 and E3 are allotted only the absolute minimum amount of time and moves. This FIGURE shows where the uncertainty is within the columns within the EP.

FIGURE 2.12 data type by columns

FIGURE 2.12 shows the FREQULM first order interaction, data type by columns, for the three scores of the FREQULM data set. $F(3, 54) = 7.34$, $p < .000325$. FIGURE 2.12 shows that the FREQU and FREQL scores have almost the same value for the column units E2 and E3, or in other words, they have a one to one relationship. The rats usually, on the average, require one second per move in these two column units. However, the ratios for E1 and E4 are three to one and four to one respectively. If the uncertainty in the maze is directly related to the ratio between these scores then E2 and E3, it must be concluded are not the source of any uncertainty. However, E1 and E4 are the major source of uncertainty.

The BFEL data

FIGURE 2.13 main effect for 12 rows

FIGURE 2.13 is the main effect for 12 rows for the BFEULM data set. FIGURE 2.13 should be compared with FIGURE 1.4.11. This curve shows that the row curvilinear effect of this experiment was much steeper and shorter than the curvilinear effect in the first study seen in FIGURE 1.4.11. The constraint set of this study produced a steep but narrow curvilinear effect, one that lasted for only one row.

The EP data: 25 growths

In the second part of the results section above, the result of the analysis of the Essential Path data for the first growth was presented. In this third part of the results section, the Essential Path data for twenty-five growths will be reviewed. In the analysis of the run data in the first section above, the effect of punishment was seen to produce avoidance or devious behaviors as shown by several run and loop scores. In the first-growth data, no punishment effect could be seen since the punishing contingency of the constraint set obviously required several growths before it was be detected by the rat and could begin to take effect. In this section, the effect of punishment upon twenty-five growths will be examined to determine exactly when and how devious behavior is manifested in the growth data but first two technical problems must be explained.

Problem one

The analysis of this multiple-growth data presents a major technical statistical problem just in terms of the massive amount of relevant data that must be processed in one analysis. For example, for each growth of this experiment each subject generated 48 repeated measures scores. There are 25 growths, which produced 1200 repeated measure scores per subject (25 * 48 = 1200). These numbers produce a technical problem because our statistical package, CSS, only permitted about 100-repeated measure scores per subject in a design. Our pragmatic solution was to treat the 25 growths for each subject as a between rather than as a within variable. While this distorts the between degrees of freedom, for it looks like the design has 75 rather than 3 subjects (3 * 25 = 75), and thus is far from a perfect solution, yet it gives us the graphs needed to evaluate the data even though the DF are technically not valid. However, the results are still probably conservative.

Problem two

Second, the selection of 25 growths itself, presents several problems: First, the data had to be divided into blocks for it is not possible to draw graphs with lines for each of the 25 growths; second, the way the number 25 can be factored (5 * 5), was meaningless. Twenty-four can be factored into 3 * 8 and 4 * 6, which raises the question: which 24 growths to analyze and which of the factors to select. The analysis of growths 1-24 turned out to be unsatisfactory for the growth-one data involved such large numbers that the effect of analyzing 1-24 was almost the same as the result of the analysis of growth one. The growth one data was the one in which the rats manifest their greatest uncertainty. Besides, **empirically, the analysis of growths 2-25 proved to be the most interesting, and the results of the 3 * 8 factoring produced the cleanest looking graphs.**

Problem three

In this section, things get complicated for three reasons: first the topic the difference between the one-growth and the twenty-five growth data. Second, the rat's behavior becomes more complex as it runs more times and thus has time to evolve an understanding of the complexity of the maze. Third, there is just so much data to be processed. The first kind of complexity, which the reader must learn to deal with, is the difference between one-growth performances and twenty-five growth performances by comparing three sets of figures.

The chart below breaks out the figures that need to be compared. FIGURES 2.08 and 2.14 are the whole effect, with 2.08 being for the first growth and 2.14 being for 24 growths. The next figures in each set are the parts, which ANOVA gives us for understanding the different causes of the complexities in the figures.

FIGURES 2.09 and 2.15 tell us about the main effect for rows, while 2.10 and 2.16 tell us about the effect due to the data types by columns. FIGURE 2.17 explains the differences due to blocks by rows.

One-growth	**Twenty-four growths**
FIGURE 2.08	FIGURE 2.14 -- Data types by units
FIGURE 2.09	FIGURE 2.15 -- Main effect for rows
FGURE 2.10	FIGURE 2.16 -- Data types by columns
	FIGURE 2.17 -- Blocks by rows

FIGURE 2.14 data types by units

FIGURE 2.14 shows the first order interaction, data type by units, for 48 units for the three scores of the FREQULM data set summed over 24 runs, 2-25. F (94, 6721) =8.51, P< .0000. The cyclical structure, which characterized this data set, was also seen in the Growth 1 data and the relationship between the three scores is still much the same. See FIGURE 2.08. However, this FIGURE has important and unique features not seen before. This FIGURE is a whole, which will be broken into parts in the next set of FIGUREs: FIGURES 2.15, 2.16, 2.17. So, we have the nonverbal mind nonverbally telling us a most complex narrative. Figure2.14 is the opening scene with lots of complexity; like a three ring circus. Gradually ANOVA takes the scene apart for us. If we follow the script carefully; finally, in FIGURE 2.17 the plot will be resolved.

However, this narrative has everything, most notably the narrative is based on one part hedonism and one part Heterostasis

If you compare Fig 2.08 and 2.14, the reader will begin to see the problem. What are the key differences between them? The differences between FIGURES 2.08 and 2.14 are an obvious difference in complexity. Where does this complexity come from and what causes it? The answers will come from examining the next figures with their counterparts from the one growth figures discussed previously.

As a preliminary, before we actually look at the data, I would call your attention to the differences in the two figures that occurs at approximately Unit 25 and the later units of the data, which suggests that a major if not exclusive change, which occurred, is due to what is happening in the second half of the maze. What is causing the differences in the second half of the maze? This will be answered in the next set of figures.

Skinner's Theory of Punishment

Skinner's (1974) theory of Operant Conditioning, in spite of making all the same mistakes that the classic behaviorists were heir to, still at times managed to study several topics in such an intense manner that his results gave us a definitive insight into some critical aspects of behavior. His study of the effect of pain or punishment on behavior is at the top of my list of his accomplishments.

However, in this context of the study of the rat spontaneously exploring a complex hierarchically organized space, only one of the critical features of his theory must be held in mind.

- **Punishment always produces fear.**
- **This fear is associated with whatever the nonverbal mind perceives as the cause of the punishment.**
- **As far as a scientist is concerned, the wild card in this situation is discovering how to identify what the nonverbal mind perceives as the cause of the punishment.**

The facts

After the fact, I discovered that the nonverbal mind of the rat enjoyed being out of its home cage; first, exploring, or second, just spending quality time in the matrix maze.

As always, my task as a scientist is to "read the rat's nonverbal mind," where the nonverbal mind does not verbally talk about what it is thinking, but nonverbally reveals its thoughts. So, in the next set of four graphs, we have sizable snippets of the rat's most eloquent nonverbal talk.

Our question is what did the rat's nonverbal mind learn to fear because of its being punished for re-entering the alpha unit?

The nonverbal mind is a rational organ, which has the power of attributing causality so powerfully that we humans may have trouble believing our eyes as the behavioral nonverbal consequence of fear evolves before them.

The rat fears the start box because it was punished for re-entering the start box. So, what follows from that?

The push-pull hypothesis

My theory proposes that on subsequent loops, the fear of the alpha unit will act in two different ways: on the way out the fear will act as a wind at the subject's back and will shorten the "Out" phase, or the fear-pressure will quicken the "out" phase. However, on the return trip, the fear-pressure from alpha will be a head wind, which slows this leg of the loop. Remember, each rat was run for 25 growths, so the two-part question is how the pressure of the winds of fear changed the rat's behavior in both directions in its growths over time.

FIGURE 2.15 is a main effect for rows, and as we might expect, the path out was a smooth curve. The second half of the curve, however, shows that the rat had considerable trouble on the way back. While all the means on the back trip were not the same value, yet all the back row values were higher than the last of the out-rows.

Some readers, however, may be concerned with the peak at Row 7. We could just make the facts up and say, for example, that the rat after running north has trouble making the transition to having to run south toward the alpha unit. The question is where did the row seven peak come from?

However, let's not guess. FIGURE 2.16 shows that part of the peak at row seven was due to differences between the three types of FREQULM scores, with unit FREQL scores being the largest scores. Note that FREQL peaks on row seven, which explains some small part of the mystery of the row seven peak.

Moreover, FIGURE 2.17, it turns out, is the major source of the seventh row peak. FIGURE 2.17 shows the block effect. Remember the rats were run for 24 growths, which produced a massive amount of data. This figure shows how the rats did when the data is analyzed into three blocks of eight growths. First, notice there is no peak on block one on the first eight growths. This figure settles our question completely. The seventh row peak only shows up on the second and third block. The fear that was conditioned to the alpha unit acted continuously but somewhat weakly; yet over time, the cumulative effect of the fear in the rat's face as it moved back or "south" had a major impact on the rat's behavior, which becomes most obvious after perhaps ten to twenty-five days.

FIGURE 2.15 main effect for rows

The first and major difference between FIGUREs 2.08 and FIGURE 2.14 can be seen in FIGURE 2.15, which shows the main effect for rows for the FREQULM data set over twenty-five growths. $F(11, 1573) = 11.66$, $P < .000000$. Compare this data with FIGURE 2.09, the data for growth 1. The curves are the same for the first six rows. The differences between these row FIGUREs is most important for it shows very dramatically, how devious behavior evolves over the twenty-five growths. While the first-growth data for runs 7-12 is a simple continuation of the data for the first six rows, on the other hand, the means for rows 7-12 for the 25 run data shows a large increase in the FREQULM data in the second half of the maze. This begins to reveal a significant difference in the animal's behavior, starting at row 7. This is going to be called the row 7 peak. After row 7, the data for the second half of the maze is very different from what was in the one growth data. This increase in the activity in the second half of the maze gives a completely new meaning to devious behavior. The punished rats not only increased the number of nested devious loops they make each run, and as we will see they not only increase their activity in the second half of the maze, but they do so in a very special way! This data suggests that the rats after they are punished for re-entering alpha develop a fear of re-entering alpha, and that fear especially affects the rat's behavior in the rows on the way back, or the rows in the second half of the matrix; specifically, the rat's fear increases the closer they get to the alpha unit. After leaving the omega unit, they start finding ways of avoiding the alpha unit.

Rats that are punished for some component of their exploratory play behavior start generating a new kind of behavior. Devious behavior is designed to meet three needs: first, a behavior that avoids being punished; second, a behavior that allows the rat to run the shortest path, and third, behavior that allows them to maximize their play behavior in the maze.

The new behavior that meets these needs is called devious behavior, and what is amazing about this behavior is how the rat finds a way to design a new set of behavior patterns that fulfills its goals in a modified way.

There are two kinds of test environments: first, the classic or ideal test environment in which the rat can design one skill that can satisfy all its needs simultaneously; and secondly, a far from ideal test environment. How many varieties of these less than ideal test environments exist, I am in no position to speculate, but one such test environment I serendipitously discovered in this experiment, one involving punishment. In these imperfect test environments, the rational mind may have to design, spontaneously, a set of very special behaviors in which a unique pattern of behavior is required for each need.

The rats of this experiment, for example, solved the problems created by the punishment associated with the alpha unit by evolving devious behaviors that evolved in several directions: first, they evolved a new form of behavior, which avoided the alpha unit, which was the punished unit. Second, this new form of behavior was modified by the rat decreasing the number of loops; three, by increasing the number of alternate growths. Fourth, by the redefining the criterion loop; and, fifth, finally the development of stalling behavior which as pure devious play behavior seems to have no purpose other than to delay the time when the rat is removed from the maze.

The fact of punishment complicates our task of understanding the rat's behavior. The rat's problem in the normal test condition is that the rat's original goal was to discover the shortest path through the maze, but the fact of punishment changes everything. Now the rat must define a new complex goal, which includes the notion of maximizing its play- time in the maze. The normal rat always had two goals, but they were confounded and now because of punishment, these two goals each require a different solutions, different behavior patterns.

The effect of punishment is, in effect, to increase the amount of uncertainty in the maze. In the first growth situation, the rat exhausted all the uncertainty in the matrix while exploring in the first half of the matrix on the way out. However, in the punishment situations, suddenly the second half of the matrix now can be used to perfect the new goal of finding how to avoid being removed from the matrix, while it perfects the shortest path under these new test conditions.

FIGURE 2.16 data types by rows

FIGURE 2.16 is the first order interaction, data type by rows, and shows the twelve-row effect for the three FREQULM scores. $F(22, 1573) = 5.0$, $P < .0000$. This FIGURE shows that the avoidance behavior in the back portion of the maze was so pronounced that row 7 was not the same for all three FREQULM scores; that the FREQL score was larger than the FREQU or FREQM data.

Comparing 2.16 with 2.10 shows the difference is due to the the FREQULM data set. Examining the mean curve in these figures, the mean in both figures is relatively comparable. If you examine the FREQU, the activity in the second half is different from what it was in the second half of the first growth. Looking at the Latency curve, it shows a significant contribution of error over the first growth. The peak 7 effect is partially due to the differences between the three kinds of FREQ scores, with the latency component contributing the largest part of the peak.

FIGURE 2.17 blocks by rows

FIGURE 2.17 is the first order interaction, blocks by rows, for the FREQULM data for 24 growths. F (22, 11573) = 5.00, p<. 000000. The 24 growths, 2-25, were divided into three blocks of eight growths each in order to be able to draw a graph that would be readable. This FIGURE shows that as the rats' devious behavior evolved and perfected, they systematically changed the way they spent their move and time allotment of the constraint set over the 24 growths. FIGURE 2.17 shows that in the first six rows the rats decreased their activity primarily between block 1 and 2 and showed little improvement between blocks 2 and 3. However, during each of the successive three blocks in the second set of six rows, the rats increased their amount of activity. As the rats were perfecting their skill of spending less and less of their activity in the first half of the maze, they were systematically spending more and more of their activity in the second half of the maze. This FIGURE shows how the devious avoidance in the second half of the maze evolved over the twenty-four runs 2-25. In general, the order of the rows in the two halves of the maze is reversed. The avoidance behavior in the first half of the maze facilitated the shortest path learning of this half of the maze. However, the avoidance behavior in the second half of the maze facilitated the devious behavior in the second half of the maze.

This FIGURE is full of amazing facts, or perhaps the results can be summarized with one relation. The whole goal of the rat reversed: In block one; the rat spent most of its moves in the first six rows and the fewest of its moves in the last six rows. In block three, the rat spent most of its moves in the last six rows and the least moves in the first six rows. The second block data fits in between these two trends. Compare the order of the lines in the first half of the maze and the second half.

This figure explains most significantly the differences between the one growth data and the 24 growth data. The runs have been divided into the three groups. These groups show the avoidance behavior of the animal towards the alpha unit was like the sailing analogy. The alpha avoidance did not act in one tornado like effect; it acted more like a steady breeze. The cumulative avoidance of the alpha unit produced over these three blocks so that you get a sum in block 1 of 8 trials for the avoidance effect to produce a difference, it is a smaller difference here. If you get to block two, the wind has had twice as long to produce a cumulative effect and you see a larger difference. In block 3, the devious behavior avoidance effect has produced a significant data effect.

Actually the simplest way to say what happened is this: In the beginning the rat did almost all its moves exploring in the first half of the matrix, but by the end the rat was spending almost all moves exploring in the second half of the matrix. Perhaps an even simpler summary would be this: Over the twenty-five growths the rat shifted its activity from the first half of the matrix to the second half; it is as if the rat were trying to find a way to spend its time as far from the alpha unit as was physically possible.

The alpha unit where the punishment occurred, in effect, had become a feared spot; a rational goal was to spend as much time and motion as possible, as far away from this unit as physically possible.

FIGURE 2.18 directions by units

FIGURE 2.18 shows the first order interaction, directions by units, for the FREQULM data set for 25 growths. Yes, this 25 unit data includes even growth one. F (24, 3600) = 3.477, p< .000000. This data supplements the antigravity row effect seen in FIGURE 2.16. This FIGURE also shows the systematic improvement in the avoidance behavior that occurred over the 25 growths that were distributed over the 50 units of the maze 25 units out and 25 units back. Again, the data shows a gradual decrease in the activity in the first half of the maze, while the activity in the second half of the maze systematically increased. In the first nine growths, the out scores were smaller than the back scores, but after growth 10, the reverse was true. In addition, FIGURE 2.18 shows, as FIGURE 2.17 did not, the cyclical nature of the rats' column behavior as these learning changes occurred. These cycles are the column effect within rows. However, this effect confirms the antigravity nature of fear.

FIGURE 2.19 blocks by directions by columns

FIGURE 2.19 shows the second order interaction, blocks by directions by columns, for the FREQULM data set for 25 growths. F (6, 429) = 8.74, p< .000000. The U or bucket shape for the column effect again held up over 25 growths. The column effect is powerful and doesn't go away after the first growth. This data again shows, as did the data of FIGUREs 2.15 and 2.16 that as the performance improved in the first half of the maze; it decreased in the second half of the maze. This again shows the antigravity nature of fear. On the way out, the primary improvement was between block one and block two, but on the way back there was deterioration between 1 and 2, and between 2 and 3. The direction of the changes between the two directions is the most interesting feature in this FIGURE.

The BFEULM (Between First Entry Unit, Latency, Mean) analysis

FIGURE 2.20 blocks by rows

FIGURE 2.20 plots the BFEULM data, which tell a completely different story than does the FEULL data. FIGURE 2.20 is the 2-way interaction blocks by rows for the BFEULM data. F (22, 1584) =2.89; p< .0000. This FIGURE is another kind of growth curve and shows how the BFEULM behavior, the retracing component of exploratory-play behavior of the rats was emitted and evolved over the twenty-four growths. Please note that this data was the first 24, 1-24, of the twenty-five growths, not the 2-25 set analyzed as was previously described. However, these twenty-four growths were also analyzed as three blocks of eight trials or growths (3 * 8 = 24). The growth of the FREQULM was noted above in FIGUREs 2.17, 2.18, and 2.19. The growth of the BFEULM behavior is different from the growth of the FREQULM data.

The BFEULM EP data is a measure of retracing, particularly retracing loop type 3, which is the response of rats to the massive uncertainty induced by anomalies. The type-three retracing loop represents a period of accommodation during which the rat is engaged in the nonverbal equivalent of what Kuhn (1962) calls the revolutionary stage of science. It is during this retracing period that rats create new paradigms in response to anomalies, as do human scientists, and during retracing rats made major revisions in their model or paradigm of the matrix maze domain. In the first experiment, each new level of the hierarchy induced a new round of retracing as the rat experienced the new anomaly of each new order of the hierarchically organized maze. In this study, however, the constraint set seems to have had the effect of somehow preventing the rats from encountering the uncertainty one grade at a time, or one order of the hierarchy at a time, thus producing a series of curvilinear retracing loops. Instead the rats in this study, suddenly in row 5 on the first growth seemed to have, all at once, become aware of the whole of the uncertainty in the all the lower levels of the several hierarchies prior to row 5. Thus, the features of row-five induced one massive dose of uncertainty, which induced one massive retrace loop. This massive period of retracing induced by row 5 can be seen in FIGURE 2.20 as the spike in block one at row 5. Actually this spike was entirely due to growth one. There was so little retracing in the next seven of the eight growths in block one that growth one dominated this curve. FIGURE 2.20 shows that the massive period of retracing seen in block one was never seen again. The behaviorists, who believed that behavior was a simple reflex phenomenon, had no idea that retracing was when the major advances in adaptation occurred. On the contrary, they always considered retracing a curse that experimentalists must endure if they would be rat runners, but it has become the practice of some rat runners to minimize retracing by using non-retrace doors. This retracing the behaviorists proposed as proof of the existence of random behavior, which justified the concept of trial and error learning.

FIGURE 2.20, however, shows that the behaviorists' entire conceptual scheme was a myth. The rat data in FIGURE 2.20 shows that rats clearly stop retracing after block one, actually growth one, and they stop retracing in both the first and second experiments without ever finding reinforcement in the goal box. Rats stop retracing, not as the behaviorist claimed because they are punished, in effect, for retracing by the delay retracing induces in their finding of food or the reinforcement. However, this study shows that the reason rats stopped retracing had nothing to do with food, neither the presence nor the absence of food. Rats stopped retracing when they had created a valid model for the hierarchically organized domain, one that removed all the uncertainty from the maze, and thus there was no longer any rational mental need for retracing. Retracing behavior is the phase during which rats' create the model that takes the uncertainty out of the anomalies of the maze. During retracing, the abductive behavior of the rat discovers the causal structure of the maze. The complex pattern induced by abduction is called Heterostatic behavior, which represents the third wisdom, the wisdom of the primary mind. The Heterostatic behavior reduces the uncertainty induced by the anomalous events of the domain. Heterostatic behavior reduces uncertainty needs, rather than any physiological or homeostatic need, and they make these changes in the absence of Homeostatic need reducing events or reinforcement.

Kinds of Uncertainty

It is necessary to distinguish two kinds uncertainty: first, between the primary or nonverbal uncertainty and; secondly, the secondary or verbal uncertainty. The classical information theorists define information as that which reduces uncertainty. However, this concept of information and uncertainty only works in the formal stage of science with a closed alphabet, and thus does not work in the revolutionary pre formal stage of science induced by an anomaly, in either the primary generic stage or the formal secondary institutionalized stage. Secondary uncertainty is a weak if not feigned kind of uncertainty, for any uncertainty that can be measured is clearly not a full blown or robust kind of uncertainty. Both primary knowledge and primary uncertainty belong to the primary generic stage of science, while secondary information and secondary uncertainty belong to the secondary normal stage of formal science.

The nonverbal form of primary uncertainty comes in two kinds, negative and positive. The former negative kind is an excessive form of uncertainty induced by an anomaly. The positive form of primary uncertainty is the degree of uncertainty, which organisms experience as not excessive, as neither too much nor too little, but as just the right or desirable degree of uncertainty. This positive degree of uncertainty induces the optimal degree of arousal on the Yerkes-Dodson law. Primary uncertainty, in other words, comes in several types, but only the positive and negative forms will be distinguished at this place in the story.

Abductive logic

The implications of the nonverbal mind are too many and too broad to all be covered in one book. The West has been wearing Stratton Glasses for over two thousand years, and perhaps it is unreasonable to expect that the West would go "cold turkey," as they say, and take the glasses off completely at one time and place.

This discussion will employ several concepts unique to Rat-runners, plus several terms unique to Kuhn, and Aristotle.

The "man-function" types of logic

However, I have been trying to avoid another topic but now it is obvious that is not possible. The general topic is one of the many topics that Aristotle (1952) first developed: the topic of logic. Aristotle distinguished two kinds, which I briefly discussed in Book 1, Chapter 5: deduction and induction. These two classic types of logic have only a limited role to play, even in human behavior, for essentially they only work in closed verbal domains. In effect, these two classical types of logic support what Plato (1952) called the "man-functions," and only qualify, therefore, as secondary forms of logic.

So what is the primary kind of logic? Charles Sanders Peirce (1878) identified a third type of logic he called abduction, whose function was to furnish the logic of human verbal hypothesizing. Peirce as I have noted before was a first generation pragmatist, while I am a third generation one. Peirce was concerned with the meaning of words--the verbal mind's logical capacity, which he claimed were ultimately defined by action. As a third generation pragmatist, however, I am concerned with bigger questions: not just the meaning of words, but the meaning of nonverbal images, but most particularly about the origin of ideas or the conceptual elements out of which the model of the maze, for example, created by F3's nonverbal mind was constructed. In her model of the maze, I contend there are four causal conceptual elements, A2, B3, C2, and D8. In my discussion of F3's behavior, I devoted a considerable amount of time explaining how these four ideas emerged i.e. how abduction worked as she solved the puzzle of the maze.

The "god-function" logic

The primary difference between Peirce (1878) and me is that he is concerned only with the possible forms of human verbal logic, while I am concerned with the possible forms of the nonverbal mind's logic. In both deduction and induction, the human verbal mind executes logic whose moves are formally restricted to a closed set of forms, which legally may be found in say Webster's dictionary. By contrast, imagine how different what is happening in F3's mind when she is exploring the maze. Remember, she had no pool of concepts from which to appeal. Her only restriction is that the conceptual structure she employs be original with her, and they advance her ability to run the maze

Female 3's exploratory behavior operationally defines the pragmatic behavior, which best characterizes abduction. F3 was placed in a very anomalous domain, very open, or totally without any preconceived form: one as full of ignorance or uncertainty as could be conceived. Yet in less than 30 minutes, her nonverbal mind abductively created a closed set of four hierarchically organized ideas, which reduced or destroyed all the uncertainty within the domain.

Adult humans engage in verbal logic behavior with a vast arsenal of verbal forms from which they can draw. This meme population is their logical tool kit. By contrast, I can't imagine how a female rat upon having just been inserted into the anomalous matrix must feel, first because she has no comparable population of nonverbal memes in her toolkit. All the forms she employs must first be personally created.

My thesis says that mammals are self-organizing systems. It is then appropriate to ask the next question in the form of a riddle. Who is the more self-organizing, an adult rat, or an adult human? Just keep in mind that a rat is, in every way possible, on its own in the hierarchically organized maze.

It is clear to me that F3 offers us a most powerful metaphor for understanding the behavior, which authors historically refer to as god-function behavior. It may seem strange, but of the four hierarchical levels of behavior, the innermost level is the most useful for talking about abduction. The innermost level of the four hierarchical levels is the node-path level. A natural node is a unit where the physical structure of a space forces the organism to turn around.

Nodes as anomalies

The emergence of the nonverbal mind changed everything. One of the major insights required by my theory of Heterostasis is that there are two kinds of nodes, the physically natural nodes such as the innermost or second level units of all the Blind paths, which are physically natural nodes. When an organism enters one of these units, there is only one response possible - turn around and leave.

Another item that makes the discovery of the nonverbal mind exciting is that now one of the major kinds of nodes can be recognized as a mentally not a physically induced node. This started with my discovery of the dependent variable BFEULM, which measures retracing; which objectively measure the uncertainty induced by an anomaly. To get to the point, none of the three units, which triggered F3's retracing were natural anomalies or natural nodes.

Choice Points, E1's, as nodes

Just as natural nodes are physical nodes, just so mental modes are what rat runners have long called "choice point" units, which usually in the matrix maze is the E1 unit, the Functional Unit Names, FUN as defined in FIGURE1.2.8 where the SRS-functional units were defined. For the male rats, all the units in the maze that were anomalous and induced retracing were E1 units, which are the unit a rat first encounters upon entering a row. Some of the females also found E1 as an anomalous unit.

Generalizing

While we are talking about the human verbal types of logic, the two human closed types, deduction and induction, and the one nonverbal or open creative type, it may be appropriate to take a moment and consider the implication of these differences.

Again, let me repeat this difference. In the case of human logic, employing deduction and induction, the human logician must conceive both types of forms deduction, and induction employing only words that are memes. The elements, the words, which can be used to generate any particular logic form, are all given as memes.

By contrast, when F3 uses abduction in the maze, the whole point of her logical behavior is to create *de novo* mental forms, which had never before been conceived. The conceptual structure, which her nonverbal mind creates, the A2, B3, C2, D8 conceptual system is a brand new mental creation, which probably never before existed. This abduction system creates forms that in some future human mind may be transformed into memes, as it was in my case, as an experimenter observing F3's behavior, but nothing is more important than understanding where Ideas come from. The nonverbal mind has the ultimate power, which I keep referring to as the God Function, for the facts are that upon entering the maze these four ideas did not exist. But 30 minutes later, F3's mind by employing abductive logic has created these four ideas by interacting nonverbally with the maze.

Life as a Heterostatic Challenge

There are two kinds of nodes: first, ordinary man-function nodes and second, the god-function nodes. This latter class of nodes can also be called the "Make or Break" nodes.

Every success and every failure can be conceived of as a node or as a block or barrier; the road up ahead is closed.

F3 was able to break through every anomaly, or we could conceive of every anomaly she encountered in the maze as a temporary node. Success is breaking through a conceptual node. A conceptual node as an anomaly induces a quantity of uncertainty, which induces retracing. If a mammal encounters an anomaly for which it cannot destroy the uncertainty it induces, then it could never resume forward motion, it could never explore past the node or obstacle. So, that node would sorrowfully become the high water mark in its life, its waterloo. While we are speculating in broad terms, then every life can be characterized by two boundaries, a low, and a high. In AA, for example, one of their core concepts is called "hitting bottom." By contrast, the high point in each field of endeavor is a node, which could be characterized as hitting our High.

The central concept of heterostasis is that the mind makes every individual mammal a self-organizing system, yet circumstances make for individual differences. The limit for each individual in every domain is simply the encounter of a node, which induces a quantity of uncertainty, which the individual cannot overcome. On the other hand, success in heterostatic terms is achieved by destroying the uncertainty, which temporally obstructs progress. The goal of life is growth, continued mastery of the uncertainties of life.

FIGURE 2.21A linear plot block by row by data type

Both FIGUREs 2.21A and B plot exactly the same set of data; the only difference us that FIGURE2.21A is a log plot, while FIGURE2.21B is a linear plot. In the first graph, it is obvious that the only difference observable in the data is the spike at Rows 4-6 in Block one. If a scale is picked that is appropriate for plotting the large values of the spike, then all the other scores are relatively so small, as to appear not as a straight line but so close to straight that on a linear graph, the data seems uneventful as seen in FIGURE2.21B.

Slowly the story of the rat's behavior is emerging so that humans can discover and formalize the entire of set of rational rules the rational nonverbal mind employs.

What motivates the rational nonverbal mind

- First, the primary purpose of this book is to focus upon Heterostasis, which is the main force that keeps the nonverbal mind awake and alert and in the B-region of the Y-D law. The primary feature of the spatial world that activates the Heterostatic regions in the mind are empirical anomalies, which induce retracing, which is measured by BFEULM scores. FIGURE2.21A shows that the rats exhausted all the anomalies on runs 4-6, which account for the spike, and then it shows that the rats keep running for 25 runs or for an additional 19 runs but there were no anomalies of the kind that induce retracing. So what is this missing motivating force?

As noted above the rat runs the maze with at least two or more goals in mind: First to discover the shortest path, and second to maximize their exploratory playtime within the constraints of the domain.

FIGURE 2.21B log plot block by row by data type

FIGUREs 2.21A and B are two different views of the 3- way interaction, block by row by data type, for the BFEULM data set. $F (44, 1584) = 2.35$; $p < .0000$. FIGURE 2.21A is a linear plot of the data and shows that an enormous spike of retracing occurred at row five of block 1 that dwarfed all the other data. However, FIGURE 2.21B is a log plot of the data that allows all the data to be seen. This data is an extension of the data of FIGURE 2.20 except the data type variable has been added.

FIGUREs 2.22A and B block by columns by data types, linear and log

FIGUREs 2.22A and B are two views of the 3-way interaction, block by columns by data type for the BFEULM data set. $F (12, 432) = 2.47$; $p < .0040$. FIGURE 2.22 A is the linear view, while FIGURE 2.22 B is the log view. The linear-view shows once again how the block-one data dwarfs all the other column data. FIGURE 2.22 B shows the column effect for all three blocks of data. This is another type of learning curve, is the column equivalent of the row effect seen in FIGURE 2.21, and shows that Heterostatic growth behavior can be seen in both the row and column effect. This shows the importance of the matrix two-dimensional description of the data.

These graphs and others show that the nonverbal mind of the rat is a complex neuronal system with the complexity revealing itself, as we shall see more clearly in the fourth experiment, as a function of the number of goals a nonverbal mind can work or play to achieve at any one time.

The still invisible behavioral goal is to maximize playtime in the maze or to maximize the time out of the home cage. Normally both of these goals can be achieved at the same time and as a function of the same behavior. However, the next remarkable discovery is that punishment forces these different goals to be independent, so that a different response is produced.

This will all become clear in the next chapter on devious behavior.

The full maze analysis

The full maze behavior will be described in the next experiment.

Discussion

Four implications of devious behavior

First, devious behavior is a revolutionary new component of rodent behavior.

Second, devious behavior makes a strong case that rodents have a rational mind.

Third, the rational mind does not exhibit the pure Platonic type of non-emotional behavior. On the contrary, the self or the ego is center stage in rational nonverbal behavior.

Fourth, in other words, devious behavior shows how punishment completely alters the nature of rational behavior: how rodents apparently become fearful or angry when punished, how they rationally alter their behavior when fearful or angry.

Individual Differences, Bonsai behavior

The major difference between the TTP (Trial Testing Procedure) and the LTP (Loop Testing Procedure) is that in the former with its harsh and complex set of constraints, especially the extreme homeostatic motivational constraint, experimenters were able to reduce the rat to the lowest common denominator of purpose. Each rat was brought, in Maslow's (1974) terms, to the absolute base of the motivational triangle. By reducing every subject to this base physiological state, effectively all the higher forms of motivation in the triangle, which might produce individual differences, were eliminated. The TTP was a most effective straightjacket keeping each organism in its most primal, elemental, and single purpose state. This made the smooth classic learning curve possible and made the statistical group mean a reliable and valid estimate of behavior, and it was in this ability of the TTP to produce uniform behavior that the Behaviorists took such great pride. It was this induced physiological motivational uniformity, which made the mindlessness animal metaphor of Thorndike (1911) work. The behaviorist's error in this case was their generalization that all animal behavior was machine-like, as if this induced abnormal behavior was somehow a representative sample of animal behavior. This study shows that while the TTP can reduce rats' behavior to this standard form this does not mean that the rats' natural spontaneously adaptive behavior normally takes this simple form. While bonsai trees demonstrate that even trees can be made to assume strange, stunted, and grotesque forms that some humans find interesting, or even beautiful, only the behaviorists could mistake bonsai behavior for normal behavior. Thorndike's learning curve is a bonsai form of animal behavior, which the behaviorists have sold for far too many years as normal animal learned behavior.

However, with the LTP the rats' physiological tissue-need state is not manipulated in any way. The behavior of the rat in the maze tested in the LTP is the natural spontaneous behavior of the rat. The Heterostatic LTP generates natural behavior that is full sized behavior.

The statistical analysis of purposive vs. multi-purposive behavior

This study has shown that the spontaneous behavior of the rat is complexly motivated, but these complex goal patterns are not normally seen in the TTP. It was our good fortune in this experiment to build a conflict into the design of these experiments that luckily forced the rats to reveal both of their goals by making these purposes mutually exclusive. The rats' first purpose is to run the shortest path. The other purpose is to play. Normally these two goals are compatible, but when, as in this experiment, rats are punished, these two purposes became mutually exclusive, and rats evolve two different learning curves, one for each of their now mutually exclusive purposes.

This experiment made it clear that the exploratory-play behavior of the rat becomes manifestly multi-purposive when the rat is punished. On any one trial, the punished rat's behavior may reflect a single purpose, but from trial to trial, the rat can shift its purpose between two or more mutually exclusive purposes. The consequence of this shift is that group means are meaningless. For example, when the rats' purpose is to maximize play behavior on any one trial, they produce large run, step, and latency scores. When their purpose is to produce the shortest path, they produce small run, step, and latency scores. If all rats were synchronized on their alternation between these two purposes then the group means would be useful. However, the fact is that their behavior is not so synchronized and thus, these two purposes produce behavioral effects, which cancel each other out in the group mean curves. One of the paradoxical virtues of the TTP was the fact, as noted above it acts as a straightjacket reducing the rats' behavior to the single purpose that the classical model of statistical science required. In this study, group means were used to illustrate their uselessness in describing multi-purposive behavior generated by the LTP. It was shown that individual curves must be used when multi-purposive behavior is being described unless some mean can be found to synchronize these purposes in all rats.

Multi-purposive Behavior; the Effect of Punishment on Exploratory-Play Behavior

The effect of the conflict upon exploratory-play behavior induced by punishment can be seen in many ways in the data. The twenty-five run data reveals how rats deal with conflicts within their purposive systems. The female 11 data of FIGURE 2.02 introduced the important first part of the complexity of the rat's purposive system. There are two conflicting trends obvious in this data. The rat's first purpose is to create a valid cognitive model for producing the shortest path form of behavior. The second purpose is to maximize its playtime in the maze. These two purposes are, in a punished rat, mutually exclusive purposes and thus the punished rat must choose on any day, which of its purposes it will pursue on that day. On some days, the rat improves its skill on the shortest path behavior (days 7, 10, 18, and 23). On other days, the rat improves its skill to maximize its playtime. The conflict arises from the fact that, on the one hand, on all those days the rat improves its mapping behavioral skills by taking the shortest path through the maze; it is punished by being removed from the maze. Thus as the rat perfects its mapping skill, it loses proportionally more and more play-time in the maze, since there is an inverse ratio between the rat's expertise at making a complete loop and its playtime in the maze. On the other hand, if the rat maximizes its playtime it must forsake its primary goal of running the shortest path. It is clear that the rat, on some days, does resolve the conflict simply by forsaking some one of its goals. On days the rats elect to run the shortest path, they forfeit all the playtime that they could have had. This is time over and above the moves and time required by the shortest-path goal, but thanks to the rats' capacity for evolving devious behavior, on the majority of the days, the rat learns partially to satisfy both goals. These days produce the devious behavior pattern.

The rat by redefining its criterion loops was able, on the days noted above, partially to achieve both of its goals on the same day and thus resolve the conflict between these two purposes. Another part of the puzzle is the fact that on some days the rat did choose one of its goals and forsook the other. Why did the rat do this? The type-one multiple-loop must fulfill a special cognitive need of the rat so that the exercise of this option is worth the price required in terms of the loss of play moves and time.

The discovery that our three-element constraint set had serendipitously included an element that constituted a punishing contingency was a major and pleasant surprise. However, for the first group of four rats, the punishment behavior was misunderstood as a form of extinction, and these rats were stopped after only seven runs, before the full sequence of the punishment phenomena had a chance to run its course. The second group of three rats was run for twenty- five growths, which allowed enough time for the rats to evolve fully the two patterns of behaviors that were their response to punishment effect.

The role of the self in Heterostasis

The effect of punishment upon exploratory-play behavior is complex, and the complexity seems to result from the fact that punishment induces a conflict into the purposive systems of the rat. The rat can be conceived as having several selves, if we may use the language of James (1950). The mapping-self's purpose is to perfect its cognitive map of the Lashley III maze in manifest maze exploratory behavior, which attempts to run through the maze with the shortest path. The excellence of the rat's manifest exploratory behavior is an expression of the validity of its map. When the map is valid, the rat explores new units of the matrix. When the map is invalid, the rat retraces. The rate of exploration is measured by the emergent-loop, which is the best behavioral index of the growing validity of the map.

The rat's play-self, however, has the purpose of maximizing the rat's exploratory-play behavior, maximizing both the rat's daily moves and time quota in the maze. The conflict between these two selves under this three-element constraint set arises because the respective behavior generated by these two selves seems, at first glance, to be mutually exclusive events. The mapping-self which generates behavior that makes a brilliant short complete loop in say 70 moves and 70 seconds has selfishly cheated the play-self out of hundreds of possible moves and hundreds of possible seconds of play.

There is, however, much more to this picture than is apparent at first glance. The fact that makes the situation interesting is this: the primary rational mind of the rat is capable of redefining the purpose of its mapping-self. Logically there are eight possible combinations of multiple loops, but only the first combination is punished. The rat's mapping self has the option, at least on some days, to settle for perfecting some one or several of these seven multiple loop types that are not punished. The advantage of this option is that the mapping self and the play self are no longer mutually exclusive.

The behavior of the rats run for twenty-five growths shows that on 85% of the days these rats were content to avoid the type one multiple loop and thus avoid the conflict between the two selves. On these two-self days, the rats imperfectly perfected both selves. First, it satisfied its mapping-self's cognitive needs by perfecting its ability to make more than one of some one of the seven multiple growths that are not punished. Second, it satisfied its play self's needs, by perfecting its ability to make these devious loops which allowed it to enjoy more and more time each day in the maze. The important point to be clarified is this: The nature of the purpose of the exploratory-play self's behavior is not some frivolous time and move-wasting behavior. The goals of the play-self and the mapping-self normally are or should be the goals of the same self. The normally unified self is only forced to split into two competing selves when some element in the constraint set is punished, which generates a conflict within the self. The punished contingency, however, forced the rat to reveal cognitive skills that otherwise would never have been seen.

The measurement of deviant behavior

Punishment produces avoidance or devious behavior, which can be measured in different ways at different levels of the hierarchically organized behavior. First, at the loop level, devious behavior is seen in the decrease in the number of loops made per run and an increase in the number of growths per run. Second, at the growth level, devious behavior is seen in the shifts in the rats' activity in the direction, row, and column variables, with the punished rat decreasing its activity in the first half of the matrix and increasing its activity in the second half of the matrix. Third, at the run level, devious behavior is seen in the conflict that develops between the rats' purposes; how the rats periodically alternate between purposes from day to day.

Chapter 3 A Belated Control Group for Devious Behavior

D87 Experiment III - Background

Heterostasis, Devious Behavior: The effect of punishment on the exploratory-play behavior of rats

In this series of experiments essentially every aspect of the behavior described in this book for the first time has been anomalous; and every concept evolved to describe these behaviors has been anomalous. The sum of this series of anomalies and concepts is a new kind of behavior, a new kind of wisdom. Heterostatic behavior of the rat is the manifestation of the third wisdom, the wisdom of the primary mind, or the wisdom of the generic nonverbal mammalian mind.

The Three Purposes:

The first purpose

The first purpose was to run a belated control group that would allow a belated formal statistical test of the hypothesis that devious behavior is the result of punishment. Since, in the last experiment when devious behavior was discovered for the first time, no control group had been planned and thus it was impossible to determine, which of the features of exploratory-play behavior observed in the experimental group were due to punishment.

The second purpose

The second purpose of this experiment is to run this control group for 25 growths, the same, as were the rats of the experimental group in the second experiment. Of course, the behavior of the control group will not be punished with the Alpha-Omega-Alpha, AOA constraint. This control group will make it possible to determine which of the several 25 growth effects of the experimental group were caused by the punishment in the second experiment and which parts due to play.

The fact that no control group was originally designed as an element of the second experiment makes an interesting point about the difference between normal and revolutionary science; only in normal science can a proper formal experiment be designed *a priori*, one with a proper control group. When an anomalous phenomenon is observed for the first time in an experiment, in this case the phenomenon of punishment, obviously it would have been impossible that a valid control group could have been included in the original design. The behavior of the rats in both phases of the second study was full of anomalies all of which I attributed to the effect of punishment, but since there was no non-punished control group, the experimental proof for this attribution was lacking. The anomalous behavior I attributed to punishment and called devious behavior was a complex new form of behavior and had the following attributes.

The attributes of devious behavior

When rats' exploratory-play behavior is punished, they do not passively accept the punishment by simply avoiding the stimulus with which fear is associated. On the contrary, the punished rat actively modifies its behavior in several ways in order to minimize the effect of the punishment on its spontaneous exploratory-play behavior. So far, I have been able to show that this devious behavior can be seen at three of the four levels of the hierarchical structure of behavior: at the run, loop, and growth levels. Yet, devious behavior may also occur at the fourth or innermost node-path level, but I have not yet devised a way to measure the possible changes at this level.

- First, at the run level, the rat achieved a behavioral conflict resolution skill by evolving two different purposes and shifting periodically between them, by either avoiding the alpha unit by redefining the nature of the criterion loop or simply occasionally running the conventional shortest path.
- Secondly, at the loop level of the hierarchy, the experimental group of rats modified their loop-behavior by decreasing their mean number of loops per run from 30 to 1. Meanwhile they increased the mean length of their loops from 20 to 600 moves, which allowed them to start making several nested growths per loop up to six complete nested loops per loop.
- Third, at the growth level, the experimental group of rats shifted their behavior in complex ways between the direction, row, and column variables and in different ways for each of the three sets of dependent variables. The most dramatic effect of punishment was the conflict it induced in the rats' purposive system at the run level. The particular manifestation of devious behavior was revealed as the rats evolved two separate skills; they alternated between running the shortest path and then secondly maximizing their play times and play moves in the maze.

The third purpose

The third purpose was to explain the anomalies of the Blind Path, BP behavior. Perhaps, a most interesting and complex purpose of this experiment was to study the Blind Path, BP, behaviors. The rat's full maze behavior consists of two sub elements, the Essential path EP and the BP element. Thus, theoretically, three kinds of growths or emergent loops can be analyzed. First, the growth in the essential path, EP, that was described in the first experiment, secondly, the growth in the full-path, FP, which is a mixture of the EP, and the blind path, BP, and thirdly, the pure BP growth, which it will be possible to study in this experiment for the first time.

The blind path behavior of rats, I will show, contains a number of anomalous properties, or surprises. First, Blind entries are rare events contrary to the classic behaviorists' position that claimed otherwise. If initially rats' behavior in the maze and in all parts of the maze is random, then rats should make a random number of Blind Path errors. Blind Path errors, however, are neither random nor plentiful. On the contrary, they are scarce events. In fact, some kinds of Blind Path errors almost never occur and this fact has a surprising consequence namely that Full-Maze Growths, FMG, and Blind-Path Growths, BPG, are rare events. Why is that? Because these non-deprived non-reinforced rats quickly learn during spontaneous play to avoid certain kinds of Blind Path units, which has an interesting consequence: It takes many EP Growths, before a rat will make one FP Growth, which means it will take many EP growths before they will make one complete BP growth. Both FM and BP growths, by definition, are not complete until all their cells, one hundred cells for the FM growth of the matrix maze, have been entered at least once. Secondly, there are different kinds of blind entries, in particular the Forward, F, type, and the Alternate, A, type. Third, BP behavior is a unique kind of behavior, which requires its own skill, which is the unique expression of its own set of laws. What this section on the rats BP behavior shows is that: rats make two relatively independent models of the matrix maze, one for the EP and another for the BP. In other words, the whole of the spontaneous behavior of rodents in the matrix maze consists of two partially independent behavioral skills, conceptual laws, images, or maps. The two sub-map elements are domain specific knowledge, laws, or behaviors, which now require the recognition that even this level of domain-specific knowledge cannot be the smallest unit of knowledge. Domains themselves must be understood as divided or subdivided into sub-units for which the primary mind must also evolve relatively independent knowledge-systems. Therefore, in the matrix maze, rats make a model of the whole FP by evolving two subsystem models, a model for the EP and another for the BP. Each of these models are required because each of these two sub-spaces of the FP domain present different problems for which the rat must evolve different rule systems, different behavioral skills, or different maps, different scripts, etc. The primary mind must evolve a different special purpose behavior for each sub-domain of the FP domain; ones that are appropriate for each element of the matrix subspace, the EP and the BP.

Each subspace-domain problem is, first, a special ontological-empirical problem. While the general problem of the primary mind is to create a conceptual representation system for the whole matrix space, yet this larger purpose can only be achieved in parts. For each of these two sub-elements of the matrix space must be differentiated because each consists of different classes of units, with a different kind of causal hierarchical structure, each of which can only be described by a unique sub-domain set of laws, and each set of laws requires a unique set of responses.

Each of the rows of the matrix maze consists of eight column elements evenly divided into two halves, four EP elements, and four BP elements. The primary task the primary-mind evolved to solve was to create domain-specific maps or domain-specific causal conceptual systems. The primary task of the map-making function of the primary-mind is to invent domain-specific causal systems of conceptual elements that can take the uncertainty out of some domain-specific empirical set of patterns of environmental energies that are experienced as independent domain-specific sets of experiences as the rodent spontaneously explores the whole hierarchically organized domain. The senses present the primary mind with complex but unique domain-specific patterns of strangeness and the primary mind was designed to give meaning to each unique syndrome of elements by evolving unique hierarchically organized scripts for each domain.

One of my primary tasks as a rat runner is to challenge Plato's (1952) original myths about the nature of the human mind, but especially to challenge his notion about the nature of an idea, which he claimed, was also always an ideal. My task is to show that nonverbal ideas start out as very concrete far-from-perfect ideals. The primary mind of the rat does all kinds of things Plato could never have imagined things, which he might have found offensive. However, the primary mind of the rat conceives ideas as nonverbal, not verbal, but moreover these nonverbal domain-specific ideas are rationally conceived. The concept of a rational nonverbal domain-specific idea would no doubt have been strange and foreign to Plato.

The domain-specific concrete map that the primary mind evolves to take the uncertainty out of just the EP (I will address the larger picture later) consists of four ontological elements, one for each of four spaces, E1-E4 for each row. While just this domain-specific sub-map is able to take the uncertainty out of the EP, it is, of course, useless in reducing the uncertainty induced by the domain-specific elements of the BP. The spatial units of the blind path consist of different classes of objects from the EP elements, and these BP elements are related variables that are unique for the BP. Thus, in order for the rat's primary mind to understand the causal structure of the hierarchically organized elements of the BP, the primary mind must evolve a completely different set of causal conceptual forms. Forms, which consist of a set of sub-domain specific ideas, which are as far from the kind of idea that Plato (1952) had in mind as east is from west. The primary mind, in other words, must not only evolve a distinct map for recognizing the four elements of the BP, but the primary mind must evolve a domain-specific causal conceptual system. A system which is unique for the blind path, and more specifically one or ones that can distinguish two classes of blind units, the Alternation units, A and the Forward units, F subset of spaces, A1-A2 and F1-F2. On the other hand, in the EP the primary mind also had to evolve a domain-specific causal conceptual system that can map the sensory potential knowledge from the four elements of the EP, E1—E4, into images of four elements of the nonverbal consciousness and into four classes of behaviors. In other words, to be very concrete, in order for the primary mind to solve the behavioral problem of the FM, the primary mind must parse this FM problem into two sub-problems: the problem of the EP, and the problem of the BP. In turn, in order to solve the BP problem, the primary mind must conceive a set of sub-maps for the four elements of the BP, which must map the potential knowledge from the four elements: A1-A2 and F1-F2 into coherent sets of knowledge.

Furthermore, the rats' primary nonverbal rational mind must generate two different maps in order to predict the distribution of the two different classes of spatial objects within the matrix domain, an EP sub-map of the domain and a BP sub-map of the domain, but the latter, itself has two sub-elements. The causal principles which predict the distribution of the probability space for the two subsets of spaces are very different cause-effect rules, and the meaning of these two ontological subsets of spatial beings are very different for rodents at different times in its exploration of the 100 element domain. Each of the two BP sub-maps must both predict the distribution of two subsets of elements and give meaning to these two subsets. The meaning of the EP subset and the BP subset are both equally exciting when the rat is first exploring the matrix. In other words, any new class of space that must be represented is initially exciting until a valid representation has evolved which can take the uncertainty out of the class of events that define the domain. This model must convert this unique precept stream generated by each space into a class of recognizable, predictable sets of ontological beings. Then the second stage of map making begins. The larger question then becomes, where and how does each of the classes of ontological beings that are the elements of this whole domain fit into the species-specific set of purposes that the exploring rodent brings into the domain. The rat has two purposes; first, to evolve a map will allow the rat to traverse the domain with the shortest path; secondly, to maintain consciousness- to generate the specific degree of novelty that will maintain the optimal degree of arousal- in short, its second purpose is to play.

Therefore, in this second stage, the heterostatic process of inventing maps that take the uncertainty out of the strangeness imbedded in the hierarchically organized classes of physical spaces in the domain must now evolve. The primary mind must possess the requisite intelligence for classifying the total population of recognizable spaces into subsets that are relevant to each of its spontaneous goals. The first question, of course, is how to parse or classify the total set of classes of ontological beings that have been empirically experienced in the domain into those that are relevant, and those that are irrelevant for each purpose. For example, which of the 100 classes of spaces are essential for its shortest path purpose, and which are irrelevant for that purpose. The EP elements are the former, while the BP elements are the latter.

This raises two major questions that it is now appropriate to discuss at this point in the book. First, how many sets of conceptual elements can the primary mind of the rat evolve? In my first attempt to understand the conceptual systems evolved by the primary mind, I only considered the possibility of the evolution of the salient conceptual elements. However, the study of the full maze and particularly the blind path behavior forced me to recognize that the primary mind must evolve two sets of conceptual elements. First, the primary mind must evolve a set of salient conceptual elements for the EP, which is the element of the matrix through which the rat must run while executing it shortest-path purpose. Secondly, the BP behavior forced me to recognize that the primary mind must also evolve another independent conceptual system to program its behavior in the BP elements of the matrix maze, which are the units OUTSIDE the EP elements.

Secondly, what kinds of reflexes are there? Heterostatic learning must be understood as mental and behavioral activity designed to deal with primary rather than the secondary reflexes. For one hundred years, learning has been conceived as a function of reflexes that Pavlov (1960) called the unconditioned and the conditioned (the US and the CS). However, Heterostatic learning is primarily motivated by reflexes which Pavlov called the Orienting Reflex, OR, and the Freedom Reflexes, FR. Pavlov recognized four reflexes, the US, CS, OR, and FR. The Behaviorists have tried to explain learning only in terms of the first two reflexes. However, as I was to discover, these latter two reflexes are the primary reflexes. The former two are the secondary or lower priority reflexes. Heterostasis growth is mediated by the primary reflexes.

Pavlov (1960) was an original scientist, who was not just seeing with someone else's concepts or Kantian molds. (Kant, 1952) For example, Pavlov tested his dog by first doing surgery to externalize one of its salivary glands. He would then put some tubing over the fistula to catch the saliva. To keep the dog from tearing out Pavlov's implant he had to constrain it in a stock. Initially, the dog tried to escape from the stock.

This reflex to get free, Pavlov called the Freedom Reflex. Pavlov found he could not condition the dog while the FR was active.

The OR was like a startle response and was triggered by a surprising event. The OR was the master reflex; it suppressed all other reflexes. (Pavlov, 1960)

At this second stage of map-making, the rats' primary mind is motivated to evolve two maps; first to invent a map that can **increase** the probability of the shortest-path behavior in the EP. Secondly, invent a map that can **decrease** or minimize the probability of its wasting time in the BP subset of spaces.

The purpose of the EP map is to increase the success of the behavior within the subset of the EP. However, the purpose of the BP map is very different; to decrease the probability of behavior in the BP-in those elements of the whole that are in conflict with its shortest path purpose.

My purpose in the results section will be to describe the anomalous empirical EP and BP behavioral data that the behavioral telescope reveals for the first time. One of my first tasks is to show the differences between the limited set of naked-eye behavioral data the behaviorists recognized and the enormous set of BP behavior, which the behavioral telescope generates.

Two Gradients: Goal and Start

The Hullian Goal-Gradient model of maze behavior recognized two elements of the rats' behavior. (Hull, 1943) First, the random component of the rat's behavior that moved the rat through the maze until the organizing effect of the reinforcing event begins to work its magic. Secondly, the organized behavior that evolved through the action of the reinforcing object that organized behavior through a goal gradient, which is behavior that evolved from the goal back toward the start box.

Our purpose in this section is to show that neither of these behaviors is seen in the spontaneous behavior of the rat in the matrix maze. The behavioral phenomena, which Hull as a naked-eye behaviorist saw in the behavior of rats, just does not exist when the spontaneous behavior of rats is observed through the behavioral telescope. Therefore, neither the behavior nor the theory, which Hull (1943) recognized, had any empirical foundation. In this experiment, I will examine the full maze behavior in general, and the blind path data in particular, and I will show that rats never exhibit random behavior. The random behavior that the whole series of classical learning theorists imagined from Thorndike (1911), Hull (1943), and Tolman (1961), etc. seen with their naked eye was a myth. The random behavior they believed they saw no more existed than the old belief that the sun goes around the earth.

Secondly, the goal gradient behavior that the naked-eye behaviorists saw is not seen in the spontaneous behavior of rats when viewed through the behavioral telescope. Goal-gradient behavior may exist in a later stage of behavioral evolution, but in the first stage of the evolution of spontaneous map-making behavior, this goal gradient behavior is not to be seen. On the contrary, Heterostatic learning is the product of a start-gradient rather than goal gradient behavior.

My point is that both the data and the theory that the classical learning theorists invented to understand the rat's maze behavior were myths. Nothing the classical theorist said or did about maze behavior had any validity; their data was an artifact, and the theory they invented to explain this imaginary data was also an artifact.

Therefore, in this section, I will describe for the first time, the spontaneous full maze behavior of the rat in the matrix maze as seen through the BT, but I must warn the reader that every element of the data will be anomalous. Furthermore, no element of the classical learning theory, neither the data nor the theory, has any validity in this context. Nothing in classical learning theory will have prepared the reader for the kinds of behavior that the BT reveals in the EP much less the BP.

There will be those who probably will refuse to look through my telescope, just as there were those that refused to look through Galileo's new scope, but hopefully some will not be afraid of behavioral anomalies, which open the mind up to new realities.

Negative transfer

As noted earlier, understanding the BP growth behavior of the rat was especially difficult for me because Hull (1943) and Tolman (1961) had generated theories of BP behavior that are not only irrelevant to understanding the BP behavior in the matrix maze, but even worse their theories pointed me in the wrong direction, i.e. they produced negative transfer.

The classical theories designed to explain TTP behavior were based on the myth of mindlessness. My new LTP theory to explain blind-unit behavior begins with the recognition that rats have a rational primary nonverbal mind. It is necessary to assume that the rat is an intelligent being capable of making models of the cause-effect constraint in a domain, both in the EP and in the BP.

Two theories of intelligence: the mindless, and the mindful predictions compared

The theory of Heterostasis makes a paradoxical prediction about the effect of intelligence upon the BP behavior of rats because rats running the matrix maze under the LTP make a far-below-chance rate of BP unit entries. This is evidence that the rats are making a model of the matrix maze that allows the rats to avoid the redundant Blind Path Units of the maze, by avoiding the exemplars of the redundant units of the Blind Path.

To understand this behavior requires a theory of Heterostasis, which picks up the other end of the stick as compared to the classical learning theorists.

The mindless prediction

The first unhelpful prediction was the classical theorist's claim that all entries into blinds were errors. These classical theorists also claimed that rats make blind entry errors for two opposite reasons; first, because they are mindless or stupid and secondly because of reinforcement learning. The latter kinds of errors are called anticipatory errors; but in both cases, according to classical theory, a blind entry is an error.

The important and interesting question is really: Are Blind path unit entries a sign of intelligence or stupidity. If we assume the former, then this assumption turns out to be the key to understanding the spontaneous behavior of rats in the matrix maze. If BP behavior is a nonverbal intelligence test for rodents, then the longer rodents delay making an entry into a blind unit, the more intelligent they are, or the smarter the rats the longer they can avoid entries into redundant BP units of the maze.

The topic of errors in an animal theory of mind takes on a complexity that never could occur in the Behaviorist's view of things. Numerous errors must be distinguished, but the concept of error itself must be called into question.

Physiological motivation as physiological error

In order to understand the blind-unit behavior of the rat, it is essential to distinguish at least two kinds of errors, physiological error, and psychological error. The classical behaviorist, of course, only recognized physiological or Homeostatic motivation, which is defined in terms of tissue needs, either physiological errors of deficiency or excess.

Psychological motivation as physiological error

The behaviorist did not recognize the confounding of the three different HHH (Homeostasis, Hedonism, and Heterostasis I) mechanisms, which generate behavior in the TTP. The LTP was designed to maximize the possibility that only the Heterostatic mechanisms are actively generating behavior. Since the rats in this series of experiments are never homeostatically deprived, their behavior is neither deprived nor reinforced; yet the rats in these experiments exhibited spontaneous exploratory-play behavior, which is designed to make models or a variety of sub-models of the hierarchically organized matrix maze. Since this spontaneous behavior is not homeostatically motivated by physiological tissue needs, the concept of physiological error is not helpful in this context in understanding the rats' behavior. The exploratory-play behavior of the rats in this experiment can only be understood in terms of psychological motivation or in terms of psychological needs or psychological errors based on the psychological, epistemological, or ontological needs of the organism that will be activated by the strange anomalous features of the maze. The psychological motivation of the primary mind induces the rats to evolve a valid cognitive map of the matrix domain in order to achieve its purpose-to run the shortest-path through the maze. The ninety-day old rats suddenly immersed in the Lashley III maze domain, after spending their entire young lives in a homogenous wire home-cage, needs no additional motivation other than the cognitive motivation generated by the anomalous stimulation of the maze domain to induce their primary mind to spring into spontaneous exploratory-play behavior. The uncertainty induced by the anomalous domain initiates the spontaneous execution of the rat's Heterostatic modeling mechanism. The hierarchically organized maze requires that several versions of each sub map of the maze be created and revised before a map of the whole 96 unit maze evolves that is valid for all levels of all variables of the hierarchically organized whole domain. The first units of each of the several levels of the hierarchy within the EP are the anomalous triggers that initiate retracing in the EP proportional to their novelty.

Cannon, the ANS, and the emotions

The Autonomic Nervous System (ANS) has two divisions: the sympathetic (SNS) and the parasympathetic (PSNS) nervous systems. The SNS is the emergency Nervous system.

Heterostasis conceives of the rat's exploratory-play behavior in the EP as motivated by the uncertainty induced by the anomalous units of the maze. In other words, Heterostatic exploratory behavior requires the support of the sympathetic nervous system, SNS. However, Heterostasis requires that the nature of the SNS be redefined. Cannon (1932) conceived of the SNS as the "Flight or Fight" or as an emergency nervous system responsible for energizing homeostatic behavior. Cannon's imagination did a great disservice to the role of the SNS. Cannon conceived the role of the SNS as too narrow in scope as serving only needs of the body.

Heterostasis on the other hand allows us to recognize that the SNS serves not only the body but also the needs of the nonverbal mind. In fact, the structure and function of the primary mind can be said to have evolved out of a new set of arousal functions imposed upon the SNS by the evolution of the primary mind.

Lindsley (1960), for example, conceived of the SNS as responsible for the arousal function as in his activation theory of emotion. With Heterostasis, however, the arousal function is assigned a more specific cognitive and generic function; the arousal function is identified specifically with the induction of the generic arousal function - surprise, which is the generic emotion of uncertainty. The SNS induces uncertainty or psychological error, and the purpose of the exploratory-play behavior of the rat is to evolve a model that reduces the uncertainty or ignorance induced by mental errors or anomalies.

A Heterostatic theory of work

The classical learning theorists' errors were that they were only able to conceive of survival in terms of physiological tissue needs. For the classical behaviorists, physiological and psychological needs were confounded. In their confounded conception of motivation, a food-deprived organism, for example, only had a physiological need. They never recognized the psychological side of deprivation. The physiologically deprived rat, however, has two problems. First, of course, it has a Homeostatic physiological need. Second, and this has always been invisible to the behaviorist, the fact that the organism is hungry is also proof that its model of the environment of its home cage, which had always supplied food and water ad libitum, has suddenly become invalid. This second side is the Heterostatic side of the picture. Hunger is always the product of two kinds of deprivation, psychological and physiological. The rat deprived of food has not only been physiologically deprived, but also has been deprived of a valid model of when, where, and how to find food. The hungry rat's old model or paradigm of its home cage domain as an ad libitum source of food, which had been valid is now suddenly obviously invalid; if this old model were still valid, the rat would not now be hungry. The theory of Heterostasis assumes that the primary role of food deprivation is to invalidate at least one element of the home cage model of the rat, specifically the food-obtaining action system model.

Heterostasis is, in a sense, based on the two sides of the cliché "knowledge is power." If true, then surely the opposite is also true, ignorance is weakness. The act of deprivation has two sides; it deprives the mind as well as the body. Hunger is the awareness of both a physiological and psychological need for food. Hunger is an emotion as well as a motivation. Hunger is the heterostatic sign of a specific kind of ignorance. The hungry organism has two needs; a physiological tissue need, and a psychological need to know how to obtain food. The hungry Heterostatic rat is first interested in making a model, a valid map by which it can always find food.

Map making or making a model of the domain is the primary kind of knowledge. However, rats as well as humans have many kinds of maps. One of the rats' primary maps is its food map. One of the many possibilities that the behaviorists dominated by their theory of mindlessness overlooked is that when evolutionary survival mechanisms looked to brains rather than brawn for solutions, that psychological errors rather than physiological errors became the priority. The brain of mammals has been recognized as being unique in several areas: language, tool use, etc., but now it is time, perhaps, to add yet another function to the list. The capacity to learn is not the product of nurture shaping a tabula rasa mind motivated by tissue needs and organized by homeostatic reinforcement. The nonverbal brain of mammals is the primary mind with a complex mapping mechanism as its primary first line of defense against ignorance. The SNS is not just a defense against bears, but especially against ignorance-of all kinds-against anomalous bears or anomalous Lashley III mazes. Hunger is an emotion. Food deprivation does not just induce a physiological need but more importantly, it induces a psychological need, a state of uncertainty. Hunger is an activating emotion, an uncertainty-inducing event.

Rats do not need to be deprived of food or any other physiological primary need in order to be induced to learn the matrix maze. Most rats when suddenly injected into the matrix maze will find the pattern of the units of the hierarchically organized matrix maze a compelling but complex and strange pattern of units. So much so, that the rat's primary mind normally will begin to spontaneously explore this maze and seek to evolve a model, which understands and predicts the spatial pattern of the maze, so that the rat can run through the maze with the shortest path. Initially the pattern of this maze is full of events that are anomalous, and each induces all the psychological error, ignorance, or uncertainty required to induce the exploratory-play behavior of heterostasis that motivates rats to evolve valid conceptual structures that are required by the rational mind to organize their behavior in a hierarchically structured maze.

The primary reflexes

Heterostatic learning is not a function of either Pavlov's (1960) Unconditioned or Conditioned Reflexes, for these are secondary reflexes. The rat learns the maze motivated by the primary or first priority reflexes such as Pavlov's Orienting Reflex, OR. When a mammal, such as a rat, is removed from its home cage, the only domain it has ever experienced and has a model of, and suddenly injected into an anomalous maze in an anomalous room; every stimulus, both in the maze and in the test room, is an anomalous event that elicits an OR. Heterostasis is a mechanism of the primary mind that creates a model that can convert all these strange and complex anomalous events into simple and familiar events. Heterostasis is a form of adaptation that is a function of the primary reflexes, not the secondary reflexes.

Transfer, assimilation, generalization, extrapolation and extrapolation errors

Subjective vs. objective error

As noted above there are many kinds of errors and now is the time to consider yet another distinction. The first distinction between the physiological and psychological conception of error was noted above. The next concept of error that must be distinguished is between objective and subjective errors. The concept of objective errors was a fundamental assumption of Watson's (1972) notion of data as being public and objective. For a behaviorist, a rats' entry into a blind-unit of a maze was both public and objective data and public and objective error-an objective error is one defined by the observer. The animal theory of Heterostasis, however, is based on the conception of a **subjective** error--that is an error from the subject's point of view. Heterostasis takes Titchener's (1912) definition of psychology seriously. Titchener defined psychology as, "The objective study of the subjective." Heterostasis recognizes several types of subjective errors. The primary subjective error-inducing event is the anomaly. In this section however, the focus is upon the rats' subjective ability to anticipate that an entry into a redundant blind unit would be an error or waste of time. When a rat enters a blind unit of the matrix maze, or any unit for that matter, the experimenter tracking the rat objectively records the unit entry on the graphic tablet. When a rat enters any unit of the blind Path, the unit-entry is recorded simply as a unit entry into a BP unit, as a public and object fact, but not recorded as an error. The central issue is whether this BP unit-entry was subjectively experienced by the rat as an error or not. The important question is this: how can an objective determination be made as to how the rat experienced the unit entry? How can a scientist objectively know that the rat's mind classifies its behavior as a subjective error? When the rat's mind experiences an event as an anomaly, the fact that the event was an anomaly is proof that the rat was surprised by the event; that the rat experienced the event as an unexpected event; as a subjective error—no one else was surprised by the event. In order that the anomalous event be experienced as anomalous: it need only be anomalous to one rat, the rat that was surprised.

How can a scientist objectively know that a rat subjectively experienced a unit of the matrix maze as an anomalous event? I will say and show many times in this book how anomalies can be objectively recognized. However, the most important answer to this question is anomalies induce retracing! Moreover, the length of the retracing loop is a measure of how anomalous such a unit was. This was carefully described using experimental data in the section of Experiment I in Chapter 1, in which Female 3's behavior is discussed. The reader can review the discussion of flat places in her behavior. Flat places in behavior tell us loud and clear in this case subjectively what is happening in her brilliant female mind.

An anomaly is a private event with a public aspect. When scientists observe the public behavior of surprise, they can infer that the surprised individual had just experienced the fact that its model was in part or in whole invalid.

Transfer in the EP

The ultimate questions are these; first, how is the model of the matrix maze evolving in the rat's primary mind affecting the rat's spontaneous moment-to-moment behavior in the units of the maze; and second, how is the behavior of the rat affecting the evolving model? These are the ultimate questions for the behavior in both the EP and the BP. In this section, the focus is on how the evolving model affects the rat's decision whether or not to enter some blind units. Unfortunately, the language of psychology is a particularly cluttered set of domain-specific terms.

First, the Associationist would call the phenomena in question as one of transfer of training, and in this context, they recognize two kinds of transfer, either positive or negative.

In Piaget's (1952) conceptual system this phenomena was called assimilation.

"Among the aggregate of external realities assimilated by the organism are some which are incorporated into the parts of physicochemical mechanisms. In the first case, there is physiological assimilation, whereas the second may be called psychological assimilation." (Piaget, 1952, P. 43)

Pavlov (1960) called this phenomena generalization. However, while any of these terms is acceptable, I prefer to use the mathematical concept of extrapolation rather than positive transfer, assimilation, or generalization.

The types of extrapolation: Extrapolation: type one

Rats with a valid model of the ninety-six-unit matrix maze domain can use that model for several purposes. As the rat's model of the maze evolves, the rat acquires, first of all, the capacity to take the uncertainty out of all the maze-specific events that had previously been anomalous, and the validity of the model within the EP produces positive transfer, assimilation, generalization, or as I prefer, extrapolation. Growth learning is defined as the changes in the animal's behavior that occurs as the model of a domain evolves. Growth-learning is defined as the changes in the behavior that occur as the knowledge in the models initially reduces and finally eliminates the primary uncertainty inducing capacity of the previously anomalous elements of the maze. The power of the knowledge in the model to reduce or eliminate the uncertainty of the behavior in any unit of the maze that initiated retracing in the essential path will allow the rat to resume exploring in the EP. Yet exploring in the EP will again stop when another anomalous element of the maze is encountered. The power of a valid EP model is to eliminate the cognitive errors, the anomalies that induced retracing. However, the extrapolation power of the model in the EP is due to the fact the knowledge created at the lower level of the hierarchy of the model will to some degree make the next grade level of the hierarchy of the maze capable of being assimilated by the model. As the model of the domain evolved and the lower levels of the EP become valid and begin to take the uncertainty out of EP units of the maze never before explored, this knowledge will generate behavior, which is the sign of positive transfer, generalization, assimilation, or extrapolation.

Type two: Transfer in the blind units

Heterostasis, however, makes a distinction the classical associationists did not make in their discussion of transfer. Heterostasis distinguishes two kinds of positive transfer: positive approach and positive avoidance. Positive approach was described above. Positive avoidance transfer will be described below. Both approach and avoidance are positive forms of extrapolation. A rat with a valid model of any unit in the EP units of the maze will show positive approach transfer in the essential path but positive avoidance transfer in its blind unit behavior. (This avoidance can also be understood in terms of retracing with two kinds distinguishable: Anomaly induced retracing, AIR, and Redundancy Induced retracing, RIR. AIR is observed in the EP, while RIR is observed in the Blind path.) In this second case, valid knowledge in a BP unit-model allows rats to extrapolate the knowledge in the unit-model to avoid redundant blind units. This theory of positive avoidance behavior produced by a valid unit-model generates several predictions about the rats' blind-unit behavior that in the context of classical learning theory would be considered paradoxes.

- First, the order of the elimination of the blind-unit entries should be very different from that predicted by the classical chaining models of Skinner (1974) and Hull (1943), which were evolved to explain transfer in the work domain.
- Second, blind elimination should occur without tissue need or reinforcement.
- Third, the quantity of the avoidance of the blind path units is measured by the fact that the amount of retracing they trigger reflects the intelligence of the rat.

The Three Predictions

Prediction one - The order of blind elimination

Classical reinforcement learning theorists such as Hull (1943) used chaining theory to explain the serial ordering of behavior predicted first, a backward elimination of blinds. Second, blind-unit avoidance is a matter of positive transfer. In other words, if the organism makes a right response to enter the goal box where it was reinforced, then on the next trial all right responses should have their probability increased, and all left responses in the approach to the goal box should have their probability decreased. These predictions are not simply wrong in this loop paradigm, they are irrelevant and for several reasons. First, there is no goal box. Secondly, there is no reinforcement. Third, there are no trials. Trials are obtrusive human interventions not designed for the benefit of the rat. There are no trials needed in order for the rat to evolve a model of the maze. When left to its own devices, the rat will simply not explore beyond its cognitive competence. The rat has the wit to promote, retain, or demote itself to the level of the hierarchically organized maze domain that it needs to explore. In the LTP, the rat does not run in only one direction within the maze. The rat is allowed to retrace as well as explore. The rat's model of the domain is rooted, or in technical terms has its point of origin in the Alpha unit, and the rat learns in a forward gradient not a backward one. It explores the maze only as fast as the validity of its model of the maze justifies. The exploring rat is very conservative; it does not risk getting lost in the maze. The behaviorists' concept of the elimination of objective blind-errors really has no place in heterostasis for it is based on the assumption that errors can only be defined by the experimenter. When rats enter blind units, there is no logical automatic reason to assume that it has made an error. The initial unit-entries for any unit including the blind-units on the contrary are considered an essential step in the empirical exploratory process of making a model of the domain. The first entry into any unit cannot logically be called an error by an empiricist. The rat, *a priori*, does not recognize the first unit-entry as an error, and there is no good reason for the experimenter automatically to score them as errors. The first blind unit-entries are logical and necessary elements of the experimental exploratory-play behavior of the rat. However, when the blind unit-entry becomes redundant and serves no cognitive or security function, then the rat and not the experimenter will consider the blind entry an error, a subjective error and eliminate it. In other words, the intelligent rat will avoid entering redundant BP units. In the context of Heterostasis, it is the judgment of the rat not the judgment of the experimenter that is important in determining whether a unit entry was or was not experienced by the rat, after the fact, as an error. Again, let me stress this point by repeating myself. Heterostasis is a far more advanced form of behaviorism than practiced by Watson (1972). By that, I mean, in order for an experimentalist objectively to score a blind unit-entry as a subjective error, public and objective data must be presented to support that judgment. Psychology is the objective study of the subjective. (Titchener, 1912) However, Heterostasis does not propose to judge an entry into a blind-unit as error based on introspective data; public and objective data must justify the statement that the blind entry was a subjective error.

The rats in this study especially avoided the forward type blind units on the back half of the maze, which suggests that they are better able cognitively to map forward than alternation blinds.

- **Prediction two Blind elimination occurs without Homeostatic reinforcement**

Heterostatic growth learning is based on the spontaneous exploratory-play behavior triggered by the novelty of anomalous events. Heterostasis is a form of adaptation concerned with eliminating the uncertainty induced by anomalies. The entire evolution of the cognitive map, for both the essential path and the blind path, is accomplished without homeostatic deprivation or reinforcement. The theory of Heterostasis is not required, of course, to abandon the concept of motivation, feedback, or reinforcement. A theory of the primary mind can adopt a symmetrical set of definitions with those advanced by Homeostatic theorists who generally define motivation in terms of physiological needs, drive induction, and reinforcement.

Heterostasis, for example, could define motivation as uncertainty induction and reinforcement as uncertainty reduction. This deceptively simple and appealing consequence of this definition is that in the creation of the model of the maze, the primary mind of the rats can be said to create primary knowledge that reduces primary uncertainty.

Galileo (1989), Locke (1914), Berkeley (1957), and Hume (1962) first raised the issue of primary and secondary qualities in philosophy. Then Müller settled the issue in physiology. (Boring, 1929) Sensations and all experienced qualities are not copies of the proximal energies of the physical world that excite our receptors. Sensations are produced by the primary mind to represent the proximal energies and distal objects and events of the physical world.

In the context of Sperry's (1971) two-mind theory, however, primary means right-brained or the first to be processed, while secondary means left brained or a second stage processing of events. In this book, I will use the word primary and secondary in the Sperry context not the context of Locke (1914).

If the events, the proximal stimuli of the Lashley III domain, which stimulate the rats' senses, were knowledge, then why are they anomalous events? Why does the rat have to create a model to take the uncertainty out of stimuli that contains knowledge? In creating a cognitive map or model of the objects and events of a domain, the rat is inventing the forms, the knowledge that informs the precepts. The primary mind in creating valid models creates knowledge that reduces the uncertainty of the anomalous, strange, and complex events of the domain. The primary mind creates primary kinetic knowledge to reduce the primary uncertainty induced by the potential knowledge of the strangeness of the anomalies of the domain.

- **Prediction three**

In the Lashley III maze, there are two types of blinds, types A and F and they have different probabilities. The rat has to choose either to enter or to avoid the A1 blind unit from the E1 unit that is the essential path unit behind the door in each row. E1 is a three way choice unit. The F1 blind unit can be chosen or avoided from the E4 unit, which is the unit before the door in each row and is a three-way choice unit. The rats in this study most easily avoided the Forward type, F, blind units. Yet paradoxically, the F2 units were always the last units to be entered in the full maze emergent loop. These F2 blind units require three times more moves to first entry than any unit in the Essential path does.

The forward blind units are the easiest to avoid because they are entered by rats continuing their forward motion in the essential path, past the door in unit E4 into the blind. Thus, all that is required to stop the entry into this forward blind unit is for the primary mind to begin to look for, stop at the door, and exit the row. The fact that the rats' quickly learned not to enter these F1 blind units is objective data that shows that the rats, not the experimenter, considered an entry into this blind a "waste of its time" or an error. The rats' avoidance behavior is the operational proof that the rats would have considered an entry into this class of F Blind Type unit an error. The rat is saying by the avoidance behavior that the entry into this class of Blind Path units would have been **redundant** and not worth the effort.

The third element of Heterostatic motivation

For over thirty years, I found the Blind Path behavior so complex and strange that I was not able to find a statistical way to analyze the data much less understand it. However, this data finally made it possible for me to recognize a new role of models that I had not previously recognized. In the first experiments, only the role of a model in organizing the behavior in the EP of the matrix was described, and this behavior was understood in terms of anomalies and news. The BP behavior, however, allows the third element of Heterostatic motivation to be observed and understood - the role of redundancy.

One of the central issues in Heterostasis, as in classical theory, is the question of motivation. The theory of Heterostasis conceives of rats and mice as intelligent self-organizing systems. The classical theorist conceived of neither animals nor humans in a complimentary way. Freud (1962), for example, in his theory of neurosis conceived of humankind as only motivated by libido or sublimated sexual energy. In other words, sex was conceived as the only force behind all-human action and achievement! Hull (1943) modified and extended Freud's libido theory with the help of Cannon's (1932) physiological theory of Homeostasis. Hull proposed that physiological deprivation or tissue-needs were the motivation behind all human and animal action. Hull proposed that sexual deprivation was not the only source of libido energy that could be sublimated. Food and water, indeed any tissue deficiency or excess, could produce physiological energy that could be sublimated.

However, Freud (1962), Hull (1943), and indeed all the behaviorists, even Pavlov (1960) attempted to understand learning in terms of parasympathetic reflexes. Heterostasis conceives of mental growth as a function of the primary reflexes of the sympathetic nervous system, especially the OR.

The paradox is that the data of this section will make it possible to recognize that not only man, but also perhaps all mammals are motivated by a force that neither Freud (1962) nor Hull (1943) ever imagined. Up to this point in this series of studies, it has been shown that rodents in the essential path are motivated by uncertainty induced by **anomalies** to create models, which destroy the uncertainty induced by the anomalies. Anomalies are the primary source of motivation and/or emotion. Anomalies are events that are surprising events and are too strong. Organisms generally initially avoid anomalies for some time until they are finally able to get mentally prepared to deal with them.

There are two kinds of anomalies: revolutionary and ordinary. Kuhn's (1962) theory forces the scientists to recognize that a critical kind of scientific data was an anomalous event, which induced the revolutionary stage of science. However, scientific revolutions are rare events that are needed perhaps less than once every hundred years or so.

However, all mammals are Heterostats and humans no less than any other mammal, which means that their mind requires their daily novelty just as their body requires their "daily bread". My question is what should this daily ration of novelty be called, or what is it now called?

The second kind of novelty has long been called news. Newspapers are historically the cooks who prepared our daily mental plate of novelty. News is the second source or the daily kind of motivation. News is a source of a milder brand of uncertainty than anomalies. Newspapers should be a source of ordinary kinds of novelty, births, deaths, baseball scores, etc. However, newspapers also report those rare events like anomalies that are revolutionary.

The uncertainty news induces is not excessive, on the contrary, news seems to be the normal source of the uncertainty required to maintain the optimal level of arousal as required by the Yerkes-Dodson law (2004). Anomalies, HOWEVER, are rare events that induce an excessive amount of uncertainty and initiate retracing behavior in the EP of the matrix maze that is a form of avoidance behavior. Anomalies are illegal events, something like a capital offense, while news is at best a petty crime or minor illegal event.

I personally learned to recognize that there are different kinds or degrees of novelty inducing events in college, where as a ministerial student I gave sermons to some of the local churches. My youthful belief or hope was that I could change perhaps not the world but at least one congregation at a time. However, I soon recognized church members generally provided immediate feedback or evaluations of sermons. What I soon discovered was that those that I considered my best sermons, those that called for changes, were to say the least not well received. On the contrary, what was expected of me was that I should tell the "old, old story" in a **new** way that breathed **new** life into their beliefs. The members of any denomination knew what they believed and they were not looking for anomalies that required any change. Every denomination had, in effect, a paradigm, and no young revolutionary preachers were welcome.

Accommodation as conceptual mutation

Why do rats retrace in the EP? During Anomaly Induced Retracing, AIR, rats are concerned empirically with reexamining the units already explored because the current model was in error. If the representation had been valid then no anomaly would have occurred. The question is this: should the units being re-explored during retracing be classified as bearers of news or anomalies? If the representation had been valid, they would not now sustain the retracing. One of the first rules about a model is this; a hierarchically organized model of the matrix that is valid at all future levels of the hierarchy will no longer be capable of inducing surprise. In other words, a valid model is one that will not allow an event in the domain to be experienced as either news or as an anomaly. Only anomalies are *de facto* evidence that the representation at whatever stage of evolution was not valid. A valid model consists of a hierarchically organized conceptual system or taxonomy that allows every event in the domain to be recognized as a unique member of the taxonomy. Any anomalous event that invalidates the current model does two things. First, it converts at least some of the old events that had been considered as represented validly, once more into novel or news events. Events that the organism reevaluates during the retracing period may become novel, or news, for a second or third time. Perhaps Charles Kelley put it best when he said:

"Thus at whatever level in the control process at which there is novelty, i.e., unexpected events, choices for which there is no habitual response, conscious processes are brought into play". (Kelley, 1968, P. 45)

In the rat's case, the conscious process brought into play is retracing. This fact holds the key to the solution of the current mystery - how does the primary mind revise a model that will take the uncertainty out of the anomaly just experienced. The primary nonverbal mind initiates retracing because it assumes that the events of the first part of the maze have the potential knowledge required to update or to create a new valid model of the matrix, one that will make every subsequent event in the domain recognizable. The experience of an anomaly implies in the nonverbal mind that the model that created the anomaly had misrepresented some previous event or events. The rat upon experiencing an anomaly begins to retrace in order to discover the event or events that were misrepresented in the current model. Once those events have been discovered and correctly represented then the anomalous event that initiated this cognitive crisis will no longer be experienced as complex and strange, an anomalous event, but will be experienced as a simple and familiar event. The retracing loops of the rodent are the occasions in which the error in the current model is being fixed. Retracing is not the waste of time that Blodgett (1929) assumed when he devised the non-retrace doors.

When a revised pragmatic model finally works, it has destroyed the power of the previously anomalous event to induce uncertainty, and then the rat is ready to resume exploring. However, two things have happened during the updating of a model. First, the power of the once anomalous object to induce uncertainty has now been destroyed so that it no longer needs to be avoided, and its removal eliminates the mental block that prevented the rat from exploring further in the maze. Secondly, the fix of the model also has had the consequence of having destroyed the news value of the events that were being reevaluated during retracing. The Yerkes-Dodson law (2004) thus requires the rodent reinitiate exploratory behavior because the events behind the organism no longer have the requisite arousal potential to sustain retracing behavior. The fix of the model, it must be noted, has destroyed two things, first the anomaly that had been blocking its forward path, and secondly the news value of the events behind the rat that had made retracing worthwhile. The revised and now valid model destroyed the anomaly before the rat and the news behind the rat. Now is the time to begin to understand the third force that motivates the heterostatic organism, redundancy. The effect of the fix of the invalid model does three things; first, the destruction of the anomaly, second, the destruction of news quality of the events that has made retracing worthwhile, and thirdly, the effects of both of the first two fixes were to create redundancy. The fix, in other words, transforms both the anomalous events and the news events into redundant events. This was the critical part of the story, which I assume Pavlov (1960) must have somehow understood when he said, "Extinction is inevitable."

Within the primary mind, the representation of every object in a domain has a mental life, a representation cycle, and the end state of this cycle for every physical event in a domain for which an organism can evolve a valid model will always be extinction, or boredom. Alternatively, to paraphrase Pavlov's (1960) point, "every light bulb always burns out." The most fundamental phenomena, the fact from which Heterostasis derives its name is the fact that the primary mind is in the business of creating a paradox; order is to the mind what order is to physics. Moreover, both in the mind as in physics entropy are inevitable. As the representation of every object in every domain ultimately becomes perfected, it becomes redundant, it dies a mental death, and the organism must necessarily seek a new source of uncertainty. Uncertainty is to the primary mind what hunger is to the body. Uncertainty is "appetite for thought" in the primary mind. In order to be awake, alert, and feel good, the domains in which our daily actions occur must have the requisite uncertainty-inducing potential. While the Lord's Prayer teaches us to pray, "Give us this day our daily bread" a biblical recognition of the homeostatic needs of the body, heterostasis would also have us pray for the requisite uncertainty that make our waking moments interesting. Heterostasis recognizes the daily needs of our mind. Mind "does not live by bread alone."

Empirical vs. Logical or rational redundancy

We are finally getting close to having assembled a set of conceptual elements that will make it possible for us to begin to understand the blind path behavior of rats. However, one more distinction must be made. In order to understand BP behavior it is also necessary to make a distinction between two kinds of redundancy, logical and empirical. Empirical redundancy is the kind Pavlov (1960) had in mind, when he said, "extinction is inevitable." Pavlov recognized this aspect of redundancy when he specified that every experience of an event would cause an accumulation of a central unit of diffuse inhibition, an inhibitory force similar to Hull's I_R (1943).

"It is thus clear that the administration of the unconditioned stimulus results in an inhibitory state of those cortical elements on which the conditioned reflex depends- these cortical cells becoming temporarily unresponsive to their normal exciting stimulus." (Pavlov, 1960, P. 243) See FIGURE 0.2E

However, Gastaut's (1957) diffuse inhibition was a central inhibitory force while Hull's (1943) I_R was a force in the peripheral nervous system such as lactic acid. Gastaut's diffuse inhibition was more like Hull's I_C, which was a permanent force and would not dissipate over time, as would I_R.

The interesting and novel aspect of the type of redundancy that characterizes the blind-unit's behavior and induces avoidance behavior, I call rational redundancy. What is interesting and strange about Blind Path, BP behavior, is the fact that some blind units are almost never entered, so that their redundancy cannot be said to be empirical redundancy. Since some of these blind units are so effectively avoided, they are the last units of the matrix maze to be entered; obviously empirical redundancy cannot be the cause of this avoidance. This blind unit redundancy, which induces avoidance behavior, forced this distinction of the two types of redundancy, logical or rational redundancy. The blind units that are the last to be entered are avoided because they are **rationally redundant not empirically redundant**. Rational redundancy is a behavioral phenomenon that can only occur in a rule-using organism. Redundant blind units were avoided because they were redundant exemplars of sets of elements or classes in the taxonomy of units that the rat had discovered as being not worthy of reentering. Rational redundancy occurs when an organism avoids a class of events, even when the organism experiences a novel member of the set. This novel exemplar will be avoided in spite of the fact that it is technically a novel event. For example, in the blind path, there are units which rats are reluctant to enter for the first time. Why, because the rat knows which units are Forward Blind units and knows that those units are blind units, i.e. they go nowhere. The avoidance of these units is a logical learned prejudice.

Empirical redundancy should be called ideographic redundancy, while logical redundancy should be called nomothetic redundancy. **Prejudice**, in spite of the fact that socially it is considered unethical, yet behaviorally, is a phenomenon, which represents a considerable mental sophistication. It has important survival value in many domains. Behavior that allows either an organism to avoid a novel object that is a member of a set, that is dangerous, that tastes bad, or that is simply a waste of time has survival value and represents a most important aspect of the behavioral contribution of the primary mind to survival.

Rational avoidance can also be understood as a form of anticipatory behavior, except the organism does not associate the avoided unit with some positive or negative reinforcing event. The organism is not anticipating food, pain, or any particular external event. Instead, the animal is expecting the avoided units do not contain anything of interest based on experience; the unit is avoided because it is redundant. The unit is avoided because it is expected to have no interesting features.

The Yerkes-Dodson Law (2004) was described earlier in book as an inverted-U function of arousal; see FIGURE0.2B. Yet this curve is full of surprising predictions, some of which I will discuss below.

The arousal mechanism summarized by the inverted U curve of the Yerkes-Dodson law, can be understood in the context of Heterostasis as being the product of three kinds of elements or forces: redundancy, news, and anomalies, with each of these three rational forces identified with a corresponding zone of the Yerkes-Dodson curve: A, B, and C respectively. In other words, Heterostatic behavior is a complex pattern of behavior, which is the product of three kinds of rational behaviors, each motivated by one of these unique rational forces: redundancy, news, and anomalies.

While I have discussed anomalies dozens of times throughout this book, yet, in spite of all that has been said, unfortunately anomalies are so important and so complex an aspect of mammalian life that I feel I have hardly begun to cover the topic. There are many kinds of anomalies, primary and secondary. With respect to the domains of science, it is reasonable to assume that every theory in every scientific domain is still conceptually flawed, and thus each science is not about to go out of business. Yet, the more advanced the science, the higher up the hierarchical ladder the critical anomalies can be found. In effect, each time an anomaly is discovered and has its mysteries understood, the rule is that the conceptual complexity of that domain advances one grade. My too simple assumption is that each anomaly within a hierarchy operationally can be conceived as a grade to solve. The mastery of each grade simply means that another anomaly has been mastered.

If it is so that the more advanced the topic, the higher up the tree the anomalies will be found, then in especially technical domains only the most advanced Heterostat will be able to enjoy the pleasures of having a first-encounter with an anomaly.

For example, with respect to Aristotle's theory of cosmology (1952), in his day, Galileo was probably the only Heterostat who could have encountered an Aristotelian-anomaly: it was not a chance event when he encountered the moons of Jupiter. Galileo was only one of the few humans in his time that could not only make but also improve a telescope to be capable of seeing Jupiter's moons. However, just as important, Galileo was probably only one of the small populations who understood Aristotle well enough to be surprised by the moons of Jupiter.

Let us just say that the moons of Jupiter were fifteenth grade anomalies. Galileo may well have been the only human fifteenth-grade cosmological scholar of his day; so the pleasure of discovering and solving the conceptual problems these moons presented to Aristotle's theory may well have been Galileo's alone.

However, the primary nonverbal anomalies are another story. The primary mind of every healthy individual of every mammalian species is constantly encountering and destroying primary anomalies every day. My rat's concept of fun was to be allowed spontaneous exploration in the matrix maze for a short period each day. The rat's primary mind was designed to digest primary anomalies and during spontaneous play, destroying anomalies was what the primary mind did.

The anomalies of the matrix maze and anomalies of Aristotle's cosmology (1952) are very different anomalies in one sense but very much alike in another. The heavens were Galileo's maze. Both Galileo and all Heterostats have the same concepts of fun; the fun of each day is the privilege of having a puzzle to solve. The Heterostat's prayer says, "...and give us this day our daily anomalies." Anomalies are the spice of the mental life. Here I am talking about mild anomalies, which are news. Galileo's fifteenth-grade anomaly was a type B event, and all the anomalies of the matrix maze are probably Type B events. Type B events only require the primary mind to update its model not to create a brand new model.

So what is a type C event on the Y-D law? Holmes and Rahe (1967) have identified a dozen or more kinds of anomalous events and graded them for their soul-shaking quality. The sudden death of a loved one is one of the most common Type C events. Such events have a profound effect upon a person's life. They may incapacitate a different individual for different amounts of time, but a type C anomaly is excessively strong to the point that it makes the individual incapable of functioning for some period. When events are maximum type C events, they may incapacitate an individual for life, unless the person gets expert help. These events typically cause Post Traumatic Stress Syndrome, PTSS.

However, type C events come in a variety of strengths, but all too strong events to some degree must be incapacitating to qualify for the type C status.

Anomalies come in a wide variety of degrees, which will be identified and discussed at several different places in this text, but for the moment, it is important to identify the weakest kind of anomaly that qualifies as a type three event.

In the matrix maze, the weakest class of type C event is incapacitating only for a moment, in the sense that the anomalies within the maze cause the rat to abandon exploration and to initiate retracing. Technically, these anomalies are incapacitating in the most important sense. They are powerful events that force the mind to recognize that the conceptual system that allowed this anomalous event was flawed, and so the rats began to retrace in order to update the concepts, which allowed this anomaly to be encountered.

The force of news

News, on the other hand, is a sustaining force yet a manifestation of a mild form of anomaly. Those units of the matrix, the rat explores, re-explores, and re-re-explores, etc. contain features at an intermediate stage of knowing, while not anomalous yet they are only partially understood; they are not yet fully understood so well they can function automatically within the rat's current purpose. Yet these units, the rat has reason to believe in the moment before it encountered the last anomaly that it had a working understanding. Yet, the fact of the encounter with the last anomaly is now proof they were only partially understood. The significant insight the reader must achieve here is that in a hierarchical structure, all knowledge in the lower grades is partial, is incomplete and that the concept responsible for the assimilation of each class of events has to evolve and grow. After an encounter with every new anomaly, the rat must once again start to retrace because as the rat slowly becomes aware of the hierarchical complexity of the matrix, the first units of the matrix must be reconceived to make provisions for each new dimension of the causal structure to be added into the evolving conceptual structure. If each unit of the matrix is even a three-dimensional vector, then in the beginning the rat conceives of each unit as having a one-dimensional causal structure. If each anomaly forces the rat to add just one new dimension at a time: then at least three retracing sessions will be required for the primary mind to evolve the full three-dimensional vector conception for each concept that a valid model of the matrix requires. Each retracing loop would allow the mind to evolve the complexity of vectors only one dimension at a time. The concept of news allows us to understand that each anomaly forces the primary mind to recognize that each unit, which it had believed was completely understood and thus having no additional capacity to surprise, forces the primary mind to discover that each of the earlier units still possess some potential causal structure, which has not been assimilated. Thus, these old units suddenly have an unrecognized potential; each unit again contains some news.

The primary mind conceives pragmatic concepts; concepts that are not designed to be ideals, but only as good as required in order for the rat to achieve its domain-specific goal. However, this said, the fact is that as the primary mind slowly becomes aware that its concepts are not yet as good as required, it must be prepared to retrace time after time until each concept achieves the basic level of competence that achieving its goal requires. However, this is not a bad thing, retracing in other words is as much a part of the fun as the exploratory-play behavior. While the rat's mind intends to evolve a pragmatically valid conceptual structure when the goal is finally complete, yet the goal is complex and part of the goal of the primary mind is the fun generated by the growth process. The primary mind is like the alimentary canal; the goal of eating is not to swallow the food as fast as possible but to enjoy savoring each bite. Each human as a child is said to be in a hurry to grow up, yet the tragedy of youth is failing to understand the fun involved in growing up.

My point is that anomalies and news are very different kinds of events. Anomalies require the rat to begin to retrace and to avoid the anomalous unit until it can approach the once anomalous unit with the conceptual skill to understand this unit with some degree of knowledge. Each anomaly forces all the earlier units of the matrix to be viewed in a new light. All of the old units, which had become boring, now suddenly acquire a new appeal, each becomes novel once again. The primary mind must now re-explore each of these now new units until it can discover how to conceive a new conceptual framework in which these new units will once again become at least for a moment, a familiar vector. When this happens, this transformation of the primary mind's capacity to view each unit in the context of a higher order vector complexity, this new complexity will not only destroy the news of the old units but will also destroy the anomaly's power to surprise. Each retracing loop allows the primary mind to add a new higher order degree of complexity to the conceptual structure of the maze, and this new complexity accomplishes two things first, it destroys the news value of the old units, and secondly, it destroys the capacity of the last anomalous unit to be an anomaly.

An invalid concept is one the primary mind cannot assimilate, and one the rat's mind must reexamine to discover the causal structure it has never before been able to understand. However, these news-containing units contain a causal structure, which must evolve or grow cycle after cycle. A valid concept that is a working member of a set of a hierarchically organized conceptual system has an "OLD SOUL." No first-grade concept, ones with only one layer of conceptual evolution, could ever have the power to function as a full-fledged member of a hierarchically organized system.

The forces of redundancy

Sometimes I wish the behaviorist's simple S-R (stimulus-response) concept of human and animal behavior had been correct. In an S-R world, each stimulus elicits one response; ideally, there is a one-to-one match between stimuli and responses.

However, the whole point of the evolution of mind was to create beings with what could be called Behavioral Freedom, BF. While creatures designed by biological evolution have a simple nature defined by their genetic material; creatures designed with a Mind have a complex nature, which means that the relations between stimuli and responses can be very complex since these creature have BF.

Concepts are partial

BF (Behavioral Freedom) is the result of the basic law of the rational mind: Every object is capable of being represented by the mind with an almost infinite number of partial or domain-specific concepts. What is paper? James (1950) noted that while each human may harbor only one partial concept for most objects, yet this only proves how little humans understand the complexity of mammalian nature.

In order to understand the nonverbal rational behavior of even rats, we must begin to understand the varieties of ways the primary mind can represent objects. In these experiments, we must understand how the primary mind represents the units of the BP, but this part of the story requires that we also begin to understand how redundancy alters the primary mind's functions.

In an S-R world, the rule would be that mammals avoid boring places. However, in the Y-D law (2004), a bedroom could be defined as a boring place, and yet it is designed to be a safe, quiet, dark place where we can retire each night for a few hours of rest. Yet, healthy normal individuals after a good night's sleep can hardly wait to get out of bed and back into the midst of the complexity of activities that are happening in other places of the home or of their community.

What is a cul-de-sac to a rat? In the matrix maze where every unit is painted black, when a rat is too excited they may retreat into these units with three-sides and no doors because such a unit may seem safe. However, for a rat that is wide-awake and eager to engage the complexity of the hierarchical structure of the matrix maze, these units are boring.

The Yerkes-Dodson law (2004) was designed to rationalize the complex nature of mammals. The rule is that mammals seek a place, which will optimize their arousal level. However, this law can be understood several ways. First, the literalist version of the law could be understood to say that mammals will always seek those places, which have just that ideal rich mix of stimuli, which will maintain the organism in an ideal state of arousal.

On the other hand, the pragmatist's version of the law understands the law to mean that a rational organism will seek whatever place at any time or place, which will have the effect of restoring the organism to its ideal state of arousal. This is what must be understood as the interaction understanding of the law. Mammals will seek places, which will arouse them to the degree required to restore their ideal state of arousal-type B. The interaction-view states that mammals seek to control their state of arousal and will move to whatever place will most likely restore their arousal level to its ideal state.

Review of Experiment II Hypothesis – A Belated Control Group

The behavior of the experimental group of rats in the second experiment revealed first to our great surprise, that the extinction which Pavlov (1960) caused us to expect, did not occur. Secondly, in its place, we observed a radical new kind of behavior, which we had not expected, devious behavior, and that rats emitted this devious behavior at three of the four behavioral levels of the hierarchical organization, which we were able to observe. However, the first two studies in this series were designed in ignorance of the phenomena of punishment, so no appropriate control group was included as a part of their design. This third study was then designed as an imperfect way to run a control group, albeit belatedly, that would allow an objective operational test of the devious behavior hypothesis. If the devious behavior seen in the experimental group was caused by the punishment, then the symptoms of devious behavior should not occur in the control group. Therefore, the purpose of this experiment is to test the hypothesis that a control group will not exhibit devious behavior. The data of this study essentially confirms this hypothesis; control rats do not exhibit devious behavior at any of these three levels, but also to my delight, one of the control rats did oblige me and show me that Extinction occurs and how it looks. However, with only one rat as a witness to this phenomenon, it is not possible to infer the causal condition for extinction.

METHODS SECTION

Subjects

Subjects 8, sixty-day old Sprague-Dawley rats, 4 female and 4 male, obtained from Charles-Rivers were the subjects of this experiment. The females mean weight was 171 grams and the male mean weight was 249 grams at the beginning of the experiment. The subjects were housed as in the previous two studies.

Apparatus The same as in the previous study

Procedures

The rats in this experiment were run on a three-element constraint set. The first constraint was a fifteen-minute (900 s) limit per run, the second was 300 moves per run, and the third was a five-minute sit (300 s). Every rat was removed from the maze after it had achieved 900 s, 300 moves, or a 300 s sit in some unit, whichever came first. However, the important difference between this and the earlier constraint sets is the fact that with this constraint set for the first time making an AOA loop was not an out element. In other words, in this study, the rats were not punished for making a criterion loop. Technically, of course, the rat could conceive each of these three elements of the constraint set as a punishment. However, rats did not seem to experience these contingencies as aversive. Alternatively, if rats did find them aversive, there was nothing to be done to alter the consequences of these contingencies. However, questions remain open, does extinction occur as an element of rational behavior under what conditions and of course, what does it look like?

The primary mind creates a new form of life. All the lower forms of life were and are a pure manifestation of biological life. However, with the evolution of the primary mind, mammalian behavior is a manifestation of both the biological and the mental forms of life. The mental life varies as a function of or in response to the laws, which govern their world.

The rats in this experiment have their being as a function of several sets of laws; the laws of nature and the laws of man, or more honestly the laws that I, as an experimentalist, chose to impose upon the world of the subjects of my experiment. The latter sets of law include two such set of laws: first, the laws, which govern the structure of the Matrix-maze space which I allow my subjects to explore, and secondly, the laws governing the conditions under which I allow them to stay within the maze.

The difference between the experimental and control groups of rats in this experiment are in this second set of human imported laws, which defines the shape and size of the world within which their time on earth is spent.

The critical lawful difference between the experimental and control groups of rats is that the control group of rats have their time in the maze defined by four man-made laws, while the control group of have only a set of three man-made laws or constraints. The critical legal difference is that the experimental rats have to live in a more complex four-law world, in which this fourth law is the criterion-loop constraint. This law requires the rats to be removed from the maze immediately upon completing a criterion loop, upon returning to the alpha unit after having made a complete loop through all the units of the EP of the matrix maze.

In this results section, I will try to describe the consequence of both sets of manmade laws: The effect of the laws of spatial constraints built into the design of the matrix maze and the effect the consequences of the four manmade laws concerning how the rat's time in the maze is defined.

Devious Behavior can only be clearly seen in individual not group data

Experiment II showed that at the run level of the organization of behavior, heterostasis produced large individual differences in devious behavior so that individual rather than group curves were required to see the effect of punishment. In other words, at the run level, group means partially masked the individual different patterns of devious behavior generated by the conflict of purposes induced by punishment. However, there are still visible peaks at rows 7 and 12. While all the rats exhibited devious behavior, yet the timing of the individual difference effects was unique, was not synchronized for each rat with the results that the individual differences were partially canceled out in the mean group effects.

In the second half of the second experiment, IIB, the individual data for rat F11 was shown in FIGURE 2.02. The run behavior for F11 showed that she did not exhibit devious behavior on every day, but in fact, on some days she did not completely suppress all signs of devious behavior. F11 had two goals on some days, for example, she was primarily concerned with running or expressing her dominant need running the shortest path, yet on other days she was exclusively concerned with practicing, or exhibiting her second concern; maximizing the amount of play she was allowed to enjoy per day. However, and this is a critical point, this complex shifting between these two mutually exclusive purposes was not observable in the control group data. The run data for several individual animals will be shown to make this point and to show, however, that each rat exhibits one of several possible styles for running and exiting the maze. Since the design of this experiment specified that the rats would be given a three-option set of exiting-the-maze constraints, it was most interesting to discover if individual rats had any strong consistent day-to-day preference for how they exited the maze. Again, I was pleased to discover that these "simple" creatures clearly had a strong preference even at this level of detail over what the rules were that governed their daily lives. I also was pleased that one rat, at least, exhibited extinction.

The three exiting-optional elements of rational behavior

Rational third-wisdom behavior, so far, we have discovered exhibits at least two kinds of behaviors, which can best be studied as individual differences, devious behavior and now the exiting options. Since, by definition, the control rats' behavior could not exhibit devious behavior, yet they did, never the less, exhibit the other kind of individual difference; each rat revealed they had a preferred way for exiting the maze. The three-element constraint set used in this experiment set clear limits on the exploratory-play behavior of rats and each of these constraints was, in effect, an option for exiting the maze. The control rats' behavior clearly showed which exiting option they preferred: 300 moves, 900s, or a 300s sit.

The first option specified that a rat could spend its allowed 300 moves if they did so before the 900s time-constraint was up. The second option, on the other hand, was that the rat could spend all its allowed 900s of time if they did so before their move-constraint was used-up. The third option was to sit in some unit for the allowed 300s of time.

FIGUREs 3.01, 3.02, and 3.03, show three different rats' performance curves for 27 runs. Each FIGURE gives an example of one rat that chose one of each of these three different exiting options.

Again, I must note that these three particular exiting options were not available to the experimental group of rats—by definition, what made the experimental rats members of the experimental group was that they had a very different exiting option; they were removed from the matrix maze immediately upon making a criterion loop.

The control and experimental rats lived in two different rule-governed, lawful worlds. The set of rules within which they had to function were two different sets of rules. The question this experiment asks is how a rat's behavior is affected by rules, which rules can they respond to, and how does each rule alter the rat's behavior.

A comparison of the three control rats' individual curves reveals that each rat had a preferred choice for exiting for 27 runs. But these differences must not only be compared with one another, but they also must be compared with the complex 24 trial curves for the experimental rats, which revealed devious behavior. The point of these curves is the difference between these three plain or boring exiting curves, and the dynamic curves for the experimental rats exhibiting devious behavior.

The move-maximizing rat

FIGURE 3.01 F3 steps and time per run

FIGURE 3.01 shows two run scores for F3 (please note: this is not the same rat referred to as F3 in experiment one in this book): steps and time per run for 27 runs. FIGURE 3.01 shows the number of steps and the amount of time the rat, F3, a move-maximizing rat, allocated for each of 27 runs. The data shows that F3 almost perfectly attained the maximum possible moves each day-note the almost perfectly straight line at 300 moves. However, her time allocation varied from 450 to 900s with the latter having been reached on the ninth run. Since both F1 and M4 were step-maximizing rats like F3, their data is not shown. FIGURE 3.01 allows me to make two additional points. First, it shows that this control rat did not develop devious behavior. While her time score did vary from day to day, the step-score did not for there was no room for improvement. This rat was not punished and as a move or step maximizer showed none of the complex variability exhibited by the experimental rats seen in FIGURE 2.02 in the last experiment. Secondly, this data illustrates one of the ways control rats chose to exit the matrix when the choice they had to resolve was how to exit the maze.

The time-maximizing rat

FIGURE 3.02 M2 steps and time per run

FIGURE 3.02 shows the run behavior of a time-maximizing rat, M2, for 27 runs for two scores: the steps and time per run. Compare FIGURE 3.01 and 3.02, these two sets of curves show that the differences between the behaviors of these two rats could not have been more different. FIGURE 3.02 makes two points. First, this control rat did not exhibit devious behavior at the run level of organization. Second, as a time maximizing organism this rat chose every day to obtain almost its maximum possible time allotment, 900 s in the maze, but only on three occasions obtained the maximum number of allotted moves, 300 moves.

No rat was observed to maximize both moves and time

The difference between F3 and M2 raises an interesting question. For example, why does F3 maximize moves, and yet seems unconcerned about all the time it is losing in the maze because of this strategy. For example, on several days she lost as much as 400 seconds of her allotted playtime. This may well be a sex effect. Females are generally high movers; that is they set a pace and continue a steadier pace throughout the maze than do males. Males, on the other hand, are distinguished by their sitting-behavior. In several studies, the only rats that sat in one unit for as long as a minute were males. And so M2 did not maximize moves but maximized the time spent in the maze. Again, note how boring this curve is compared to the experimental devious behavior trial effect in the second experiment.

Extinction

Finally, the behavior of this third rat, M1 clearly answers at least two of my basic questions: does the phenomenon of Extinction occur in cases of pure Heterostatic behavior, and if so what does it look like?

FIGURE 3.03 M1 extinction behavior

FIGURE 3.03 shows the run behavior of M1 for twenty-three runs. This rat's behavior is both sad and beautiful, at the same time. It is beautiful in that it is a textbook example of what extinction should look like, yet from the personal point of view, if the reader can empathize with a rat, this behavior is very sad. This rat's behavior tells a story. M1 clearly did not have a good time in my maze. On only four days was the rat able to endure remaining in the maze for the full-allotted time of 900s. Generally, the rat simply chose to take the fastest and easiest way out; it simply had only to sit still in one spot for 300s, and by law E was obliged to immediately remove it from the maze and take it back to its home cage, and that is what M1 most often chose to do.

Now the question is why. Why did this rat choose to perform as it did? Was the rat stupid or smart? Who was at fault here: the rat or me? If I were a normal mentor, teacher, or parent, I would probably blame the rat that is the easy way out. But I am not normal, and I am too well aware of several good reasons for blaming myself.

First, the data shows that for, at least, the first day M1 took full advantage of the opportunities afforded by the matrix maze; on day one, it spent over 200 of its allotted moves, and it spent all 900s of its allowed time, but it was downhill after that. M1 obviously found nothing about the space that could even initially gain its full attention much less sustain its interest over twenty-seven days. Notice that on day twenty-five M1 had perfected the skill of escaping from the maze as quickly as the law allowed—M1, took no steps: M1 sat for 300s, and was removed from the maze. M1 extinguished with perfect rational efficiency. The record shows that he made no moves on three other days, for a total of four days. If this behavior isn't extinction, I don't know what else it might be.

Group means

FIGURE 3.04 Group means moves and times

FIGURE 3.04 shows the group means for moves and times for four subjects for 25 runs. This FIGURE shows that the average moves became stable at about 250 moves per day, however the average time decreased from about 850 s to somewhere between 600 and 700 s. $F(24,192) = 2.182, p < .001$. The control rat's priority, in general, seems to have been to maximize moves not time. Furthermore, they simultaneously decreased the time spent in the maze without experiencing this as a loss or punishment.

The control group in this study, made it possible to understand more completely, the devious behavior of the rats seen in FIGURE 2.02 in the second study. Devious behavior at the run level of organization had a V shape, see FIGURE 2.01 and FIGURE 2.02, or in other words, devious behavior at the run level has essentially two phases. In the first phase, the run data seems like a traditional learning curve with the run-scores, run steps, and run latency all improving by getting shorter. Several points must be distinguished here. First, during exploratory-play behavior, once the induction phase has passed and the rat has found and accepted a worthy challenge, the challenge of making a model of the anomalous matrix maze, the rat will, from that point on, always be learning or growing. However, some animals do not find the maze a worthy challenge, as for example M1 seen in FIGURE 3.03. These animals do not seem to find any challenge capable of generating the degree of arousal that is necessary in order to qualify as having fun in this particular domain. Second, once a rat finds and accepts the challenge to solve the puzzle of the hierarchically organized structure of the matrix maze and is learning or making a model of the maze, punishment complicates the rat's options. There are several mutually exclusive options from which the rat can choose, but the presence of the punishment-induced-conflict seems to facilitate the rats' interest in the maze. The first phase of the V in which the rat is initially seen as improving its behavior, as indicated by both of the run scores, is in fact induced by the punishment contingency. That is, this first phase does not occur in the control rats even though they too are learning exactly the same thing the experimental group is learning. The difference is that the control rats also learn to run the shorter path and thus they learn to make growths that are more efficient. However, the control rats are not taken out after completing each of these ever more efficient growths as is the experimental group, and so the experimental run data will show only the first phase of the V curve. Both groups have the same purpose: to learn to run the shortest possible or the most efficient path, but the control group of rats can make as many growths as they like on any run for they are not punished after each of these AOA loops. The experimental rat starts the second phase of the V when it finally realizes that it is being penalized more and more as it perfected the efficiency of its path. The beginning of the second phase of the V signals the beginning of devious behavior. The second phase of the V signals the beginning of the rat's awareness of the conflict the punishment induces into its purposive system, the conflict between the exploring-self's purpose to perfect the shortest path behavior, and the play-self's purpose to maximize its time and moves in the matrix. Once rats accept the challenge of the puzzle of the hierarchical matrix, they understand how the options are modified by the punishment contingency; only then can they finally reveal their capacity for devious behavior. The second phase of the V is a new kind of growth phase, a second growth phase. The first phase of the V, the first growth phase, is a relatively simple phase of adaptation compared to the second phase. In the second phase, the devious behavior reveals the complex ways the primary mind of the rat is rational. During punishment, the rational nature of the function of the primary mind enables the mind to be especially agile. Mental agility is seen in the swiftness and the varieties of adaptive alternatives the primary mind is able to invent as a means to its end of attaining its goal in spite of the punishment. With devious behavior, the primary mind is capable of altering behavior in order to satisfy, as much as possible, both of its mutually exclusive purposes, to learn to run the most efficient path through the matrix and to maximize its playtime in the matrix.

The devious behavior of experimental rats whose spontaneous behavior in the matrix maze was punished shows that the rat's primary mind functions in some ways as a general-purpose batch-processing computer; but one capable of both writing and executing its own programs. However, the data shows that the primary mind effectively can program for only one purpose at a time. So as a batch processing system, the primary mind must alternate between designing a devious act to negate the effect of punishment upon play-behavior, and then on a different run solve the problem induced by punishment upon its goal of seeking the shortest path. There are two goals, two effects of punishment, and two devious solutions that must be solved one puzzle at a time. The complex devious behavior pattern of the devious rat reveals the rat's primary mind as a batch-processing computer, which shifts between two mutually exclusive purposes.

Devious behavior at the loop-level of organization

FIGURE 3.05 loop scores for 25 runs

The experimental group of punished rats in the last study revealed a complex pattern of devious behavior at the loop level that evolved over twenty-five growths. Since the control groups of rats in this study were not punished, this pattern of devious behavior should not and did not evolve.

FIGURE 3.05 shows the first order interaction between four loop scores for 25 runs. The basic phenomenon of interest in this FIGURE is the inverse relationship seen in the interaction between the two curves at the bottom. This interaction is between the number of loops per run, the red curve, and the number of growths per run, the blue curve. As the number of loops decreased from a mean of eight to a mean of about four, the number of growths increased from a mean of .2 to a mean of about two. But it is most important to understand the rest of the story, which the other two curves tell: as the number of loops decreased, the other two curves show that while the rats were making fewer loops, yet these loops were becoming longer in steps, and time.

As the experimental rats avoided the punished and feared alpha, by definition the number of loops went down, but now by necessity the length of these fewer loops increased in steps and time. So what were the rats doing with this additional activity in these fewer loops? The answer is that they were doing their devious thing: they were perfecting their ability to make a criterion using one of the other seven kinds of nested loops.

This same general relationship was also seen in the experimental group in the last study but this inverse relation was extreme in that experiment. See FIGURE 2.04C.

The other two variables in FIGURE 3.05 show that as the numbers of loops decreased per run, the size of the loops increased in both moves and time.

FIGURE 3.06 growths per run

FIGURE 3.06 shows the number of growths per run, but it is plotted by itself so that this important effect can be seen more clearly than was possible in FIGURE 3.05.

Devious behavior at the growth-level of the organization of behavior

The primary purpose of this third experiment was to run a control group for the experimental group that was punished in the second experiment. The growth data for the two experiments, II and III, were combined and analyzed, as if the data were from one experiment. The design is a mixed ANOVA with three between variables: year, data-type, sex, and then the several possible variations on the repeated measure variables: blocks, units, directions, rows, and columns. The difference between the experimental and control group at the growth level of the behavioral hierarchy, however, unambiguously supports the devious behavior hypothesis.

The FREQULM data set

FIGUREs 3.07A, B and C experiment by block by data type by unit

FIGUREs 3.07A, B, and C are the 4-way interaction, experiment by block by data type by unit (2*3*3*48), for each of the elements of FREQULM data set for three blocks for the growths 2-25 for the two groups $F(188,19458)=1.33; P< .0017$. The data is plotted as three separate, year by growth by units effects, one for each of the three elements of the FREQULM data set; A is the FREQU data, B is the FREQL data, C is the

FREQM data. FIGURE 3.07A confirms the prediction of the devious behavior hypothesis, the control group did not show the growth of avoidance behavior for the three blocks of growths as is seen in the experimental group data. This data proves that the increased amount of activity of the experimental group behavior in the second half of the maze was due to punishment. The control group of rats, which were not punished, continued to improve their performance in the second half of the maze: their FREQU scores got smaller while the experimental group got larger over blocks. This data shows that the effect of the third element of the constraint set, the removal of the rats after they made a criterion, AOA loop, had for the experimental group of rats, a punishing effect. This FIGURE is, however, somewhat deceptive for it seems to suggest that the avoidance behavior was confined to the second half of the maze; that the avoidance force was unidirectional, acting only in the second half of the maze, but this is not correct. The avoidance force acted in both directions, improving the performance in the first half of the maze, while at the same time degrading the behavior in the second half of the matrix maze. This effect will be described in FIGURE 3.08 below.

FIGURE 3.07A experiment by block by data type by unit

The FIGURE3.07A is a most complex and important figure; complex for its data bears witness to two different aspects of Heterostatic behavior: both the direction and the devious behavior effect.

The direction effect: The alpha-omega model distinction

Block one of FIGURE3.07A so effectively reveals the direction effect that it forces me to make a distinction between two EP models: The Alpha and the Omega model. The difficulty of both groups of rats on the way back shows that they were trying to use the Alpha model on the way back, which no longer works, and they have to start over and conceive a new model, the Omega to use on the way back.

The Matrix maze has several asymmetries, which force the rat to conceive more than one model, a universal model, for the whole maze. The first asymmetry is between the EP and the BP units of the maze. The second asymmetry is the difference between the out and the back half of the EP model. The EP model breaks down into two models: the alpha and the omega models. My first task is to describe the asymmetry that forced this distinction between the alpha and the omega models.

The direction effect: the direction asymmetry

The data in the first block of FIGURE 3.07A requires a comment. Note in Block 1, that the behavior of both groups of rats shows that they had difficulty in the second half of the maze. This difficulty, however, was not due to avoidance. A careful examination of the turn sequence required by the two directions will show that the back sequence is opposite of the out sequence. In other words, one of the asymmetries of the matrix maze is the direction effect. The sequence on the way out requires the rats evolve this first sequence as say the alpha model, and the sequence on the way back as the omega sequence. Part of the rat's problem is that it naturally assumes that the omega unit is another exemplar of the alpha unit, and thus they leave omega fully prepared to make the same sequence of turns that characterized the alpha unit. However, they quickly get feedback and no sooner do they emit the first element of the alpha sequence of turns than they wind up in units they did not expect to occupy. Technically, the Omega sequence of turns usually isn't more difficult than the alpha sequence. However, the data shows that some rats required as many sets and time to master the omega as was required by the alpha model.

As an aside, let me note that the two directions sequences of turns are different because the matrix maze is an even number of rows. Since the matrix maze has an even rather than an odd number of rows, the row sequence in the second half of the maze is reversed over the row sequence in the first half of the matrix, and both groups require time to master this reversal. This figure shows that the direction effect is a product of one of the asymmetries of the matrix maze.

This problem can be partially solved by using a 7-9 row maze; see FIGURE2.02E.

The 7-Row, symmetry, problem

What is the 6-Row, back problem? Why am I proposing that rats need to be run in a 7-Row maze? In the second half of Block 1 of FIGURE3.07A, for one example, we can see that both groups of rats have about the same amount of difficulty getting started on the way back. This difficulty is in the second half of the maze, on the return leg of the shortest Path through the EP to the Alpha unit is due, I propose, to the fact that the six-row maze is not symmetrical.

For example, look at FIGURE 2.02D and compare what the rat must do when leaving the Alpha unit. In FIGURE 2.02C note that the omega unit is Unit 62, and is opposite the alpha unit that is they both are in column six. So what would happen if one of our rats turns right when they leave Omega? The answer is that they would enter unit 55 which is A1 not E1. This asymmetry is a basic structural problem generated by having a six-row maze.

FIGURE3.07A shows that the fact that an even unit maze not being symmetrical is a universal problem that all rats must learn to live with. The fact is that the asymmetrical hierarchical structure, the D2, Rep 3. RT2, Units 8 maze could be said to have a direction dimension, only because it is asymmetrical. As long as there are two directions, technically there is a direction effect in the design. However, the D effect will not be significant unless there is something for the rats to master, but more simply nothing can qualify as a factor that can't be factored.

The paradox: there is no perfect solution to the matrix maze problems.

My 7-9 maze solution is really only a thought experiment for Plato is partially correct, matter sometimes is obstinate. Sometimes, there are no solutions to a problem without the solution creating new problems.

A twelve-row matrix maze hypothesis

A simple test would be to run rats on a loop testing procedure in a maze with a twelve row matrix and determine whether they experience any special problem at the transition between the six and seventh row. The difficulty my rats experienced could have been due to the number of rows not the direction effect.

Devious behavior: The punishment effect

The second and most important problem, which FIGURE3.07A calls attention to, is the devious behavior aspect of the nonverbal mind. The data of Block 2 and 3 clearly show that the punished experimental group emitted an enormous amount of data in the second half of the maze that can only be described as devious behavior. The data in Blocks two and three show that the experimental rats were perfecting their ability for staying as far away from the alpha unit as the maze allowed. Yet in the meantime, the control group was working just as hard to overcome the initial difficulty of the direction effect in the second half of the maze. In block three, the control group behavior in the second half of the maze finally looks like a smooth continuation of their behavior in the first half of the maze. The large difference in block three between the behaviors of the two groups in the second half of the matrix maze reflects the full avoidance effect induced by punishment. The difference between the two groups in block three show just how powerful a force the fear induced by punishment was.

FIGURE 3.07B experiment by block by data type by unit

FIGURE 3.07B shows that the avoidance force had the same effect on FREQL data, as it had in the FREQU data, however, the separation of the two groups does not seem as clean as in the FREQU data. Nevertheless, this is an illusion created by the greater range of the column cycles in the FREQL data as opposed to the FREQU data. A careful examination of the data will show that while the range of the in the FREQU data were from a mean of 1 to 10, the range for the FREQL data is from .7 to 60. So while the differences are actually larger in the FREQL data, yet, because of the scaling factor differences, these extreme differences do not show up as effectively in FIGURE 3.07B.

FIGURE 3.07C experiment by block by data type by unit

FIGURE 3.07C shows the FREQM data and the avoidance effect did not have any obvious effect on this element of the data set. The avoidance force does not affect the total average time a rat spends in each unit of the maze per visit. Avoidance does not change the duration of the unit entries but the number of unit entries. Avoidance affects the number of visits made to each unit in the second half of the maze, the FREQU score and therefore, the cumulative latency, FREQL. The effect of punishment upon avoidance is, in other words, not to make rats move more slowly. The punished rat keeps moving at the same speed as the control rat, but they allocate more moves to the second half of the maze.

FIGURE 3.08 groups by blocks by rows

FIGURE 3.08 is the 3-way interaction, groups by blocks, by rows (2*3*12), for the FREQULM data set for three blocks for two groups. $F(22, 4554) = 2.15$; $P < .0014$. This FIGURE contrasts the block effect for the three blocks for the 12 rows for the two groups with the experimental or punished group on the left and the control group on the right.

This FIGURE has less clutter than FIGUREs 3.07A, B, and C, and thus allows the changes that occurred in both halves of the maze to be compared. Both groups' improvement can be seen in the first half of the maze over the three blocks with the greatest improvement for the experimental group occurring in the first block, while the control group improvement occurs between the first and second block.

However, the control group's behavior on the right forms the baseline against which the experimental group's behavior must be contrasted. Several features of this data must be noted. First, note that for the control group the change, over blocks 1, 2, and 3, are in the same direction in both halves of the maze with the larger scores always in the first block in both halves and the smaller scores in the third block in both halves of the maze. These rats' improvement was in the same direction in both halves of the maze in all three blocks. For the control group, clearly no avoidance occurred. The second half of the matrix was more difficult than the first, but by block 3, the control rats had perfected their performance as much as was possible; the mean for the green block three was down to one.

However, the experimental group's behavior may seem very complicated; yet every feature of its behavior can be explained with a simple rule. The first difference to note is that the order of the block effects is reversed with the size of the scores in the first half of the maze inversely related to block number. Block 1 scores were the largest and block 3 scores the smallest. However, in the second half of the maze, the relation between size and block score is a direct relation, the block 1 scores are the smallest and the block 3 scores are the largest. The Devious Behavior hypothesis states that because the Experimental group of animals was punished for entering the alpha unit after each growth, they began to avoid reentering the alpha unit, but they did not stop making complete loops but instead redefined the criterion loop by picking one of the other possible seven patterns.

The control group in this experiment was run to determine whether punishment was the cause of the avoidance behavior. The results of the control group's data confirm this hypothesis. The control group showed no signs of avoidance behavior. However, the control group of this experiment proved the point. The devious behavior did not occur in the control groups of rats.

However, the data in Experiment II clearly supports the push-pull theory of the deviousness hypothesis. First, while on the first block the Experimental animals got off to a slow start--with a large initial value of 25, yet on the first block they improved down to the value of 2, which was better than the control means on block one. Block one shows that fear was at their back and it moved them through the first half of the maze faster. Second, on block three they, in effect, had the wind in their face, which slowed them down and induced them to take more moves. This Figure thus adds its support to my push-pull theory of devious behavior.

The full maze data

My understanding of the spontaneous behavior of the rat in the matrix maze evolved in many steps, but the focus in this section is upon describing how the evolution of the growth level of behavior differed as it evolved in different parts of the full maze. In particular, my understanding of the emergent loop behavior evolved in three steps.

First, in Experiment 1, I described the Emergent Loop, EL, and behavior only for the EP data: since with only the data for one EL available there were many cells in the Full maze, which the rat never entered. A statistical analysis of the full maze cannot be made until every rat has entered each unit of the matrix at least one time. Since the rats, in Experiment I, were only allowed to run one growth and there were so many empty cells, it was impossible to analyze the blind data. In this section, for the first time I will be able to describe the EL for the Full maze, FM, for finally in Experiment II and III, the rats were allowed to make enough growths to fill all the cells of the full maze matrix. The paradox is that the shape of the curves, the EP and the FM, for these two sets of data are generally opposite. The reason the EP and the FM data are so different is due to the difference between the EP data, and the Blind Path, BP, data. The full maze, FM, data looks so different from the EP growth data because the BP growth data is so large it completely dominates the Full maze data. In other words, it overwhelms the EP growth data. First, I will describe the full maze data, and then I will show that the distortion of the full maze data is due to the effect of the BP data. The ultimate point of this section is the recognition that the full maze data is the sum of the EP and the BP data, but the latter is so large that it dominated the EP component.

Historically the BP behavior of the rat has always been considered as tangential to the rat's goal to learn the EP. While that perspective is useful, the fact is that the behavior of the rat is just not that simple. The behavior of the rat in the BP units of the matrix involves principles that are unique to the BP, thus the BP behavior requires its own special analysis and its own special theory, and proving these points is the point of this section.

The BP behavior of the rat was the most difficult element of the rat's behavior for me to understand, perhaps partially because I was not expecting the complexity it revealed.

The data of Experiments II and III have been combined in this analysis because the primary interest of this section is to describe the full maze growth behavior of the rat. While there are significant differences in the full maze behavior between the experimental and control groups of rats and between the sexes for the two groups, these differences are of secondary importance at least for the moment. The primary focus in this section is upon the general or basic exploratory-play behavior of rats in the FM and about how the EP and the BP components of the FM differ. After, the basic phenomena have been explained, the subtleties can then be considered in a later section.

Just as each of the three sets of growth-level dependent variables scores (FEUL, FREQUL, and BFEUL), each told an important part of the EP story of behavior, just so they tell an important part of the BP story, but they tell a very different story about the full maze and the blind-behavior. The rat's primary nonverbal mind weaves a different solution for the BP puzzle. The rat's primary mind enjoys a good puzzle and is most proficient at solving the ones posed by the Matrix Maze for both the EP and the BP puzzles are solved in less than an hour's time while exploring the matrix maze. However, it took me a scientific lifetime, forty years, to understand how they solved these puzzles. In this section, I will group the FIGUREs in order to make a direct comparison between the comparable elements of the three data sets for the BP behavior, but I will ask the reader to refer to the comparable FIGUREs for the EP data in the earlier sections.

The three data sets normally each contain three scores for nine total scores. However, in this section, I will be using abbreviated data sets with only seven scores: The FEU and the BFE data sets will each have only two elements, FEUL and BFEUL. The third element in each case was dropped because it did not serve any useful purpose in this context.

Our statistical package of choice, CSS, had a constraint, which established how many repeated measure scores were allowed in a design. This constraint was no problem as long as only EP data was being analyzed for one growth since the EP data was only a forty-eight-unit path. But now that our purpose is to analyze the full maze data, a 96-unit path, this CSS constraint makes it impossible to do the appropriate analysis. So, in this section I had to modify our design to deal with this reality. Ideally, the next three FIGUREs should be plotted as a single 96-unit effect. However, because of the CSS constraint, the data will be plotted as two 48-units paths, in other words, the full path FIGUREs will be broken up into two 48-unit effects; one for each of the two directions. This is not an ideal solution, but the compromise is not as bleak as it might sound for the focus of our attention in the study of the FM data is on the full column eight-unit effect and not on the units or the rows effects. What is unique about the FM analysis as opposed to the EP analysis is what is happening in the BP units of the maze, so my purpose in this section is to focus the reader's attention upon the 8-unit column effect instead of a 4-unit column effect.

FIGUREs 3.09, 3.10 and 3.11 full maze data types by units

FIGUREs 3.09, 3.10, and 3.11 are three full-maze first order interactions graphs, data types by units for forty-eight units. Each graph is a plot of one of the three data sets: FEUL, FREQULM, and BFEUL, respectively. The data in these FIGUREs is the average for the two halves of the matrix ignoring the direction effect. The data contained in these three data sets include the data from both the EP and the BP. Our task is to understand these ANOVA designs in order to learn to recognize the rat's solution to the blind-unit problem.

In FIGURE 3.09, the FEUL data set shows the EL data for the FM for six of the eight unit rows for a total of forty-eight units; the average for the two directions, out and back. The first entry data, FEU, for the FM emergent loop data, EL, is the average of the FEU unit scores for both directions for both groups, while the FEL score is the average of the first latency in both directions for both groups. The series of units 1-48 on the way out, were averaged with the series of units 49-96 on the way back.

The FE scores, by definition, are a monotonic series of scores, and the sizes of the scores are roughly related to the rank order of the unit in question within the matrix. The data is plotted as a log-linear graph in order that the differences in the FEU scores, which are so much smaller than the FEL scores, will be visible. Two obvious effects are responsible for the variability in these two sets of data; the long-term effect will be shown in a moment to be **the row effect**, and the higher order variability is **the column effect.** Vertical lines should have been added to mark each set of eight units in order to facilitate the recognition of the full maze column effect.

So what does the FM 48 unit curve loop look like? In both the FEU and the FEL data, the 48 unit exhibit an eight unit-cycles, or 6: 2-4-2 cycles: see FIGURE3.16 for the basic cycle. FIGURE3.9 consists of 6 cycles as seen in FIGURE3.16.

FIGURE 3.10 full maze data types by units

FIGURE 3.10 shows the FREQUL data set with each of its three scores having its own unique rhythm or pattern. However, while the size of the FEU score was directly related to the unit order, in this graph the size of each of the FREQUL scores is inversely related to the unit order. This inverse relationship is one of the most well established facts in animal maze learning. Animals exploring a maze for the first trial are notorious for spending most of their time retracing in the first units of the maze. The reader should divide the data into sets of eight in order to be able to note that for each of the three scores the higher rate rhythms in the data are a function of the full maze eight-column effect within rows. Each of the three scores shows the inverse effect with units, even the FREQM scores. The learning effect is very similar for each of these sets of scores even though each set has its own rhythm.

The Behaviorists for all their talk about studying behavior never found the need or the time to measure or objectively describe any form of behavior in the detail it deserves, but especially they never studied this retracing behavior, which was at the heart of their theory of behavior. The consequence was their loss; they never observed this behavior, which they claimed was random with the degree of resolution required objectively to test their theory. If they had, they would have observed as I have observed that this behavior is complex and reveals in the matrix maze, the beautiful patterns, and rhythms. These patterns can only be measured by a technology such as what I call the behavioral telescope, BT.

FIGURE 3.11 BFEUL full maze data types by units

FIGURE 3.11 shows the BFEUL data set for 48 units. This data also should be divided into sets of eight in order that the full maze column pattern or rhythms could be observed. However, the complexity of the full maze pattern in this FIGURE may not be obvious to everyone's eye. But that is why ANOVA is useful. However, in general, what is new and different about the full maze BFEUL scores is that the size of these scores are also roughly but directly related to their unit position within the matrix. The column effect, however, in this data set has a much greater variability than in the two previous FIGUREs. The direct relation seen in the full maze data between the size of the BFEULM scores and the unit position shows that retracing in the full maze data does not extinguish as does retracing in the EP. The question is this: why does the FM BFEULM data behave so differently from EP data?

The parts of the whole

The three data sets seen in the FIGUREs above are complex wholes, each of which consists of many parts. These parts will be identified in this section, and the breakdown of this data will show that in spite of the apparent complexity of these

FIGUREs, the FM exploratory-play behavior of rats has a relatively simple, yet unexpected order.

The row effect

FIGURE 3.12 main effects for rows

FIGUREs 3.12, 3.13, and 3.14 are the main effect for rows for the three previous data sets, respectively. Notice how very different each of these three row main effects is. FIGURE 3.12 is the row effect for the FEUL data of FIGURE 3.09, and shows the direct relation between the FEU data and the unit order that was obvious in FIGURE 3.09. In other words, the first entry scores of the emergent loop for the FM keep getting larger with each successive unit, which is exactly what has to happen given the nature of the score. However, the row effect seen in FIGURE 3.12 shows that the rate of the progression through the twelve rows was much slower in rows five and six than in the other rows. This row main effect shows that the rate of exploratory progress is not smooth and continuous throughout the maze. Rows 5 and 6 were rows in which the rats' progress was very slow for reasons, which will become clear in a moment.

The means of ANOVA

ANOVA is fun and powerful for me for I understand it nonverbally rather than mathematically. The statistic is the sum of d^2, which I think of as a measure of cheating, or bad matter. If every exemplar were perfect, was an exact copy of its "ideal," then Plato would have had no need for an out-of-this-world-world, and all F-ratios would be zero, and no significant effect could be seen. When exemplars differ from their ideal, it is because either some Independent variable or some error factor is acting on the data.

A significant F-ratio indicates that some independent variable is causing the replicates to be different from their mean.

FIGURE 3.13 row main effect for FREQULM

FIGURE 3.13 is the row main effect for the average of the three scores of the FREQULM data set. This data confirms once again as was seen in FIGURE 3.10 that the rats in this study spend the majority of their allotted moves and time exploring or retracing in the first units of the maze. This is conventional behavior for all rats run in a maze using the work, TTP, or play test procedure, LTP. The shape of this curve is the same for both the EP data and the FM data. The log-linear plot makes it possible to see the relatively minor difficulty rats have in the first row of the second half of the maze, row seven, because of the reversal of the order of the rows in the matrix maze. The log-linear plot was chosen because otherwise on a linear plot this small effect would not have shown.

FIGURE 3.14 Row effect for BFEUL

FIGURE 3.14 and 3.15 are two different ways of describing the row effect for the BFEUL data. The former shows the row effect for 12 rows; the latter shows the row effect for six rows. The twelve-row effect allows the change of rate of exploration that occurred in row five and six to be seen and explains the extreme amount of retracing triggered by the units of the fifth and sixth rows seen in FIGURE 3.11. However, since this is row data it is not possible to tell which column units within these two rows, five and six, induced the retracing or if the retracing was induced by the EP units or by the BP units of these two rows, and this is the critical point.

FIGURE 3.15 row effect for both directions

FIGURE 3.15 shows the average row effect for both directions. In other words, the data is the average of the two directions and thus this six-row effect shows more clearly than did the 12-row effect that the retracing induced by the rows increased in all six rows in both directions. This is one of the novel aspects of the full maze data, which must be explained. However, once again, the question remains did units of the EP or BP induce the retracing? The BFEU data seen in FIGURE 3.14 and 3.15 most effectively reveals the phenomenon that distinguishes the BP exploratory-play behavior from the EP data. BFEUL data is always a measure of retracing behavior. The BFEULM data set for the EP for the first growth showed that retracing in the EP usually stopped somewhere within the first half of the maze. This behavior proves that rats were able to make a valid model of the EP units of the matrix so quickly that they stop retracing even before all the units of the EP in the first half of the maze had been explored. However, the BFEUL data of FIGURE 3.15 shows the full maze behavior of the rat to be very different from its EP behavior. This FIGURE shows that in the full maze data, which includes the BP as well as the EP data, retracing did not extinguish half way through the maze, but continued to increase all the way to the last row! This extreme difference between the retracing behavior of the rat in the EP and the full path shows that the BP component in the FM data must be very powerful. And so, this data raises the next question: How does the evolution of the rat's model of the maze affect the behavior of the rat in the BP units? Why do rats not stop BP retracing the same way they stop EP retracing?

In order to understand this FM retracing data, it is important to recognize that the BFEUL retracing data for the full maze includes two different components of retracing; first, the retracing induced in the EP by **anomalies** and **news**, and secondly, the retracing induced in the BP units by **redundancy**. In other words, there are several similarities and differences between the EP and the BP retracing behavior. The first difference--and this is a major difference--is this: EP retracing is induced by anomalies while BP retracing is induced by redundancy. Anomalies and redundancy are on opposite ends of the same continuum, and yet the BFEUL score for the full maze data is a measure of the retracing induced by both of these forms of retracing. The problem with the full maze data is that these two forms of retracing are confounded in the Full maze data.

FIGUREs 3.14 and 3.15 make it possible to gain a new appreciation of the BFEULM data of FIGURE 3.11. Once the full maze BFEULM row effects for the 12 or 6-row main effect have been recognized, the same effect can be seen even more dramatically in FIGURE 3.11—except at the time we first looked at this FIGURE, this feature failed to draw our attention. In other words, the overall increasing monotonic shape of these curves in FIGURE 3.11 reflects the effect of the BFEULM BP data, not the

EP BFEULM data. Therefore, the first big conclusion is that the big difference between the EP and the FM exploratory-play-retracing behavior is the difference between the shapes of the retracing functions. EP retracing, as was shown in earlier experiment, is a curvilinear function that hits a peak in the first half of the maze, while the FM retracing, or by implication, the BP retracing function is a monotonic function. While EP retracing extinguishes even while the rat is still exploring the first half of the matrix, yet the FM retracing, or the BP component gets larger row by row in both sections of the maze, both out and back. The BP retracing does not extinguish at least on the time scale reported in this study.

The simplest way to understand these facts is to recognize what is common to both.

In a real sense, the whole point of rationality, the whole point of biological evolution's creation of the primary mind was to create a machine that could, in effect, take over the role of evolution and do kinds and rates of evolution that were impossible for biological evolution.

Homeostasis was, in Cannon's concept (1932) of physiology, the wisdom of the body. Cannon, however, did not recognize any other kind of wisdom, nor even suggest that there was any other kind, or even what wisdom is, in the first place. This book is designed to take off where Cannon stopped. I define many kinds of wisdom with Heterostasis the third wisdom, the wisdom of the primary mind. The third wisdom invents mental things not biological things, mental species not biological species.

The rat during spontaneous play within the matrix maze is exercising this third wisdom; it is inventing sets of species of mental things, classes of mental things which allow the primary mind to understand where it is within the matrix maze at all times. The primary mind of the rat, I am suggesting, finds it so simple to evolve a valid set of classes to represent the EP units, that it designs a valid conceptual system for the EP during the first few rows. This model is so powerful that the primary mind of the rat can extrapolate this conceptual system and run the last few rows of the EP without making any errors.

However, the primary mind finds that evolving a set of rational concepts that allows it to understand and extrapolate a path through the BP a much more difficult task. However, there are some units of the BP, which the primary mind of the rat can conceptualize very easily. What I must do next is answer this question: which units of the BP are easy, and which ones are difficult to conceptualize? The rate of Mental Evolution is a function of the conceptual difficulty of a unit. What makes a unit difficult to conceptualize?

One of the major advantages conferred by a valid model of the matrix domain is that it makes it possible for rats to avoid both nonessential or redundant units, but this advantage works differently in the EP and in the BP units. In the EP, the rat continues to retrace within those units of the matrix having potential knowledge or undigested strangeness that has not yet been assimilated by the lower orders of the hierarchical models of the maze. Once the rat has exhaustively explored a maze unit and has been effectively able to incorporate all of its salient strangeness into its model, the rat stops retracing and begins exploring new units of the maze. In the blind-units, the same phenomena also occurs, except, and this is the paradox, while the avoidance of redundancy in the EP stops the retracing in the EP, the avoidance of redundancy in the blinds has the opposite effect and initiates retracing. The sooner the primary mind evolves a model that is able to recognize redundant blind units, the greater will be the size of the BP retracing scores. However, the sooner the rats' model is able to detect redundancy in the EP, the smaller will be the retrace scores. This point is strange, but the difference between EP and the BP behavior is tied up in this point.

While the concepts of retracing in the EP and retracing in the BP are technically the same, yet they have opposite meanings or consequences on retracing behavior. The BFEULM scores measure retracing by definition. Yet, in the EP, retracing is induced by anomalies and news; the BFEULM scores measure how much retracing activity the rat must make before its model of the domain is able to assimilate the potential knowledge contained in the uncertainty-inducing units that initiate or sustain retracing. However, in the blind units the rule is very different: As the model of more and more elements of the BP units becomes valid, the primary mind of the rat is able to anticipate redundant blind units, but now notice what happens. While in the EP when a unit becomes redundant, the rats stop retracing into that unit. However, in the BP when the rat detects that a model is redundant, it refuses to enter that unit for the first time and by avoiding that redundant unit, initiates BP retracing. The rule is that the FM or the BP BFEULM retracing scores measure how much activity is required before the rat for some reason finally enters the redundant blind units in spite of the redundancy. The BP retracing score measures how many steps a rat takes before it enters a BP unit after it has refused or failed to enter that unit for the first time.

FIGUREs 3.14 and 3.15 both furnish the data that allows the problem to be stated which must be explained: how does the evolving validity of a model affect the retracing behavior of rats in the blind units? A valid model in the BP allows rats to avoid blind units, which are redundant exemplars of sets of spaces that the model is competent to recognize and predict, while the size of the BP BFEUL scores reflects which elements of the blinds are first experienced as redundant. The size of the BP BFEUL scores are the best estimate of the amount of redundant activity the rats were able to generate due to the evolving validity of the model of the maze to recognize the redundant blind units of maze.

Understanding the exploratory-play behavior of the rat in the matrix maze is not easy because as I said, the understanding of this behavior must evolve. In this next step of mental evolution, I will show how the mystery deepens and requires yet another step before the story begins to get clearer.

The column effects

FIGURE 3.16 full maze 8 column effect

The particular data that is unique for the FM can be seen to be a member of the eight-unit column effect as opposed to the four-unit column effect that characterized the EP data. FIGUREs 3.16, 3.17, and 3.18 are graphs of the full maze, eight-unit column effects; one for each of the three data sets and explains the higher frequency cycles seen in FIGUREs 3.09, 3.10 and 3.11, respectively.

My fondest hope, or perhaps illusion, is that the readers of this book will include those who are not authorities on ANOVA. So I am going to explain some of these sections in more detail, and more often than would be required if I were writing for a pure scientific audience. In the paragraph above, I stated that there is a special relation between the 2 sets of FIGURES 3.16, 3.17, and 3.18 and FIGURES 3.09, 3.10, 3.11.

One cycle six cycles
1. 3.16------------------3.09
2. 3.17----------------- 3.10
3. 3.18----------------- 3.11

In this section, my concern is with calling attention to the behavior of mammals in columns. The behaviorist in the TTP attended exclusively upon what I would call the row effects, so my task is to call attention to the column effect, which seems to be the more important component.

In the whole maze there are six rows out and six row back or 12 rows and each of these has 8 column units and the behavior of the rat in these 8-units of each row is the column effect. ANOVA works with averages or means, and there are many ways the column effects can be averaged, but in the first column above, the one-labeled one cycle; the effect shown in this column is the average of the column effect for all twelve rows. The question remains are the column effects all the same or are they different?

Metaphorically, Plato might call this mean of the twelve columns the "ideal" mean, for it is the more stable than the six particular row means. Again in Plato's world, we could be asking how exactly the six exemplar column means matches the ideal, mean column effect. Plato imagined that matter was both good and bad. It was good in that it strove to replicate the ideal. Supposedly, there was a perfect eternal model of a cat, and every exemplar cat ever born had tried to duplicate the perfection of the ideal cat. Matter was bad for every exemplar failed to capture the perfection of the ideal.

And roughly, ANOVA looks at the world in much the same way.

In our case, the mean of the six columns or the main effect can be thought of as Plato's (1952) ideal and each of the six columns is an exemplar, which strove to copy the ideal. But matter is bad, and none of the six row means every achieved the perfection of the main effect of the column mean.

Next, the question is what is the relation between this one ideal cycle in this context and the six cycle curves? Plato would say that each column cycle in the six column or six cycle effects is trying to take on the ideal form of the one cycle mean. But what I want the reader to do is to see that each of the complex curves consists of six eight-unit column effect curves. The computer asks for us if each of the six curves is the same or different from the ideal form, which is an average of them all.

With FIGURE 3.16, the FM data begins to get very interesting. FIGURE 3.16 is the main effect for the eight-unit column effect for the FEUL data set and shows that the eight-column effect produces a very different column effect than the one seen in the EP. The EP four-unit column effect is, of course, still a component of this data and can be seen in the E1--E4 units in FIGURE 3.16. What are new and strange about this eight-unit BP data, are the four scores for the F and A-types of BP units. The most striking effect is the data for the forward-type blind units, F1 and F2, which are column levels 7 and 8 in FIGURE 3.16. The data in FIGURE 3.16 shows that F1 and F2 had, on the average, the largest FEUL scores, FEU and FEL: which means that they were the last units of the matrix to be entered. Additionally, this FIGURE shows that the A1 and A2 were also, on the average, among the last four units of the maze to be explored by rats, these are units 1 and 2 in FIGURE 3.16. This data shows that all four of the BP units had larger FEUL scores than any of the four units of the EP.

The FM data is only being described in this study for the first time because as I have said, it takes many growths before all the BP units finally have at least one first-entry score. This FIGURE reveals, in other words, another remarkable property of rats' behavior, their power to avoid redundant blind units. FIGURE 3.16 shows that the BP units of the matrix are the last units to be entered, with the Forward BP units most effectively avoided by rats. The Forward BP units of the matrix are the units that the rats' model is most powerful at recognizing and predicting as redundant. These BP units are judged by the rats' primary mind as having no Potential Knowledge, PK, worth exploring or gleaning. Why then, do rats finally ever enter these redundant BP units? Eventually, I can only guess, after the primary mind has harvested the entire PK from all the other units of the maze, there must finally come a time when even the avoided Forward BP units are judged as deserving a look. Even these redundant Forward BP units, which had the lowest priority for attention, must finally achieve relatively enough PK to deserve some attention that can only be had by finally making a FE.

FIGURE 3.17 data type by columns

FIGURE 3.17 is the first order interaction, data type by columns, for three FREQULM scores for the FM maze eight-unit column effect. $F(14,126) = 4.85$; $p<0000$. Each of the three elements of this data set has its own unique pattern, and each looks very different from the EP four-unit column bucket effect, which still can be seen in the four units E1-E4. The additional two units added on each side of the EP data has a Gestalt embedding effect and make the EP bucket data, which we have seen several times by now almost unrecognizable. Each of these three shapes explains one component of the high frequency cycle's effect for one of the curves seen in FIGURE 3.10. The relation between these three scores is important to understand, and their relation to the other growth data set scores is important to understand. Contrast the FREQU scores in FIGURE 3.17 with the FEUL score in FIGURE 3.16.

However, in all the complexity of FIGURE 3.17, the one curve that first deserves our attention is the FREQU score. Compare this curve with the FIGURE 3.16. The FREQU curve of FIGURE 3.17 is roughly a rolled over version of FIGURE 3.16. Together these two curves tell an important part of the story. For example, compare the value of the F2 unit in the two curves. F2 in FIGURE 3.16 was the last unit of the matrix to be entered by the rat, and FIGURE 3.17 shows that F2 was also the least entered unit in the maze. So the rule is, roughly, in the BP there is an inverse relation between the numbers of times a unit was entered, and the time it was first entered. In other words, the unit that was entered last was the least entered. The BP units that were entered first had the most unit entries, while the units that were entered last had the least number of unit entries. The rule holds for both the EP and the BP. In the EP, those units that were the first entered were the most entered units. However, in the BP, the units last entered were the least entered units of the maze.

These two FIGUREs show that column eight, the F2 unit, was the last unit entered on the average and the least entered unit of the matrix. The rats' behavior shows that the F2 unit of the matrix was the unit most easily perceived by rats as a redundant unit of the maze. The rats' behavior shows that this unit was the easiest BP unit of the matrix for the rats' model of the maze to recognize, predict, and thus avoid. Moreover, the latency of the rats within this unit when it was finally entered was on average over 10 s; this latency is a relatively large one, as we shall see.

Each of these three curves in FIGURE 3.17 is the prototypical FM 8 column shaped curve. If the reader will look at FIGURE 3.10, they will see that each of these prototypical curves accounts for the high frequency component for one of the three curves in FIGURE 3.10. Each of the three curves in FIGURE 3.10 consists of a series of one of the prototypical curves. Each of the three curves of FIGURE 3.10 is an eight cycle series of one of these curves.

FIGURE 3.18 column effect for BFEULM data

Each of the FREQU data points in FIGURE 3.17 tells us how much retracing occurred in each unit of the maze, but the FREQU data does not say what unit of the maze triggered the retracing. FIGURE 3.18 shows the column effect for the BFEULM data and reflects the amount of retracing triggered by each unit. FIGURE 3.18 shows the eight-unit column main effect for the FM BFEUL data. What an amazing set of data! The scores for the EP, the E1-E4 scores are, of course, the same as the data in FIGURE 1.4.13. The EP data tells the first part of the story. The EP data shows that the only units that triggered EP retracing were E1 and E2. In other words, these two EP units were the only two anomalous units in the EP of the maze; the only units in the EP whose novelty was so excessive the rat could not assimilate them on its first encounter. These two units were unique in that they forced the primary mind to retrace until the model had evolved sufficiently so assimilation of the features of these units could occur.

But now, the blind-units tell the second half of the story. The F1 and the A1 units are to the BP what the E1 and E2 were to the EP. That is, just as E1 and E2 triggered all the retracing in the EP, just so F1 and A1 triggered all the retracing that occurred in the blind units! F1 is to E1 as A1 is to E2. However, while there is this important similarity between these units, there is on the other hand a major difference between them. The mechanism that caused the retracing in the two cases could not be more different but more about mechanisms later.

The FIGURE 3.18 and FIGURE 3.11 relation

In the discussion of FIGURE 3.16 above, I included a little table showing how three pairs of Figures were related. In the first column, I listed three one-cycle figures, and in the second column, I listed three six-cycle figures. The last pairing was FIGURES 3.18 and 3.11, and I explained that ANOVA has the statistical power to measure how alike or different the Six-cycles were from the "Ideal" one cycle curve. You have to look closely, but FIGURE 3.11 consists of a series of six of the one-cycle curves seen in FIGURE 3.18. Can you see FIGURE 3.11 as six instances of FIGURE 3.18 if you view them side by side?

The row by column interaction

The Group main effect for columns for the FEU data of FIGURE 3.16 only tells the first part of the story, but it gives the first clue to understanding the rat's BP avoidance behavior. While the F1 class of BP matrix units was, in general, the most avoided class of matrix units, it is now important to look at how the avoidance of the F1 unit varied with rows. In which particular row was F1 most effectively avoided? FIGURE 3.19 tells this part of the story. FIGURE3.19 is twelve cycles of FIGURE 3.16, one for each row. But FIGURE 3.19 shows that the large values of the wings for the BP units did not show up until row 6. The avoidance of the Blind units did not get serious until each rat had a chance to make a model of the Blind path units.

FIGURE 3.19 rows by columns for FEUL data

FIGURE 3.19 is the first order interaction, rows by columns, for the FEUL data set and is another one of the many new kinds of Heterostatic growth curves that the matrix conception of the maze makes possible. This FIGURE gives a detailed picture of how the BP column-avoidance for each of the eight column units evolved over the 12 rows. The data in FIGURE 3.19 shows that the FEUL main effect for rows seen in FIGURE 3.16 effectively only takes form in the second half of the maze. The data of FIGURE 3.19 shows the rate which the blind units of each of the columns became redundant over rows and thus were avoided by the rats. The blind units did not become distinct from the EP units until long after all the EP units had been entered for the first time in the first growth; in fact, the BP units were still largely empty cells at the completion of the first growth. The redundant BP units were finally entered only after the rats made several additional growths. The data in FIGURE 3.19 shows that the last unit of the maze to be entered was F2 of row six at a mean score of over 6000.

I am going to stop describing the BP behavior for these rats at this point in this Chapter and pick up this topic in the Experiment IV when describing the BP behavior of mice. Since the BP behavior of rats and mice are sufficiently alike, it will save some time to describe their BP behavior together for even the two species' differences are instructive.

Discussion

The anomalies of this experiment afford an opportunity to gain new insight into the workings of the primary mind in the two dimensions of the spatial structure of the matrix maze. Two sets of anomalies will be considered: first, the anomalies associated with the direction dimension of the matrix maze and secondly, the anomalies associated with the BP dimension of the matrix.

The direction-dimension anomalies

So far, my analysis of the rat's behavior has led me to assume that on the way out some rats evolved a set of nonverbal laws that allowed them to understand and predict the spatial pattern of the matrix structure of the whole maze with a single set of laws called the Alpha Model of the matrix. Or perhaps, one that should be called an Alpha-Omega model, which was valid in both directions. This Alpha-Omega model was conceived as a set of nonverbal laws, which would support extrapolation and take the uncertainty out of the maze in both directions. If you recall, some animals evolved a model that proved valid for all the units of the whole maze while exploring only the first half of the matrix, often even before they were half way to the omega unit.

Yet, the data of this experiment shows that this assumption is not always true. In other words, there is still another aspect of rat's behavior I have not yet discussed; one more set of anomalies or asymmetries that must be studied and understood. These anomalies can be seen in both FIGUREs 3.07A and B in the behavior of both the control and the experimental groups. What puzzles me is that if you look closely, you will see that according to the data in these FIGUREs that the rats, in effect, had to solve the mystery of the structure twice, once on the way out and again on the way back.

Under these new test conditions, the Alpha model, which was conceived on the way out, proved valid only for the first half of the maze. The data of FIGURE3.07A and B shows that on the way back these rats had to conceive a new model, the Omega model, which removed the uncertainty in the second half of the maze for use on the way back from Omega to alpha.

The question is why wasn't one model enough? I can only assume that there is some difference between the out and back structure of the maze that was obvious to the rat but not to me. What was this originally unrecognized difference? The rat's behavior must provide an answer to this question.

The first clue we have to the rat's problem is the quantity of exploratory steps and time the animal required in the first few rows on the way back. If their alpha model, their first solution, had been valid for the whole maze, then why did these animals upon leaving the omega unit on the way back to alpha, as FIGURE 3.07A shows; all of a sudden have to start over? The data shows they had to evolve a completely new set of nonverbal laws for the second half of the matrix. The mystery is obvious: why did they have to evolve two sets of laws? Obviously, the laws evolved on the way out didn't work on the way back.

In order to take the next step in the evolution of our thinking about the nature of the rational machine, which is the primary mind, we must find a better way to understand and talk about the hierarchical structure of the matrix maze. Perhaps, the rat's primary mind sees features of the matrix, which are beyond our current conceptual capacity to recognize. In experiment M87 of Chapter 2, I proposed several alternative conceptual frameworks, which time had proven necessary and useful to understand the features of the rat's behavior that I had recognized based on the data of that study. However, this study, D87, forced me to take in new realms of mental functions; ones we encountered, but for which we were not adequately conceptually prepared. Every anomaly says the same thing; anomalies cannot happen when the mind is prepared. The primary mind is an anomaly engine, a machine that burns anomalies in the fire of its mind and only anomalies. Anomalies fuel the primary mind, and a wise person rejoices at the pain each anomaly induces, for anomalies are always two-edged swords, as the oriental god Shiva symbolized, the god of death is the god of life; a new life can only emerge upon the ashes of the old, a new paradigm cannot come until an old one dies.

So now, it is time for me to take the next step and propose a new model of the matrix, which will bring into our working consciousness features of the matrix, which have heretofore been invisible. It is now time to go back to school and enter the next grade. When we have mastered this next level of the hierarchy, we will be able to see things that heretofore were invisible. We become wiser; as we climb the mountain, for every grade gives us new conceptual perspectives that allow us to see more and further, both quantitatively and qualitatively than ever before.

Figure3.07 2nd half problem

Figure3.07A is one of the most beautiful and interesting figures within the set of figures generated by the four experiments described in this book. What makes it interesting is that it reveals two separate anomalies, the direction, and the devious behavior effect.

However, the data in FIGURE 3.07A shows that the mastery of the second half of the matrix was not as difficult as was the first half. Yet, the curve clearly shows that the rats almost had to start over; they had to evolve two models, one for the first half of the matrix, the alpha model and another for the second half of the matrix, the Omega model.

A new Frame of Reference for the matrix maze

In the second experiment, I proposed several alternative conceptual frameworks, which time had proven necessary and useful to understand the features of the rat's behavior that I had recognized based on the data of that study. However, the new data of this experiment demonstrates that there are structural features of the matrix, which I have overlooked until now. The primary mind is a two edged sword with respect to anomalies. The primary mind is both an anomaly detecting and an anomaly-destroying machine. The primary mind is a bird dog that seeks anomalies and a metabolic system that burns anomalies. Anomalies fuel the primary mind. It is time to take the next step and propose a new model of the matrix, which will bring into our working consciousness features of the matrix, which were invisible in the first model.

With the evolution of each kind and instance of wisdom, by definition, life expands its mental/symbol capacity to interact with the real world, not Plato's (1952) "real spirit world," but the real material world. The Third Wisdom was the first rational nonverbal wisdom to emerge, and it expands or grows in response to spatial/temporal anomalies.

So each new anomaly, once again demands that the rats expand their nonverbal wisdom, or, in effect, to go back to school, to enter into the next grade of complexity of the hierarchically structured maze. When rats or their observers mastered each new level of the hierarchy, they will be able to see things that heretofore were invisible.

The nonverbal mind evolved to execute a Kantian function; one Kant (1952) attributed to God. Kant proposed that the mind supplies sets of hierarchically organized concepts, the mental tools by which the mind grasps or gives meaning to precepts, every perception was the product of the synthesis of a precept and concept, yet for Kant the origin of these concepts were innate-i.e. they were the gift of god.

Both Plato (1952) and Kant (1952) conceived concepts as ideal spiritual beings: eternal and perfect, which had no need to change or grow. However, our thesis is that while the primary nonverbal mind is the author of all nonverbal concepts, yet the primary nonverbal mind cannot create by issuing verbal orders, "Let there be concepts." Rather the primary mind evolves nonverbal concepts, but this non-Darwinian evolutionary function takes only minutes. On the contrary, centuries or millenniums are required by biological evolution to create new sets of domain-specific concepts, which are far too slow.

The nonverbal rational mind machine, however, endowed mammals with a revolutionary new adaptive power or wisdom, the mental technology to evolve coherent sets of causal concepts in only a matter of minutes. My task is to describe how the nonverbal mind executes the god function, how it conceives mental evolution.

Figure 3.07A, B out and back problem

The rat's behavior can only be explained by recognizing that the primary mind creates models, which must embody the conceptual structure of such a spatial world; the primary mind was designed to function at this level of spatial complexity. All I can say is look at the data, look at FIGUREs 3.07A and 3.08. When rats first encounter this break in the sequence on the way back, they expect that the laws they conceived on the way out will work on the way back. However, in row seven or row one in the back direction immediately upon leaving the Omega unit, they discover that the structure of the spatial world on the way back is not the same as it was on the way out; this anomaly forces them back to the drawing board. Now they must rewrite the rules. Two sets of rules are required: an "out-set" and a "backset" of rules, or alpha set and an omega set.

(For more on this topic, the reader can refer to FIGURES 2.02C, D, E in experiment two of this book)

Both FIGUREs show that the control rat's behavior suffered a setback between the out and the backset of six rows. However, FIGURE 3.07A shows that the control group of rats had effectively solved the problem of running the second half of the matrix by the end of the experiment, so that their performance in the second half of the matrix was as efficient as was physically possible. But the fact is unmistakable, the asymmetry of the rules in the two directions in an even number of row maze was a considerable challenge, which the primary mind of the rat not only detected but also solved. However, it was a considerable problem and took a considerable amount of time to solve.

FIGURE3.07A, however, is not only both beautiful and interesting but very complicated. Two major effects are observable: first, the direction effect, but then secondly, the devious behavior effect is also most obvious beginning in block two.

The BP behavior: The relative or functional theory of place

This experiment presents two sets of anomalies, which allow critical new insights into the workings of the primary mind. In the section above, the anomaly associated with the direction effect was discussed. In this section, the anomalies associated with the BP behavior of rats will be considered.

However, in the paragraph above, I talk as if the two effects were independent of each other, while nothing could be further from the truth.

The BP behavior of rodents has many implications for understanding how the primary nonverbal mind works, but one of its most interesting properties is its pluralistic capacity. The primary mind evolves or creates rational behavior by inventing sets of causal concepts and then recognizing things or units of the matrix in this case, as exemplars of these sets. James (1950) noted that, at least, in humans concepts are partial, which today would be called pluralistic; in other words, there is no one universal domain-general concept for anything. Each thing may be experienced as a member of many sets in different contexts, and have as many concepts as it has functions. Every animal's nonverbal concepts are partial, domain specific functions, and action-specific functions. The primary mind has the capacity to represent every physical thing that it can represent by as many partial concepts as it has functions. Indeed, animal nonverbal concepts may achieve the status of an abstract universal as Plato imagined with intense laboratory training.

The partial nature of concepts, however, is as valid for nonverbal and verbal human concepts as for animals. This rule may be a universal for mammals simply because all mammals share the primary nonverbal mind. My task now is to illustrate how rodents' conceptual systems are pluralistic, or if you like, partial, or relativistic and how this conceptual relativism evolves how concepts become differentiated. In rodents, conceptual relativism is seen in a form of behavior I call the functional representation of space.

The paradox of BP behavior can best be described by observing the behavior of the rat in BP unit 55, which is an F1 unit in row 6 in the out direction. What the reader must understand is that this unit conceptually is two units, one unit in the out direction, and another unit in the back direction. The computer model shows that on the very first loop this unit first gets a FEU score on the back half of the maze, but not for the out portion of the loop. Rat 8141 enters this unit time after time, but the reader needs to understand that each of the entries into this unit 55 occurs when the rat is on the way back, and yet this unit of the matrix on the way out is never entered. Time after time when I keep watching this behavior, I kept thinking that the program was not working, until I finally understood that the rats recognize this unit of the matrix functionally, not structurally. Unit 55, which is an F1 unit in row 6 in the first half of the matrix, however, is an A1 unit of the matrix for row one in the second half of the matrix. On the way back, the rat enters unit 55, but only when it is an A1 unit, and almost every possible occasion, why? Perhaps, because when this rat leaves the omega unit it usually makes a right turn, which would be the correct response if this were alpha not omega, if this were alpha one not alpha two. From alpha one, a right turn puts the rat into E1, but from Omega or alpha two, a right turn put the rat into an A1 unit. And yet, this unit is always the same unit 55, which is, on the way out, the F1 unit of the sixth row and is one of the last units in the maze to be entered. On the way out, long before the rat approaches the F1 unit in the sixth row, the rat has made a valid model of the Forward BP units F1 and F2 and learned to avoid these redundant units. Yet on the way out, they have not made a valid model of the role this unit plays on the way back that is in the second half of the maze.

If the matrix maze were symmetrical, then the turn sequence would be the same in both directions, but the back sequence is more difficult than the out sequence, as discussed in the last chapter.

And so, rats ran the maze and entered unit 55 or A1 in row 1 on the way back, repeatedly, and yet the FEU score for the out-unit 55 of the maze remains empty. The puzzle does not go away until the functional definition of the matrix maze units is fully understood.

A unit of the matrix maze is not an absolute place. While a physicist may consider Unit 55, simply as a single unit in space but to the mind of the rat this unit is not a single place but two different places. Unit 55, when it is F1 in row 6 is the last unit of the matrix the rats enter, and yet unit 55 when it is A1 of row 1 on the way back is usually entered. An entry of the rat into unit 55 does not have a single meaning. An entry into unit 55 means one thing when the rat in running in one direction and has a second and very different meaning when the rat is running in the opposite direction. The two possible meanings a rat may experience upon entering a particular unit of the matrix maze depends upon which direction the rat is running in the matrix. Unit 55 of row 6 on the way out is an exemplar of the blind type that the rat first successfully models, an F1 unit. However, Unit 55 is on the way back an A1 unit, and thus is the only unit in the first half of the maze whose redundancy all rats are able to learn to avoid.

In the second half of the maze, functionality works the other way around. For example, Unit 15, which in the first half of the maze is an A1 and is generally the first blind unit entered- is in the second half of the maze an F1 unit and is one of the last units of the maze the rat enters. In the first half of the maze, Unit 15 is full of novelty, and rats cannot seem to get enough of it. And yet, in the second half of the matrix, this unit is perceived as most redundant, and rats will have nothing to do with it.

In Heterostasis, growth consists of the evolution of a set of recognition schemes, or an understanding mechanism, and rats evolve the 2*8*8 taxonomy of the maze structure that defines a maze-place functionally not structurally. Every matrix element has at least two meanings and the behavior of the rat within a unit depends upon which meaning the rat experiences before entering that unit. The perceived stimulus property of any unit is not an objective public event, but a subjective private event determined by which element of the 2*8*8 taxonomy the primary mind uses to assimilate the precept patterns generated by the unit.

The EP and the BP data ultimately both tell two different sides of the same story. The EP data could not be understood; indeed the EP data could not even be effectively analyzed by ANOVA, until I learned to transform the row data so that the data was entered into ANOVA designs in the Functional rather than the Natural or Structural form. Only the functional analysis of the EP data produced the significant effects that are the empirical foundation of the theory of heterostasis. Both the EP and the BP data show that rodents represent the units of the matrix maze functionally, partially, pluralistically, or relatively.

Both sets of data, both the EP and the BP data, force the recognition that heterostatic growth is essentially the evolution of sets of conceptual schemes for representing the units of the maze, one for the EP and another for the BP. The behaviorists conceived of learning as the simple acquisition of an association, which Hull (1943) conceived as Habit strength, as an association (H) between S and R (Stimulus and Response), $_SH_R$. There were three elements in a Hullian association: S, H, and R. According to Hull, what changed during learning were solely H. S and R were fixed elements or fixed forms, fixed elements of the association.

However, all this changes in the context of Heterostasis, the behaviorist held the tacit assumption that concept formation was not a part of learning, concept acquisition is a function of one test procedure, and learning is a function of another. However, heterostasis views this classical concept of learning as a myth, for it was conceived as a Homeostatic function. Mental growth is a function of Heterostasis and essentially an evolution or differentiation of a complex hierarchically organized conceptual system capable of allowing the organism to recognize where it is within the 100-unit space of the matrix maze. The poor behaviorist had no concept of mind and no concept of mental growth. Learning was a nonsense term that served as a placeholder until someone discovered how, effectively, to study adaptive behavior, which in over one hundred years the behaviorist made not a drop of progress.

The behavior of a mammal, a being with a rational nonverbal mind, is determined not by habits but by the validity of the conceptual systems evolved by its mind. A valid conceptual system allows the rat to move about in the EP without surprise and achieve its self-defined goals. A valid conceptual system allows the rat to recognize where it is within the hierarchically organized domain or where it is within the whole 100-unit structure. An organism that knows where it is because of having a valid conceptual model can respond appropriately, presumably for any momentary goal that it may conceive in the matrix.

The effect of punishment upon the blind behavior

The control and experimental groups of rats each have a different set of problems to solve as they run the matrix maze. The control rats have of course, the simpler set of problems, and they are very effective at avoiding the redundant BP units in fact, they are significantly better than the experimental rats at avoiding the redundant blind units. However, this comparison is not entirely fair for the experimental rats have a dual avoidance problem, which complicates their behavior, one that the control rats do not have, and this dual avoidance problem produces in the E group, a conflict that particularly degrades their performance in the avoidance of the redundant blind units. For example, the experimental rats are punished for entering alpha after completing an AOA loop. However, after making such a loop as they approach alpha and need to avoid alpha, they also cognitively need to avoid the redundant units 15 and 16, and so the experimental group of rats has fewer options available to them than the control rats. A control rat can avoid the F1 and F2 blind units in the sixth row in the second half of the matrix by simply entering alpha, but the experimental rats do not have that simple option. The experimental rats, on the other hand, have a double avoidance problem that does not have a simple solution. So, it should not be too surprising that the experimental rats' blind avoidance behavior is less efficient than the control rats.

This experiment had two purposes. First, to run a control group of rats that would make it possible to determine which of the behaviors of the experimental group in the three hierarchical levels of behaviors were caused by punishment. Also, to be able to describe the Full maze growth behavior of the rat in the matrix maze, while distinguishing the EP and the BP behavior.

The significance of the control group

The belated control group of rats run in this experiment did make it possible to test the hypothesis that the behavior of the experimental group of rats in the second experiment was in fact avoidance behavior. The control group of subjects in this experiment did not exhibit the devious behavior seen in the experimental group. This proved that the devious behavior that took twenty-five growths to evolve in the experimental group was indeed due to the punishment induced by removing the experimental rats from the maze upon entering the alpha unit after having made a criterion loop. The control group of rats did not experience this element of the constraint set and did not develop devious behavior.

The full maze growth behavior

The rats' behavior in the blind-units makes it possible to distinguish two different kinds of avoidance behavior. The first kind is the devious behavior in the EP of matrix maze. The second kind occurs in the BP units.

The full maze avoidance in the essential path

The development of avoidance behavior in the EP gives us new insight into the nature of exploratory-play behavior. EP avoidance behavior can only be understood if it is assumed that exploratory-play behavior is a positive experience for the organism, as pleasurable or even as fun. Exploratory-play heterostatic behavior occurs when non-deprived and non-reinforced rats spontaneously encounter anomalous events within an anomalous domain. The avoidance behavior in the EP is seen in the direction effect. Avoiding rats shift the distribution of their activity within the matrix maze over blocks so that they spend less time in the first section of the matrix and more time in the second section of the maze, the one the most distant from the alpha unit which is punished. Punished rats make fewer loops, since they refuse to enter the alpha unit, which is required to end one loop and start a new loop; and they make more growths per loop; they spend more time in the maze; etc. The evolution of this complex avoidance pattern of behavior reflects the evolution that is also occurring in the rats' play self-behavior and intent induced by punishment.

The Full maze, avoidance-of-redundancy in the blind units

The second kind of avoidance behavior evolves in the blind units of the matrix maze. This second type of avoidance is seen in the degree of avoidance of the redundant blind units. The effect of punishment on the spontaneous behavior of rats in the matrix maze creates behavioral phenomena, which suggests that rodents possess powers of rationality that even I was reluctant to believe at first.

Devious Behavior

First, in devious behavior, the rats alternate their days between the shortest path goal and the devious behavior goals, which show that rats do not need an extrinsic reward for they are capable of defining their own spontaneous goals; each purpose has its own unique set of subjective rewards. And thus, each purpose is responsible for creating its own transformation or mental software or virtual conceptual machine for generating each unique pattern of behavior.

Reinforcement

Second, today I recognize after over fifty years that the first of Hull's (1943) experiments I replicated, the drive discrimination study, has a new interpretation within the context of heterostasis. Hull showed that when rats are run in a T-maze with food on the right and water on the left, that when the rats' deprivation is shifted from food to water and back, rats will run to the appropriate arm of the maze. Hull only understood this experiment as showing drive discrimination in rats, but today I realized that it shows that reinforcement is not an objective property of an object, but a subjective property, both food and water are only reinforcing when the rat is hungry or thirsty. Thus, reinforcement is a function of the animal's subjective state. Reinforcement is not an objective property of either food or water. The same principle is also true of the spontaneous heterostatic behavior in the maze when rats are punished. On the days when the rat is running the shortest path goal, the consequence that is reinforcing is a set of events, S, but on any day, when the rat is running or motivated by the devious behavior goal, another set of events are reinforcing.

Chapter 4 An Experiment to Test for Devious Behavior in Mice

Experiment IV 90 Punish

The four purposes of this experiment:

This experiment has four purposes: first, to study the effect of punishment on mice behavior. It is important to note that one of the hazards associated with original research is how difficult it is to design a proper experiment. To note, while this was the fourth experiment in the series, yet it was the first one designed as a proper study, specifically, in this case with a proper control group for studying the effect of Punishment. In Experiment I, the rats were run for only one growth so the phenomenon of punishment was never observed. In Experiment II, a control group for the effect of punishment was not included for the possibility of punishment in exploratory-play behavior had not yet been recognized. In Experiment III, a belated control group was run. In this fourth study of the series, finally enough was known about the phenomena of punishment on the spontaneous behavior of rodents being tested in the LTP so that a proper experiment could be designed. This experiment was designed with two groups, a control group that was not punished and an experimental group that was punished; secondly, to study the exploratory-play growth behavior of mice both in the FM and in its parts, both in the EP but especially in the BP of the maze. Finally, third so I can compare the FM behaviors of the mice of this study with the rat's data from Experiments II and III. While I have run rats and mice in the Lashley III maze for almost forty years, mice have always seemed strange for their behavior in the maze differs from the rats' behavior in many ways that I have never been able to understand. The issue is this are rats maze bright and mice maze dull, or is it the other way around? Different criterions favor one species over the other. Hopefully this experiment will furnish some new data that will illuminate this issue and advance the understanding of the differences between these two species. But in any case, the point is that the two species employ different strategies for making models of the maze. Fourth, to demonstrate some of the many ways that the matrix maze data can be partitioned into mixed ANOVA designs; designs which, in effect, allow almost any conceivable experimental question to be put to an exact statistical test. The Matrix Maze has a unique set of features that makes it possible to partition the whole of the maze domain into various combinations of parts.

The formal partitioning skill, which is critical for conceiving alternative Anova designs, is also a critical element of the generic exploratory-play behavior of mice. Of course, the verbal skill of humans and the nonverbal skills of rats differ in many ways. Classical science has historically been conceived as exclusively a function of verbal and mathematical intelligence. This conception, however, is but an element of the myth that there is no Primary mind. The nonverbal mind is the foundation for all the verbal and mathematical functions of the human mind.

This experiment is only possible because I was able to invent a way to marry the hierarchical structure of the matrix maze and the hierarchical character of

ANOVA. Hierarchical structures are the primary feature of the domain the primary mind was designed to solve. The data generated by the behavioral telescope consists of an overwhelming quantity of potential knowledge that in its raw form consists of a complex set of inputs, a precept string with a quantity of complexity and strangeness that the primary mind must either assimilate or accommodate. Assimilation occurs when the primary mind imposes upon a set of precepts, a set of forms, which give them valid meaning; ones, which elicit appropriate responses. Under these conditions, no anomaly will be encountered and the animal will be seen as having great skill in dealing with this domain. But if the complexity and strangeness are not met with a corresponding preformed conceptual system in the primary mind, one ready to take the strangeness and complexity out of the precept potential knowledge, then accommodation must occur.

The problem that I faced as an experimenter, when confronted with the strings of hundreds of thousands of data points or precepts generated by the behavioral telescope is similar to the problem the rat faces as it explores the matrix maze. The precept is a string of nonrandom data points, a string of data points with a considerable quantity of potential knowledge. My primary mind and the rat's primary mind both face a similar challenge. In my case, my nonverbal mind had to identify a pattern within any string of repeated measures that would allow my mind to construe the replication of an invariant pattern of events. My mind and the rat's mind must create a consciousness of the invariant pattern of the domain that will allow us to interact successfully with the objects and events of our respective domains by being able to anticipate and thus be prepared to interact with them in real time.

The rodent and the human problem are the same problem except, of course; the rodent problem consists exclusively of a set of nonverbal precepts, while the human form of the problem is framed within the context of both nonverbal and verbal/mathematical precepts. But the primary mind of both the rat and the human faces the same problem. While I am trying to invent a conceptual system capable of understanding the rat's behavior in the matrix maze, the rat is trying to invent a conceptual system capable of understanding the matrix maze itself.

Ultimately, my problem as a human explorer of the rat's behavior and the rat's problem as a rodent explorer of the matrix maze both reduce to the same form. The problem both species ultimately face is a string of neural impulses, a precept string, which either of our primary minds must either assimilate or accommodate. And my primary mind and the rat's primary mind must both invent a valid set of forms, which can give meaning to the invariant precepts patterns, which allow both primary minds to reconstruct the events of their respective domains.

My mind and the rats' mind must somehow identify a set of independent and dependent variables that are causally related, not in any absolute sense, but only causally related to the specific goals of our species. Both of our minds must search the precept string and find a way to parse these strings to identify reproducible sets of events, functions, and relations between the Independent Variables, IV, and the Dependent Variables, DV, which are relevant to our goals. The task of the primary mind is to construe the replication of the events of the domain within which it is embedded and thus learn to be prepared for and to interact successfully with these invariant patterns relevant to some specific species goal. Only the successful organization of these strings of precept data into a stream of consciousness will allow the organism to achieve successfully its goals, to enjoy its time in the domain, and to discover the shortest task.

The task, which the gene designed the primary mind to solve, was the revolutionary task of transforming the machine language precept code generated by the senses into a higher nonverbal-language code of consciousness, a phenomenological code, which describes the objects and events of the domain as classes of things relevant to some purpose. The primary mind was evolved for transforming the machine language neural code into the object language or class code of consciousness.

The primary mind of the rat has the problem in the matrix maze of identifying the classes of things, which will allow the neural precept code being generated by its senses as it explores the matrix maze to be experienced as exemplars of classes of things relevant to some purpose. The primary mind has the task of inventing a set of reliable and valid classes of things that populate the matrix maze. The rat's primary mind and my primary mind both face the same task. The question in both cases is: what variables, what independent and dependent variables can we attend to that will allow us to discover a relevant invariant function, which characterize the unique but invariant causal features of a hierarchically organized domain.

My task as a human scientist is first, to select a test procedure, which will cause a set of variables that will measure the relevant causal functions of the domain to become visible. ANOVA allows me as a human scientist to have a computer do for me what the rat's primary mind must automatically do; to identify the independent variables, which explain the statistically significant relevant components of the variance in the data strings.

The primary role of the primary mind is the identification of these relevant causal sets of variables, IVs and DVs that will objectively prove to explain the variance in the data strings. The rat's primary mind creates these variable sets and tests and confirms them in its direct action especially during those occasions in the maze when it is retracing. The proof of the validity of a set of variables, which define a function in some domain, can only be objectively demonstrated in generic science within the context of the organism's overt or manifest behavior in the domain in question. The ultimate test in generic science and even in human science is the pragmatic test. The pragmatic test is the ultimate test of the validity of any idea. The pragmatic test consists of the organism acting upon the elements of a conceptual system and discovering whether the embedded set of ideas works or not. The primary meaning of any idea is this: either when it is acted upon in some appropriate domain the function works, is valid, or it doesn't and is invalid. The rat's primary mind has several sets of behaviors, which are elements of its spontaneous exploratory-play behavior, each of which are designed to compute the specific kinds of unique discovering/inventing functions which the specific character of the problem of the moment requires. These are Explore, Sit, and Retrace functions.

One of the many problems with twentieth-century experimental psychologists was that they never knew how to put their conceptual systems to a proper test. Behaviorism and cognitive psychology, for examples, consisted of a large set of conceptual beliefs that were never capable of being put to a valid pragmatic test. Twentieth-century psychology never recognized that the primary component of science is the nonverbal task of designing sets of unconfounded IVs and DVs that can be put to the pragmatic test. The primary validating test for the fruits of the primary, generic, scientific mind is not some verbal or mathematical test but a nonverbal test. The first step of science is the creation and validation of an unconfounded set of variables to use to represent any domain. To attempt to proceed to the formal stage of science to begin to make mathematical models of confounded data is ultimately a waste of time, but the fact is that the favorite myth of twentieth-century Psychology was that the formal mathematical treatment of the data would magically make Psychology a science. However, the rule still is "garbage in garbage out." The fact is that the formal mathematical treatment of confounded data will not clean up the data. Psychology must learn to put first things first. The first task of any would-be science is to solve the nonverbal problem for which the primary mind was designed. The first skill of a scientist is the mastery of the craft of designing sets of unconfounded variables and this is just as important for a rat to survive as it is for human science.

The purpose of this fourth experiment is to demonstrate how the matrix maze allows the experimentalists to master this primary skill; the design of alternative sets of scientific causal variables that can then be tested against the data, until a set of variables that pass the test are identified.

Really, for over fifty-years, I have devoted my scientific life to the task of identifying a set of valid variables that describe the growth learning behavior of the rat, to find a new set of variables that I always knew were there, but just out of reach in the behaviorist's scheme of things. The TTP variables, somehow, I knew were not valid pragmatically; they were not describing the mental behavior of the rat but were describing some artifact. There are two kinds of scientific skills not one, as myth would have it. The primary skill of both primary and secondary science is the nonverbal skill of identifying a set of unconfounded variables. Only a set of unconfounded variables will generate the higher order data language that can usefully be transformed formally by mathematical functions.

A critical skill that a revolutionary experimentalist must master is conceptual-alternativism. The primary skill of the primary mind is the skill of identifying alternative ways of partitioning the potential knowledge in the data strings. The primary mind must invent different nonverbal conceptual systems and use them to test the data strings and thus pragmatically identify, which of the factorial designs prove themselves more useful.

The primary generic scientific skill is the generic skill of the primary mind, the intelligence to identify invariant patterns in the machine language neural codes of the data strings and to evolve sets of forms, conceptual systems that can anticipate and prepare the various centers in the cortex to impose meaning, to form these patterns of potential knowledge.

A human experiment is a well-formed nonverbal act emitted by the primary mind objectively put to nature; a monistic secular prayer put to the pantheistic god or nature, which will be tested objectively as a data stream by statistical methods such as ANOVA. Variable sets, which generate useful pragmatic functions, will generate effects that are statistically significant assuming the data is not confounded.

One of the myths of experimental science that must be exposed is that the experimental method generates causality. The experimental method only yields causality when the data is not confounded. Experimental science is generally aware of the fact that partial correlation can make correlation data meaningless, but they have not recognized that the same is true of the experimental method. The data generated by the TTP was experimental data, but this data cannot yield causal relations because the Independent variables of the TTP were confounded, so in effect, the causal variables the behaviorist believed they observed were artifacts.

However, it is only possible to solve the problem of confounding in an experimental context. The purpose of this particular section is to demonstrate the role of the primary mind in the creation of the nonverbal unconfounded conceptual systems. Such conceptual systems can be understood as analogous to factorial or mixed ANOVA designs, which are put to nature, and which allow statistics to have the last word, to say whether the conceptual system was or was not isomorphic with some constraint of the domain.

The EP designs

Various types of EP designs were demonstrated in each of the previous experiments. The set of independent variables consists of sets of between and within variables from which various sorts of mixed designs can be constructed.

The between independent variable list is sex, age, species

The within independent variable list, which can be used to partition the repeated measures are Row, Column, Trials, block, directions, data types,

The dependent variable list consists of the many types of variables;

The growth dependent variables
The Run dependent variables
The Loop dependent variables
The node\path dependent variables

The full maze designs

The full maze designs include all the variables of the EP design, except now the number of levels for the column variables is doubled from four to eight. In the EP designs, only half of the matrix's data is being analyzed, only the data from the four innermost cells or columns of the matrix are included in a design. In the FM design, the data from all eight columns in the matrix can be included in the ANOVA design.

The advantage of the full maze designs are that they allow the direct comparison of the four units from the EP and the four units from the BP.

Purpose 2: Full Maze, FM, and Blind Path, BP, designs

The FM behavior consists of two parts, the EP, and BP behavior components. The recognition of this distinction was one of the most difficult distinctions I had to learn to make, although in retrospect it seems obvious. The behaviors of rodents in these two subsections of the maze are opposite forms, two sides of same behavior. There are several points, both technical and theoretical, that must be developed concerning this distinction before the full maze behavior can be understood.

There are three kinds of growths, FM, EP, and BP where a growth is defined as the adaptations that occur in the outer-most of the four hierarchical levels in the maze. Rodents are run until they make a criterion loop, a complete round trip through the maze, an Alpha-Omega-Alpha, AOA loop, through the maze. Each AOA loop is a growth; the mice in this experiment were run until they all made at least seven growths. As noted above, however, these growths can be scored several ways. First, they can be scored as either FM or 96 unit growths, or secondly and thirdly, as growths in one of the two forty-eight sub-units of the maze, either the EP or the BP. In the last experiment, the FM, 96 units, growth for rats was described but only a limited analysis was made of the data. In this experiment for the first time, I will describe the other kinds of full and partial maze analysis of the data that are possible.

During their spontaneous concept-system-making behavior, rodents in the matrix-maze make not one but two sub-maps of the matrix, the EP and the BP sub-maps. The evolution of the EP sub-map has been partially described in the earlier experiments in this series, but now for the first time it is possible to begin to describe the manifest spontaneous Blind Path mapmaking behavior of mice.

The two first-order goals

When rats are punished during exploratory-play growth learning, it is possible to force the rats to reveal that they have two goals, which normally cannot be seen because they are compatible. But punishment makes these two goals, the shortest path goal and the play goal, mutually exclusive and visible.

The two second-order conceptual sub-goals of the shortest-path goal

When rodents are spontaneously exploring the matrix maze and evolving a valid conceptual system, they conceive of the matrix domain as consisting of two sub-domains, the EP and the BP, for which they design a sub-map. The shortest-path purpose requires a separate sub-map for each sub-domain.

For the forty-eight units of the EP, the primary mind evolves a conceptual subsystem, capable of recognizing all forty-eight elements of the EP as elements of the EP. Such a sub-map allows rodents to make appropriate responses in these units, responses consistent with their sub-goal of running the shortest-path in this sub-domain of the matrix.

For the forty-eight units of the BP of the matrix, the primary mind must also evolve a conceptual subsystem capable of recognizing all the forty-eight elements of the BP. Such a sub-map will allow the rodent to make appropriate responses in the units.

However, while in principle the goal of the primary mind in each sub-domain, both the EP and the BP, are the same, yet in fact, the tasks of the primary mind in these two sub-domains are opposites. In the EP, the purpose of the conceptual subsystem is to generate behavior that allows the rat to approach, to explore, and to navigate efficiently through this sub-domain of elements. Yet in the

BP, the purpose of the conceptual subsystem is to generate behavior, which allows the rodent to avoid the elements of this sub-domain. For example, the BP consists of elements all of which are by definition not elements of the EP and are therefore elements of the maze to be avoided when rodents are executing their goal of running the shortest-path, which only includes the EP units. The BP units are units the rat must avoid if it is to run the shortest path. However, in order for these BP elements to be avoided the primary mind must have a conceptual subsystem capable of recognizing, anticipating, and avoiding the elements of the BP. The overall goal of the rodents in the matrix maze is to create a map that will allow them to move efficiently through the maze. Yet, the fact is that this general goal has opposite meanings in the EP and the BP, to run through all the EP units and to avoid as many of the redundant BP units as possible. The EP and the BP are two sides to the same coin.

A Growth is defined as the evolution of a valid conceptual taxonomy

The theory of Heterostasis defines mental growth as the evolution of a valid conceptual system, one capable of permitting the organism to recognize where it is within the matrix, to be capable of emitting appropriate behavior, either to approach or avoid each of the one hundred units of the matrix. The evolution of this one hundred unit conceptual system evolves by Causal Conceptual Conditioning in several stages of evolution of a factorial design as described above.

The evolution of a valid conceptual system occurs in stages. First of all, rodents learn to discriminate the EP and the BP, secondly they must learn to discriminate the units of the BP as belonging to one of two classes of element F and A. The F-Class of blind-path units is entered when rodents ignore the exit door in each row in unit E4. If the rodent misses the opportunity to exit the row accidentally or deliberately, its continued forward motion will bring it into the two blind unit elements of the F class of units, F1, and F2.

The A-class of units are the most difficult units of the matrix for rodents to predict and thus to avoid. The A class of units requires rodents to learn to move around the exit door with the correct spin. The classical theorists, Lashley (1963) in particular, conceived of rodents either making right or left responses, a 90-degree turn, at a choice point. However, when Lashley designed the Lashley III maze, I have no way of knowing whether he was aware of the fact that he designed the placement of the alpha unit off to the right side of the exit door. Because of this placement of the unit, it then became necessary for a rodent exiting the alpha unit to learn to make, in effect, a 180-degree turn. The correct exiting response that rodents must learn was not a right or 90-degree response. The natural behavior of the rat exiting the alpha unit is to move in a straight line. Because of the special size of and the placement of the alpha unit relative to the door and the placement of the rodents within the alpha unit; the normal rat upon exiting the alpha unit in a straight line will end up in the A class of blind units. What rodents have to learn is to make a 180-degree clockwise turn when exiting the alpha unit door in order to stay in the EP.

However, the first rule that must be recognized is that rodents exploring the matrix maze adapt to the blind units of the matrix by evolving a set of classes, which allow them to recognize each blind unit as a member of not one, but two classes. The primary mind must predict class membership. The primary mind must predict to which class each unit belongs as a member of a blind path unit or a member of an essential path unit. Valid conceptual knowledge allows the rat to avoid those BP units, which are redundant and outside of the EP. The sooner a rodent can recognize a unit as a redundant member of the A or F class of unit, the sooner they can avoid those units. This knowledge greatly advances its goal to master the shortest path.

The differences between the BP behavior and the EP behavior are described by different dependent variable data sets and different sets of independent variables.

Method

Subjects

The subjects were six male and six female, 60-day old C57/10J mice, bred and reared at The William Paterson University of New Jersey. They were housed in individual cages with food and water ad libitum. The females averaged 24 grams and the males averaged 26 grams at the beginning of the experiment.

Apparatus

The mice were tested in a Lashley III Matrix maze under the same conditions described in the earlier studies

Procedures

Eight experimenters each tested two randomly assigned mice, one of each sex, one experimental and one control subject. Each mouse was tested at the same time each day and was weighed just before each run, carried in its home cage to the test room, and placed in the alpha box. The timer was started, the computer graphics tablet initialized, and the non-retrace door was removed to begin a run. The mice were immediately returned to their home cage in the colony room upon completing a run.

The constraint set for the control group had a single element; they were allowed to explore the maze for ten minutes, without any constraint on the number of moves. The experimental mice had a two-element constraint set. In addition to the ten-minute constraint, they were removed from the maze whenever they made a complete alpha, omega, alpha loop (an AOA criterion loop). Each mouse was run for a minimum of thirteen runs. Some were given additional runs, with the number depending upon the experimenter's schedule. Some were run for twenty-five runs.

Results

The results of this experiment will be described in several sections.

Mice Devious behavior as seen in group-effect data

Devious behavior has been operationally defined as behavior modified in some special way in order to minimize the effect of the punishment. This devious behavior can be seen in each of the first three of the four levels of heterostatic behavior growths, runs, loops, and node\path respectively. The purpose of this first section will be to describe the group effects of punishment in mice upon the second and third levels, the run and loop hierarchical levels of behavior. In the previous experiment, it was shown that at the group level of organization, the rats' devious behavior was modified in several dimensions. They decreased the number of loops per run, increased the length of these loops, and increased the number of multiple loop growths per run. In this section, the question is whether mice will also generate devious behavior with these same characteristics. While mice behavior is different from rat behavior in many ways, at least in this area of punishment and devious behavior, the data shows that mice and rat behavior are remarkably similar.

The run and loop analysis.

The run-loop data consists of the following seven variables:

1. Number of steps per run: FIGURE 4.01, and 4.02
2. Time per run: FIGURE 4.01, and 4.02
3. Number of loops: FIGURE 4.03
4. Number of steps per loop: FIGURE 4.05
5. Time per loop: FIGURE 4.05
6. Mean time per step per loop: FIGURE 4.06
7. Number of growths per loop: FIGURE 4.04

FIGURE 4.01 groups by runs

FIGURE 4.01 shows the result of the multivariate two-way interaction, groups by runs, for the mean of the run data for two variables: moves and latencies for thirteen runs. This FIGURE shows that the mean of the two behaviors of the two groups of mice was significantly affected by the difference between these two constraint sets. The control mice data shows how much daily activity mice will emit when they are not punished. The control group of mice, which were not punished, gradually increased their average score for thirteen runs. On the other hand, this FIGURE shows the experimental group, which was punished, paid a heavy price as a punishment for quickly making a complete loop. The E-group's behavior improved only for one day; in fact, on day three the average score for runs for the experimental group began to decrease. Since the experimental subjects were taken out of the maze on making a growth, they were punished and the immediate effect of this punishment was to decrease the total amount of their allotted daily activity per run they were able to spend as measured in both moves and time. However, since this data is an average of two scores, effectively it tells us nothing about the detailed character of mice devious behavior, but it demonstrates that the punished mice are paying a considerable price in terms of the decreased activity per day they are able to enjoy due to punishment. If mice are capable of devious behavior, this decreased activity most clearly shows that the need for devious behavior to evolve is clearly present.

FIGURE 4.02 groups by runs by data types

FIGURE 4.02 shows the 3-way interaction, groups by runs by data types. This FIGURE breaks the average data in FIGURE 4.01 down into the two scores, steps and latency per run for each of the two groups, the control group on the left and the experimental group on the right. The control group had only a one-element non-punishing constraint; they were allotted ten minutes, 600 s., for exploratory-play behavior in the maze but with no limit on the number of moves. FIGURE 4.02 shows even on run one, they spent their full allotment of 600 s. The move data, however, shows that for the first twelve days they increased their capacity to explore more units of the maze each day. FIGURE 4.02 shows that the number of steps per run increased over the 13 runs. On day one, the average control mice used only 325 steps, but by day thirteen, they had learned to make almost 600 steps per run. These control mice obviously enjoyed their stay in the maze and adapted to this constraint set by becoming explorers that are more efficient by improving their skill to maximize the number of moves per run. The control mice were both time and move maximizers. While initially they only made about 300 moves in 600 seconds on run one, by run 13 they were making almost 600 moves in 600 seconds. In other words, their average speed per step in the maze had almost doubled during the thirteen runs from one move every two seconds to one move per second.

The experimental data in FIGURE 4.02 however, shows that the mice whose behavior was punished were forced to reduce measures of behavior, time, and moves over the 13 runs. FIGURE 4.02 shows that on day one, the experimental mice with the two-element constraint set were very similar to the control mice data on both the numbers of steps and time spent in the maze with about one move every two seconds. However, for the experimental group, which learned to depress both scores over the next twelve days, the latency score became even more depressed than the step scores.

FIGURE 4.03 groups by runs

FIGURE 4.03 shows the first order interaction, groups by runs, for thirteen runs for both the experimental, and control groups for the number of loops per run. $F(12, 192) = 2.49$, $p<.0047$. The number of loops is operationally defined as the frequency of the alpha unit entries. FIGURE 4.03 shows that the control group data fluctuated from an average high of nine loops per run on day two to an average low of 4.5 loops per run on day eight and back to nine on day 11. The Experimental groups, however, decreased from an average of better than five nested loops per day for the first two days to one loop a day by the end of the thirteen days. This experimental group behavior shows that the experimental group of mice responded to punishment just as rats do; in other words, mice learned to decrease the number of nested loops per day during the thirteen runs, as did the rats. Since a new loop begins whenever a rat or mouse leaves the alpha unit, this decrease in the number of loops shows that the experimental mice learned, as did the rats of the previous studies, to avoid completely the alpha unit, as evidenced by the fact they made only one loop per run. The data so far presented show that mice modified their behavior in this situation the same way the rats did under the same conditions.

FIGURE 4.04 groups by runs

FIGURE 4.04 shows the first order interaction, groups by runs, for the number of growths per run for the two groups. $F(12, 192) = 4.02$, $p<.0000$. This FIGURE shows that both groups of mice improved the number of growths throughout the 13 runs. However, FIGURE 4.04 shows that the control group increased at a faster rate than did the punished mice, so that by day thirteen the average control mouse was making 4.5 growths per day, while the experimental group was averaging 2.6 growths per run. Moreover, since FIGURE 4.03 showed that the experimental group of mice learned to make only one loop per day, the 2.5 growths they made per run were necessarily nested within that one loop per run. The average control mouse made better than six loops per run, as seen in FIGURE 4.03 and an average of 4.5 of these were growths, so obviously the control group's growths were not all nested growths as were the experimental group's growths. This technical point is important in understanding the exact nature of devious behavior. The control group of mice had only one contingency or only one constraint upon their behavior with which to contend. They were able to evolve their play behavior almost without constraint, and they increased their behavior in three different dimensions that have so far been described; they made more moves, ran faster, and made more growths per run. The Experimental mice had an additional constraint within which they were forced to adapt their evolving play behavior, and this additional constraint, which was perceived as a punishment, robbed them of 300 moves and 300 seconds per run. In order to compensate partially for their losses, they learned to avoid the alpha unit, which was punished, and thus they gained some additional time, which they spent learning to make more nested growths per run. Considering they had fewer moves and time with which to work compared to the control mice, their improvement in the number of growths per run is impressive.

FIGURE 4.05 groups by runs

FIGURE 4.05 shows the first-order interaction, groups by runs, for thirteen loops for the average of the two variables: moves, and latency per loop. $F(12, 384) = 4.20$, $p<.0000$. This effect shows that the average control group's loop size, as measured by both steps and latency decreased over the thirteen trials until their average loop became about 90 steps and 90 seconds long. However, the experimental group's average on these two loop variables increased from about 150 moves in 150 seconds per loop to close to 300 moves in 300 seconds per loop. This is one of the signs of devious behavior.

FIGURE 4.05 shows the average number of moves and steps per run. Since FIGURE 4.03 showed that the experimental group decreased their average number of loops to one loop per run on run 13, and FIGURE 4.04 showed that their average number of nested growths increased to about 2.6 on run 13, therefore, FIGURE 4.05 shows that the inevitable happened. As the number of loops per run went down and the number of nested loops per run went up, the average loop size increased to 300 steps and 300 seconds per run. These are the same numbers seen in FIGURE 4.02, the number of run steps and run latency, while the control group's loop lengths were decreasing to less than 100 steps and seconds per loop. FIGURE 4.05 shows the punished mice evolved devious behavior as did the rats in the earlier experiments. They also learned to decrease the number of loops, increase the size of those loops, and learned to make several nested growths within this loop, which was now twice its original length. Control mice did not show this type of behavior. This pattern of behavior is the devious pattern of behavior, which only evolves in punished mice and rats during exploratory-play behavior.

FIGURE 4.06 main effect for latency per move per loop

FIGURE 4.06 shows the main effect for the mean latency per move per loop. $F(12, 192) = 22.89$, $p<.000$. This FIGURE confirms directly what we inferred from FIGURE 4.04 the control and experimental mice both decreased from an average of two seconds per step on run one to one second per step on run thirteen. The group by run effect for this variable is not significant. It is interesting that the experimental mice in spite of their punishment were able to maximize their running speed in the maze and get the most growth possible under these less-than-ideal test conditions. The devious behavior of the experiment, in other words, was very effective in maximizing their play behavior given the unfavorable contingency set with which they were forced to contend in this domain.

These first FIGUREs, 4.01-4.05, show that the punished experimental mice modified their behavior relative to the control mice in several ways. First, they decreased the number of loops per run. Second, they decreased the number of steps and latency per run from 1:1 at 600 seconds to 1:1 at 300 seconds. Third, they increased their growths from one growth per run to a mean of 2.5 growths per run. Fourth, they increased the number of steps and the latency per loop from 145 to 300. Fifth, they decreased the mean latency per step.

This five-element syndrome of behavioral change by the experimental group of mice shows that when punished, mice can spontaneously make a complex set of changes in their behavior, which supports the idea that they have the ability to change their means to attain their ends. This five- element set of changes confirms that mice also exhibit devious behavior almost the same way rats do.

Devious behavior at the run level seen in Individual Differences

Mice, of course, exhibit individual differences just as do rats. For example, they do not all share the same set of priorities, and this can be seen in the way the different individual mice allocate their resources over the thirteen runs. While FIGURE 4.01-4.05 demonstrates that devious behavior can be seen in group means, yet unfortunately the individual difference aspects of devious behavior gets lost in the group means. In the above section, the attention was focused upon the way the devious behavior could be seen in the group behavior.

The purpose of this section is to focus upon the ways devious behavior can be seen in the individual differences of mice at the run level of behavior. One of the most interesting and complex manifestations of devious behavior that can be seen in the devious behavior of both species are the conflicts that develop within their run behavior when they are forced to choose between their goals. By being punished, the mice must choose between their desire to run the shortest path and their desire to maximize their playtime within the maze. At the human level, this conflict would be explained as a conflict between the serious scholarly side of an individual and the frivolous carefree side.

The exact daily pattern of conflict that develops is unique to each mouse subject; it is these individual difference details, which are washed out in the group data. However the next FIGURE shows how the conflict induced by punishment is expressed in the individual behavior of one most amazing mouse, subject 8150. But first let us look at the individual behavior of a control mouse.

FIGURE 4.07 run data for M8141

FIGURE 4.07 shows the run data for control male mouse, subject 8141, for two scores, steps, and latency per run for twenty-five runs. This mouse was never punished in the maze, so no devious behavior should be expected and none was observed. The scores of this mouse on run one were 400 moves and 650 s latency. The latency data for this control mouse showed that it obtained all the allotted 600 s. for each of 25 days. Notice the almost perfectly straight line of the latency data. The move data, however, was not an uneventful, straight line. By run 10, this mouse had learned to attain over 700 moves in 600 s., but then it settled down and finally its behavior became stable at about 450 steps in 600 s. after twenty-five runs. It always spent all its allotted time each day, and since there was no limit on the number of steps, this mouse almost doubled the number of moves it made per day. By day eleven, it settled down to a daily allotment of 450 moves. The data on the left half of FIGURE 4.07 shows no devious behavior. It shows a relatively simple set of curves.

The data of the right half of FIGURE 4.07 however, reveals a very different pattern of behavior, it shows the number of moves and time per run for 25 runs for experimental mouse 8150. This data shows that the punishment constraint complicated the exploratory-play behavior of the experimental mouse as it did with the behavior of the experimental rats. Compare the behavior of the punished mice on the right half of FIGURE 4.07 with the behavior of the control mouse seen on the left half of FIGURE 4.07. Also, compare the Experimental mice data of FIGURE 4.07 with the data of FIGURE 2.02 for rat F3 in the second study. The data of FIGURE 4.07 shows that this punished mouse evolved two mutually exclusive goals and switched between them on different days just as F3 did.

Mice and rats both have a strong primary need to run the shortest path through the matrix maze even though on the days they run the shortest path, they pay a considerable exploratory-play penalty

FIGURE 4.07 is one of my proudest accomplishments; the data in this figure makes its point with awesome perfection. What is Devious Behavior? This figure gives the ultimate nonverbal or operational answer to that question. This figure is pure pragmatism, pragmatism at its best!

Mouse 8150

FIGURE 4.07 shows the performance data for 2 mice, 1 control subject 8141, and 1 experimental subject 8150. The difference between the data of these two subjects makes it unmistakably clear what punishment does to behavior. Look at the data for 8150--on those days where its priority was the shortest path, days 9,

17, and 24, the mouse ran the maze in 100 moves in 100 seconds and therefore lost about 500 s per day play time and approximately as many moves.

This Figure shows how mice modify their behavior when punished. This behavior is devious behavior and can be seen evolving in FIGURE 4.07 on the days when the mouse is not running the shortest path. However, on these play-days the mouse's behavior evolved until it was able to make a remarkable 700 moves within the allotted 600 s by day 23. Mice that are punished for making the AOA loop show a great capacity for devious behavior. They learn to run the maze to avoid punishment with considerable skill. However, their need to execute the shortest path occasionally cannot be inhibited and on these days, mice run the shortest path in spite of the considerable negative behavioral consequences.

The experimental mouse of FIGURE 4.07, finally, laboriously maximized its playtime, as did the control mouse but only after a much longer period and only after having made a much greater number of moves. There was three of these avoidance or devious growth periods: runs 6-8; runs 10-16; and runs 18-23, which can all be seen in the data of FIGURE 4.07. This punished mouse, Experimental female 8150 obviously had two innately defined goals and FIGURE 4.07 shows the complex pattern of behavior that emerged in the individual data as the conflict between these two mutually exclusive goals evolved and interacted.

The difference between the experimental and the control mice is that the control mice since they are not being punished are able to satisfy each day both of these two sets of needs as did the control rats, without any conflict. They were able to make as many shortest path loops each day as they desired, spend as much time in the maze as was allotted, and make as many moves as they desired.

Only the experimental mice, by contrast, are punished each time they make a shortest-path loop. However, this gets really interesting; in order for this mouse to attain each day as much satisfaction for both of its goals as possible, it must modify its shortest path goal by accepting some one of the seven alternative nested loops in order to avoid being punished! Mice, in this respect, also show the same cognitive power to modify their purposes in such a manner as to avoid punishment, as was seen in rats.

This does not mean, however, that mice adopt exactly the same devious means to achieve their end as do rats. Only a comparison of the type of nested loops used by rats and mice can answer this question, and then there is the possibility of a difference in efficiency between mice and rats. But, at least, the first step of the comparison, so far, shows that mice and rats both have the same capacity for generating devious behavior, whether they have the exact qualitative capacity remains to be seen.

In FIGURE 4.07 it is interesting to note that generally there is a close to 1 to 1 relationship between the two scores, which shows that the mice were generally moving at a high constant speed of one move per second no matter which behavioral priority they were executing. Mice punished during exploratory-play behavior exhibited behavior that is almost the same as the conflict induced behavior exhibited by rats under the same circumstance.

That is both rats and mice, when in conflict, alternate between different purposes on different days. Some days their purpose is to run the shortest path, and they seem to ignore the severe consequence this purpose imposes. This purpose has such a high priority in both rats and mice that the need to achieve this goal can only be deprived for so long, and after that interval they do what they must and accept the consequence.

However, both species are able to switch between two purposes: to run the shortest path is one but not their only purpose. Their secondary purpose is the need to play, which in this context translated into their need to spend as much time in exploratory-play behavior directed at evolving growths, which are a modified form of the ideal AOA growth; even their play behavior has a very serious purpose. Their need to perfect growths is so great that they are willing to settle for growths that are not the ideal growth, the AOA growth. Both mice and rats prefer to make the ideal AOA growth and control animals make this ideal growth almost exclusively, but when punishment enters the picture, both mice and rats are pragmatic organisms that are able to settle for something less than the ideal. Neither mice nor rats are idealists.

The set of eight nested essential-loop patterns

FIGURE 4.08T 8 essential path types

Figure 4.08T shows eight possible essential path patterns for both nested and criterion loops. The eight are divided into two subsets: the first growth subset and the nested growths subset. The nested set is what rats and mice use when they are being devious.

The normal criterion loop set of moves is a type 8, an alpha-omega-alpha loop. The purpose of this section is to show the other types of nested growth and nested loops, which characterize devious behavior. Punished rats and mice avoid any unit that is punished. Skinner (1974) made an important scientific discovery when he reported that the one consistent consequence of punishment was the fear that was associated with the object or event that caused the punishment.

Devious behavior in rodents validates Skinner's (1974) principle. In this case, the punished mice avoid the alpha unit by redefining the nature of their behavior by choosing to run one of the other types of nested loops.

Only the type 8 loop qualifies as a criterion loop and criterion growth, the only one that is punished. The other types of nested loops and nested growths can be run by the mouse with impunity. The other seven types of nested loops within a nested growth do not qualify as a criterion loop but rather, qualify as a devious nested loop. Punished rodents chose to run one of the nested loops when their goal is to maximize their playtime in the matrix maze.

Types 1-7 qualify as a nested loop within a growth, **ONLY** when the mouse **does not** make a criterion (Type 8) loop on its first trip through the maze. On the first trip through the maze, the mouse must decide which of the seven nested loop types it is going to make.

Nested loop types are broken into two sub-sets. Sub-set one includes loop types 5-7. These nested loop types can be chosen only on the first nested growth. When the mouse does not make a criterion loop on the first run through the maze, it will become one of the types shown on the table as 5-7. For example, the mouse could start in alpha, enter omega and on the return trip, **not** enter alpha a second time. So the rat has made a type seven initial growth or loop. Any subsequent nested loops now must be one of the type 1-4 options.

It is easy to look at this table and feel that the mouse cannot make any type 1-4 nested loops since the mouse is always placed in the alpha unit to start its run. The thing to remember is that the mouse can only make types 1-4 as a nested loop within a criterion loop. So for example, let's say that in this experiment, a mouse is placed into the alpha box for his daily run. The mouse runs out to omega and comes back towards alpha, but the mouse turns around in unit 14. Now the mouse has chosen not to enter the alpha unit and the nested loop it is about to run will be a choice between types 1-4. So this nested loop that begins when the mouse does not enter alpha to complete the criterion, type 8 loop, will be a nested loop of types 1-4 depending on the order of the units the mouse now chooses to enter. So if the mouse runs a type 1 nested loop, it will make a 48 step run through the essential path without entering alpha-omega-alpha at any time in the nested loops. If a type 2-nested loop was chosen, the mouse would not have entered alpha on its return trip, will run out but not enter omega and will return to enter alpha. However, please note, if the mouse chose this type of loop, once it enters alpha on its return from the nested type 2 loop, it will have completed a criterion loop.

Let's look at this more closely. If the mouse started with a loop type 7 on its initial run through the maze, it has started in alpha and entered omega on its first trip. When the mouse returned it did not enter alpha (no criterion loop was achieved), so it is beginning a nested loop. On this nested loop it chose type 2, it avoided alpha on its first trip back and turned around in unit 14. It now runs back to omega, does not enter omega but instead it turns around in unit 54. When it returns to alpha this time, it enters alpha. Now the mouse is removed because it has completed a criterion loop. It has now gone from alpha to omega and returned to enter alpha a second time. So while the mouse did not make a criterion loop on its first trip through the maze, it has now completed a criterion loop, it has entered alpha-omega-alpha and is removed from the maze.

It is important to note that the nested loop scoring was all figured by computer, all the tracker had to do was note when the animal entered the omega unit after leaving alpha and then track when it entered alpha a second time and remove the animal. The nested loop scoring was then figured by the statistics package based on the number of steps and order of units entered by the animal. The statistics package used the first entry data to compute these criterion loops and nested loops. Looking solely at the table may not give the reader sufficient insight into this behavior, but it was included to help the reader see how devious behavior develops in the loop and growth behavior.

The first four of the MLE types represent the simplest solution to the punishment problem. These four types allow subjects to make as many of these four MLE type growths as nested loops as they have time for and as long as they avoid Alpha, they can enter omega or not as often as they chose. Type 3 is the simplest of all choices. The rule is, avoid Alpha but not Omega. The most complex MLE growth is type six. This involves the strategy of avoiding Omega instead of Alpha. This solution, if it were consistently used, would imply that the organism has the cognitive power to understand the causal contingency of this constraint set in a most sophisticated way, the ability to take advantage of this type six solution, which allows the most ingenious form of devious behavior. Only this type allows the organism to enter the alpha unit all it wants and without ever being punished. This solution has all sorts of behavioral implications and one important theoretical one. How does a rodent detect causality contingencies? Does the delay effect the detection of causality in play behavior the same way as delay effects reinforcement in work behavior? The interesting and complicating fact is that the causality consequence of entering, or not entering omega, can never be immediate, as Learning requires for Reinforcement. The consequence of not entering Omega can only be detected after the subject returns to alpha after not entering omega and this time delay is a function of the omega-alpha latency. The latency for traversing these twenty-four units will be 24 seconds at best if the mice travel one unit in one second, and even twenty-four seconds is a long delay. But Garcia has shown that mice can infer causality even after long delays in food aversion studies. (Garcia et al, 1966)

FIGURE 4.09 groups by types of multiple loops

FIGURE 4.09 shows the first order interaction; group by types, for the two groups of mice for the eight multiple loop types. $F(7, 112) = 5.426$, $p<.000022$. This data was computed on the percentage of each type for each of the subjects in both groups. The experimental mice's devious behavior resulted in their making 33% of their growths in a type 3 nested loop— FIGURE 4.08T shows that this is a nonalpha1, omega, nonalpha2 type of multiple loop. FIGURE 4.09 show that the experimental mice used the first four types 61% of the time. Notice that the type six multiple loop was the least used multiple loop, only 13 type six loops were made out of the total 572 multiple loops for both groups combined. This inability to make these type six loops suggests that mice have a weak ability to detect the complex causal structure of the third element of the constraint set if any at all. Male rats learn to make more of this type of loop as a part of their devious behavior, which may suggests the presence of a powerful causal attribution mechanism in rats. The type six multiple loop pattern is a particularly powerful form of devious behavior for it allows the organism to freely enter what would otherwise be the dreaded alpha unit. For an organism to use this type of multiple loop, it would represent a powerful calculating cognitive ability especially for an organism without a second signaling system and supposedly even without a mind. Type 6 is the opposite of type three, the mice's favorite multiple loop. FIGURE 4.09 show that the control mice used type 7 and 8 49% of the time as opposed to 36% for the experimental group.

Mice first growth Behavior for the EP

In this section, the heterostatic pattern of behavior of the first growth of mice in the EP of the Matrix maze will be described. Our purpose is to show how mice and rats' first-growth behavior differs, and how these differences can be seen in each of the three data sets, FEULL, FREQULM, and BFEULM.

FIGURE 4.10 data types by units

FIGURE 4.10 shows the first order interaction, data types by units for FEU, FEL, and FL for the first growth. This is the average for the two groups of mice for this study. This FIGURE shows that the mice of both groups made the first essential path growth in 719 steps and 1229 s. Mice FEULL growth data differs from rats' data in several respects: first, mice are considerably faster than rats on completing the first growth. Secondly, the plateaus so prominent in rat data are missing. Third, the phenomenon of extrapolation is missing. Only one short retracing period is seen in this FIGURE between units four and five. The FL data shows mice learn to reduce the time it takes to explore each unit but only for some units. Also, note the column cycle effect. Compare this figure to Figure1.3.3 to the see the distinct differences between the mice and rat plateaus and retracing behavior.

The FREQULM data

FIGURE 4.11 data types by units

FIGURE 4.11 shows the first order interaction, data types by units, for the FREQULM data set for the 48 units of the essential path of the Lashley III maze. F (94, 1974) = 6.279, p<.00. The three variables of this data set show where the mice of both groups spent their allotted moves and latencies for the first growth and their average unit latencies during the first growth. This effect is a whole whose significant parts will be identified below. The relation between the three data sets is very similar to the rat data. All three scores appear to cycle as if they were sine waves. However, FREQL while having the same frequency as FREQU has a larger amplitude with higher highs and lower lows than the FREQU score. While, the mean latency score per move also cycles with the other two scores and with a column cycle eight units long that is the high frequency component of FIGURE 4.11; the FREQM score does not show the same low frequency components that reflects growth or avoidance in FIGURE 4.11.

FIGURE 4.11 shows the rhythmic sine-wave-like multi-dimensional nature of the spontaneous behavior of mice in the Lashley III matrix maze. The classical behaviorists in their trial paradigm conceived of behavior as one-dimensional from the start to the goal box, or what is called the row effect in this design, except even this dimension they generally reduced to a single error and a single latency score. FIGURE 4.11 shows the spontaneous complexity of the Lashley III maze behavior. The essential path growth behavior was, of course, circular in nature, but in FIGURE 4.11, it was opened up and appeared as a 51 unit linear phenomenon. The analysis of this data with ANOVA will allow this complex rhythm to be broken down into its component parts: the first order interaction, a unit by data type, for the FEULL data type is significant. F (94, 1974) = 6.279, p<.00.

FIGURE 4.12 main effect for rows

FIGURE 4.12 reveals an important difference between the way rats and mice make their model of the matrix maze. This FIGURE is the main effect for rows and is the first of the significant components of the FREQULM data structure of FIGURE 4.11. F (11, 462) = 20.2975, p<.000000. This is a linear plot of the data. This type of curve is seen in rat data only when it is plotted with a log-linear plot. The log-linear plot is required with rats because the effect is so small it is not noticeable on a linear plot. The mouse effect is large enough to be seen on a linear plot. What is important about this curve is the fact that it is not the simple curve observed with rats in FIGURE 2.09, which showed a single smooth continuous improvement over all twelve rows, but this mice curve is a two-part curve. This two-part curve shows a primary smooth learning curve on the way out but then a secondary smooth learning curve on the way back. The second part of this two-part curve shows that mice have much more difficulty getting started on the way back than rats. It is not immediately clear why mice have difficulty getting started in the second half of the maze. Two possibilities need to be considered. First, it is possible that mice are shortsighted explorers who on the way out take no thought of how to return home. Secondly, the problem may be that there is something novel about the second half of the matrix maze, which rats but not mice are able to master on the way out. In particular, it should be noted that the Lashley III maze is not symmetrical out and back. The out sequence is RLR..., while the back sequence is LRL... However, an odd number of rows would convert the maze into a symmetrical pattern both out and back. If mice exhibited this same difficulty on a maze with an odd number of rows then only the first hypothesis would be left. Mice, according to this hypothesis, require so much time exploring the units of the first rows of the maze on the way back because on the way out they do not create a model that is valid in both directions as do rats. Rats are maze bright in the sense that on the way out they take their time and learn the path in both directions. Rats are conservative, when they explore, they do not explore only in one direction. Rats never explore on the way out without creating a model that will allow them to make an efficient return trip. Mice, according to this first hypothesis, seem to take no thought on the way out, how they will get back. Notice the difficulty mice show at row 7 in FIGURE 4.12.

FIGURE 4.13 rows by data types

FIGURE 4.13 shows the first order interaction, rows by data types, which compares the main effect for rows for each of the elements of the FREQULM data set $F(22,528)=5.54; p<.0000$. An inspection of these three curves shows the FREQM row learning effect was fundamentally different from the FREQU and FREQL effect in two ways: first, the FREQM learning curve shows only a small improvement over the twelve rows. Second, the difficulty on the way back was exclusively a function of FREQU and FREQL variables. The FREQM data shows that the average latency was a smooth curve over twelve rows. The FREQM row learning curve is a one part rather than a two-part curve.

FIGURE 4.14 data types by columns

FIGURE 4.14 shows the first order interaction, data type by columns, for the FREQULM data set. $F(6,144)=53.89$, $p<0.0000$. This effect shows, once again, the column effect in the learning behavior of the Matrix maze is just as important for mice as it is in the behavior of rats. The classical behaviorist essentially ignored the role of the column dimension of learning. However, the amount of variance explained by the column effect in ANOVA is at least as large as the variance of the row effect. In other words, the column effect and its interactions are at least as important as the row main effect and its components and only a matrix representation of the maze allows these components of learning behavior to be analyzed.

All three curves for each element of the FREQULM data set have the bucket or U shape for the column effect but not to the same degree. The FREQL creates the deepest bucket, with FREQU an intermediate depth and with the FREQM curve very shallow. The column levels 2 and 3, which are the E2 and E3 units respectively, received the least amount of activity in all three curves. These are the two middle units of the essential path of each row and are both two-way choice units. The column one and four units are the E1 and E4 units, which are the first and the fourth units in each row respectively. The E1 and the E4 unit are door units and both are three-way choice units and thus should be expected to generate more activity than two-way units did. FIGURE 4.14 show that the E1 unit generates more activity than the E4 unit does. The column variable with its bucket effect is responsible for generating the high frequency cycle effect seen in FIGURE 4.11. Each cycle of the high frequency cycle was generated by the column effect for each row.

It is interesting to note that this data resembles the data shown for the rats in FIGURE 2.12. The same units generated more activity for both the rats and mice in these experiments.

The BFEULM data

FIGURE 4.15 data types by units

FIGURE 4.15 shows the first order interaction, data types by units, for the three scores of the BFEULM data set: BFEU, BFEL, and BFEM, for the 48 units for the first growth of the essential path. FIGURE 4.11 and FIGURE 4.15 are both FIGUREs of complex data structures, but each for a different set of scores. Both sets of data have very different shapes. While FIGURE 4.11 plots the FREQULM data set, which shows where the mice spent their allotted moves and time in the matrix maze, FIGURE 4.15 shows the amount of activity that was triggered by each unit of the maze as measured by the BFEULM data set.

The organism experiences each exemplar of a new set as an anomaly. Knowledge is that which reduces uncertainty, but an anomaly is opposite of knowledge, it is an event that induces uncertainty. Anomalous-units and news-units are alike in that they both require the current model of the maze be modified. If events are anomalous, neither mice nor rats can assimilate the event. Accommodation is the process during which a model is created that will allow the organism to understand these events, the organism will continue to retrace until it can create a model that is sufficiently valid to reduce the anomalous event so that it is no longer avoided.

FIGURE 4.15 shows: first, that some units of the maze were more powerful in initiating retracing than others; second, the retracing was initiated in cycles; third, that the retracing cycle behavior is obviously also multi-dimensional as was seen in FIGURE 4.11 for the FREQULM data set. However, the rhythms of FIGURE 4.11 and 4.15 are very different. Note: the BFEM is the simplest of the rhythms of FIGURE 4.15 and BFEU and BFEL parallel each other.

Again, this data is very similar to the rat data from FIGURE 3.11 showing that both rats and mice find the same units anomalous in the maze and retrace to find the cause of the latest anomaly.

FIGURE 4.16 row main effect for 12 rows

FIGURE 4.16 shows the row main effect for 12 rows for the BFEULM data set. The amount of retracing initiated by each row was an increasing growth function for the first six rows with the exception of row four. The second six rows also showed roughly an increasing growth function, although not so steep. This curve is not exactly the same as was seen in either of the punished groups of rats, but it is still a modified curvilinear shape function as was seen in the rats' data, except mice have more retracing behavior in the second half of the maze.

FIGURE 4.17 data types by rows

FIGURE 4.17 shows the first order interaction, data types by rows, for the three elements of the BFEULM data set. This FIGURE shows that the row effect of FIGURE 4.16 was primarily due to the two scores BFEU and BFEL since the BFEM score does not parallel the other two scores. Again, notice that just as in FIGURE 4.15 the unit and latency score of the data set have the same frequency but differ in amplitude. The ratio between the two scores reflects the relative difficulty of each move or the quantity of uncertainty associated with each unit of the maze. For example, on row 6 the uncertainty ratio is 2 to 1. The BFEM latency score is only weakly related to the other two scores.

FIGURE 4.18 main effect for columns

FIGURE 4.18 shows the main effect for columns for the BFEULM data set and shows that the E1 and E2 units were both major triggers for retracing with E2 being somewhat more powerful than E1. The E3 and E4 unit, columns 3 and 4, were both weak triggers of retracing. The E1 and E2 units of the maze are the two anomalous units of the rows that induce uncertainty and thus retracing. The quantity of the retracing that mice exhibit, however, relative to the retracing of rats is minimal, although the shape of the column effect is the same for both species.

Heterostatic growth is seen in the changes in the EP growth data for thirteen runs or seven growths

Frank Beach (1947) a comparative psychologist spent his life studying the sexual evolution of mammals. He showed that as sexual behavior evolved, there were systematic changes taking place in two components of sexual behavior: the arousal mechanism, and the consummatory mechanism. Beach showed that for over fifty million years sexual behavior evolved dramatically. Let me be clear that although I have only studied the nonverbal mind of rats and mice, I would not think of saying that the nonverbal mind did not evolve in the over fifty million years from rats to humans, nor would I say that there were no differences between the nonverbal mind of rats and mice.

In fact, in this section, I must report several differences between the way rats and mice perform in the maze. Yet that is not to say that mice don't have a heterostatic mind. In some respect, rats perform faster in the maze, while in other places mice perform the best. One of the ways the two species differ is that mice don't create models on the way out as do rats, yet mice exhibit devious behavior just as effectively as do rats.

Mice may not be Heterostatic creatures like rats that make models of the events in a domain so that they are never surprised. The purpose of this section is to show how the self-organizing behavior of mice evolves over time. There are two main goals of this section; first to describe the multi-dimensional character of the growth learning curves of mice in the matrix maze domain, which the Behaviorists ignored. The exploratory-play growth behavior in the matrix maze allows the evolution of the mice behavior to be described in several dimensions with the column dimension particularly the novel dimension of the mice and rats' behavior to be seen for the first time. The second purpose is to describe how punishment and sex affects the evolution of this behavior.

The number of the growths mice made in thirteen runs varied, but all mice made at least seven of the alpha, omega, alpha, type 8-MLE growths. In this section, the evolution of the devious exploratory-play behavior of mice will be described for thirteen runs and seven type 8 growths.

In the third experiment of the series, the devious behavior of individual rats evolved over twenty-five runs in a curvilinear fashion. The moves-per-run behavior of rats showed an increase in the total amount of activity for the first thirteen days until a max of 700 steps per day was achieved, and then gradually the moves settled down to a daily level of about 450 units of activity a day. The experimental mice of FIGURE 4.07, in this study however, did not achieve 700 units of exploratory- play activity until day twenty-five. Unfortunately, in this section the data available only allows the behavioral analysis for thirteen days so that the full range of the evolution of the devious behavior in the experimental mice could not be analyzed.

FIGURE 4.19 groups by directions: the push-pull hypothesis

FIGURE 4.19 is the groups by directions, first order interaction, for the FREQULM data set. $F(1,360) = 18.86$, $p < .0000$. This FIGURE shows that direction is a major dimension, which affects the devious behavior for both rats and mice.

In the rat's data, I proposed the push-pull hypothesis, which said that on the "out" half of the maze the fear-factor that was acting as a wind at the animals back would hasten or shorten the "out" portion of the loop. However, on the "Back" half of the loop, the fear-factor was acting as a Head Wind and would slow and lengthen this half of the loop. This hypothesis was confirmed for the rats, and now I must show that it is also confirmed for the mice.

This Group by Direction interaction effect confirms the prediction of the push-pull hypothesis for mice:

- First, The blue line of FIGURE 4.19 shows that the base line or the control group clearly spent the majority of their moves and time in the first half of the matrix and much less time in the second half of the maze.

- Second: The red line of data in FIGURE4.19 also shows that the Experimental group, which was punished, significantly modified, or actually reversed, the pattern of behavior exhibited by the control group.

- The fear produced by the particular kind of Punishment administered in this experiment generally has two effects. First, it acts as a wind on their back that moves them through the first half of the maze, which decreases the amount of behavior in the first half of the maze. The figure shows that the mean decreased from 5 to 3.9. Secondly, on the way back, the fear acted as a wind in their face, which slows them down, and increases the amount of the behavior emitted in the second half of the maze. The increase was from 3.25 to 4.75. The effect of punishment is to motivate mice to get as far away from the punished alpha unit as possible, which has a positive effect in the first half of the maze and the opposite or negative effect on the second.

FIGURE 4.20 group by rows

FIGURE 4.20 shows the groups by rows interaction for the FREQULM data. F (11, 3960) =11.18), p<.00000. This FIGURE shows how the group by direction effect described above looks at the row level. The same rule applies; the effect of punishment was to make the experimental group's scores for the first six rows smaller than the control group's scores and larger than the control group scores in the second six rows.

FIGURE 4.22 blocks by data types by columns

FIGURE 4.22 shows the learning curve for another major dimension of exploratory-play behavior, which the classical experimentalists ignored - the column effect. FIGURE 4.22 shows the second order or three-way interaction, blocks by data types by columns, for the three elements of the FREQULM data set. F (12, 1080) =13.95, p<.0000. The FIGURE contains three parts, one for each data type. Each part shows the growth curve for the three blocks for one of the three types of data. For example, on the left, the FIGURE shows the column growth effect for the three blocks for the FREQU score. The column effect for each block for the first two scores is a bucket shaped curve, and the growth effect is seen as if the bucket is falling from block to block. The middle portion of the FIGURE, which is for the FREQL data shows that the first bucket for block one was a deep bucket and the change between block one and two was larger than the change between the next two blocks. The third section shows that the FREQM score was exceptional and did not follow the path of the other two FREQ scores. The FREQM scores per block did not fall but climbed and the bucket shape was not achieved until the third block.

FIGUREs 4.23A data types by blocks by rows by columns

FIGURE 4.23A, B, and C are elements of a 4-way interaction, data types by blocks by rows by columns. FIGURE 4.23 A is the block by row by column effect for the FREQU data set; FIGURE B is for FREQL and FIGURE 4.23 C is for FREQM. Actually, the main point of these FIGUREs is to call attention to the FREQM data. Mice are unique, relative to rats, in the way they adapt to the matrix maze. Rats, especially female rats keep moving or exploring in the maze at a constant rate per unit of the maze. When something special happens like punishment, rats exhibit more avoidance behavior but do not change their rate of motion. What is especially important about this FIGURE is that it shows that mice do not work this way. Mice have the capacity to change their rate of motion from unit to unit, and they evolve a complex set of rules governing their rate of motion that takes into account several variables, direction, rows, and columns. FIGURE 4.23A shows the FREQU data, which shows how mice evolve the way they spend their allotment of moves over the three blocks for direction and rows and columns. Notice the column effect first. The column effect is typically what I have called a bucket shaped curve with E1 and E4 receiving most of the mice's allotment of both moves and time, while E2 and E3 receive the absolute minimum attention possible. In this style graph, a row effect is plotted for each column, so that there are four curves one for each column. The column effects for E1 and E4 are always the two top curves in this FIGURE and the two E2 and E3 curves are always the lower two curves. What mice learn in the first six rows can be seen in the fact that they decrease the size of the scores between blocks. They decrease from an average of about 20 in block one for row one to less than 10 in block two for row one to about 5 in block three row one. The rank order of the four-column effect is stable.

FIGURE 4.23B FREQL scores over blocks

FIGURE 4.23B shows how the FREQL score evolved over blocks. The point to note in this FIGURE is how the rank order of the four column effects changes over row from block to block. What distinguishes the FREQL data is the separation between the four column effects. The four column effect obviously consists of two subsets with E1 and E4 belonging to the high score subset, while E2 and E3 belong to the low score subset. And the difference between these subsets is what evolves between blocks.

FIGURE 4.23C blocks by rows by columns

FIGURE 4.23 C shows how the FREQM data evolved over blocks and over rows and columns over blocks. In block one, the row by column effect showed no consistent order, but the overall FREQM score was small, averaging about 1.5 seconds. In block two, the average FREQM score increased to about 2 seconds, but the order of the column effect still had no pattern. But in FIGURE 4.23C the mice really became organized and the behavior in this block is incredible as well as unique. Notice how the two subsets of column effects separate into two unique sets of latency scores. The high sets are the E1 and E4 subsets, which in the second half of the maze have unit latency scores of greater than three seconds. The fast latency subset of column effects is at its lowest point in row six with an average of seven seconds. Mice are, in my experience, unique in this ability to shift gears and dynamically change their unit latency in the complex pattern seen in FIGURE 4.23 C.

The Emergent Loop, EL, Full Maze, FM, behavior of mice

When rats and mice are tested on the Matrix maze with the LTP, both species are so efficient at making models of the matrix maze that many BP units are never entered on the first growth or even on the first several growths. These empty cells make it impossible to analyze the FM behavior for each EP growth. However, after several EP growths, rodents eventually enter each of the BP units and then a FM analysis of the data is possible. In this section, for the first time an analysis of the Emergent Loop behavior for the full maze (ELF) will be described.

A comparison of Rat and Mice FREUL full maze behavior

In this section, a comparison of rats and mice FREUL data for a full maze growth will be presented. This will save time since the full maze blind behavior for rats and mice are remarkably similar. However, this comparison is important because the difference between the primary mind of rats and mice is one of our major concerns.

There are two kinds of extrapolations - EP and BP extrapolation. Rats exhibit both kinds, while mice only exhibit the latter. It would have been convenient if the comparison of the blind behavior of the two species of rodents would have simplified the differences between the behaviors of these two species in the matrix maze, but it now seems that the BP behavior of mice has only deepened the mystery. The extrapolation of mice in the BP shows that the primary mind of a mouse is capable of exhibiting extrapolation behavior in the BP so why does it not extrapolate in the EP. The simple solution would have been for rats to exhibit two kinds of extrapolation but mice none. For mice to be able to extrapolate in the BP but not in the EP complicates the picture.

I will pick up where I left off describing the Full maze behavior in the last experiment. I stopped showing the result of the rats' full maze behavior because I wanted to include the mice data from this study in the analysis.

The FREUL analysis of the full-maze rat and mice data combined

The FREUL data set analyzed in this section was created with several special between-type variables in order to allow various kinds of ANOVA comparisons. First, a new independent variable was created with three levels, E II, E III, and E IV, that allowed these three experiments to be compared. Secondly, a species variable with two levels, mice and rats, was created that allowed a comparison of the two species. A comparison was also made involving sex, constraint conditions, and data type variables.

The full-maze data changes the appearance of every ANOVA effect for several reasons. First, behavior is now being described in 100 units of the maze instead of just the 50 units of the EP. Second, it takes many EP growths to complete a FM growth. Third, the FM column effect becomes an eight-unit effect rather than a four-unit effect for the EP growth. Understanding this eight-element column effect is a major challenge for it represents a special new kind of avoidance behavior.

Avoidance Behavior can be roughly classified two ways; first, into three pure motivational types homeostatic, hedonistic, and heterostatic, and their interactions. The first two kinds of motivation are supposedly generally well recognized. The overall purpose of this book is to define heterostasis and to distinguish it from the other two, but the point in this section is to distinguish between heterostatic and other types of avoidance behavior. Secondly, Heterostatic avoidance behavior can be classified into two types Anomaly induced avoidance and Redundancy induced Avoidance. The BP avoidance behavior is of the RIA type, while the EP is of the AIA type.

Understanding the RIA amounts to answering the question why are spontaneous blind entry errors so rare. In the next several FIGUREs, I will demonstrate the rarity of these BP errors.

FIGURE 4.24 eight unit main effect for columns

FIGURE 4.24 is the eight unit main effect for columns for the three experimental groups of rodents for the FREUL data set. $F(7, 280) = 81.78$; $p<.0000$. The abscissa shows the functional unit names, FUN, for the eight column units with the two A units on the left and the two F units on the right. This FIGURE shows that both types of blind units are the last units of the maze to be entered and show that both mice and rats avoided the F type units more effectively than the A blind type. This point will be proven in a later analysis. This BP behavior for the two types of blind unit shows that both rats and mice quickly and most efficiently learn to avoid blind units and this avoidance occurs in rodents that are neither deprived nor reinforced. The primary mind of rodents exploring the matrix maze spontaneously makes a sub model of the blind units of the matrix, the BP-model, that allows them to predict which novel elements of the BP are exemplars of redundant classes of space and thus not worth entering for many EP growths. The rarity of the BP unit errors is, in other words, due to the rodents' spontaneous rational avoidance of the redundant units of the BP made possible by the existence of the BP-model.

FIGURES 4.25A-4.25B Linear and log plots for rows by columns

FIGURE 4.25A linear and 4.25B log are two versions of the two way interaction, rows by columns, for the average of the two FEUL scores, FEU and FEL, for the three experiments, and together they show how the column effect changes over the 12 rows of the maze $F(77, 3696) = 11.07$; $p<0.0000$. This effect is a learning curve and shows that the blind units in the last seven rows are the last units of the ninety-nine units of the matrix maze to be entered. This FIGURE shows how the avoidance of the blind units grows over the twelve rows. FIGURE 4.25B, which is a log plot of the data, shows the first five row's column data in detail, and shows that the blind path avoidance for these rows was minimal for this average FREUL score. FIGURE 4.25A shows it took several rows before the rodents were able to evolve a model with enough primary knowledge for the recognition of the redundancy of the sets of the blind path.

FIGURE 4.26A rows by data types

FIGURE 4.26A is the 3-way interaction, experiments by rows by data types, for two scores of the FREUL data set. $F(22, 484) = 5.95$; $p<.0000$. This is the first FIGURE that shows how the three experimental groups' FREUL behavior differed. The FIGURE shows that the FEL score for the three groups from Experiments II,

III, and IV are very different, but notice that the FEU Scores for all three groups is very similar in one respect-the final FEU score for all three groups is close to 1000 moves. However, there is a most important difference between the two rat curves II, and III, and the one mouse, IV, FEU curve. The two rat FEU curves show that these two groups of rats hit 1000 moves by row six and then ran the last six rows of the maze with almost perfect efficiency, while the mice data shows that mice only hit 1000 at row 12. This FIGURE is just another way of making the point that one of the most fundamental Heterostatic differences between rat and mice behavior is the way they learned the maze.

The simplest solution would have rats conceiving an alpha model on the way out which also proves valid on the way back. However, the mice must evolve two models - an alpha and an omega model. Mice spend almost as much time conceiving or evolving the Omega model as they did in designing the alpha model.

Rats were able to make a valid model of the matrix maze in both directions while exploring in the first half of the maze so that they stop retracing in the first half of the maze. They then extrapolate and explore the rest of the maze with almost maximum efficiency. However, mice did not make a valid model of the matrix maze while they were exploring the first half of the maze. In fact, mice spend more time exploring the second half of the maze than the first half. Rats spend better than 85% of their total time exploring the first half of the maze, while mice spend better than 50% of their time exploring the second half of the maze.

The FEU data shows that mice on average complete an emergent loop for the full maze in a thousand moves as do rats. However, the FEL data shows that the mice complete the full maze growth in only 2000 s. on the average, which is only 33 minutes, while rats require better than 4000 s in 1 case, or 12000 s in the other case, 66 minutes vs. 200 minutes, respectively. Obviously if time is a major consideration, an experimenter should consider running mice rather than rats.

FIGURES 4.27A, 4.27B experiments by rows by columns by data types

FIGURES 4.27A and B are the 4-way interaction, experiments by rows by columns by data types, for the FREUL data set. $F(154, 3388) = 1.36$; $p < .0025$. This set of FIGUREs gives a much richer picture of the FM FREUL growth behavior than the earlier FIGUREs. FIGURE 4.27A and B compares the row by column effect for the three groups for the two elements of the FREUL data set. FIGURE 4.27A shows the mean FEU score, while 4.27B shows the mean FEL score. This effect is a learning curve that is unique for heterostatic behavior in the matrix maze, and it shows how the column effect seen in FIGURE 4.24 evolves over twelve rows. All three groups of rodents evolve the same kind of curve, which reflect the avoidance of the BP units, particularly the avoidance of F type BP classes of units. In FIGURE 4.27A showing the FEU data, the extreme example of FEU avoidance of the F type can be seen in the experiment III group of rats for column 8 row 6. The average first-entry for this unit was on step-number 2400. These numbers may seem strange in the context of the data of FIGURE 4.26A, which was just discussed above. However, it must be remembered that FIGURE 4.26A shows that the average total FEU for all three groups is 1000 moves. However, the BP data in

FIGURE 4.27A shows there were many BP units that were entered after step 1000; for example, the last unit entered was on step 2400.

FIGURES 4.27 A and B avoidance behavior shown

Both FIGUREs 4.27 A and B for the experiment II group deserve a comment for the data for this group of rats reveal, in the most striking form, both types of avoidance behavior, both AIA and RIA. The AIA type of avoidance for Experiment II is seen in the large empty central region of the FIGURE between row 4 and 5 especially in FIGURE 4.27 B. However, the RIA is seen in all three groups in the FEU data as noted above. The AIA-type of avoidance occurs in the EP and induces the EP retracing behavior. In the EP, a rat's retracing is induced by anomalies and continued until rodents evolved a valid model of the EP, a model that destroys the anomalies of the domain. However, the BP retracing is induced by redundancy that generates avoidance behavior, RIA that lasts for many EP growths.

While there are some obvious differences between the BP FEU scores for the three groups of rodents, the important point is the fact that the mice row by column effect nevertheless shows that the mice BP avoidance behavior has the same pattern as the rat BP avoidance pattern. While the degree of the avoidance is somewhat less intense, it is clear that mice exhibited the capacity to use the model of the matrix maze they evolved to avoid the redundant class of units of the matrix maze.

The FIGURE shows that mice exhibit Redundancy Induced Avoidance, RIA behavior even though they do not exhibit the Anomaly Induced Avoidance, AIA, type behavior. Why do mice exhibit one kind of avoidance and not the other?

The column type analysis of the FM BFEULM data

The full maze data in the section above was analyzed with the eight unit column data in its natural order. In this section, the data was rearranged in order to test the differences between the two column types of data, the EP and the BP and to make exact tests of several important hypotheses about the behavior that characterizes these two types of BP units. In the above section in the analysis of the full maze 96-unit data, it was obvious that there were important differences between the EP and the BP behavior of rodents, but statistical proof of those apparent differences were not possible. In this section, the full-maze 96-unit column data was sorted or rearranged so that a formal comparison of EP and BP data effect could be tested with ANOVA designs.

Two possible column arrangements for the full maze data are shown below; first, the natural eight-column order of the full maze data, and secondly, the sorted order is shown:

The natural column order: A2, A1, E1, E2, E3, E4, F1, F2
The sorted column order: A1, A2, F1, F2, E1, E2, E3, E4

In the sorted arrangement of the data, three rules for ordering the data were used: first, all the BP data was on the left; second, the A type was placed before the F type; third, all three types of data was sorted from left to right.

FIGURE 4.28 main effect for 8 columns

FIGURE 4.28 is the sorted main effect for the eight column BFEULM data. Compare this with FIGURE 4.24. The difference is that the F data now occupies position 3 and 4 instead of position 7 and 8, but this arrangement of the data makes possible the new analysis to be described.

FIGURE 4.29 column type by order

FIGURE 4.29 is the 2-way interaction, column type by order, for the BFEULM data seen in FIGURE 4.28 above. $F(3, 126) = 29.39$; $p < .0000$. The analysis of the sorted data shows that almost all the AIR for the EP was induced by orders 1 and 2, Units E1 and E2. The RIR induced by the A1 and A2 units in the blind path were almost the same magnitude as the AIR for the EP. The major difference was in the extreme amount of RIR induced by the F1 and F2 units of the BP compared to E3 and E4 units of the EP.

FIGURE 4.30 column types by rows

FIGURE 4.30 is the 2-way interaction; column types by rows, for the BFEULM data set and this analysis makes possible a direct comparison of the row learning curve for the two column types, EP and BP. This data shows that most of the mice's retracing was emitted in the first six rows for the EP units of the matrix maze. The BP unit avoidance was greater than the EP avoidance for all the rows beginning at row 4. The BP redundancy induced avoidance, RIA, shows a linear row effect in the first six rows and another linear affect in the second half of the maze.

FIGURE 4.31 column types by rows by data type

FIGURE 4.31 shows the 3-way interaction, column type by rows by data type, for three BFEULM data types. $F(22, 464) = 1.77$; $p < .0176$. This FIGURE shows that the data of FIGURE 4.30 masked the BFEM score. The FIGURE 4.30 curve only represented the two scores BFEU and BFE rather than BFEM. In addition, FIGURE 4.31 shows that BFEL was always larger than BFEU. FIGURE 4.31 shows that in spite of all the complexity seen in the BFEU and BFEL scores, the BFEM score was very stable, and in general, the BFEM score improved so the mice were running faster constantly over rows. But again this curve shows that while mice continue to retrace in the second half of the maze in the EP data, yet the amount of this retracing is much less than the quantity of their retracing in the second half of the maze seen in the BP behavior.

The mice evolved two sub models of the maze, one for the EP and another for the BP. The BFEU and BFEL score are in both cases the best estimate of the degree of retracing induced while the two models were evolving. While the mice are learning to run the shortest-path, in other words, they are acquiring a distinction between two different things: first, to evolve a conceptual system for the EP and second, to evolve another conceptual system of the BP. The conceptual system for the EP has the task of moving the mice through the EP units of the matrix with the fewest possible steps. However, the conceptual system for the BP has the task of promoting the avoidance of the BP units. The mice are most successful at avoiding the Forward type of Blind type unit.

FIGURE 4.32 path types by rows

FIGURE 4.32 is the first order interaction, path types by rows, for the FREUL data type. F (11,308) =12.58; p<.0000. This data is a comparison of the learning curve for the mean of the FREUL row effect data for the EP and the BP units. Note that the hump in the BP data at the row 6 level is not in the EP data. In order to understand and appreciate FIGURE 4.32, it is necessary to compare it with FIGURE 4.26, which also has this hump in the third column on the right of the FIGURE, which shows the data for Experiment IV. However, since the two curves in FIGURE 4.26 are the mean of the two curves in FIGURE 4.32, the hump in the data seen in FIGURE 4.26 can now be attributed to the BP not the EP data. Recognizing this makes the point of FIGURE 4.26 even clearer. The first difference between mice and rats is that rats learn in the first half of the maze and stop retracing in the first half of the maze. Mice do not create a rule-based behavioral system in the first half of the maze for the EP that allows them to extrapolate and stop retracing in the first half of the maze. The hump, in other words in FIGURE 4.26 in the column on the right for Experiment IV, makes the mice look better in the first half of the maze than they deserved to look. The data of FIGURE 4.32 clarifies this picture and shows that the mice row effect for the EP path is a smooth effect. The hump in the mouse data is in the BP not the EP behavior.

FIGURE 4.33 column types by rows by orders

FIGURE 4.33 shows the 3-way interaction, column types by rows by orders, for the FREUL data type. F (33,924) =11.72; p<.0000. This FIGURE compares the EP and the BP data for eight rows by four orders of columns and shows that the evolution of the avoidance was confined to the BP units.

Blind type analysis

In this section only the BP data will be analyzed; only the 48 unit data of the BP of the full maze 96-unit data will be analyzed. This analysis will allow exact tests of the differences between the two BP types, the Alternate A, and the Forward F, to be made.

FIGURES 4.34A- 4.34B blind type by row

FIGURES 4.34 A and B show the same first order interaction, blind type by row, for the BFEU data type plotted two ways. A is linear and B is log. F (11,462) =6.30, p<.0000. FIGURE 4.34A shows the absolute difference between the A and F BP types of avoidance and show the extreme difference that occurs at row 6 with the F BP type showing such an extreme degree of avoidance compared to the A type. The linear plot suggests that the A-type showed no avoidance. However, the log plot shows that even the A units for the BP were also avoided. While the F type of redundancy triggered the most retracing especially at row six, yet the A type also triggered redundant avoidance behavior.

Figure4.34A and B work together to induce the reader to make just the proper inference about this signifance of the interaction. Figure A the linear plot by itself overstates the case for the forward BP type error. Yet then Figure 4.3.4A provides us with data to soften the blow. While it is correct that the effect is significant, the reader should realize that the effect is less than suggested by the log plot.

FIGURES 4.35 A, B, and C blind types by rows by orders by data types

FIGUREs 4.35 A, B, and C are elements of the 4-way interaction, blind types by rows by orders by data types, for the FREQULM data type. F (22,462) =3.40, p<.0000. FIGURE 4.35A is for FREQL, 4.35B is for FREQU, 4.35C is for FREQM, in other words, each of these three FIGUREs shows a blind type by row by order interaction for one of the three FREQULM data types. Order 1 is for the F1 and A1 blind path units and Order 2 is for the F2 and A2 types of blind path units, essentially explaining the depths of the cul-de-sac units. The most important point of these three FIGUREs is to call attention to the FREQM score. The A FIGURE shows that the order effect for the twelve rows for both BP types is almost the same, but the B FIGURE shows that the FREQU score for the two orders is different with the second order having always the fewest entries. So FIGURE C shows that the FREQM score, which is the mean of these two scores and shows a constant large difference between the latencies for the two orders with the second order having the larger latencies. This C FIGURE reveals an interesting attribute of mouse behavior. While mice avoided the F type of blinds more than the A type and avoided the second order more than the first order, still the fact is that paradoxically when mice finally get around to entering the second order of both of the blind types, their long latency is in the second order. This shows that they took their time and savored their rare one-time visit to this class of BP units.

I should make two technical points here - patting myself as it were on the back. Of the several innovations I introduced in this series of studies, two of the most controversial proved themselves in this figure. They are first, the Data Type Dependent variable, with three elements: FREQL, FREQU, and FREQM. Each of these three kinds of scores is involved in this significant effect, but this effect validates the data type variable as a dependent variable. Several times throughout these experiments, I have simply dropped the use of the FREQM score when it was obviously not useful but here FREQM proved its worth.

Second, the original unit size of the matrix chosen by Lashley was one foot. One of the decisions I made was to reduce the unit size to six inches, which doubled the number of units.

This set of three FREQ scores validates the implementation of the decision to change the unit size from 12 to 6 inches.

Discussion

Mice are, and I am still not certain about the exact choice of words here, bolder explorers than rats. By that, I mean they seem almost rash in their exploratory attack upon the Lashley III matrix maze. Rats, on the other hand, are cautious explorers. Rats, for example, never explore at a rate faster than the validity of their model justifies; rats never get themselves lost in the maze, that is, they never explore into parts of the maze without knowing how to return immediately to the alpha unit. The most remarkable aspect of rats' exploratory behavior is the fact that they quit retracing generally by the fifth or sixth row. Mice, however, are so bold as to appear imprudent. They move so quickly or rashly through the first half of the maze that the model of the maze they evolve in this half of the maze is not valid in the second half of the maze. The result is that they are not able to return immediately to the alpha unit from anywhere in the second half of the maze. Rats do essentially all of their retracing on the first few rows of the maze and refuse to explore maze units that are beyond the range of convenience of their cognitive map. Rats develop models of the maze in the first half of the maze that are valid for both halves of the maze. Mice do most of their retracing and most of their learning on the back half of the maze. They appear to get momentarily lost on the way back suggesting that on the way out they gave no thought to the problem of having to return, much less to making a model on the way out that is valid for both directions. Yet, in spite of these differences, mice are clearly equal to the rats as explorers by at least one criterion. Mice can make a complete loop in the EP on the average in fewer steps than a rat, but they require the same number of moves to make a full growth. Rats and mice also differ in their distribution of the frequency scores in the maze with the rats having a greater frequency in the first half. Mice have a much greater frequency of activity in the second half of the maze. Two hypotheses about their behavior are possible, between which the data of this experiment does not allow me to choose. First, mice are high-risk explorers, exploring faster than the validity of their cognitive map of the maze justifies. Or secondly, mice are particularly sensitive to the asymmetry between the out and back half of the Lashley III maze. Only a Lashley III maze with an odd number of rows-which is symmetrical- such as a 7 or 9-row maze would allow the back series of turns to have the same pattern of turns as the out sequence and thus would allow a test of the second hypothesis. In an even number of rows maze, the first turn out of alpha is right, but the first correct turn out of Omega is left. A matrix maze with an odd number of rows would generate the conditions so that the first turn out of alpha and Omega would both be right-turns. The mice's problem may be either that they don't make as good a map on the way out as rats do, or they may not be as capable of reversing the sequence as are the rats.

The effect of punishment upon the behavior of rats and mice are the same in general, but in mice, punishment increases the frequency of the activity in the second half of the maze, which is as was noted above already greater in mice than rats.

Rats and mice differ in that rats make a model in the first half of the maze that allows them to extrapolate and run the second half of the maze almost without error and without retracing. Mice are not able to make a map that allows extrapolation, but both mice and rats make a model of the BP element of the matrix that allows extrapolation. Both mice and rats avoid the redundant BP elements of the matrix; the redundant Forward blind type is more effectively avoided than the Alternation blind type.

The ability of rodents to adapt so quickly, efficiently, and cleverly to complex constraint sets shows that they possess a primary mind; a mechanism with a set of cognitive modeling skills with properties that no one ever dared before to assume. Rodents in the matrix maze spontaneously exploring the maze under a particular constraint set quickly learn to estimate the quantity of time and moves allowed by the constraint set even on their first runs through the maze. They are then able to recognize what they have lost in term of times and moves when a punishing contingency is introduced. Devious behavior is the change in behavior induced in animals aggrieved at the loss they encounter in play times and moves contingent upon the punishing constraint. The fact that the control mice continued to make a considerable number of moves in the back units of the maze is a puzzle. Rats treat the maze as a whole, which is exhibited in the FREQULM data set as a single continuous curve that includes both sides of the maze. The mice behavior, on the other hand, shows two curves in the FREQULM data, first the curve for the first half of the maze and a secondary smaller curve for the second half of the maze. The bold hypothesis doesn't explain the fact that the mice continued to exhibit the two-element learning curve for seven growths. The asymmetrical hypothesis also fails to explain why the mice didn't master this technicality after one or two growths. Mice may not be capable of conceiving of the maze as a whole, so their conception of the shortest path comes in two parts, one for each half of the maze.

Rats and mice that are punished for making a type eight or criterion loop evolve devious behavior, which takes the form of their dropping from the usual 20 loops per run to one loop per run but then learning to make several nested loops in that one loop. In this study, the same devious behavior was observed in mice. The experimental mice also dropped their number of loops and started making nested loops. This is one consequence of the avoidance of the alpha unit.

Devious mouse behavior, however, has another dimension. This new dimension is seen in the abnormally large FREQULM scores for the second half of the maze in mice. These scores give the play-growth curve the appearance of being two curves; one curve the larger and primary curve for the out units and another smaller secondary curve for the back units. The avoidance of the alpha unit reflects the fact that fear gets associated with the alpha unit because of punishment, and this fear thus produces the avoidance behavior in the second half of the maze.

The third aspect of avoidance is the way the rats and mice avoid the alpha unit. Their strategy seems to come in two parts. The first part seems to be to run as far away and as fast as they can from the dreaded alpha unit, which produces a pressure which takes them directly out to the omega unit. Thus, this first phase of the avoidance produces clean efficient performance on the first half of the maze. Then in the second phase, a conflict arises. In the first phase, both their desire to traverse the maze with the shortest path and their desires to avoid the alpha unit are compatible. In the second phase, their desire to run the shortest path and their desires to avoid the alpha unit are in conflict. So, the second part of the two-part learning curve is the product of the avoidance behavior generated by the punishment of alpha. The abnormal quantity of FREQULM behavior in the second half of the maze simply reflects the fact that the experimental mice are not in a rush to get started back to the alpha unit, as are control mice.

Hypothesis III

So, perhaps it is now possible to develop yet a third hypothesis as to why control mice show this second component of the two part growth curve, even up to the max of the twenty five runs, which some of the control mice were tested on. It could be possible that mice are more fearful of the experimenters than are rats. The control mice behavior seen in the second part of the two-part play-growth curve simply reflects an element of avoidance of the experimenters, which are at the same end of the maze as the alpha unit. This hypothesis requires that the location of the experimenters relative to the position of the maze be made an independent variable. But this hypothesis may enlarge the meaning of devious behavior. Another solution could be to sit the Experimenter behind a one-way mirror so the mice could not see them when they are running back to the alpha unit, but this was not available during these experiments.

The hybrid maze, Work-play combinations

The four combinations of work and play;
1. Pure work
2. Pure play
3. Work-play
4. Play-work

Devious behavior can only be sustained when the player is willing to pay to play.

The phenomena of devious-behavior make it necessary to enlarge our concept of work and play. Work is generally associated with behavior that only can be sustained by extrinsic reward. While play is behavior independent of extrinsic reward, or to put it the other way around, play is behavior that is dependent upon intrinsic reward. However, as I noted above, devious behavior complicates this picture.

Consider the pure-work human case, in the case of human work the employer pays the worker. Presumably, the work is contingent upon the pay: no pay no work.

Type III, Hybrid; work-play

What then do we call the hybrid contingency where the worker enjoys their work, so that the situation has both an intrinsic and extrinsic component? How will these two components interact? Will the performance be different in this hybrid situation from either the pure work or the pure play situation?

Type IV, Hybrid; play-work

Type IV must be distinguished from the type III; work-play type. The type of learning that was unintentionally created in this study was the play-work type.

In the work-play type, both extrinsic and intrinsic motivation is involved. However, in the play-work type of behavior, the organism does not work to obtain some extrinsic reward; rather the organism must work to continue to obtain some intrinsic reward! At the human level, when we pay to play, this is not pure play but a play-work situation. In a work-play situation, the workers are paid to do something they enjoy and the employer pays. In the play-work situation, the player pays. This animal experiment is not, of course, an exact analogue of the human play-work situation. For example, play-work requires a player to pay to play like a carnival where it requires money to play the games, or if visiting a prostitute, it is necessary to pay to play. However, it may be fair to note that the rodents may be required to pay to play, if additional cognitive, emotional, and behavioral efforts must be spent in order to continue to play and are recognized as a part of the cost of play that rodents seem willing to pay.

Summary – the function of or the purpose of the nonverbal mind

Which species are the smarter, rats or mice? Based on their performance in the Lashley III maze using the TTP, I long ago convinced myself that rats were smarter than mice. But judging them based on the LTP data; it is probably safer to say simply they are different. Mice master the first growth in fewer steps and in less time than rats, but mice do not exhibit extrapolation in their growth behavior in the EP, but mice do exhibit extrapolation of their growth behavior in the BP. Why if they are capable of extrapolation in learning of the BP, should mice not exhibit extrapolation in the learning of the EP?

Ontology, knowledge, and learning as conceptual evolution

Heterostasis is the wisdom of and the sign of the primary nonverbal mind. Historically philosophers, psychologists, and behaviorists have conceived of mind as a human verbal or mathematical function, rather than as a nonverbal animal function. This hubris had the effect of not only demeaning animals, but locked the concept of the mind in the limited form in which it has been conceived in the philosophy of Plato (1952). Plato rightly recognized that the central problem of philosophy was the problem of mind, which he reduced to what he called the problem of the forms. But Plato had no understanding how forms could evolve naturally. Plato had no idea of how mind even the human mind much less an animal mind could naturally invent a domain specific set of forms in every domain that would allow even a nonverbal mammal to adapt to an originally anomalous domain. Plato fell back on the ancient notion of reincarnation, which simply is the assumption that the mind magically comes with unconditioned or nativistic sets of forms, which if I may borrow a term from Pavlov (1960), sets of forms, which presumably are built into the soul. Plato and Pavlov both employed the same trick but in different realms, Pavlov was a physiologist, while Plato was a philosopher, yet they both assumed that the mind contained an unconditioned set of forms. Plato's forms were classes of things or ideas, while Pavlov's unconditioned stimuli and unconditioned responses were nonverbal sensory and motor forms: food and salivary response to food respectively or Homeostatic sets of reflexes.

Information is one of the many terms, which the classical theory of mind polluted and which I must make every effort to recover. Information can be defined two ways:

- The information theorists define information in the context of a closed system of alphabetic characters, A through Z. The Bell Labs further defined information as "that which reduced uncertainty in the context of Bell Telephone communication. (Shannon, 1951)

The challenge of the nonverbal mind is to define information without reference to verbal communication. For Female 3, for example, upon her introduction into the maze she was essentially for some period, overwhelmed with the uncertainty induced by the anomalous character of the maze. The role of her nonverbal mind was to create not information but to create knowledge of the hierarchical structure of the matrix maze that would allow her finally to run a complete loop through the maze without error. Knowledge is empirically derived from the interaction of the rat with the maze while the rat is exploring the maze during exploratory play. The nonverbal mind was designed because there had to be a better way than for the forms of life that were nature-defined to rely upon the knowledge supplied by the genome.

The nonverbal mind evolved so that life could survive by being a SOS which could conceive whatever knowledge it required by probing the environmental context within which the uncertainty arose in the first place.

Historically, however, everyone has exhibited the same failure of the human imagination; all have been unable or unwilling to assume that the primary mind is capable of inventing its own domain specific set of forms. These forms make it possible for the mind to organize behavior and take the uncertainty out of the strange objects and events of an anomalous domain. The primary mind is an evolution-designed machine, which has for every species, a species-specific potential for conceiving forms.

Rodents bring to the anomalous hierarchically organized domain a genetically given set of potential forms that their primary mind can use to create a representation of this domain. These form potentials evolve as the primary mind is exercised by the anomalous events generated by their exploratory-play behavior in the matrix maze. Heterostasis is a theory of how the primary mind evolves causal conceptual systems, an organized set of forms. These forms allow the organism to recognize each element of the matrix and know where it is within the hierarchically organized spatial maze and thus are able to know how to respond to achieve its own species-specific goals.

Heterostasis is a theory of how conceptual systems or organized sets of forms created by the mind come into being in order to take the uncertainty out of a domain that is originally complex and strange, and convert these initial events into simple and familiar events.

Heterostasis is a theory of how the domain specific sets of mental forms evolve in a specific domain and how domain-specific forms can evolve in various dimensions, for example, into domain general forms. The forms of the primary mind evolve in several dimensions from concrete to abstract and from short term to long term etc.

Primary mental growth evolves not as associationists claim by evolving associations between fixed concepts, but by evolving concepts themselves.

The experiments of this book have shown that both approach and avoidance behaviors are critical elements of the spontaneous Heterostatic behavior that evolves in the matrix maze. Approach and Avoidance behavior are both essential elements of any theory capable of explaining EP behavior. But there are two kinds of approach behavior. Rat's approach to behavior is rule-governed behavior. This can be seen in many ways, but one of the most interesting is the extrapolation behavior of rats in the EP. Mice, however, do not extrapolate during EP acquisition and thus while they are adapting they seem not to be using rule-generated behavior in the same way as rats. However, both rats and mice exhibit avoidance extrapolation behavior in the BP. Mice have the capacity to evolve and use rule-governed behavior, but why they do not use rule-governed behavior in the EP, I have no clue.

The primary mind invents two different cognitive maps to solve the problem of the matrix maze. First, it evolves a sub-map for the EP and another for the BP. For rats, the EP sub-map is a set of concepts that allows rats to adapt so efficiently to the EP that the best rat can create a valid map for the whole hierarchically organized domain after exploring only four of the 12 rows. The validity of the sub-map is revealed in the fact that rats stop retracing after exploring the first four or five rows. They are then able to extrapolate the map evolved in the first few rows to the rest of the matrix and explore the rest of the matrix in one exploratory loop without any need for retracing.

The blind sub-map allows both rats and mice to avoid certain redundant maze units so effectively that many growths are required before these conceptually redundant units are ever entered for the first time.

Causal conceptual conditioning

The big question, of course, is how does the species-specific causal-conceptual potential that any species is born with actually become realized? This issue can be stated many ways; first, how does the primary mind grow, how is the conceptual potential of a species finally achieved. **Learning theorists have since Pavlov (1960) and Thorndike (1911) believed that homeostatic or physiological deprivation and reinforcement were the essential elements of learning. In this series of studies, however, the rodents were never physiologically deprived nor reinforced and yet their mental growth evolved with surprising efficiency. Conceptual evolution obviously does not depend upon those physiological conditions that the classical learning theorists exclusively used.** A completely new concept of conditioning is required. Our theory is called Causal Conceptual Conditioning, 3C Theory. 3C theory consists of several paradigms, which were not recognized by Pavlov (1960) or Skinner (1974), but all the classical paradigms have to be redefined.

3C theory is a theory of how the form generating potential of the primary mind achieves its potential. 3C theory is a technical theory of how the evolution of ideas, or knowledge, but in the most basic terms, how knowledge, in the primary mind occurs, how the form potential of the primary mind takes forms. Knowledge is that which reduces uncertainty. Primary Knowledge is the nonverbal form in the primary mind that makes it possible for rodents to extrapolate within the EP and the BP. Extrapolation or predictions are a function of information or knowledge within the rule system or the conceptual system of the primary mind.

Glossary of Terms

A2, B3, C2, D8	breakdown of the nested concept of the maze, where D is 8 units per row, C is two types of rows, even and odd, B is replication, three sets of the two row types and A is directions - two (out and back)
AA	Association Area - also known as the sensorium; place where consciousness resides
abductive	the kind of logic inference described by Peirce as generating hypotheses
accommodation	adjustment to new or different circumstances or ideas; an event modifies the class
action	responding somatic behavior to content or objects of the Life Space, actions are the real CS
AIA	Anomaly Induced Avoidance - rodent does not enter this unit until all the uncertainty has been reduced in their model of the matrix maze
AIR	Anomaly Induced Retracing - uncertainty so strong that the rat must go back and retrace to correct their map of the matrix maze before proceeding
Alpha Unit	also can be called the start box, all rats begin their exploration of the maze in this box and return to this box to complete a criterion loop
anomaly	any occurrence or object that is strange, unusual, unexpected or unique, an illegal event, surprise, turns on the god function, what Heterostats seek and destroy, measured by the BFEULM data set scientifically
ANOVA	A form of statistical analysis used for the data in the experiments in this book - enables you to look at the variables and their effects
ARAS	ascending reticular activating system - vertical pathways known as 3+ and 3-; diffuse activity
arousal mechanism	generic name for the objects and events that will excite sexual behavior
assimilation	the process of taking new ideas into an existing cognitive structure; no change to the class by the event
associationism	idea in philosophy that states an idea or mental process operates with its successor states
B&B	Brown and Blodgett
Behavioral Telescope	technology designed by White to measure heterostatic behavior of the rat in the maze
BFE	Between first entry a unique measure of heterostasis which measures tracing and retracing of the animal
BFEbp	between first entry elements for the blind path only
BFEep	between first entry elements for the essential path only
BFEfm	Between first entry elements for the full maze behavior

BFEL	Between first entry latency - time between first entries into two units
BFEM	between first entry mean - computed by dividing the between first entry of a unit by the latency between units
BFEU	between first entry of a unit
BFEUL	between first entry unit & latency - two of the variables of data for this set
BFEULM	between first entry unit, latency and mean - all three figures compose this three data type set
BP	Blind path - units of the matrix maze the rats do not need to enter to complete their criterion loop in the shortest path; two units on each side outside the essential path, there are two types, Forward and Alternation
Brocca's area	area of the human brain. Lower left corner of the Rolandic motor strip in the left hemisphere (13th station of Aristotle's Model) where human speech is processed
BSRF	Brain Stem Reticular Formation - most ancient vertical area, diffuse, receives signals from vertical pathway 3+ and -
CB system	cognitive behavioral system of an organism - the activity of the organism is dynamic and reflects the momentary uncertainty status being experienced
CCC	Conceptual Causal Conditioning - general theory of conditioning in which classical and operant conditioning are special classes
celestial	of or relating to the heavens or sky
CNS	Central Nervous System
confounded	confused or perplexed - for purposes in this book - results that indicate two different answers for the same variables
consciousness	certified by the nonverbal mind as being real because it is in front of your eyes, is Life space, is what is real, certified by the Sixth Sense
consummatory mechanism	generic name for physiological mechanism that controls the pattern of sexual behavior
contralateral	taking place or originating in a corresponding part on the opposite side
corpus callosum	arched bridge of tissue which connects the two cerebral hemispheres and allows communication between the right and left side of the brain
cortex	bark or outer layer of the brain; area of the human brain which houses all the millions of neurons; only mammals have a cortex
cortical	derived from or consisting of the cortex of the brain
criterion loop	rat's path from alpha out to omega and back to alpha - a complete loop through the essential units plus 2 alpha and 1 omega for 51 units, this is the goal the rodents all chose as their primary goal
CS	conditioned stimulus
CS bell	conditioned stimulus - bell sound
CS light	conditioned stimulus - light
CUC	certainty-uncertainty continuum
CV	control variables are the variables the experimenter holds constant
desynchronization	to cause to go on, or work apart at different rates, for example: all the dendrites working separately of each other in the cortex

DF	degrees of freedom
DV	Dependent Variable are the variables the experimenter measures
effector	muscle, gland or organ capable of responding to a stimulus
electrophysiological	electrical activity associated with a body part or function
EMb	External Milieu of the body
EMm	External Milieu of the mind
empiricism	theory of knowledge that asserts knowledge arises via sense experience rather than knowledge or ideas being innate
engrams	hypothetical means by which memory events are stored
EP	essential path - those units of the matrix maze the rat must enter to complete a criterion loop by the shortest path; 4 units per row for 8 rows, totalling 48 units
Epicurus	first defined hedonism as seeking pleasure and avoiding pain
EV	Essential Variable
EVb	Essential Variable of the body (like temperature is 98.6 degrees for a human)
EVm	Essential Variable of the mind
experiment	the experimenter manipulates a set of independent variables in order to determine their effect upon a set of dependent variables given the conditions of the control variables
extrapolation	inference of one or more unknowns based on what is known or has been seen
external inhibition	idea that a novel or new stimulus has the power to surprise to interupt or disrupt all behavior
F3	Female 3, a rat who demonstrated even she can perform the god function and demonstrated a prototypical heterostatic growth curve
F11	Female 11, a rat who demonstrated redundancy induced avoidance in her blind path modeling, devious behavior, and the direction effect of the matrix maze
factorials	of or pertaining to the factors of a number
FE	first entry
FEL	first entry latency
FEU	first entry unit
FEUL	first entry unit & latency
Figure	that portion of the scene that is in front - all the figure is visible
FEULM	first entry unit, latency, & mean - composed of all three types of the data set
FR	Freedom Reflex - animal attempts to escape when restrained, such as when Pavlov's dogs were put into the stock they attempted to escape and conditioning could not start until the animal oriented
FREQ	frequency
FREQL	frequency latency
FREQM	frequency mean
FREQU	frequency unit
FREQUL	frequency unit & latency
FREQULM	frequency unit, latency & mean - all three data types for this data set

FRS	Functional representation system - scoring of the data of the experiments of this book are done in the order the animal would experience the events
Given	classical meaning for word data is given - passively received like a gift
god	source of information, what is real
god functions	those ideas that Plato and others believed only god could do, man was not capable of conceiving of ideas such as ethics and laws, however F3 proves even a rat can execute the god functions
Graph53	computer program used to track and view the rat's movements within the
Ground	the surrounding area or background of a scene - only part is viewed
GNMM	generic nonverbal mammalian brain
H	one wisdom creatures - the wisdom of homeostasis
HH	two wisdom creatures - homeostasis and hedonism
HHH	three wisdom creatures - homeostasis, hedonism & heterostasis - rats and all mammals are included in this category
HHHH	four wisdom creatures, homeostasis, hedonism, heterostasis and verbal - humans are the only mammals with all four wisdoms
Habit	a recurrent often unconscious pattern of behavior that is acquired through frequent repetition
habituation	a gradual decline of the response to a stimulus resulting from repeated exposure to the stimulus
hedonism	seeking pleasure and avoiding pain
hierarchical	an organization of habits or concepts in which simpler components are combined to form increasingly higher complexity levels
heterostasis	Wisdom of the mind - mechanism for reducing surprise - seeking the anomaly and then seeking the information to destroy the anomaly, rhythm of seek/destroy
heterostat	wisdom of the mind, triggered by change
heterostatic	described as using the wisdom of the mind
homeostat	creature with only one wisdom (the wisdom of the body), the wisdom to control their internal variables automatically - no thought is required to exercise this wisdom
homeostatic	being of the wisdom of homeostasis
IMb	Internal milieu of the body
information (nonverbal)	reduces uncertainty caused by an anomaly, measured by BFEU, nonverbal behavior like retracing
information (verbal)	reduces uncertainty, measured by the probability of which alphabetic character chosen
inhibition	the weakening of a learned response usually as a result of extinction or because of the presence of a distracting stimulus; called localized sleep by Pavlov
IMm	Internal milieu of the mind
intelligence	ability to create an idea, the god function, activated when an anomaly is encountered

interaural	of or relating to sound perceptions of each ear in relation to the other or separately
invariant	that which does not change during change
IV	independent variable - variable in scientific experiment that is manipulated by experimenter and are caused to vary
Life Space	created in consciousness by the Sixth Sense station of the horizontal mind, allows animal to have valid model of their domain
LTP	loop testing procedure; used in these experiments, the animal was run both directions out and back with no deprivation and no homeostatic reinforcement
meme	meaning of a word; item transmitted by culture and repetition rather than by genes or from the genome
meaning	of words - the objects in consciousness, objects that have been certified by the sixth sense as real
Mind	special function generated by the reorganization of the cortex found only in mammals, mechanism for creating the additional information required for a system to be self organizing, used to deal with change by the technology embedded within it
model	set ideas, hierarchically organized to understand the physical structure of any given domain
MSR	Maximum Steps per Run
MTR	Maximum Time per Run
nativism	the belief that innate ideas exist - were given from god at birth
nativistic	believing ideas were given from god at birth
neocortex	largest and newest part of the cortex only found in mammals - part where a large portion of higher functions are executed
NF	Node Frequency
NFM	Node Frequency Mean
Node	place in a path where the animal turns around, easily identified by same unit before and after a nodal unit
nonrational	doing the right thing for the wrong reason for example
novelty	a new or unusual experience or occurrence
Omega Unit	rats enter this unit when they complete the out portion of the criterion loop, then turn around here and return to Alpha to complete the Loop Testing Procedure
ontology	study of the nature of existence; study of what is; study of what is "god"
OR	Orienting Reflex "what is it" reflex, instinctive reflex that forces the animal to orient to novelty
paradigm	an example serving as a model or pattern; invented by Kuhn - an agreed upon way to do something in science
plasticity	the ability of tissue (such as brain tissue) to re-organize itself and is usually lost by the time an individual is 5-6 years old
PLT	Path length in time per unit
PLTM	mean path length in time
PLU	path length in unit's steps

PLUM	mean path length in units
PNS	Peripheral Nervous System
PPA	Primary Projection Area - area of the cortex where the afferent nerves of the 5 senses dump their information to be turned into sensations and put into the nonverbal consciousness
Pragmatic	of or pertaining to a practical point of view; of or pertaining to the state of a community/uniquely American philosophy started by Peirce and James
psychophysiology	branch of physiology which deals with the interrelation of mental and physical phenomena
QUAL	quality
QUAN	quantity
rational	ability to conceive models or structures of a space or time, to indentify the parameters of the space so you can make a model and extrapolate (nonverbal activity)
rationality	heart of mind, designed to deal with change, create ideas, the god function
RIA/RIR	Redundancy Induced Avoidance/Retracing - behavior in which the rodent avoids a unit, mostly the blind path because it knows it is not a part of its goal of the shortest path because it made a model of the EP and BP
Response	any behavior emitted by an organism
sensorium	part of the brain regarded as the seat of the senses - relating to the 5 senses/ area of where Aristotle's 6th sense resides (station 7)
Sprague Dawley	type of rats used in these experiments
SRS	Structural representation system - in each society there is a different standard, in West we read left to right and top to bottom
startle reflex	response to being surprised, scared, involuntary response
stimulus	physical energy that is capable of exciting a receptor
Statistica	computer program used to help extrapolate the data for the experiments in this book
synchronization	to cause to go on, or work together at exactly the same rate - for example: the dendrites of the brain firing together
Tabula Rasa	term used to describe the mind at birth as a receptive blank slate - used initially by John Locke
Takens	means what is measured in scientific experiment is product of your own domain specific experimental procedures and is an active process
thing	made of matter and real
TRF	thalamic reticular formation - area of the brain; focal; that receives input from the vertical pathway 2+ and -
transcortical	functions of the mind are in the cortex and move back and forth within this area uninhibited
transduce	to convert energy from one form to another
TTP	trial testing procedure; animals were run only one direction and reinforced with food after being deprived

US	unconditioned stimulus
US food	unconditioned stimulus food
USM	Unit speed mean
Wisdom	domain specific heterostasis; set of symbols unique to a specific function - for example musical notes for music, letters for the alphabet, numbers for counting, etc.

References

Andrews, G., & Harlow, H. F. (1948). *Performance of macaque monkeys on a test of the concept of generalized triangularity*. Berkeley, CA: University of California Press.

Aristotle. (1952). The works of Aristotle, volumes i and ii. In M. J. Adler, C. Fadiman & P. Goetz(Eds.), *Great Books of the Western World* (8 ed.). Chicago: Encyclopaedia Britannica, Inc.

Ashby, W. R. (1956). *An introduction to cybernetics*. London: Methuen & Co.

Ashby, W. R. (1960). *Design for a brain*. London: Chapman & Hall Ltd.

Beach, F. A. (1947). Evolutionary changes in the physiological control of mating behavior in mammals. *The Psychology Review*, *54*, 297-315. Retrieved from DOI: 10137/h0056549

Berkeley, G. (1957). C. M. Turbayne (Trans.), *A treatise concerning the principles of human knowledge* Forgotten Books.

Bernard, C. (1957). *An introduction to the study of experimental medicine*. New York, NY: Dover Publications.

Blodgett, H. C. (1929). The effect of the introduction of reward upon the maze performance of rats. *University of California Publications in Psychology*, *4*, 113-134.

Brown, W. L., Benson, R., Bitter, H., Carr, R. M., McDowell, A. A., & Wicker, J. (1959). Primate in space progress report no. 1. *Bioastronautics Advances in Research*, (March), 58-71.

Brown, W. L., Pickering, J. E., Stallings Jr., H. D., Benson, R. E., Zellmer, R. W., Bitter, H. L., Carr, R. M., McDowell, A. A., & Wicker J. (1959). Primate in space progress report no. 2. *Bioastronautics Advances in Research*, (March), 72-94.

Boring, E. G. (1929). *A history of experimental psychology*. New York, NY: Appleton-Century-Crofts, Inc.

Breland, K., & Breland, M. (1972). The misbehavior of organisms. In D. Robinson (Ed.), *Readings In the Origins and Principles of Psychology* (pp. 443-449). Encino, CA: Dickinson Reading company, Inc.

Bruner, J. (2004). Life as narrative. *Social Research*, 691-710. Retrieved from http://search.proquest.com/docprintview/209669779/record/133D1CFCE77692D97B4/4

Bryan, W. L., & Harter, N. (1897). *Studies in the psychology of the telegraphic language*. Retrieved from http://grants.hhp/coe.uh.edu/clayne/historyofmc/bryan&harter1897.pdf

Buck, R. (1988). *Human motivation and emotion*. New York, NY: John Wiley and Sons, Inc.

Calvin, W. H. (2005). *A brief history of mind*. New York, NY: Oxford University Press.

Cannon, W. B. (1932). *The wisdom of the body*. New York, NY: W.W. Norton & Company Inc.

Chomsky, N. (1979). *On language*. New York, NY: The New Press.

Chomsky, N. (1997). Written transcript of Noam Chomsky's language and mind. *Princeton University Lecture*, 11-22.

Conkin, P. (1976). *Puritans & pragmatists eight eminent American thinkers.* Waco, TX: Baylor University Press.

Cruickshank, W. M., Bentzen, F. A., Ratzeburg, F. H., & Tannhauder, M. T. (1961). *A teaching method of brain-injured and hyperactive children*. Syracuse, NY: Syracuse University Press.

Csikszentmihalyi, M. (1990). *Flow the psychology of optimal experience*. New York, NY: Harper Perennial.

Cubberly, E. (1948). *The history of education*. Cambridge, MA: Houghton Mifflin Company.

Darwin, C. (1859). *On the origin of species by means of natural selection, or the reservation of favoured races in the struggle for life*. London, England: John Murray.

Dawkins, R. (1989). *The selfish gene*. New York, NY: Oxford University Press.

Descartes, R. (1946). Two meditations. In J. H. Randall Jr., J. Buchler & E. Shirk (Eds.), *Readings in Philosophy* (pp. 26-38). New York, NY: Barnes and Noble Books, a division of Harper & Row Publishers.

Dewey, J. (1916). *Democracy and education*. New York, NY: The Macmillan Company.

Dewey, J. (1916). *Essays in experimental logic*. Chicago, IL: The University of Chicago Press.

Donald, M. (1991). *Origins of the modern mind*. Cambridge, MA: Harvard University Press.

Eckblad, G. (1981). *Scheme theory*. New York, NY: Academic Press.

Erdmann, E., & Stover, D. (1991). *Beyond a world divided human values in the brain-mind science of roger Sperry*. Boston, MA: Shambhala Publications, Inc.

Eysenck, H. J. (1971). *Readings in extraversion-introversion. fields of application*. New York, NY: Thomas Y. Crowell Company.

Freeman, K. (1970). *Ancilla to the pre-Socratic philosophers*. Cambridge, MA: Harvard University Press.

Freud, S. (1962). J. Strachey (Trans.), *Three essays on the theory of sexuality* New York, NY: Basic Books.

Galileo, G. (1989). A. Van Helden (Ed.), *Sidereus nuncius or the Sidereal messenger* (pp. 64-86). Chicago, IL: University of Chicago Press.

Garcia, J., Ervin, F. R., & Koelling, R. A. (1966). Learning with prolonged delay of reinforcement. *Psychon. Sci., 5*, 121-122. Retrieved from http://fitelson.org/woodward/garcia.pdf

Gastaut, H., Jus, C., Morrell, F., Storm-VanLeeuwen, W., Dongier, S., Naquet, R., Regis, H., Gazzaniga, M. S., & Sperry, R. W. (1972). Effects of split brain on language behavior. In F. Mcguigan & P. J. Woods (Eds.), *Contemporary Studies in Psychology* (pp. 161-172). New York, NY: Appleton-Century-Crofts, Inc.

Gazzinaga, M. S. (1998). The split brain revisited. *Scientific American, 279*, 50-55.

Gerschwind, N. (1979). Specialization of the human brain. In *The Brain A Scientific American Book* (pp. 108-117). San Francisco, CA: W.H. Freeman and Company.

Gould, S. J. (2002). *The structure of evolutionary theory*. Cambridge, MA: The Belknap Press of the Harvard University Press.

Graham, D. W. (2002). *Lep.utm.edu.* Retrieved from http://www.iep.utm.edu/heraclit/

Gupta, A. S., van der Meer, M. A. A., Touretzky, D. S., & Redish, A. D. (2010). Hippocampal replay is not simple function of experience. *Neuron, 65,* 695-705. doi: 0.1016/j.neuron.2010.01.034

Hampden-Turner, C. (1981). *Maps of the mind*. New York, NY: Macmillan Publishing Company.

Hanouch, Y., & Vitouch, O. (2004). When less is more information, emotional arousal and the ecological reframing of the yerkes-dodson law. *Theory and Psychology, 14,* 427-452. doi: 10.1177/0959354304044918

Harlow, H. F., & Hicks, L. H. (1937). Discrimination learning theory: Uniprocess vs. duoprocess. *APA.org*, Retrieved from http://psycnet.apa.org/journals/64/2/104.pdf

Heidbreder, E. (1933). *Seven psychologies*. New York, NY: Appleton-Century Company incorporated.

Helson, H. (1964). *Adaptation-level theory*. New York, NY: Harper & Row, Publishers.

Herbart, J. F. (1901). *Outlines of educational doctrine*. Norwood, MA: Norwood Press.

Hilgard, E. R., & Bower, G. H. (1975). *Theories of learning*. New York, NY: Appleton-Century-Crofts, Inc.

Holmes, T. H., & Rahe, R. H. (1967). The social readjustment rating scale. *Journal of Psychosomatic Research, 11,* 213-218. Retrieved from http://www.testandcalc.com/richard/resources/Teaching_Resource_Holmes_Rahe_social.pdf

Hull, C. L. (1943). *Principles of behavior*. New York, NY: Appleton-Century-Crofts, Inc.

Hume, D. (1962). *A treatise of human nature. book I: Of the understanding*. Cleveland, OH: The World Publishing Company.

James, W. (1950). *Principles of psychology volumes I and II complete*. United States of America.

Jaynes, J. (1976). *The origin of consciousness in the breakdown of the bicameral mind.* Boston, MA: Houghton Mifflin Company.

Kamin, L. J. (1969). Predictability, surprise, attention and conditioning. In B. Campbell & R. Church (Eds.), *Punishment and aversive behavior* (pp. 279-296). New York, NY: Appleton-Century-Crofts.

Kant, I. (1952). The critique of pure reason, the critique of practical reasons and the critique of judgment. In M. J. Adler, C. Fadiman & P. W. Goetz (Eds.), *Great Books of the Western World* Chicago, IL: Encyclopaedia Britannica, Inc.

Kelley, C. R. (1968). *Manual and automatic control a theory of manual control and its application to manual and automatic systems*. New York, NY: Academic Press.

Kelly, G. A. (1955). *The psychology of personal constructs, volumes one and two*. New York, NY: W.W. Norton & Company Inc.

Klein, N. (2007). *The shock doctrine the rise of disaster capitalism*. New York, NY: Metropolitan Books Henry Holt and Company, LLC.

Kluver, H. (1951). Functional differences between the occipital and temporal lobes with special reference to the interrelations of behavior and extracerebral mechanisms. In L. A. Jeffries (Ed.), *Cerebral Mechanisms in Behavior. The Hixon Symposium* (pp. 147-182). New York, NY: John Wiley & Sons, Inc.

Kohler, W. (1929). *Gestalt psychology*. New York, NY: Horace Liveright.

Kuhn, T. (1962). *The structure of scientific revolutions*. Chicago, IL: The University of Chicago Press.

Lashley, K. S. (1963). *Brain mechanisms and intelligence a quantitative study of injuries to the brain*. New York, NY: Dover Publications.

Lashley, K. S. (1951). The problem of serial order behavior. In L. A. Jeffries (Ed.), *Cerebral Mechanisms in Behavior. The Hixon Symposium* (pp. 112-135). New York, NY: John Wiley & Sons, Inc.

Levy, J. (1985). Right brain, left brain: fact and fiction. *Psychology Today*, (May), 37-44.

Lewin, K. (1943). Defining the "field at a given time." *Psychological Review, 50*, 292-310.

Lewin , K. (1951). Field theory and learning. *APA.org*, Retrieved from http://psycnet.apa.org/books/11335/006.pdf

Lieberman, P. (2000). *Human language and our reptilian brain*. Cambridge, MA: Harvard University Press.

Lilienthal, D. (1944). *Democracy on the march*. New York, NY: Pocket Books Inc.

Lindsley, D. (1960). Attention, consciousness, sleep and wakefulness. In J. Field, H. Magoun & V. E. Hall (Eds.), *Handbook of Physiology Section 1: Neurophysiology Volume 3* (pp. 1553-1594). Washington D.C.: American Psychological Society.

Locke, J. (1914). *An essay concerning human understanding*. Oxford, England: Oxford at the Clarendon Press.

Magoun, H. W. (1963). *The waking brain*. Springfield, IL: Charles C. Thomas Publisher.

Maslow, A. H. (1974). A theory of human motivation. In R. J. Mueller, D. Ary & C. McCormick (Eds.), *Readings in Classroom Learning and Perception* (pp. 223-245). New York, NY: Praeger Publishers.

Menninger, K. (1963). *The vital balance the life process in mental health and illness*. New York, NY: The Viking Press.

Mill, J. S. (2002). *A system of logic, ratiocinative and inductive*. Honolulu, HI: University Press of the Pacific.

Mitchell, W. (1907). *Structure and growth of the mind*. London, England: Macmillan and Co., Limited.

Montessori, M. (1965). *Dr. Montessori's own handbook*. New York, NY: Schocken Books.

Moser, M., & Moser, E. I. (1998). Distributed encoding and retrieval of spatial

memory in the hippocampus. *The Journal of Neuroscience, 18*, 7535-7542.
Olds, J., & Milner, P. (1976). Reinforcement by septal stimulation. In F. M. Levine (Ed.), *Theoretical Readings in Motivation: Perspectives on Human Behavior* (pp. 402-408). Chicago, IL: Rand McNally College Publishing Company.
Palmer, D. C. (2000). Chomsky's nativism. *NCBI.NLM.nih.gov*, Retrieved from http://www.ncbi.nlm.nih.gov/pmc/articales/PMC2755455/pdf/anverbbehav00028-0039.pdf
Pascal, B. (1665). The arithmetic triangle. In R. Pulskamp (Trans.), Retrieved from http://www.cs.xu.edu/math/sources/pascal/sources/arith_triangle.pdf
Panksepp, J. (1998). *Affective neuroscience the foundations of human and animal emotions*. New York, NY: Oxford University Press.
Pavlov, I. (1960). *Conditioned reflexes an investigation of the physiological activity of the cerebral cortex*. New York, NY: Dover Publications, Inc.
Peirce, C. (1878). How to make our ideas clear. *Popular Science Monthly*, (January), 286-302. Retrieved from http://www.peirce.org/writings/p119html
Pestalozzi, J. H. (1859). Pestalozzi and pestalozzianism. H. Barnard (Ed.), New York, NY: F.C. Brownell.
Piaget, J. (1970). *Psychology and epistemology*. New York, NY: Grossman Publishers.
Piaget, J. (1968). *Structuralism*. New York, NY: Harper & Row, Publishers.
Piaget, J. (1952). *The origins of intelligence in children*. New York, NY: International Universities Press, Inc.
Pinker, S. (1997). *How the mind works*. New York, NY: W.W. Norton & Company.
Pinker, S. (1999). *Words and rules the ingredients of language*. New York, NY: Perennial.
Plato. (1952). The dialogues of Plato. In M. J. Adler, C. Fadiman & P. Goetz(Eds.), *Great Books of the Western World* (8 ed.). Chicago: Encyclopaedia Britannica, Inc.
Pribram, K. H. (1960). The intrinsic systems of the forebrain. In J. Field, H. Magoun & V. E. Hall (Eds.), *Handbook of Physiology Section 1: Neurophysiology Volume 3* (pp. 1323-1344). Washington D.C.: American Psychological Society.
Redish, A. D. (1999). *Beyond the cognitive map from place cells to episodic memory*. Cambridge, MA: The MIT Press.
Redish, A. D., & Touretzky, D. S. (1996). Cognitive maps beyond the hippocampus. *Hippocampus, 7*, 15-35.
Rescorla, R. A., & Wagner, A. R. (1972). A theory of Pavlovian conditioning: variations in the effectiveness of reinforcement and nonreinforcement. In A. Black & W. Prokasky (Eds.), *Classical Conditioning II, current research and theory* (pp. 64-99). New York, NY: Appleton-Century-Crofts.

Rogers, A. (1996). *Amazing grace: the story of John Newton*. Retrieved from http://www.anointedlinks.com/amazing_grace.html
Roger, A., Bekkering, D., Kamp, A., & Were, J. (1957). A topographic study of

electroencephalographic CR's in man. *Electroencephalography and Clinical Neurophysiology, 9,* 1-34.

Ryan, R. M., & Deci, E. L. (2000). Intrinsic and extrinsic motivations: classic definitions and new directions. *Contemporary Educational Psychology, 25,* 54-67. doi: 10.1006/ceps.1999.1020

Santayana, G. (1941). The unknowable. In C. Fadiman (Ed.), *Reading I've Liked* New York, NY: Simon and Shuster.

Schank, R. C., & Abelson, R. P. (1977). *Scripts, plans, goals and understanding. An inquiry into human knowledge structures.* Hillsdale, NJ: Lawrence Erlbaum.

Shannon, C. E. (1951). Prediction and entropy of printed English. *The Bell System Technical Journal, 30,* 50-64.

Skinner, B. F. (1974). *About behaviorism.* New York, NY: Alfred A. Knopf.

Smith, A. (1993). *An inquiry into the nature and causes of the wealth of nations.* New York, NY: Oxford University Press.

Sperry, R. W. (1971). The effects of hemisphere deconnection on conscious awareness. In J. Kagan, M. M. Haith & C. Caldwell (Eds.), *Psychology Adapted Readings* (pp. 182-190). New York, NY: Harcourt Brace Jovanovich, Inc.

Sperry, R. W. (1961). Cerebral organization and behavior. *Science, 133,* 1749-1757.

Spinoza, B. (1933). Philosophy of benedict de Spinoza. R. Elwes (Trans..), New York, NY: Tudor Publishing Co.

St. Augustine. (2002). *Augustine confessions.* Retrieved from http://www.ourladywarriors.org/saints/augcon8.html

Sternberg, R. J. (1998). *In search of the human mind.* Orlando, FL: Harcourt Brace & Company.

Stratton, G. (1896). *Some preliminary experiments on vision without inversion of the retinal image.* (Master's thesis, NYU)Retrieved from http://www.cns.nyu.edu/~nava/courses/psych_and_brain/pdfs/stratton_1896.pdf

The holy bible. authorized king James version. Cleveland, OH: The World Publishing Company.

Thorndike, E. L. (1911). *Animal intelligence.* New York, NY: Macmillan.

Titchener, E. B. (1912). Prolegomena to a study of introspection. *The American Journal of Psychology, 23,* 427-448. Retrieved from http://www.jstor.org/stable/1413427

Tolman, E. C. (1961). Cognitive maps in rats and men. In R. A. King (Ed.), *Readings for an introduction to psychology* (pp. 148-158). New York, NY: McGraw-Hill Book Company, Inc.

Touretzky, D. S., Wan, H. S., & Redish, A. D. (1994). *Neural representation of space in rats and rodents.* In Zurada, J.M. and Marks, R.J. (Eds.), *Computational Intelligence: Imitating Life* (pp. 1-12). IEEE Press.

Touretzky, D. S., & Redish, A. D. (1996). Theory of rodent navigation based on interacting representations of space. *Hippocampus, 6,* 247-270

Underwood, B. J. (1949). *Experimental psychology an introduction.* New York, NY: Appleton-Century-Crofts, Inc.

Watson, J. B. (1972). Psychology as the behaviorist views it. In D. N. Robinson (Ed.), *Readings In the Origins and Principles of Psychology* (pp. 105-119). Encino, CA: Dickenson Publishing Company, Inc.

Watson, R. I. (1978). *The great psychologists.* New York, NY: J.B. Lippincott Company.

White, R. (1976). Motivation reconsidered: the concept of competence. In F. M. Levine (Ed.), *Theoretical Readings in Motivation: Perspectives on Human Behavior* (pp. 323-338). Chicago, IL: Rand McNally College Publishing Company.

Whitney, C. A. (1984). Generating and testing pseudorandom numbers. *Byte, 11,*

Wilson, E. O. (1998). *Consilience the unity of knowledge.* New York, NY: Vintage Books.

Wilson, E. O. (1975). *Sociobiology the new synthesis.* Cambridge, MA: The Belknap Press of the Harvard University Press.

Wollstonecraft, M. (1976). *Mary and the wrongs of women.* New York, NY: Oxford University Press.

Wollstonecraft, M. (1792). *A vindication of the rights of woman.*

Yerkes, R. M., & Dodson, J. D. (2004). The relation of strength of stimulus to rapidity of habit-formation. *Journal of Comparative Neurology and Psychology, 18,* 159-182. doi: 10.1002.cne.920180503

Index

About the Author

Dr. Robert White is a scientist and rat runner with 55 years' experience. He obtained his doctorate in Psychology from The University of Texas. He has worked at Balcones Research Center in Austin, TX, the AFRI Lab in Maryland, and Bellcomm in Washington D.C. and was a professor at Texas Tech and William Paterson College in New Jersey. He has previously co-authored articles published in *The Journal of Genetic Psychology, Radiation Research,* and *Perceptual and Motor Skills.*

Dr. White *has* devoted his life to the completion of this work to share with the public how their nonverbal mind works. Every day, books are written and published on the verbal mind, but this is groundbreaking new material on the "quiet" or "nonverbal" mind. The study of animal minds is a new and growing field and Dr. White is at the forefront of this field.

Dr. White currently resides in New Jersey with his son and daughter-in-law and looks forward to publishing his next book.

Made in the USA
Monee, IL
08 November 2026

67971460R00339